Springer Series in
CHEMICAL PHYSICS 83

Springer Series in

CHEMICAL PHYSICS

The purpose of this series is to provide comprehensive up-to-date monographs in both well established disciplines and emerging research areas within the broad fields of chemical physics and physical chemistry. The books deal with both fundamental science and applications, and may have either a theoretical or an experimental emphasis. They are aimed primarily at researchers and graduate students in chemical physics and related fields.

69 **Selective Spectroscopy of Single Molecules**
By I.S. Osad'ko

70 **Chemistry of Nanomolecular Systems**
Towards the Realization of Molecular Devices
Editors: T. Nakamura, T. Matsumoto, H. Tada, K.-I. Sugiura

71 **Ultrafast Phenomena XIII**
Editors: D. Miller, M.M. Murnane, N.R. Scherer, and A.M. Weiner

72 **Physical Chemistry of Polymer Rheology**
By J. Furukawa

73 **Organometallic Conjugation**
Structures, Reactions and Functions of d–d and d–π Conjugated Systems
Editors: A. Nakamura, N. Ueyama, and K. Yamaguchi

74 **Surface and Interface Analysis**
An Electrochmists Toolbox
By R. Holze

75 **Basic Principles in Applied Catalysis**
By M. Baerns

76 **The Chemical Bond**
A Fundamental Quantum-Mechanical Picture
By T. Shida

77 **Heterogeneous Kinetics**
Theory of Ziegler-Natta-Kaminsky Polymerization
By T. Keii

78 **Nuclear Fusion Research**
Understanding Plasma-Surface Interactions
Editors: R.E.H. Clark and D.H. Reiter

79 **Ultrafast Phenomena XIV**
Editors: T. Kobayashi, T. Okada, T. Kobayashi, K.A. Nelson, S. De Silvestri

80 **X-Ray Diffraction by Macromolecules**
By N. Kasai and M. Kakudo

81 **Advanced Time-Correlated Single Photon Counting Techniques**
By W. Becker

82 **Transport Coefficients of Fluids**
By B.C. Eu

83 **Quantum Dynamics of Complex Molecular Systems**
Editors: D.A. Micha and I. Burghardt

84 **Progress in Ultrafast Intense Laser Science I**
Editors: K. Yamanouchi, S.L. Chin, P. Agostini, and G. Ferrante

85 **Progress in Ultrafast Intense Laser Science II**
Editors: K. Yamanouchi, S.L. Chin, P. Agostini, and G. Ferrante

D.A. Micha I. Burghardt (Eds.)

Quantum Dynamics of Complex Molecular Systems

With 99 Figures, 9 in Color and 7 Tables

Springer

Dr. David A. Micha, Professor of Chemistry and Physics
University of Florida
P.O. Box 118435, 2318 New Physics Bldg., Gainesville FL 32611-8435, USA
E-Mail: micha@qtp.ufl.edu

Dr. Irene Burghardt
Département de Chimie, Ecole Normale Supérieure
24 rue Lhomond, F-75231 Paris cedex 05, France
E-Mail: Irene.Burghardt@ens.fr

Series Editors:
Professor A. W. Castleman, Jr.
Department of Chemistry, The Pennsylvania State University
152 Davey Laboratory, University Park, PA 16802, USA

Professor J.P. Toennies
Max-Planck-Institut für Strömungsforschung, Bunsenstrasse 10
37073 Göttingen, Germany

Professor W. Zinth
Universität München, Institut für Medizinische Optik
Öttingerstr. 67, 80538 München, Germany

ISSN 0172-6218

ISBN-10 3-540-34458-6 Springer Berlin Heidelberg New York
ISBN-13 978-3-540-34458-2 Springer Berlin Heidelberg New York

Library of Congress Control Number: 2006928272

Springer is a part of Springer Science+Business Media.

springer.com

Typesetting by the authors and SPi using a Springer LaTeX macro package
Cover concept: eStudio Calamar Steinen
Cover production: WMX Design GmbH, Heidelberg

Printed on acid-free paper SPIN: 11502227 54/3100/SPi - 5 4 3 2 1 0

Preface

Quantum phenomena are ubiquitous in complex molecular systems, and yet remain a challenge for theoretical analysis. A complex molecular system is composed of many atoms and may for example constitute an assembly of molecules, a cluster, a polymer, a chromophore-protein complex, or an adsorbate-surface structure. The system may be isolated, or more likely in contact with some physical environment. Its properties and behavior usually depend on the way it interacts with external fields or with other molecular species, and typically involve excited atomic and electronic states, which must be described in terms of quantum mechanics. From the point of view of quantum theory, one is dealing with a system with many quantized degrees of freedom, a subject that has been formally explored for a long time. But molecular systems are special, in that they involve particles (electrons and nuclei) with very different masses leading to interactions with very different time scales. Therefore, quantum molecular dynamics can often be described in terms of potential energy surfaces within the Born–Oppenheimer approximation – even though it is the breakdown of this approximation, at avoided crossings or conical intersections, which is at the root of many reactive and photochemical processes. Further, molecular systems are subject to thermodynamical constraints when they interact with a medium, which in turn dynamically evolves as a result of the interaction with the molecular subsystem. The subsystem's quantum dynamics is thus entangled with the nonequilibrium evolution of the environment. Due to these many facets of dynamical behavior, the quantum dynamics of molecular systems, including statistical effects, has become one of the most challenging and active areas of molecular science.

Much current activity is directed at developing methods to tackle quantum dynamics in many dimensions, including quantum coherence and dissipative phenomena, often with the aim of interpreting and predicting experimental observations based upon detailed molecular scattering experiments or ultrafast spectroscopic techniques. Indeed, the direct, femtosecond scale, observation of molecular phenomena ("femtochemistry") has given a strong impetus to the theoretical and computational developments in quantum dynamics. Applications and comparisons with experiments demand theories that can be implemented numerically to calculate measurable properties. Straightforward numerical methods for solving the differential equations of quantum mechanics, based on basis set expansions or discretization of variables on a grid, are restricted to small systems and are not practical for complex molecular systems. Relevant and useful treatments include self-consistent field methods for atomic motions and their multiconfiguration extensions, path integral methods for molecular motions, semiclassical and mixed quantum–classical approaches, various trajectory based methods, and density matrix methods describing both population relaxation and decoherence.

The present book grew out of a workshop organized in May 2005 in Paris, France, to bring together workers in the field of quantum dynamics of molecular systems, to discuss applications of present theories to a variety of phenomena, along with new theoretical concepts and methods. The following chapters have been contributed by some of the workshop participants and their collaborators, and have been grouped in what follows into Part I, with applications to complex molecular systems, and Part II, on new theoretical and computational methods. In fact, method development and applications are closely interconnected and related work is found in both parts.

Much can be done to explain phenomena in systems excited by light or through atomic interactions, extending from the molecular scale to nanoscales and even to macroscopic dimensions. The following chapters show that promising new methods are now available for those purposes. They demonstrate how one can tackle the multidimensional dynamics arising from the atomic structure of a complex system, and address phenomena in condensed phases as well as phenomena at surfaces. The chapters on new methodological developments cover both phenomena in isolated systems, and phenomena that involve the statistical effects of an environment, such as fluctuations and dissipation. The methodology part explores new rigorous ways to formulate mixed quantum–classical dynamics in many dimensions, along with new ways to solve a many-atom Schrödinger equation, or the Liouville-von Neumann equation for the density operator, using trajectories and ideas related to hydrodynamics.

The workshop leading to this book was made possible by sponsors from the University of Florida in the USA and by several institutions in France. We thank in connection with the University of Florida: the Paris Research Center, the Vice President for Research, the Quantum Theory Project (an Institute for Theory and Computation in the Molecular and Materials Sciences), and the Chemistry and Physics Departments. On the French side, we thank the Centre National de la Recherche Scientifique (CNRS), the Ministère de l'Education Nationale, the Ecole Normale Supérieure, Paris, and the Ecole Doctorale 388 "Chimie Physique et Chimie Analytique." The workshop greatly profited from the support of the Director of the Paris Research Center, Dr. Gayle Zachmann, and from the help of Rachel Gora. We appreciate their enthusiasm and hospitality.

Gainesville (Florida), USA
Paris, France
July 2006

David A. Micha
Irene Burghardt

Contents

List of Contributors

J.A. Beswick
Laboratoire Collisions Agrégats
Réactivité, IRSAMC
Université Paul Sabatier
Toulouse, France
beswick@irsamc.ups-tlse.fr

E.R. Bittner
Department of Chemistry and Center for Materials Chemistry
University of Houston
Houston, TX, USA
bittner@uh.edu

S. Bonella
NEST Scuola Normale Superiore
Piazza dei Cavalieri 7
It-56126 Pisa, Italy
s.bonella@sns.it

A.G. Borisov
Laboratoire des Collisions
Atomiques et Moléculaires
UMR CNRS-Université Paris-Sud 8625
Bât. 351 Université Paris-Sud
91405 Orsay Cedex, France

B. Brüggemann
Chemical Physics, Lund University
P.O. Box 124 SE–22100
Lund Sweden
Ben.Bruggemann@chemphys.lu.se

I. Burghardt
Département de Chimie
Ecole Normale Supérieure
24 rue Lhomond
F–75231 Paris, France
irene.burghardt@ens.fr

G. Ciccotti
Dipartmento di Fisica
Università "La Sapienza"
Piazzale Aldo Moro 2
00185 Rome, Italy
ciccotti@roma1.infn.it

D.F. Coker
Department of Chemistry
Boston University
590 Commonwealth Avenue
Boston, MA 02215, USA
coker@bu.edu

C.F. Craig
Department of Chemistry
University of Washington
Seattle, WA 98195-1700, USA

E. Deumens
University of Florida
Gainesville, FL 32611-8435, USA
deumens@qtp.ufl.edu

W.R. Duncan
Department of Chemistry
University of Washington
Seattle, WA 98195-1700
USA

J.P. Gauyacq
Laboratoire des Collisions
Atomiques et Moléculaires
UMR CNRS-Université Paris-Sud
8625
Bât. 351 Université Paris-Sud 91405
Orsay Cedex, France
gauyacq@lcam.u-psud.fr

B.F. Habenicht
Department of Chemistry
University of Washington
Seattle, WA 98195-1700
USA

G. Hanna
Chemical Physics Theory Group
Department of Chemistry
80 St. George St.
University of Toronto
Toronto, Canada M5S3H6
ghanna@chem.utoronto.ca

K.H. Hughes
Department of Chemistry
The University of Wales Bangor
Bangor Gwynedd, LL57 2UW, UK
keith.hughes@bangor.ac.uk

R. Kapral
Chemical Physics Theory Group
Department of Chemistry
80 St. George St.
University of Toronto
Toronto, Canada M5S3H6
rkapral@chem.utoronto.ca

S.V. Kilina
Department of Chemistry
University of Washington
Seattle, WA 98195-1700
USA

H. Kim
Chemical Physics Theory Group
Department of Chemistry
80 St. George St.
University of Toronto
Toronto, Canada M5S3H6
hkim@chem.utoronto.ca

I. Kondov
Department of Chemistry
Technical University of Munich
D-85748 Garching, Germany
kondov@ch.tum.de

H. Köppel
Theoretische Chemie
Universität Heidelberg, INF 229
D-69120 Heidelberg, Germany
horst.koeppel@pci.uni-heidelberg.de

A. Leathers
Departments of Chemistry
and Physics, University of Florida
Gainesville, FL 32611, USA
leathers@qtp.ufl.edu

V. May
Institut für Physik
Humboldt-Universität zu Berlin
Newtonstraße 15
D-12489 Berlin, F. R. Germany
may@physik.hu-berlin.de

C. Meier
Laboratoire Collisions Agrégats
Réactivité, IRSAMC
Université Paul Sabatier
Toulouse, France
chris@irsamc.ups-tlse.fr

D.A. Micha
Departments of Chemistry
and Physics, University of Florida
Gainesville, FL 32611, USA
micha@qtp.ufl.edu

S. Miret-Artés
Instituto de Matemáticas
y Física Fundamental
Consejo Superior de
Investigaciones Científicas
Serrano 123
28006 Madrid
Spain
s.miret@imaff.cfmac.csic.es

K.B. Møller
Department of Chemistry
Technical University of Denmark
2800 Kgs. Lyngby, Denmark
klaus.moller@kemi.dtu.dk

Y. Öhrn
University of Florida
Gainesville FL 32611-8435
USA
ohrn@qtp.ufl.edu

E. Pollak
Department of Chemical Physics
Weizmann Institute of Science
76100, Rehovoth, Israel
eli.pollak@weizmann.ac.il

O.V. Prezhdo
Department of Chemistry
University of Washington
Seattle, WA 98195-1700
USA
prezhdo@chem.washington.edu

J.G.S. Ramon
Department of Chemistry and Center
for Materials Chemistry
University of Houston
Houston, TX, USA

A.S. Sanz
Chemical Physics Theory Group
Department of Chemistry
80 St. George St.
University of Toronto
Toronto, Canada M5S3H6
asanz@chem.utoronto.ca

B. Thorndyke
Radiation Oncology
Stanford Medical School
Palo Alto, CA
thorndyb@stanford.edu

M. Thoss
Department of Chemistry
Technical University of Munich
D-85748 Garching
Germany
thoss@ch.tum.de

D.G. Truhlar
Department of Chemistry
and Supercomputing Institute
University of Minnesota
Minneapolis, MN 55455-0431, USA
truhlar@chem.umn.edu

D. Tsivlin
Institut für Physik
Humboldt-Universität zu Berlin
Newtonstraße 15, D-12489 Berlin
F. R. Germany
tsivlin@physik.hu-berlin.de

H. Wang
Department of Chemistry
and Biochemistry, MSC 3C
New Mexico State University
Las Cruces, NM 88003, USA
whb@intrepid.nmsu.edu

Part I

Complex Molecular Phenomena

I.1 Condensed Matter and Surface Phenomena

Photoexcitation Dynamics on the Nanoscale

O.V. Prezhdo, W.R. Duncan, C.F. Craig, S.V. Kilina, and B.F. Habenicht

Summary. The chapter describes real-time ab initio studies of the ultrafast photoinduced dynamics observed in quantum dots, carbon nanotubes, and molecule-semiconductor interfaces. The theoretical modeling of these nanomaterials establishes the relaxation and charge transfer mechanisms and uncovers a number of unexpected features that explain the experimental observations. In particular, the ultrafast electron injection from alizarin into TiO_2 surface occurs via strong coupling to a few surface states rather than through the commonly assumed interaction with multiple TiO_2 bulk states. The injection does not require high densities of acceptor states and, therefore, can function close to the edge of the conduction band, avoiding energy losses and maximizing voltages attainable in Grätzel solar cells. The phonon-induced electron and hole relaxation in the PbSe quantum dots is symmetric and slow. As a result, the carrier multiplication that generates multiple electron–hole pairs and increases solar cell efficiency becomes possible. In contrast to quantum dots, the relaxation of charge carriers in carbon nanotubes is mediated by the high frequency phonons and is, therefore, fast. Substantial contribution of the low frequency breathing modes to the dynamics of holes, but not electrons rationalizes why holes decay slower and over multiple timescales, even though they have been expected to decay more rapidly due to their denser state manifold. The systems considered here are representative of a wide spectrum of problems and contribute to the general framework for control and utilization of the novel nanomaterials.

1 Introduction

Rapid advances in chemical synthesis and fabrication techniques generate novel types of materials that exhibit original and often unforeseen properties and phenomena. These are immediately studied by physical detection tools that probe material response to a variety of perturbations. The experimental data generated in such measurements demand understanding and interpretation that are greatly facilitated by theoretical modeling and simulation. The current chapter presents three closely related theoretical studies of charge transfer and relaxation phenomena recently observed in novel nanoscale materials using ultrafast laser spectroscopies. The materials under investigation are the molecule–semiconductor interface and the two types of quantum confinement devices, including the quasi-zero-dimensional quantum dots (QD) and quasi-one-dimensional carbon nanotubes (CN). The motivation for the studies largely stems from the search for alternative energy sources. The materials under investigation have the potential to replace the existing solar cells with more efficient designs and to generate chemically stored energy, such as hydrogen obtained by splitting water. The questions raised by the experimental observations and elucidated by the described simulations bear on a wide range of problems encountered in molecular and nanoscale electronics, spintronics and quantum information processing, biological imaging and detection, etc.

The first problem addressed below deals with the photoinduced charge transfer across a molecule–semiconductor interface. The system originates from the Grätzel type solar cell and provides an excellent example of numerous issues that arise when two fundamentally different types of systems are brought together. Molecules, typically studied by chemists, show finite sets of discrete, localized quantum states. Bulk semiconductors, on the other hand, are studied by physicists, and are characterized by continuous bands of delocalized orbitals. The intrinsic difference in the quantum states of the two systems, as well as the often disparate sets of theories and experimental tools used by chemists and physicists, create challenges for the study of the molecule–semiconductor interface. Similar issues arise in nanoscale electronics, where small molecular objects are sandwiched between bulk electrodes.

The second and third projects tackle charge and energy relaxation in recently created materials showing quantum confinement effects. Originally considered to be nanoscale derivatives of bulk materials with related properties, QDs and CNs have taken on a life of their own and are now often regarded as artificial atoms and nanowires, due to their close resemblance to traditional molecular objects. Yet QDs and CNs are in neither the bulk nor the molecular regime, and each exhibit an entirely new range of properties placing them squarely in between the two traditional types of materials. The study of the electron and hole relaxation in the QDs reported below is prompted by the recent experimental detection of multiple charge carriers generated upon absorption of only a single photon. The study investigates the mechanisms for increasing the current and voltage in photovoltaic cells and also directly relates to spintronic and quantum computing applications of quantum dots, in particular by establishing the limits on vibrationally induced dephasing that must be avoided. The electron and hole relaxation facilitated by phonons results in CN heating and is critical to understand for successful development of nanotube-based miniature electronic devices. The relaxation plays a key role in CN- and fullerene-based photovoltaic designs.

The simulations described below became possible with the development of the state-of-the-art theoretical tools designed to tackle the specific problems, which resulted in theoretical advances important in their own regard. The theoretical approaches are explained in Sect. 2, which is followed by the sections on the molecule–semiconductor charge injection, the excitation dynamics in

QDs and the electron and hole relaxation in CNs. The chapter concludes with a summary and a broader prospective of the key results.

2 Theoretical Approaches

The simulations are performed using the time-dependent (TD) Kohn–Sham (KS) density functional theory (DFT) for electron-nuclear dynamics, where the electrons are described quantum-mechanically, while the much heavier and slower nuclei are treated classically. Three variants of the theory are used. They share the same equations for the electronic evolution and differ in the implementation of the nuclear dynamics that is chosen depending on the problem under consideration and computational simplicity. DFT provides a modern and versatile means for the investigation of molecular and solid state structures, reaction pathways, thermochemistry, dipole moments, spectroscopic response, and many other properties [1, 2]. It is accurate, flexible, and computationally efficient compared to the Hartree–Fock and post-Hartree–Fock methods [3]. The electron-nuclear TDKS theory is implemented within the VASP code that provides a commercially available distribution of time-independent DFT [4, 5].

2.1 Time-Dependent Kohn–Sham Theory for Electron-Nuclear Dynamics

The electron density is the central quantity in DFT. It is represented in the KS theory [6] as the sum over single-electron KS orbitals $\varphi_p(x,t)$

$$\rho(x,t) = \sum_{p=1}^{N_\mathrm{e}} |\varphi_p(x,t)|^2 , \tag{1}$$

where N_e is the number of electrons. The time-evolution of $\varphi_p(x,t)$ is determined by application of the Dirac TD variational principle to the KS energy

$$E\{\varphi_p\} = \sum_{p=1}^{N_\mathrm{e}} \langle\varphi_p|K|\varphi_p\rangle + \sum_{p=1}^{N_\mathrm{e}} \langle\varphi_p|V|\varphi_p\rangle + \frac{e^2}{2}\iint \frac{\rho(x',t)\rho(x,t)}{|x-x'|}\mathrm{d}^3x\mathrm{d}^3x' + E_{xc}. \tag{2}$$

The right-hand side of (2) gives the kinetic energy of noninteracting electrons, the electron-nuclear attraction, the Coulomb repulsion of density $\rho(x,t)$, and the exchange-correlation energy functional that accounts for the residual many-body interactions. Application of the variational principle leads to a system of single-particle equations [1, 2, 7–9]

$$\mathrm{i}\hbar\frac{\partial\varphi_p(x,t)}{\partial t} = H(\varphi(x,t))\varphi_p(x,t),\ p = 1,\ldots,N_\mathrm{e}, \tag{3}$$

where the Hamiltonian H depends on the KS orbitals. In the generalized gradient approximation [10] used in the current simulations, E_{xc} depends on both density and its gradient, and the Hamiltonian is written as

$$H = -\frac{\hbar^2}{2m_e}\nabla^2 + V_N(x;\mathbf{R}) + e^2\int\frac{\rho(x')}{|x-x'|}\,\mathrm{d}^3x' + V_{xc}\{\rho,\nabla\rho\}\,. \tag{4}$$

The KS energy (2) may be related to the expectation value of the Hamiltonian with respect to the Slater determinant (SD) formed with the KS orbitals [6]

$$H = \langle\varphi_a\,\varphi_b\,\cdots\,\varphi_p|H|\varphi_a\,\varphi_b\,\cdots\,\varphi_p\rangle. \tag{5}$$

The single-electron density (1) is obtained from the SD by tracing over $N_e - 1$ electrons.

$$\rho(x_1) = N_e\,\mathrm{Tr}_{x_2,\ldots,x_{N_e}}|\varphi_a(x_1)\varphi_b(x_2)\cdots\varphi_p(x_{N_e})\rangle\langle\varphi_a(x_1)\varphi_b(x_2)\cdots\varphi_p(x_{N_e})|. \tag{6}$$

The TD KS orbitals $\varphi_p(x,t)$ are expanded in the basis of adiabatic KS orbitals $\tilde{\varphi}_k\,(x;\mathbf{R})$ that are the single-electron eigenstates of the KS Hamiltonian (4)

$$\varphi_p(x,t) = \sum_k^{N_e} c_{pk}(t)|\tilde{\varphi}_k\,(x;\mathbf{R})\rangle. \tag{7}$$

The adiabatic KS orbital basis is readily available from a time-independent DFT calculation [4,5] and provides a preferable representation for one of the nuclear dynamics approaches described below. The TDKS equation (3) transforms in the adiabatic KS basis to the equation for the expansion coefficients

$$\mathrm{i}\hbar\frac{\partial}{\partial t}c_{pk}(t) = \sum_m^{N_e} c_{pm}(t)\left(\epsilon_m\delta_{km} + \mathbf{d}_{km}\cdot\dot{\mathbf{R}}\right). \tag{8}$$

The nonadiabatic (NA) coupling

$$\mathbf{d}_{km}\cdot\dot{\mathbf{R}} = -\mathrm{i}\hbar\langle\tilde{\varphi}_k\,(x;R)|\nabla_{\mathbf{R}}|\,\tilde{\varphi}_m\,(x;R)\rangle\cdot\dot{\mathbf{R}} = -\mathrm{i}\hbar\langle\tilde{\varphi}_k\,|\frac{\partial}{\partial t}|\,\tilde{\varphi}_m\rangle \tag{9}$$

arises from the dependence of the adiabatic KS orbitals on the nuclear trajectory and is computed from the right-hand-side of Eq. (9) [11]. Similarly to (7), the time-evolving SD (see (5)) evolves into a superposition of adiabatic SDs

$$|\varphi_a\,\varphi_b\,\cdots\,\varphi_p\rangle = \sum_{j\neq k\neq\cdots\neq l}^{N_e} C_{j\,\ldots\,l}(t)\left|\tilde{\varphi}_j\,\tilde{\varphi}_k\,\cdots\,\tilde{\varphi}_l\right\rangle \tag{10}$$

with the many-electron coefficients $C_{j\,\ldots\,l}(t)$ expressed in terms of the single-electron coefficients

$$C_{j\,\ldots\,l}(t) = c_{pj}(t)\,c_{qk}(t)\,\cdots\,c_{vl}(t). \tag{11}$$

The evolution of $\mathcal{C}_{j\cdots l}$ follows from (8)

$$i\hbar\frac{\partial}{\partial t}\mathcal{C}_{q\cdots v}(t) = \sum_{a\cdots p}^{N_e} \mathcal{C}_{a\cdots p}(t)\left[E_{q\cdots v}\delta_{aq}\cdots\delta_{pv} + \mathbf{D}_{a\cdots p;q\cdots r}\cdot\dot{\mathbf{R}}\right]. \tag{12}$$

$E_{q\cdots v}$ is the many-electron eigenenergy, and the many-electron NA coupling

$$\mathbf{D}_{a\cdots p;q\cdots r}\cdot\dot{\mathbf{R}} = -i\hbar\langle\widetilde{\varphi}_a\,\widetilde{\varphi}_b\,\cdots\,\widetilde{\varphi}_p\,|\frac{\partial}{\partial t}|\widetilde{\varphi}_q\,\widetilde{\varphi}_r\,\cdots\,\widetilde{\varphi}_v\,\rangle. \tag{13}$$

is nonzero only if the determinants differ in a single KS orbital.

2.2 The Classical Path Approximation

The equations above define dynamics of the electronic subsystem evolving in response to the nuclear degrees of freedom that determine the electron-nuclear potential V in the Hamiltonian (4). The nuclear trajectory $\mathbf{R}(t)$ has yet to be defined. While it is common and straightforward to define the effect of the classical nuclei on the quantum electrons through the $\mathbf{R}$-dependence of the electron-nuclear potential, the back-reaction of the electrons on the classical nuclei is not straightforward. Numerous prescriptions have been proposed, each with its own merits [11–18, 20–45, 66]. All quantum-classical approximations, however, violate some essential properties seen in a fully quantum electron-nuclear dynamics. The classical path approximation (CPA) provides the simplest solution by ignoring the back-reaction and assuming that the classical path is predetermined [13, 14]. The CPA is the simplest and computationally most efficient approximation, and is a valid approach if the nuclear dynamics are not sensitive to changes in the electronic subsystem. The classical nuclear trajectory associated with the electronic ground state is often used in cases where excited state potential energy surfaces (PES) are similar to the ground state PES, and where the nuclear kinetic energy and thermal fluctuations of the nuclei are large compared to the differences in the PES.

2.3 The Ehrenfest Nuclear Dynamics

The mean-field or Ehrenfest [46] approximation is the simplest form of the back-reaction of electrons on the nuclei. Here, the classical variables couple to the expectation value of the quantum force operator [12–14]

$$M\ddot{\mathbf{R}} = -\langle\varphi_a\,\varphi_b\,\cdots\,\varphi_p|\nabla_{\mathbf{R}}H|\varphi_a\,\varphi_b\,\cdots\,\varphi_p\rangle. \tag{14}$$

The gradient $\nabla_{\mathbf{R}}$ is applied directly to the Hamiltonian according to the TD Hellmann–Feynman theorem [14]. Thoroughly investigated by many authors, the Ehrenfest method remains valid under the conditions similar to those needed for the CPA and requires modification when electron-nuclear correlations [16] and detailed balance must be taken into account [17, 47, 48]. Advanced versions of the Ehrenfest approach include "quantum fluctuation variables," alleviating some problems [47–57]. More radical solutions are provided by other techniques; one of the most popular and efficient is the trajectory surface hopping (SH) approach [11, 14–18, 20–31, 66].

2.4 Surface Hopping

In SH, the nuclear trajectory responds to the electronic forces by stochastically "hopping" between electronic states [11, 14–18, 20–31, 66]. Analytical and numerical arguments have indicated that SH should be performed in the adiabatic representation (7). Among many flavors of SH, the fewest-switches (FS) SH is designed to minimize the number of hops and satisfy a number of other key physical criteria [16]. The nuclear trajectory in SH propagates adiabatically

$$M\ddot{\mathbf{R}} = -\langle \tilde{\varphi}_a \tilde{\varphi}_b \cdots \tilde{\varphi}_p | \nabla_{\mathbf{R}} H | \tilde{\varphi}_a \tilde{\varphi}_b \cdots \tilde{\varphi}_p \rangle \tag{15}$$

rather than in the mean-field, (14). The probability that the nuclear trajectory hops to another adiabatic state over time interval dt is

$$\mathrm{dP}_{a\cdots p;q\cdots r} = \frac{B_{a\cdots p;q\cdots r}}{A_{a\cdots p;q\cdots r}} dt, \tag{16}$$

where

$$B_{a\cdots p;q\cdots r} = -2\,\mathrm{Re}\big(A^*_{a\cdots p;q\cdots r} \mathbf{D}_{a\cdots p;q\cdots r} \cdot \dot{\mathbf{R}}\big); \quad A_{a\cdots p;q\cdots r} = C_{a\cdots p} C^*_{q\cdots r}. \tag{17}$$

If the calculated $\mathrm{dP}_{a\cdots p;q\cdots r}$ is negative, the hopping probability is set to zero. After the hop, the nuclear trajectory continues adiabatically in the new state $q \cdots r$. In order to conserve the total electron-nuclear energy after a hop, the nuclear velocities are rescaled [11, 16] along the direction of the electronic component of the NA coupling $\mathbf{D}_{a\cdots p;q\cdots r}$. If a NA transition to a higher energy electronic state is predicted by (16), and the kinetic energy available in the nuclear coordinates along the direction of the NA coupling is insufficient to accommodate the increase in the electronic energy, the hop is rejected. The velocity rescaling and hop rejection produce detailed balance between upward and downward transitions [17].

The CPA can be adapted to SH in order to achieve computational speed-up and improved statistical sampling. The SH probabilities can be computed based on the ground state nuclear trajectory if the following assumptions hold (1) that the electronic PES are similar, (2) that the electronic energy dumped after a hop rapidly distributes among all vibrational modes. The detailed balance that is achieved in the original FSSH by the nuclear velocity rescaling performed after each transition is restored by multiplying the probability of transitions upward in energy by the Boltzmann factor.

While several SH procedures have been derived using the partial Wigner transform techniques [23–25], most SH approaches remain ad hoc. SH can be viewed as a quantum master equation with the transitions probabilities that are computed nonperturbatively and on-the-fly for the current nuclear configuration. In contrast to the traditional quantum master equations, SH is capable of describing the short-time Gaussian component of quantum dynamics that is responsible for the quantum Zeno effect and related phenomena [30, 58–61].

3 Ultrafast Photoinduced Electron Injection in Dye-Sensitized TiO_2

Electron transfer (ET) at organic/inorganic interfaces plays a key role in many areas of research, including molecular electronics [62–66], photo-electrolysis [67], photo-catalysis [68–71] and color photography [72]. ET at semiconductor interfaces constitutes the primary step in novel photovoltaic devices comprised of dye-sensitized semiconductors [73–82], assemblies of inorganic semiconductors with conjugated polymers [83–86], and quantum confinement devices [87, 88]. The exact mechanistic details of the interfacial ET in these materials are an issue of practical importance and theoretical debate.

The alizarin–TiO_2 interface is a particular example of the photoinduced charge separation component in the Grätzel cell, where highly porous nanocrystalline titanium dioxide is sensitized with transition metal or organic dye molecules [73–75]. Grätzel cells offer a promising alternative to the more costly traditional solar cells. Absorption of light excites the dye-sensitizer molecules from their ground state, which is located energetically in the semiconductor band gap, to an excited state that is resonant with the TiO_2 conduction band (CB) (Fig. 1). The electron is then transferred on the ultrafast timescale to the semiconductor, which is in contact with one of the electrodes. Upon carrying an electric load and reaching the second electrode, the electron enters an electrolyte that carries it back to the chromophore ground state. Ultrafast laser techniques have shown that electron injection can occur in less than 100 fs [76–82], making it difficult to invoke traditional ET models, which require slow ET dynamics to allow for redistribution of vibrational energy [77].

Direct modeling of the ultrafast electron injection processes between dyes and semiconductors observed in laser experiments has been performed with reduced models and a full quantum-mechanical description of electrons and nuclei [81, 89, 90] and at a detailed atomistic level using a quantum description of electrons and classical treatment of nuclei [91–98]. The first real-time ab initio atomistic simulation of the interfacial ET were carried out in our group [92–94]. The isonicotinic acid dye was chosen to have an excited state well within the semiconductor CB, since the researchers usually assume that a high density of semiconductor states is needed for fast and efficient ET [73–76, 99–101].

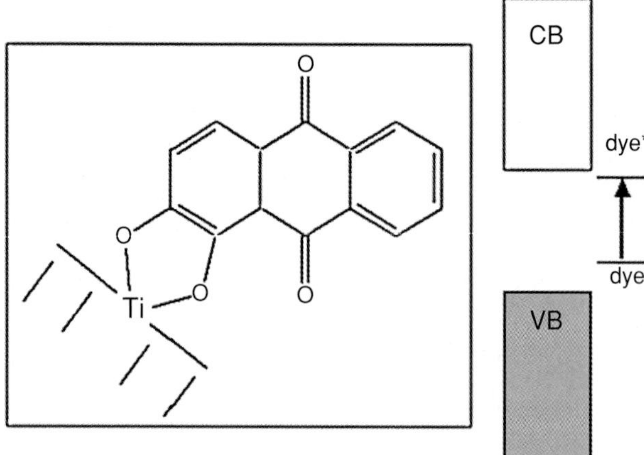

Fig. 1. Dye-sensitized TiO_2. Upon photoexcitation, alizarin chemisorbed onto the TiO_2 surface transfers an electron into the semiconductor. The ground state of alizarin is in the gap between the valence band (VB) and conduction band (CB). The excited state of alizarin is energetically near the edge of the semiconductor CB, and nontrivial electron injection dynamics ensues as the state crosses in and out of the band

The alizarin/TiO_2 system investigated in our group most recently represents an interesting and novel case in which the photoexcited state is positioned near the band edge. The experiments show that electron injection from the alizarin excited state near the TiO_2 CB edge is no less efficient than ET from chromophores with excited states deep in the CB. Moreover, the injection is extremely fast with a record 6 fs transfer time [80]. Efficient ultrafast injection from photoexcited states near the CB edge is both fundamentally interesting and practically important. The fundamental question is: what mechanisms make the ET so fast in the absence of a high density of acceptor states? On the practical side, injection at the CB edge has the potential to aid in the design of cells with higher maximum theoretical voltage, since energy will not be lost by rapid relaxation to the bottom of the CB [102]. We have modeled the injection dynamics by the classical path approximation in the ab initio TDKS theory described in the previous section. The simulation has uncovered a number of novel features of the injection process that are not observed in the previously studied cases [73–82, 89–94].

3.1 Nuclear Dynamics

Nuclear dynamics have a twofold influence on the ultrafast electron injection process. On the one hand, thermal fluctuations of the nuclei create an ensemble of initial conditions with slightly different geometries and photoexcitation energies. On the other hand, upon photoexcitation, nuclei drive ET by moving

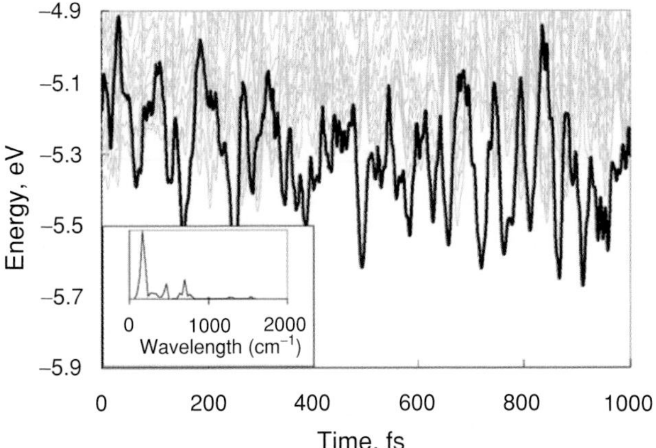

Fig. 2. Evolution of the photoexcited (*thick line*) and CB (*thin lines*) state energies in alizarin-sensitized TiO_2. The energy of the photoexcited state fluctuates by about 0.15 eV due to atomic motions. The fluctuation is small relative to the 2.5 eV excitation energy, but it moves the dye state into and out of the TiO_2 CB. *Insert*: Fourier transform of the photoexcited state energy shows low frequency peaks associated with alizarin and surface atoms motions up to the 1,600 cm^{-1} frequency of the C–C stretching

along the reaction coordinate and, alternatively, by inducing direct quantum transitions between the donor and acceptor states. The evolution of the photoexcited and CB state energies is presented in Fig. 2. The fluctuation of the energies at room temperature is sufficient to move the photoexcited state into and out of the TiO_2 CB, generating two ET regimes. Outside of the band the coupling of the chromophore excited state to the semiconductor states is small. Inside the band the density of states (DOS) grows substantially with increasing energy, and the chromophore excited state can therefore interact with a larger number of TiO_2 states. The Fourier transform (FT) of the photoexcited state energy is shown in the insert of Fig. 2. The main contributions to the energy fluctuation are seen at the frequencies below 700 cm^{-1}, corresponding to bending and torsional motions. Small peaks are seen up to 1,600 cm^{-1}, characteristic of the C–C and C=O stretches. Vibrations above 1,600 cm^{-1} do not contribute to the oscillation of the photoexcited state energy, although they do contribute to the fluctuation of the photoexcited state localization and, therefore, dye-semiconductor coupling [96].

3.2 Distribution of Initial Conditions for ET

Thermal fluctuations of atomic coordinates produce a distribution of the photoexcited state energies and localizations that creates an inhomogeneous ensemble of initial conditions for the electron injection. Near the CB edge the TiO_2 DOS is low, and there is very little mixing between the alizarin excited

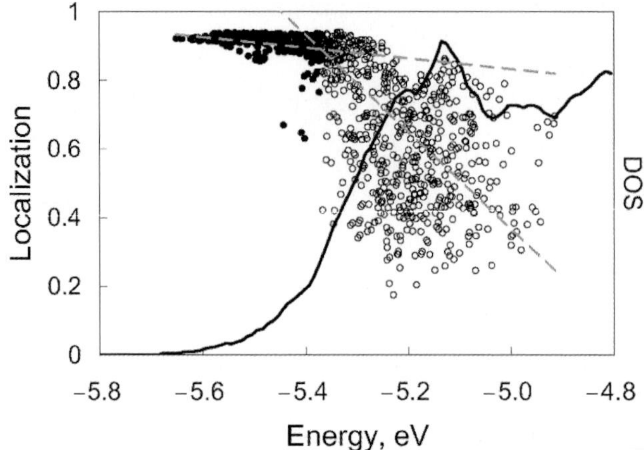

Fig. 3. Localization of the photoexcited state on alizarin (*circles*) and the TiO_2 DOS (*solid line*) as functions of energy. Below the CB (*filled circles*) the photoexcited state is localized on the dye. Above the CB (*empty circles*) the state is significantly delocalized into the semiconductor. The delocalization parallels the increasing TiO_2 DOS. The large spread of the localizations inside the CB is due to fluctuations in the chromophore–semiconductor interaction

state and the semiconductor, Fig. 3. The localization of the photoexcited state on the alizarin molecule (filled circles) is therefore close to one. As the energy increases, progressively more CB states couple to the chromophore. Under these circumstances the localization decreases (empty circles) and significant amounts of ET occur already during the photoexcitation. The large spread in the localization data at higher energies is due to fluctuations in the surface that cause changes in the energies, spatial extent and localization of the semiconductor surface states. The number of semiconductor states that the chromophore can couple to at a particular energy varies with the atomic configuration. Even if the density of acceptor states is the same, the spatial overlap between these states and the chromophore excited state vary substantially, depending on the current geometry of the docking region. Despite the spread of the localization data, there is a clear difference between the photoexcited states below and above the CB edge.

3.3 The Mechanism of Electron Injection

Two competing ET mechanisms have been proposed to explain the observed ultrafast injection events [76]. These mechanisms have drastically different implications for the variation of the interface conductance and solar cell voltage with system properties. In the adiabatic mechanism, the coupling between the dye and the semiconductor is large, and the ET occurs through a transition state (TS) along the reaction coordinate that involves a concerted motion of nuclei. During adiabatic transfer, the electron remains in the same Born-Oppenheimer (adiabatic) state that continuously changes its localization from the dye to the semiconductor along the reaction coordinate. A small TS barrier relative to the nuclear kinetic energy gives fast adiabatic ET. NA effects decrease the amount of ET that happens at the TS, but open up a new channel involving direct transitions from the dye into the semiconductor that can occur at any nuclear configuration. The NA transfer becomes important when the dye–semiconductor coupling is weak. Similar to tunneling, the NA transfer rate shows exponential dependence on the donor–acceptor separation.

The dynamics of the electron injection from alizarin to TiO_2 are presented in Fig. 4. The ET is determined by the portion of the electron that has left the dye. The timescales and relative amounts of adiabatic and NA ET are computed by separating the overall ET evolution into the contributions that are due to changes in the localization of the initially occupied state and populations of the initially empty states, respectively. In order to obtain the ET timescale, the total ET is fit by the equation,

$$\mathrm{ET}(t) = \mathrm{ET_f}(1 - \exp\left[-(t + t_0)/\tau\right]), \quad (18)$$

where ET_f is the final amount of ET, and τ is the timescale. The fact that the photoexcited state is initially delocalized onto the surface is reflected by the t_0 term of the fit. The t_0 fitting constant can be interpreted as the time the ET is advanced by the photoexcitation. Due to the delocalization of the photoexcited state onto the semiconductor, Fig. 3, about 25% of the electron is already on the surface after the photoexcitation. The adiabatic and NA ET are fit with a similar equation, but without the t_0 term. The adiabatic

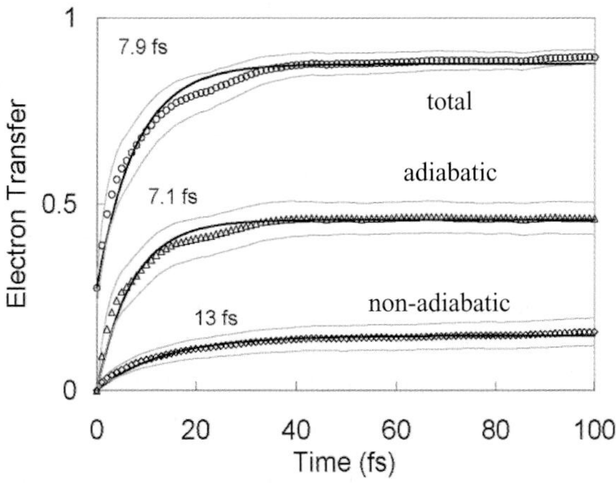

Fig. 4. ET dynamics in the alizarin–TiO_2 system averaged over 900 initial conditions and separated into the adiabatic and NA components. The thin grey lines represent 20% of the variance in the ET data. The thick black lines are fits by (18) with the timescales τ shown on the figure

mechanism dominates the dynamics and is not only faster but also reaches a much higher amplitude than the NA component. The thin grey lines show 20% of the variance of the data. The variance is quite large, which indicates a large diversity in the individual electron injection events, depending on the initial condition. The small oscillations in the total and adiabatic ET data, relative to the fit line, are similar to those observed by Willig and co-workers with perylene [82], and are due to coherent nuclear vibrations.

The ET events that originate from the photoexcited states above and below the CB, Fig. 3, show significant differences [96]. The ET dynamics starting in the photoexcited states above the band gap are qualitatively similar to the average ET dynamics shown in Fig. 4. The photoexcited state is more delocalized and the ET proceeds faster at energies above the CB. Both adiabatic and NA transfer components are faster at the higher energies. Because the DOS increases with energy, Fig. 3, there is a shorter wait until a surface state that is strongly coupled to the dye state is found and adiabatic ET takes place. A larger DOS provides more semiconductor states to interact with, leading to faster NA ET. The electron injection dynamics at high initial energies are even more dominated by the adiabatic mechanism than the dynamics averaged over all initial conditions. In contrast, the ET coordinate and its adiabatic component are markedly different for the initial states that are below the CB edge [96]. The ET is not exponential during the first 8 fs and is best fit by an inverted Gaussian, reflecting the fact that the donor state must enter the CB before crossing with an acceptor state. Once the dye state has moved into the CB, the ET can be fit with an exponential. The state crossing is not required by NA ET, which behaves exponentially even for the lower energy initial conditions. Both adiabatic and NA ET components are slowed down for the initial states below the CB. It is quite remarkable that photoexcitation below the CB can lead to fast and efficient electron injection [95, 96].

4 Excitation Dynamics in Quantum Dots

QDs have the potential to substantially improve the conversion of solar energy into electric current, thereby producing more efficient solar cells. The tunability of the absorption spectrum of QDs with their size circumvents the need for sensitizer chromophores as in the Grätzel cell, Sect. 3. The control of the charge carrier relaxation pathway with QD type, size and surface passivation creates additional tools for improving photovoltaic devices.

Conversion efficiency is one of the most important parameters to optimize in order to implement photovoltaic cells on a truly large scale [103]. The maximum thermodynamic efficiency for the conversion of unconcentrated solar irradiance into electrical free energy in the radiative limit assuming detailed balance and a single threshold absorber was calculated by Shockley and Queisser [104] in 1961 to be slightly above 30%. QD solar cells have the potential to increase the maximum attainable thermodynamic conversion efficiency of solar photon conversion up to about 66% by utilizing hot photogenerated carriers. There are two fundamental ways to utilize the hot carriers for enhancing the efficiency of photon conversion: by enhancing either the photovoltage or the photocurrent. Enhanced photovoltage requires that the carriers be extracted from the photoconverter before they cool [105]. Enhanced photocurrent requires the energetic hot carriers to produce two or more electron–hole pairs through impact ionization [106] – a process that is the inverse of an Auger process whereby two electron–hole pairs recombine to produce a single highly energetic electron–hole pair. In order to achieve higher voltages, the rates of photogenerated carrier separation, transport, and interfacial transfer across the contacts to the semiconductor must all be fast compared to the rate of carrier cooling [107]. Achieving larger currents requires that the rate of impact ionization be greater than the rates of cooling and other relaxation processes of hot carriers.

Over the past several years many investigations have been published that explore hot electron and hole relaxation dynamics in QDs. The results are controversial. It is quite remarkable that there are so many reports that both support [88, 108–120] and contradict [121–127] the prediction of the existence of a phonon bottleneck to the hot–electron cooling in QDs, defined as a strong reduction in the efficiency of electron–phonon interaction. A number of groups have investigated QDs created with III–V semiconductor materials, such as GaAs, InAs, and InP, and reported slowed charge-carrier cooling due to the QD quantization effects [108–113]. Relaxation of both hot electrons [108–111, 128] and holes [112, 113] was considerably slowed down relative to the bulk materials. The studies of QDs of the II–VI type, and CdSe in particular, found two relaxation time scales, whose relative weights depended upon the molecules capping the QDs [114–119]. A phonon bottleneck was observed similar to the III–V QDs. In addition, a new, faster relaxation component was seen and attributed to the Auger mechanism for electron relaxation, whereby the excess electron energy is rapidly transferred to a hole, which then relaxes rapidly through its dense spectrum of states. If the hole is removed and trapped by the molecules capping the QD surface, the Auger mechanism for the hot electron relaxation is inhibited and the overall relaxation time increases. However, there are many investigations that indicate no phonon bottleneck. Such results were reported for both III-V QDs [121–123] and II–VI QDs [124, 125]. In some cases [126, 127] hot-electron relaxation was found to be slowed only slightly.

The breakthrough in the studies of carrier multiplication came with the observations of multiple electron–hole pairs in PbSe QDs upon absorption of high energy photons [88, 120]. The observations raise questions over why certain relaxation pathways fail to quench impact ionization in PbSe QDs, when they are so effective in QDs composed of other materials. Using the FSSH approach implemented within DFT as described in Sect. 2, we investigate the relaxation mechanisms and establish that both electron and hole relaxation in

PbSe QDs is slow, allowing time for the carrier multiplication and eliminating the Auger relaxation pathway that transfers the electron energy to the rapidly relaxing holes.

4.1 Electronic States of Quantum Dots

The electronic structure of QDs is intimately related to their quasizero-dimensional nature that makes them closer to atoms than bulk materials. For this reason, QDs are often called "artificial atoms," and assemblies of QDs are referred to as "artificial molecules." The reduction in the system dimensionality that accompanies the transition from bulk semiconductors to QDs is associated with a dramatic transformation in the energy spectra, which become discrete and atomic-like. At the qualitative level, the quantum energies of the electron and hole states can be understood by regarding the QD as a spherical potential that confines the noninteracting particles. The lowest states of both electrons and holes have an approximate spherical symmetry and are labeled as S-states. The next three levels show P-like character and are polarized along the x, y, and z-directions. [87, 111]

The electronic structure of QDs is exemplified in Figs. 5 and 6 with the 32 atom PbSe QD. The simple cubic lattice of bulk PbSe allows one to create the small roughly spherical nanocrystal of about 10Å in diameter that preserves the bulk symmetry. A structural relaxation of the 32 atom PbSe QD relative to the bulk does occur even at zero Kelvin. Temperature induced fluctuations

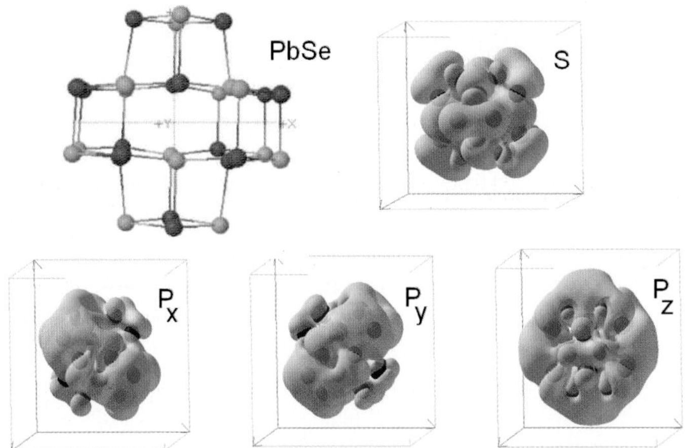

Fig. 5. Geometric structure of the PbSe QD under investigation and the spatial densities of its 4 lowest electron states. The simple cubic lattice of PbSe creates a stable 32 atom QD that preserves the bulk structure. The quantum energy levels of electrons and holes can be qualitatively understood by considering a particle in a spherical well. The lowest energy level of both electron and hole is S-like, the next three levels are P-like, etc.

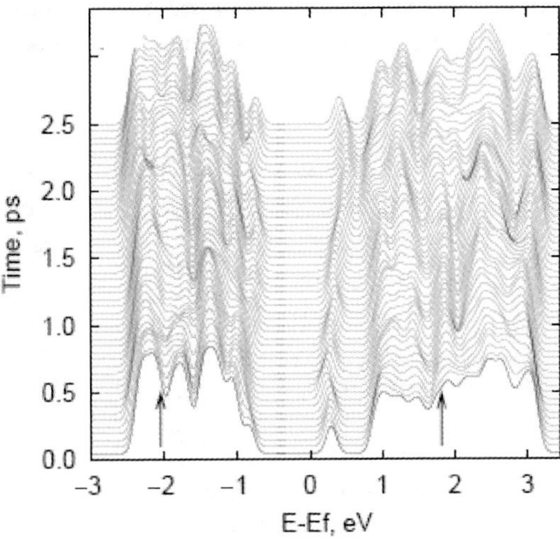

Fig. 6. DOS of the PbSe QD. The DOS fluctuates over time due to thermally induced nuclear motions. The arrows indicate the energies of the initial electron and hole excitations, which are set up to match the triple energy gap as in the experiments [88]

further distort the dot, but cause neither surfaces to reconstruct nor bonds reconnect. The four lowest electronic states shown in Fig. 5 exhibit the expected S- and P-like symmetries, which are significantly modulated by the local atomic structure. The energies of the hole and electron states shown in Fig. 6 fluctuate over time due to thermal nuclear motions. The DOS shown in Fig. 6 is constructed by the broadening of the energy levels with Gaussians of 0.01 eV width. The S-like lowest electron and hole states are clearly isolated from the rest of the states. The arrows in Fig. 6 indicate the energies of the electron and hole excitations. The energy range is set three times larger than the QD energy gap, in correspondence with the experiments. [88] The initial excited states for each nuclear configurations are chosen based on the largest transition dipole moments among the states close to the energies indicated by the arrows.

4.2 Phonon-Induced Relaxation of Electrons and Holes

The quantum confinement effects in QDs strongly affect not only the electronic spectrum, but also the rates and pathways of electron–phonon and hole–phonon relaxation. The reduced availability of pairs of electronic states that satisfy energy and momentum conservation can lead to a strong reduction in the efficiency of electron–phonon interactions in QDs, i.e., the phonon bottleneck. This effect dramatically slows down energy relaxation in zero-dimensional QDs in comparison to systems of higher dimensionality. Other, nonphonon mechanisms for energy relaxation in QDs include interactions with defects and Auger-type electron–hole interactions involving transfer of the electron excess energy to a hole, with subsequent fast hole relaxation through its dense spectrum of states. The relaxation effects are to be minimized in order to achieve the desired enhancement of the solar photon conversion efficiency [87].

The DOS of holes calculated for the 10Å PbSe QD shows a denser spectrum, compared to the DOS of electrons, Fig. 6. The separation of the S-like from the main state manifold is more pronounced for the electrons than for the holes. The difference in the electron and hole DOS is not as dramatic in PbSe QDs as, for instance, in the extensively studied CdSe QDs. The simulated relaxation dynamics are slightly faster for the holes than for the electrons, Fig. 7. The difference is minor, which explains why the Auger-type electron relaxation through energy transfer to the hole is not efficient in PbSe. The relaxation times for both holes and electrons is around 1 ps. This is orders of magnitude longer than the electron injection time in the alizarin–TiO_2 system considered in the previous section, and is several times longer than the closely related electron and hole relaxation times in CNs considered next. It may be expected that the simulation overestimates the relaxation rates, since in experiments the QD surfaces are passivated, decreasing the number of states in the relevant energy range. Comparing the DOS of Fig. 6 with the relaxation dynamics shown in the top panel of Fig. 7, one observes that the initial photoexcitation peak vanishes and reappears directly at the final P- and S-states. Although multiple states are visited by electrons and holes during the relaxation, none of the intermediate states play any special role. The slow and nearly symmetric electron and hole relaxation in the PbSe QD leads us to conclude that the observed carrier multiplication takes place due to the low rates of the other, unfavorable relaxation mechanisms.

5 Electron and Hole Relaxation in Carbon Nanotubes

Discovered in 1991 by Iijima [129], CNs continue to be at the forefront of scientific research. Their unique structural, mechanical, and electronic properties [130,131] prompt a variety of applications ranging from chemical sensors to computer logic gates and field-effect transistors [132–138]. Advancements in the synthesis and purification of CNs have enabled the study of size-selected tubes as well as rudimentary separation of metallic and semiconducting CNs [139, 140]. Developments in the spectroscopic techniques have accompanied the progress in the nanotube preparation. Numerous time-resolved experiments have addressed the electronic structure of CNs, revealing intriguing features in the nanotube response to electronic and optical excitations [141–149].

Motivated by the time-resolved experimental observations we performed the first real-time ab initio simulation of the electron and hole dynamics in a CN. The simulated dynamics agree with the experimental timescales, establish

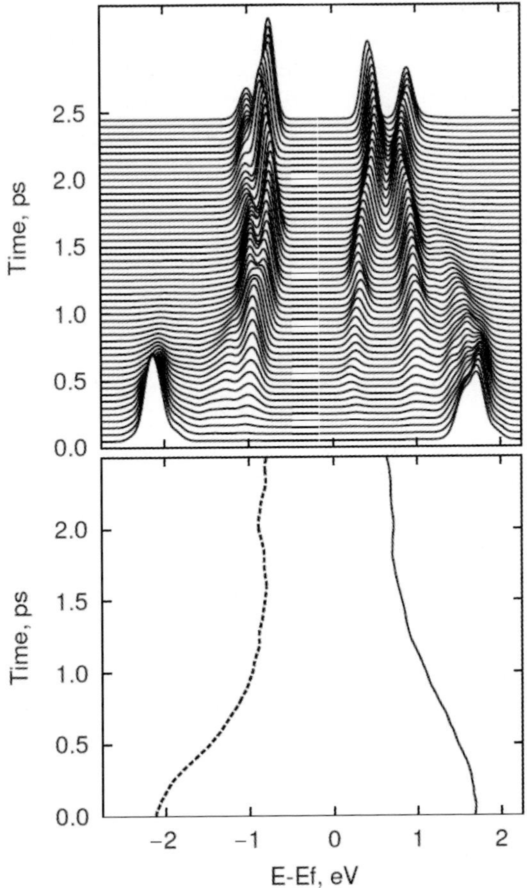

Fig. 7. Evolution of the electron and hole excitations in the QD averaged over 500 initial conditions. The top panel shows DOS multiplied by the time-dependent populations of the excited states. The bottom panel gives the average electron and hole energies

the electron and hole relaxation pathways, characterize the electronic states and phonon modes that facilitate the energy dissipation, and reveal a number of intriguing details of the relaxation processes. In particular, the simulation shows for the nanotube under investigation that the holes relax more slowly than electrons, even though the holes have a denser manifold of states facilitating the relaxation. The electrons show a single exponential decay, while holes relax by a Gaussian and then an exponential component. Both electron and hole relaxation is promoted primarily by the C–C stretching G-type phonons with frequencies around 1,500 cm^{-1}. However, holes, but not electrons, additionally couple to the lower frequency breathing modes.

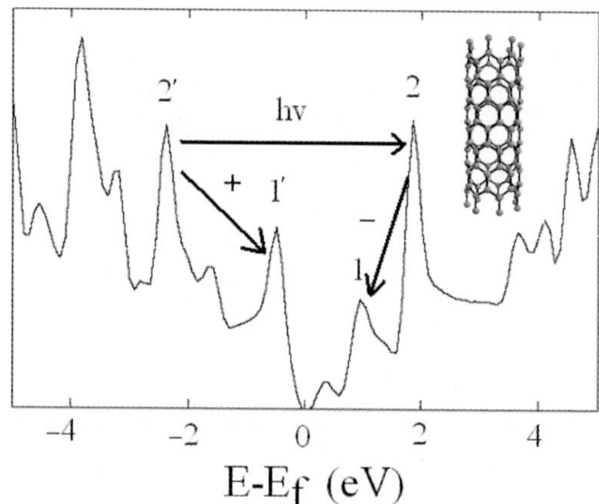

Fig. 8. DOS of the (7,0) zig-zag CN. The electron is optically excited between the second van-Hove singularities $2' \rightarrow 2$. The electron and hole then decay to the first singularities $1'$ and 1 on a subpicosecond timescale

The study described below is performed on the smallest semiconducting (7,0) zig-zag CN, since semiconductors can be simulated with fewer basis functions, and a zig-zag tube has a frequently repeated periodic pattern that helps to reduce the size of the simulation cell. The study is carried out with the SH approach described in the Theory section.

5.1 Electronic Structure of Carbon Nanotubes

The nanotube DOS exhibits characteristic van Hove singularities (vHs) due to the folding of the 1D Brillouin zone of a graphene sheet [130], Fig. 8. These singularities dominate the electronic spectrum of CNs. The curvature of the nanotube, together with electron-correlation effects, alters the DOS by creating an asymmetric distribution of states across the Fermi level with the holes having a denser manifold of states than the electrons. The electron and hole relaxation under investigation is initiated by an excitation from the second vHs below the Fermi level to the second vHs above the Fermi level, as in the recent ultrafast laser experiments [143–145]. The states within the singularities were chosen based on the strongest transition dipole moment at a given initial time. The electronic densities of the two most optically active electron and hole states are shown as inserts in Fig. 9. Upon the photoexcitation the electrons and holes relax nonradiatively through the first vHs to their corresponding band edges.

5.2 Phonons Facilitating Electron and Hole Relaxation

Figure 9 establishes the types of phonon modes that couple to the electrons and holes and promote the relaxation. The two pairs of states whose densities

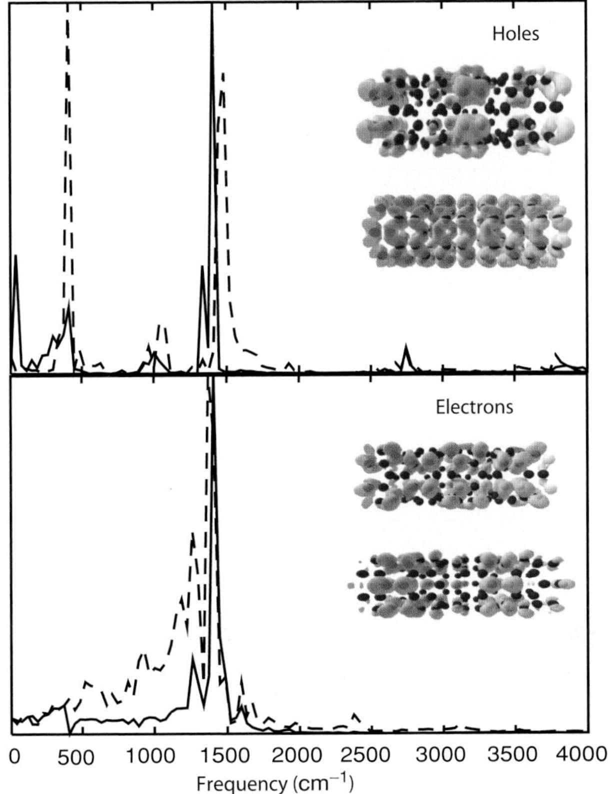

Fig. 9. Fourier transforms of the energies of the two most optically active states of electrons and holes in the CN. The insert shows the charge densities of these states. The C–C stretching G-modes around 1,500 cm^{-1} strongly couple to both electron and holes. The breathing modes at and below 500 cm^{-1} have fewer nodes and, as a result, better couple to the holes, whose states are lower in energy and also have fewer nodes than the electron states

are shown in the inserts account for 80% of the photoexcitation intensity. FTs of the phonon induced dynamics of the energies of these states are shown in Fig. 9. The FTs identify the modes that modulate the properties of the electron and hole states and create the NA coupling (13). The electron–phonon and hole–phonon coupling occurs over a broad range of frequencies starting at the C–C stretching G-type modes around 1,500 cm^{-1} down to the breathing modes below 500 cm^{-1}. The G-modes give the largest contribution to both electron and hole relaxation. In contrast to the electrons, holes also strongly couple to the breathing modes. The coupling of the holes to the lower frequency modes can be understood by considering the energies and densities of the electron and hole states. The lower energy valence band (VB) states supporting the holes have fewer nodes than the higher energy

CB states supporting the electrons. The hole states with fewer nodes couple to the lower frequency breathing phonons that also have fewer nodes. The stronger coupling to the breathing modes slows down the hole relaxation dynamics relative to that of the electrons, counteracting the effect of the denser manifold of hole states, Fig. 8, that facilitates the relaxation. Similarly, it can be expected that coupling to lower frequency phonons slows down the hole relaxation in QDs, Sect. 4, decreasing the rates of phonon heating and Auger relaxation and allowing the carrier multiplication.

5.3 Electron and Hole Relaxation Dynamics

Figure 10 details the electron and hole relaxation dynamics in CNs by showing the average electron and hole energies as functions of time. The energy of the electrons is fitted with a single exponential. The energy of the holes gives a poor single exponential fit and is described by a sum of the Gaussian and exponential components. The Gaussian component can be hardly distinguished in the electron relaxation. The exponential component of the hole relaxation is noticeably slower than that of the electrons, which is rather surprising since the holes have a larger DOS, Fig. 8, that facilitates relaxation. The Gaussian component accounts for nearly half of the hole energy relaxation and occurs while the hole spreads within the second vHs and before the Boltzmann weighting produces the exponential decay from the second to the first singularity. The existence of a smaller maximum in the second vHs of

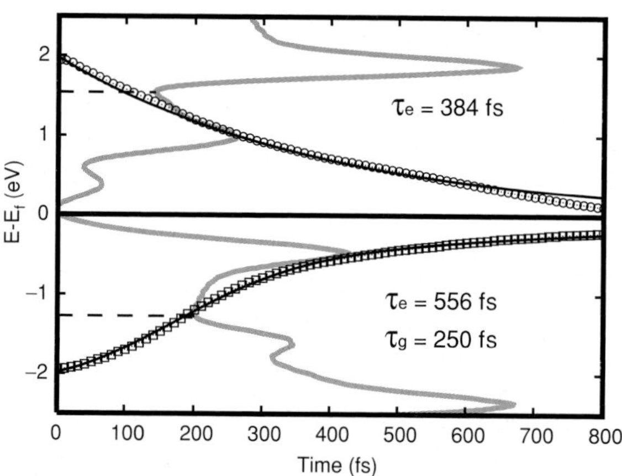

Fig. 10. Relaxation of the average energy of electrons and holes in the CN. The holes show Gaussian and exponential regimes, while the electrons follow a single exponential. The hole exponential decay is slower than that of the electrons, in spite of the higher density of hole states, Fig. 8, that facilitates faster relaxation. The slow dynamics of the holes can be attributed to the coupling with the low frequency breathing modes, Fig. 9

the hole DOS may additionally contribute to the Gaussian relaxation component. Minor deviations from the exponential fit seen with the electron energy toward the end of the relaxation are most likely related to the small peak in the electron DOS near the Fermi level, Fig. 8.

The timescales for the electron and hole relaxation from the 2 and $2'$ singularities to the band edge computed for the (7,0) tube are in good agreement with the reported ultrafast spectroscopy experiments. The experimental results for this type of process vary from tens of femtoseconds [142–144] to picoseconds [145–147] depending on sample preparation, size homogeneity, photoexcitation energy, intensity, and type of experiment. The simulation gravitates toward the slower end of the experimental data and provides an upper bound on the relaxation time, since other relaxation mechanisms, most notably charge–charge scattering and electron–hole annihilation, have not been included in the simulation.

6 Conclusions

The three case-studies described above provide a sampling of the exciting phenomena observed with nanomaterials in the very recent past. The state-of-the-art theoretical tools developed in our group allowed us to characterize the excitation dynamics in these nanomaterials, establish the mechanisms of charge transfer and relaxation, and uncover a number of interesting and practically important features that are accessible only from simulation and that explain the unexpected experimental observations.

We showed that the ultrafast electron injection takes place in the alizarin-TiO_2 system not through the commonly assumed coupling to multiple bulk states of the semiconductor, but through a strong coupling to a few surface states. The established injection mechanism does not require a high density of acceptor states and, therefore, can function at the energies close to the edge of the conduction band. Electron injection at those energies avoids energy losses, helping to preserve the maximum voltage attainable in the Grätzel solar cell.

We found that the phonon-induced electron and hole relaxation in the PbSe quantum dots is almost symmetric and occurs slowly, on a picosecond timescale. The slow phonon-assisted relaxation allows for the other productive processes to occur. The carrier multiplication that generates multiple electron–hole pairs and increases the current attainable in a solar cell becomes possible, since both the direct electron and hole cooling and the Auger assisted electron relaxation through hole states are slow.

We determined the pathways of relaxation of free charge carriers in carbon nanotubes. The simulations agreed with the available experimental data and provided important insights into the decay mechanisms of the excited electrons and holes. The nontrivial observations included the dominant role of the high frequency phonons in both electron and hole relaxation, and the substantial contribution of the low frequency breathing modes to the dynamics of holes, but not electrons. These facts rationalized why holes decayed more slowly and over multiple timescales, despite the denser manifold of states which was expected to facilitate faster relaxation.

The simulations we performed with specific systems, which address concrete experimental observations and practical questions, bear on a much wider spectrum of problems. The interfacial charge transfer is generic to molecular electronics, where the contacts between molecular conductors and bulk electron leads remain very poorly understood. The slow charge relaxation in the QD suggests a phonon bottleneck that can be used to achieve not only carrier multiplication and larger solar cell currents, but also better voltages through delayed carrier cooling. The hole–phonon and electron–phonon interaction timescales seen in our studies establish limits on vibrationally induced dephasing that must be avoided for spintronic and quantum computing applications of quantum dots. The heating mechanisms seen in the simulations of carbon nanotubes are critical for successful development of nanotube-based miniature electronic devices. The systems and problems considered here contribute to a general framework for control and utilization of the novel phenomena that become possible on the nanoscale.

Acknowledgments

The financial support of NSF CAREER Award CHE-0094012, PRF Award 150393, and DOE Award DE-FG02-05ER15755 is gratefully acknowledged. The authors are thankful to Dr. Kiril Tsemekhman for fruitful discussions. OVP is an Alfred P. Sloan Fellow and is grateful to Dr. Jan Michael Rost at the Max Planck Institute for the Physics of Complex Systems, Dresden, Germany for hospitality during manuscript preparation.

References

1. M. A. L. Marques and E. K. U. Gross, Annu. Rev. Phys. Chem. **55**, 427 (2004)
2. T. Frauenheim, G. Seifert, M. Elstner, T. Niehaus, C. Köhler, M. Amkreutz, M. Sternberg, Z. Hajnal, A. DiCarlo, and S. Suhai, J. Phys. Cond. Matter **14**, 3015 (2002)
3. A. Szabo and N. S. Ostlund, *Modern Quantum Chemsitry*, 1st revised ed. (McGraw-Hill, New York, 1989)
4. G. Kresse and J. Furthmüller, Comput. Mater. Sci. **6**, 15 (1996)
5. G. Kresse and J. Furthmüller, Phys. Rev. B **54**, 11169 (1996)
6. W. Kohn and L. J. Sham, Phys. Rev. A **4**, 1133 (1965)
7. R. Baer, T. Seideman, S. Ilani, and D. Neuhauser, J. Chem. Phys. **120**, 3387 (2004)
8. I. Franco and S. Tretiak, J. Am. Chem. Soc. **126**, 12130 (2004)
9. M. A. L. Marques, X. López, D. Varsano, A. Castro, and A. Rubio, Phys. Rev. Lett. **90**, 258101 (2003)

10. J. P. Perdew, K. Burke, and M. Ernzerhof, Phys. Rev. Lett. **77**, 3685 (1996)
11. S. Hammes-Schiffer and J. C. Tully, J. Chem. Phys. **101**, 4657 (1994)
12. A. D. McLachlan, R. D. Gregory, and M. A. Ball, Mol. Phys. **7**, 119 (1963–1964)
13. G. D. Billing, Int. Rev. Phys. Chem. **13**, 309 (1994)
14. J. C. Tully, in *Classical and Quantum Dynamics in Condensed Phase Simulations*, edited by B. J. Berne, G. Ciccotti, and D. F. Coker (World Scientific, Singapore, 1998), pp. 489–514
15. J. C. Tully and R. K. Preston, J. Chem. Phys. **55**, 562 (1971)
16. J. C. Tully, J. Chem. Phys. **93**, 1061 (1990)
17. P. V. Parahdekar and J. C. Tully, J. Chem. Phys. **122**, 094102 (2005)
18. M. F. Herman, J. Chem. Phys. **103**, 8081 (1995)
19. D. F. Coker and L. Xiao, J. Chem. Phys. **102**, 496 (1995)
20. Y. L. Volobuev, M. D. Hack, M. S. Topaler, and D. G. Truhlar, J. Chem. Phys. **112**, 9716 (2000)
21. M. D. Hack and D. G. Truhlar, J. Chem. Phys. **114**, 9305 (2001)
22. A. W. Jasper, S. N. Stechmann, and D. G. Truhlar, J. Chem. Phys. **116**, 5424 (2002)
23. S. Nielsen, R. Kapral, and G. Ciccotti, J. Chem. Phys. **6543**, 112 (2000)
24. M. Kernan, G. Ciccotti, and R. Kapral, J. Chem. Phys. **116**, 2346 (2002)
25. O. V. Prezhdo and V. V. Kisil, Phys. Rev. A **56**, 162 (1997)
26. R. W. Wyatt, C. L. Lopreore, and G. Parlant, J. Chem. Phys. **114**, 5113 (2001)
27. F. A. Webster, P. J. Rossky, and R. A. Friesner, Comput. Phys. Commun. **63**, 494 (1991)
28. E. R. Bittner and P. J. Rossky, J. Chem. Phys. **103**, 8130 (1995)
29. O. V. Prezhdo and P. J. Rossky, J. Chem. Phys. **107**, 825 (1997)
30. O. V. Prezhdo, J. Chem. Phys. **111**, 8366 (1999)
31. C. F. Craig, W. R. Duncan, and O. V. Prezhdo, Phys. Rev. Lett. **95**, 163001 (2005)
32. A. Donoso and C. C. Martens, Phys. Rev. Lett. **87**, 223202 (2001)
33. C. L. Lopreore and R. E. Wyatt, Phys. Rev. Lett. **82**, 5190 (1999)
34. E. R. Bittner and R. E. Wyatt, J. Chem. Phys. **113**, 8888 (2000)
35. E. Gindensperger, C. Meier, and J. A. Beswick, J. Chem. Phys. **113**, 9369 (2000)
36. O. V. Prezhdo and C. Brooksby, Phys. Rev. Lett. **86**, 3215 (2001)
37. A. S. Sanz and S. Miret-Artes, J. Chem. Phys. **122**, 014702 (2005)
38. K. Runge and D. A. Micha, Phys. Rev. A **62**, 022703 (2000)
39. D. A. Micha and B. Thorndyke, Adv. Quant. Chem. **47**, 293 (2004)
40. E. Deumens and Y. Ohrn, J. Phys. Chem. A **105**, 2660 (2001)
41. R. Cabrera-Trujillo, J. R. Sabin, E. Deumens, and Y. Ohrn, Adv. Quant. Chem. **47**, 253 (2004)
42. W. H. Thompson, J. Chem. Phys. **118**, 1059 (2003)
43. I. Burghardt, K. B. Moller, G. Parlant, L. S. Cederbaum, and E. R. Bittner, Int. J. Quant. Chem. **100**, 1153 (2004)
44. I. Burghardt, J. Chem. Phys. **122**, 094103 (2005)
45. S. Bonella and D. F. Coker, J. Chem. Phys. **122**, 194102 (2005)
46. P. Ehrenfest, Z. Phys. **45**, 455 (1927)
47. T. N. Todorov, J. Hoekstra, and A. P. Sutton, Phys. Rev. Lett. **86**, 3606 (2001)
48. A. P. Horsfield, D. R. Bowler, A. J. Fisher, T. N. Todorov, and C. G. Sanchez, J. Phys. - Condens. Mater. **16**, 8251 (2004)

49. O. V. Prezhdo and Y. V. Pereverzev, J. Chem. Phys. **113**, 6557 (2000)
50. C. Brooksby and O. V. Prezhdo, Chem. Phys. Lett. **346**, 463 (2001)
51. O. V. Prezhdo and Y. V. Pereverzev, J. Chem. Phys. **116**, 4450 (2002)
52. E. Pahl and O. V. Prezhdo, J. Chem. Phys. **116**, 8704 (2002)
53. O. V. Prezhdo, J. Chem. Phys. **117**, 2995 (2002)
54. C. Brooksby and O. V. Prezhdo, Chem. Phys. Lett. **378**, 533 (2003)
55. D. Kilin, Y. V. Pereverzev, and O. V. Prezhdo, J. Chem. Phys. **120**, 11209 (2004)
56. E. Heatwole and O. V. Prezhdo, J. Chem. Phys. **121**, 10967 (2004)
57. E. Heatwole and O. V. Prezhdo, J. Chem. Phys. **122**, 234109 (2005)
58. O. V. Prezhdo and P. J. Rossky, Phys. Rev. Lett. **81**, 5294 (1998)
59. O. V. Prezhdo, Phys. Rev. Lett. **85**, 4413 (2000)
60. A. Luis, Phys. Rev. A **67**, 062113 (2003)
61. P. Exner, J. Phys. A - Math. Gen. **38**, L449 (2005)
62. A. Nitzan and M. A. Ratner, Science **300**, 1384 (2003)
63. P. A. Derosa and J. M. Seminario, J. Phys. Chem. B **105**, 471 (2001)
64. X. Y. Zhu, J. Phys. Chem. B **108**, 8778 (2004)
65. M. W. Holman, R. Liu, and D. M. Adams, J. Am. Chem. Soc. **125**, 12649 (2003)
66. F. F. Fan, Y. Yao, L. Cai, L. Cheng, J. M. Tour, and A. J. Bard, J. Am. Chem. Soc. **126**, 4035 (2004)
67. D. L. Jiang, H. J. Zhao, S. Q. Zhang, and R. John, J. Catal. **223**, 212 (2004)
68. W. Zhao, W. H. Ma, C. C. Chen, J. C. Zhao, and Z. G. Shuai, J. Am. Chem. Soc. **126**, 4782 (2004)
69. T. Hirakawa, J. K. Whitesell, and M. A. Fox, J. Phys. Chem. B **108**, 10213 (2004)
70. P. D. Cozzoli, E. Fanizza, R. Comparelli, M. L. Curri, A. Agostiana, and D. Laub, J. Phys. Chem. B **108**, 9623 (2004)
71. H. G. Kim, D. W. Hwang, and J. S. Lee, J. Am. Chem. Soc. **126**, 8912 (2004)
72. D. Liu, G. L. Hug, and P. V. Kamat, J. Phys. Chem. **99**, 16768 (1995)
73. B. Oregan and M. Grätzel, Nature **353**, 737 (1991)
74. O. Schwarz, D. van Loyen, S. Jockusch, N. J. Turro, and H. Duerr, J. Photochem. Photobiol. A: Chem. **132**, 91 (2000)
75. R. D. McConnell, Renew. Sustain. Energy Rev. **6**, 273 (2002)
76. J. B. Asbury, E. C. Hao, Y. Q. Wang, H. N. Ghosh, and T. Q. Lian, J. Phys. Chem. B **105**, 4545 (2001)
77. T. Hannappel, B. Burfeindt, W. Storck, and F. Willig, J. Phys. Chem. B **101**, 6799 (1997)
78. N. A. Anderson, E. Hao, X. Ai, G. Hastings, and T. Lian, Phys. E. **14**, 215 (2002)
79. R. Huber, S. Spoerlein, J. E. Moser, M. Grätzel, and J. Wachtveitl, J. Phys. Chem. B **104**, 8995 (2000)
80. R. Huber, J. E. Moser, M. Grätzel, and J. Wachtveitl, J. Phys. Chem. B **106**, 6494 (2002)
81. S. Ramakrishna and F. Willig, J. Phys. Chem. B **104**, 68 (2000)
82. S. Ramakrishna, F. Willig, V. May, and A. Knorr, J. Phys. Chem. B **107**, 607 (2003)
83. E. Kucur, J. Reigler, G. A. Urban, and T. Nann, J. Chem. Phys. **120**, 1500 (2004)

84. W. Greens, T. Martens, J. Poortmans, T. A. J. Manca, L. Lutsen, P. Heremans, S. Borghs, R. Mertens, and D. Vanderzande, Thin Solid Films **451–452**, 498 (2004)
85. E. Hao, N. A. Anderson, J. B. Asbury, and T. Lian, J. Phys. Chem. B **106**, 10191 (2002)
86. P. Ravirajan, S. A. Haque, D. Poplavskyy, J. R. Durrant, D. D. C. Bradley, and J. Nelson, Thin Solid Films **451–452**, 624 (2004)
87. A. J. Nozik, Phys. E **14**, 115 (2001)
88. R. D. Schaller and V. I. Klimov, Phys. Rev. Lett. **92**, 186601 (2004)
89. L. X. Wang and V. May, J. Chem. Phys. **121**, 8039 (2004)
90. M. Thoss, I. Kondov, and H. Wang, Chem. Phys. **304**, 169 (2004)
91. L. G. C. Rego and V. S. Batista, J. Am. Chem. Soc. **125**, 7989 (2003)
92. W. Stier and O. V. Prezhdo, J. Mol. Struct. (Theochem) **630**, 33 (2002)
93. W. Stier and O. V. Prezhdo, J. Phys. Chem. B **106**, 8047 (2002)
94. W. Stier and O. V. Prezhdo, Isr. J. Chem. **42**, 213 (2003)
95. W. Stier, W. R. Duncan, and O. V. Prezhdo, Adv. Mater. **16**, 240 (2004)
96. W. R. Duncan, W. M. Stier, and O. V. Prezhdo, J. Am. Chem. Soc. **127**, 7941 (2005)
97. W. R. Duncan and O. V. Prezhdo, J. Phys. Chem. B **109**, 365 (2005)
98. W. R. Duncan and O. V. Prezhdo, J. Phys. Chem. B, **109**, 365–373 (2005)
99. J. Kruger, U. Bach, and M. Grätzel, Adv. Mater. **12**, 447 (2000)
100. A. C. Arango, L. R. Johnson, V. N. Bliznyuk, Z. Schlesinger, S. A. Carter, and H. H. Horhold, Adv. Mater. **12**, 1689 (2001)
101. A. F. Nogueira, J. R. Durrant, and M. A. D. Paoli, Adv. Mater. **13**, 826 (2001)
102. M. Grätzel, J Photochem. Photobiol. C: Photochem. Rev. **4**, 145 (2003)
103. M. A. Green, *Third generation photovoltaics* (Bridge Printery, Sydney, 2001)
104. W. Shockley and H. J. Queisser, J. Appl. Phys. **32**, 510 (1961)
105. D. S. Boudreaux, F. Williams, and A. J. Nozik, J. Appl. Phys. **51**, 2158 (1980)
106. S. Kolodinski, J. H. Werner, T. Wittchen, and H. J. Queisser, Appl. Phys. Lett. **63**, 2405 (1993)
107. F. Williams and A. J. Nozik, Nature **311**, 21 (1984)
108. H. Yu, S. Lycett, C. Roberts, and R. Murray, Appl. Phys. Lett. **69**, 4087 (1996)
109. M. Sugawara, K. Mukai, and H. Shoji, Appl. Phys. Lett. **71**, 2791 (1997)
110. B. N. Murdin, A. R. Hollingworth, M. Kamal-Saadi, R. T. Kotitschke, and C. N. Cielsla, Phys. Rev. B **59**, R7817 (1999)
111. R. J. Ellingson, J. L. Blackburn, P. Yu, G. Rumbles, O. I. Micic, and A. J. Nozik, J. Phys. Chem. B **106**, 7758 (2002)
112. R. Heitz, M. Veit, N. N. Lebentsov, A. Hoffmann, and D. Bimberg, Phys. Rev. B **56**, 10435 (1997)
113. F. Adler, M. Geiger, A. Bauknecht, D. Haase, and P. Ernst, J. Appl. Phys. **83**, 1631 (1998)
114. P. Guyot-Sionnest, M. Shim, C. Matranga, and M. Hines, Phys. Rev. B **60**, R2181 (1999)
115. V. I. Klimov, A. A. Mikhailovsky, D. W. McBranch, C. A. Leatherdale, and M. G. Bawendi, Phys. Rev. B **61**, R13349 (2000)
116. M. Achermann, J. A. Hollingsworth, and V. I. Klimov, Phys. Rev. B **68**, 245302 (2003)
117. R. G. Ispasiou, J. Lee, F. Papadimitrakopoulos, and T. Goodson, Chem. Phys. Lett. **340**, 7 (2001)

118. M. B. Mohamed, C. Burda, and M. A. El-Sayed, Nano Lett. **1**, 589 (2001)
119. P. Yu, J. M. Nedeljkovic, P. A. Ahrenkiel, R. J. Ellingson, and A. J. Nozik, Nano Lett. **4**, 1089 (2004)
120. R. J. Ellingson, M. C. Beard, J. C. Johnson, P. R. Yu, O. I. Micic, A. J. Nozik, A. Shabaev, and A. L. Efros, Nano Lett. **5**, 865 (2005)
121. M. Lowisch, M. Rabe, F. Kreller, and F. Henneberger, Appl. Phys. Lett. **74**, 2489 (1999)
122. X.-Q. Li, H. Nakayama, and Y. Arakawa, Jpn. J. Appl. Phys. **38**, 473 (1999)
123. I. Gontijo, G. S. Buller, J. S. Massa, A. C. Walker, and S. V. Zaitsev, Jpn. J. Appl. Phys. **38**, 674 (1999)
124. U. Woggon, H. Giessen, F. Gindele, O. Wind, B. Fluegel, and N. Peyghambarian, Phys. Rev. B **54**, 17681 (1996)
125. V. I. Klimov and D. W. McBranch, Phys. Rev. Lett. **80**, 4028 (1998)
126. T. S. Sosnowski, T. B. Norris, H. Jiang, J. Singh, K. Kamath, and P. Bhattacharya, Phys. Rev. B **57**, R9423 (1998)
127. R. Heitz, M. Veit, A. Kalburge, Q. Zie, and M. Grundmann, Phys. E **2**, 578 (1998)
128. K. Mukai and M. Sugawara, Jpn. J. Appl. Phys. **37**, 5451 (1998)
129. S. Iijima, Nature **354**, 56 (1991)
130. R. Saito, G. Dresselhaus, and M. S. Dresselhaus, *Physical Properties of Carbon Nanotubes* (Imperial College Press, London, 1998)
131. R. H. Baughman, A. A. Zakhidov, and W. A. de Heer, Science **297**, 787 (2002)
132. Y. Wang et al., Science **85**, 2607 (2004)
133. J. Kong et al., Science **287**, 622 (2000)
134. N. Mason, M. J. Biercuk, and C. M. Marcus, Science **303**, 655 (2004)
135. M. S. Strano et al., Science **301**, 1519 (2003)
136. S. G. Rao, L. Huang, W. Setyawan, and S. Hong, Nature **425**, 36 (2003)
137. S. J. Tans et al., Nature **386**, 474 (1997)
138. M. S. Dresselhaus, Nature **432**, 959 (2004)
139. R. Krupke, F. Hennrich, H. v. Lohneysen, and M. M. Kappes, Science **301**, 2003 (2003)
140. M. J. O'Connell et al., Science **297**, 787 (2002)
141. Y. C. Chen et al., Appl. Phys. Lett. **81**, 975 (2002)
142. T. Hertel and G. Moos, Chem. Phys. Rev. **320**, 359 (2000)
143. Y.-Z. Ma et al., J. Chem. Phys. **120**, 3368 (2004)
144. C. Manzoni, A. Gambetta, E. Menna, M. Meneghetti, G. Lanzani, and G. Cerullo, Phys. Rev. Lett. **94**, 207401 (2005)
145. G. N. Ostojic, S. Zaric, J. Kono, M. S. Strano, V. C. Moore, R. H. Hauge, and R. E. Smalley, Phys. Rev. Lett. **92**, 117402 (2004)
146. L. Huang, H. N. Pedrosa, and T. D. Krauss, Phys. Rev. Lett. **93**, 017403 (2004)
147. F. Wang, G. Dukovic, L. E. Brus, and T. F. Heinz, Phys. Rev. Lett. **92**, 177401 (2004)
148. H. Htoon, M. J. O'Connell, S. K. Doorn, and V. I. Klimov, Phys. Rev. Lett. **94**, 127403 (2005)
149. S. Roche, J. Jiang, F. Triozon, and R. Saito, Phys. Rev. Lett. **95**, 076803 (2005)

Ultrafast Exciton Dynamics in Molecular Systems

B. Brüggemann, D. Tsivlin, and V. May

Summary. The theory of subpicosecond Frenkel exciton dynamics in molecular systems is reviewed with emphasis on a stepwise improved description of the coupling to intra- and intermolecular vibrations. After introducing the concept of multiexciton states the motion of electronic Frenkel excitons as they appear in light harvesting antennae of photosynthetic organisms is discussed. The description is based on a multiexciton density matrix theory which accounts for the exciton–vibrational coupling in a perturbative manner. Some improvements of this density matrix theory as suggested in literature are shortly mentioned. Afterwards, vibrational Frenkel excitons as found in polypeptides are considered. By utilizing the multiconfiguration time-dependent Hartree method an exact description of the coupling to longitudinal vibrations of the peptide chain becomes possible. The discussion of the computed transient infrared absorption spectra is supported by the introduction of adiabatic single- and two-exciton states.

1 Introduction

With the dawn of femtosecond spectroscopy the study of vibrational wave packet dynamics in molecular systems became a main topic of molecular physics and physical chemistry. And immediately this new type of spectroscopy was applied to the investigation of electronic excitations in molecular systems known as *Frenkel excitons* (for recent introductions into this field see [1–4]). Frenkel excitons are spatially delocalized excited states with the basic excitations completely localized at individual molecules within the complex. Furthermore, their excitation energy is much larger than the thermal energy (at room temperature conditions). If the coupling of these excitations to vibrational coordinates remains weak the excitation energy transfer may proceed coherently up to some 100 fs. This makes the study of exciton dynamics and vibrational wave packet dynamics complementary to each other. In the contrary case of strong coupling to vibrations the excitation jumps as a localized state from molecule to molecule. One arrives at the case of incoherent excitation energy transfer named after Förster. In recent years, however, experimental data have also been obtained characterizing the intermediate case where the excitation energy becomes delocalized but couples strongly to the vibrational coordinates (see, e.g., [5]).

The description of the quantum dynamics of excitons in organic crystals and molecular systems like dye aggregates or polymer strands has a long tradition (cf. [6–10]). In the late forties and early fifties of the last century the field has been pioneered by Förster [11] and Dexter [12]. Later Davydov [6] and Agranovich [9] established the theoretical basis for the description of Frenkel excitons, their optical properties and of the related excitation energy transfer dynamics. Over the past 15 years work has been concentrated on the formulation of models adequate for the description of ultrafast dynamics of excitonic systems and related nonlinear spectra (see [4,13,14] for an overview). This required the inclusion of multiple electronic excitations (cf. Fig. 1) and the utilization of techniques of dissipative quantum dynamics.

The present article reports on recent theoretical achievements in this field with the specific account on exciton–vibrational coupling. Starting at a description of exciton dynamics for the case of a weak coupling to molecular vibrations, different possibilities will be touched to go beyond this weak-coupling approximation. We conclude by an exact consideration of the exciton–vibrational coupling. And, a theoretical description is presented which can be applied to electronic as well as vibrational Frenkel excitons.

1.1 The Multiexciton Concept

When doing nonlinear spectroscopy on aggregates and chromophore complexes higher excited states of single molecules as well as those of the whole complex have to be taken into account. The resulting *multiexciton* scheme displayed in Fig. 1 found various applications. Besides different nonbiological complexes one striking example is given by photosynthetic antenna systems (cf. Figs. 2 and 3 and the remark in [15]). Recently, the multiexciton scheme has been also applied to understand nonlinear infrared spectra of polypeptides [16] (cf. also Fig.7) where Frenkel excitons are formed by localized high-frequency vibrational excitations.

In order to introduce multiexciton states for all mentioned systems the respective states of the single molecules are denoted by φ_a with a referring to the ground-state ($a = g$), the first excited state ($a = e$), and a higher-excited state ($a = f$). If electronic Frenkel-excitons are considered the φ_a refer to electronic excitations of the individual molecules (cf. Fig. 2) whereas the states φ_a describe distinct vibrational excitations if vibrational Frenkel-excitons are of interest (see Fig. 8).

To characterize the possible states of the whole complex of excitable units (the molecular complex) one may introduce product states $\prod_m \varphi_{ma_m}$, where m counts the individual molecules of the complex. Of course, such an ansatz is only reliable if different φ_{ma} do not overlap. It is advisable to order the

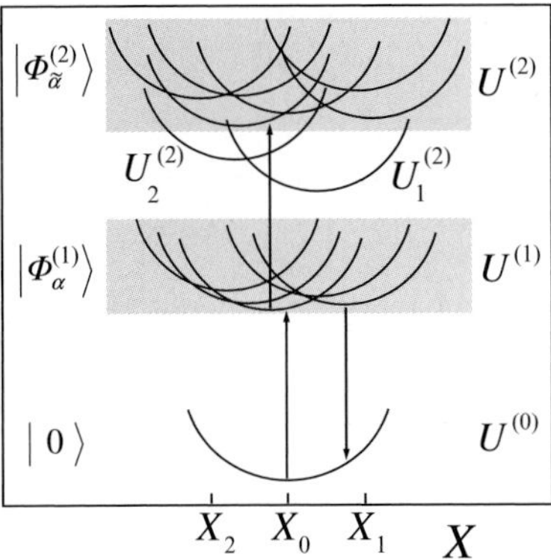

Fig. 1. Energy level scheme of a molecular complex covering the groundstate $|0\rangle$ as well as the manifold of single-exciton states $|\Phi^{(1)}_{\alpha}\rangle$ and of two-exciton states $|\Phi^{(2)}_{\tilde{\alpha}}\rangle$. Shown are respective potential energy surfaces (PESs) $U^{(N)}$, $N = 0, 1, 2$ (vs. vibrational coordinates). The PES follow when calculating adiabatic multiexciton states. X_0 indicates the vibrational equilibrium configuration of the unexcited complex, and X_1 and X_2 label the local energy-minimum configurations for particular single- and two-exciton PES, respectively. The *vertical arrows* indicate different transition processes initiated by external optical or infrared fields and observed in the experiment. (The two PESs $U^{(2)}_1$ and $U^{(2)}_2$ separated from the majority of two-exciton PESs correspond to self-trapped states described in the Sect. 4.)

product states with respect to the number of basic excitations (the single excitation from φ_g to φ_e). The overall ground-state reads

$$|0\rangle = \prod_m |\varphi_{mg}\rangle \ . \tag{1}$$

The presence of a single excitation at unit m in the complex is characterized by

$$|\phi_m\rangle = |\varphi_{me}\rangle \prod_{n \neq m} |\varphi_{ng}\rangle \ , \tag{2}$$

whereas a double excited state covers a double excitation of a single molecule as well as the simultaneous presence of two single excitations at two different molecules:

$$|\phi_{mn}\rangle = \delta_{m,n}|\varphi_{mf}\rangle \prod_{n \neq m} |\varphi_{ng}\rangle + (1 - \delta_{m,n})|\varphi_{me}\rangle|\varphi_{ne}\rangle \prod_{k \neq m,n} |\varphi_{kg}\rangle \ . \tag{3}$$

The ordering scheme can be easily continued. However, to describe existing experiments (from various different fields), so far, does not require the introduction of triple or quadruple excitations.

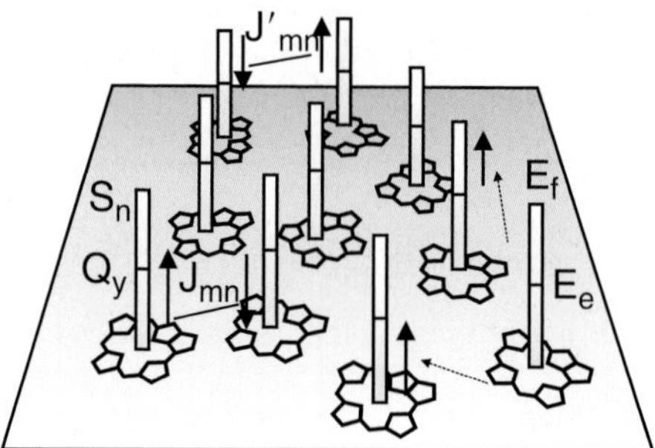

Fig. 2. Three-level model of a planar chromophore complex (first excited state with energy E_e and higher excited state with energy E_f). Shown are the tetrapyrrole rings of chlorophylls with the first excited, so-called Q_y-level and a higher excited singlet state together with all relevant interactions (interchromophore couplings J_{mn} and J'_{mn}) as well as optical excitations (cf. also the note in [22])

An expansion of the complete Hamiltonian with respect to these different excited states forming so-called *exciton manifolds* yields

$$H = \mathcal{H}_0 + \mathcal{H}_1 + \mathcal{H}_2 - \mathbf{E}(t)\hat{\mu} \,. \tag{4}$$

The $\mathcal{H}_N$ $(N = 0, 1, 2)$ are obtained as $\hat{\Pi}_N H \hat{\Pi}_N$ with the $\hat{\Pi}_N$ projecting on the different manifolds. Moreover, the $\mathcal{H}_N$ describe intramanifold dynamics including the coupling to all relevant vibrational coordinates abbreviated in the following by X (cf. Fig. 1). If necessary intermanifold transitions beside those induced by the external field $\mathbf{E}(t)$ (which couples to the complex via the dipole operator $\hat{\mu}$) may be introduced into (4) (for the description of exciton–exciton annihilation see, for example, [17, 18] and references therein). The ground-state Hamiltonian $\mathcal{H}_0 = T_{\text{vib}} + U_0(X)$ includes the vibrational kinetic energy operator T_{vib} and the potential energy surface (PES) U_0. The overall minimum of the latter defines the nuclear equilibrium configuration which will be abbreviated by X_0 in the following. For the first and second excited manifold, respectively, we get

$$\mathcal{H}_1 = \sum_{m,n} \Big(\delta_{m,n} H_m(X) + (1 - \delta_{m,n}) J_{mn}(X) \Big) |\phi_m\rangle\langle\phi_n| \,, \tag{5}$$

and

$$\mathcal{H}_2 = \sum_{kl,mn} \Big(\delta_{k,m}\delta_{l,n} H_{mn}(X) + (1 - \delta_{k,m}\delta_{l,n}) J_{kl,mn}(X) \Big) |\phi_{kl}\rangle\langle\phi_{mn}| \,. \tag{6}$$

Here, H_m and H_{mn} denote vibrational Hamiltonians referring to the respective intra-molecular excitations. The inter-state coupling matrices J_{mn} and $J_{kl,mn}$ have to be deduced from the inter-molecular Coulombic coupling (see, e.g., [2]). If a multipole expansion is possible the matrices reduce to well

Fig. 3. Water wheel like spatial structure of the LH2 (light harvesting complex 2 of purple bacteria, cf. [15]) with 27 chlorophyll molecules. The α-helical part of the carrier protein is also shown. Those 18 chlorophylls forming the water wheel like part absorb at 850 nm (transition into the single-exciton band formed by the Q_y-excitations) whereas the other (lying in the figure plane) absorb at 800 nm (for other parameters see the listing in [25])

known expressions of the inter-molecular transition–dipole transition–dipole coupling (in the case of $H_{mn}(X)$ and $J_{kl,mn}$ one has take care of the correct normalization).

Standard single and two-exciton states are obtained by diagonalizing $\mathcal{H}_1$ and $\mathcal{H}_2$, respectively, but in removing the vibrational kinetic energy operator and by fixing the X at the ground-state equilibrium configuration X_0. These exciton states are denoted as

$$|\alpha\rangle = \sum_m C_\alpha^{(1)}(m)|\phi_m\rangle \ , \tag{7}$$

and as

$$|\tilde{\alpha}\rangle = \sum_{m,n} C_{\tilde{\alpha}}^{(2)}(mn)|\phi_{mn}\rangle \ . \tag{8}$$

The respective multiexciton Hamiltonian $H_{\rm mx}$ is obtained from H, (4) and covers $H_{\rm ex}^{(N)} \equiv \mathcal{H}_N(X_0) - T_{\rm vib}$ ($N = 1, 2$, the ground-state energy of the complex has been set equal to zero). In detail we have $H_{\text{ex}}^{(1)} = \sum_\alpha \hbar\Omega_\alpha |\alpha\rangle\langle\alpha|$ and $H_{\text{ex}}^{(2)} = \sum_{\tilde{\alpha}} \hbar\Omega_{\tilde{\alpha}} |\tilde{\alpha}\rangle\langle\tilde{\alpha}|$ with related single- and two-exciton energies $\hbar\Omega_\alpha$ and $\hbar\Omega_{\tilde{\alpha}}$, respectively.

By removing the restriction of the vibrational coordinates to X_0 while diagonalizing the Hamiltonians $\mathcal{H}_1$ and $\mathcal{H}_2$ (minus T_{vib}) one arrives at so-called adiabatic exciton states (cf. Fig. 1 and [13, 19, 20]):

$$|\Phi_\alpha^{(1)}(X)\rangle = \sum_m C_\alpha^{(1)}(m;X)|\phi_m\rangle \,, \tag{9}$$

and

$$|\Phi_{\tilde{\alpha}}^{(2)}(X)\rangle = \sum_{m,n} C_{\tilde{\alpha}}^{(2)}(mn;X)|\phi_{mn}\rangle \,. \tag{10}$$

Now, the expansion coefficients $C_\alpha^{(1)}$ and $C_{\tilde{\alpha}}^{(2)}$ depend on the actual vibrational configuration.

The description introduced so far carries out an expansion of the Hamiltonian and any observable with respect to localized multiple excitations of the complex (or delocalized multiexciton states). The use of such a type of eigenstate representation is mainly motivated by a correct description of excitation energy relaxation in the framework of dissipative quantum dynamics. Other approaches have been directly based on Pauli-operators for Frenkel excitons [9] or on the introduction of the so-called anharmonic oscillator model [21].

1.2 Regimes of Exciton Dynamics

The electronic interchromophore coupling described by the J_{mn} as well as the $J_{kl,mn}$ (cf. (5) and (6), respectively) and the multiexciton vibrational coupling are the two basic interaction mechanisms determining the concrete character of the exciton transfer. The interchromophore coupling will be characterized by a representative J and the coupling to vibrational coordinates by a related reorganization energy $\hbar\lambda$. It equals the amount of energy which is set free if the vibrational coordinates X in an excited state of a single molecule or in a multiexciton state change from the vibrational ground-state equilibrium configuration X_0 to the actual equilibrium configuration in the chosen excited state. Additionally, this process of vibrational coordinate reorganization can be also characterized by a representative relaxation time τ_{rel}.

The regime of weak exciton–vibrational coupling is reached if $\hbar\lambda \ll J$ is valid. The formation of delocalized (or partially delocalized) single and two-exciton states becomes possible and the exciton dynamics appears coherent on a time scale less than or comparable to τ_{rel}. For this regime a density matrix description is most appropriate (see Sect. 2).

In the reverse case where $J \ll \hbar\lambda$ the dynamics after ultrashort photo excitation are dominated by vibrational reorganization and relaxation. J affects the excitation energy transfer only weakly, thus, the regime of localized excitation energy transfer well characterized by the so-called Förster theory is reached (cf., for example, [3, 4] and Sect. 3). Of actual interest but less investigated are those regimes of exciton dynamics where both fundamental couplings compete against each other (see Sect. 4).

2 Electronic Frenkel-Excitons: Weak Exciton–Vibrational Coupling

The treatment discussed hereafter is based on the introduction of (unrelaxed) exciton states (7) and (8) referring to the vibrational equilibrium configuration in the unexcited complex X_0. Multiexciton–vibrational coupling $H^{(N)}_{\text{ex-vib}}$ ($N = 1, 2$) is obtained as $H^{(N)}_{\text{ex-vib}}(X) = \mathcal{H}_N(X) - \mathcal{H}_N(X_0) - H_{\text{vib}}$. The expression $\mathcal{H}_N(X_0)$ defines the multi-exciton levels at the vibrational equilibrium configuration (plus vibrational kinetic energy operator). Their contributions to $\mathcal{H}_N(X)$ are removed to arrive at the multiexciton vibrational coupling. Moreover, $H_{\text{vib}} = \mathcal{H}_0(X)$ represents the ground-state reference vibrational Hamiltonian. Thus, the difference expression $\mathcal{H}_N(X) - H_{\text{vib}}$ includes the deviations from the ground-state PES (except the contributions given by the $\mathcal{H}_N(X_0)$) which act as the exciton–vibrational coupling.

If the $H^{(N)}_{\text{ex-vib}}(X)$ are linearly expanded with respect to the vibrational coordinates and the latter undergo a change to normal-mode vibrations one arrives at the standard expression for exciton–vibrational coupling [2]. A restriction to single-exciton states yields:

$$H^{(1)}_{\text{ex-vib}} = \sum_{\alpha,\beta} \sum_{\xi} \hbar\omega_\xi g_{\alpha\beta}(\xi) Q_\xi |\alpha\rangle\langle\beta| \,. \tag{11}$$

The Q_ξ are dimensionless normal-mode coordinate operators with mode-index ξ forming a set of otherwise decoupled harmonic oscillators. Neglecting any X dependence of the J_{mn} the exciton–vibrational coupling constant follows as

$$g_{\alpha\beta}(\xi) = \sum_m C^*_\alpha(m) g_m(\xi) C_\beta(m) \,. \tag{12}$$

The exciton coefficients $C_\alpha(m)$ are defined in (7) and the $g_m(\xi)$ follow from a linear expansion of $H^{(1)}_{\text{ex-vib}}(X)$. The coupling Hamiltonian introduced in (11) has been used in many studies (see [4] for a an overview and also the recent applications in [17, 18, 23, 24]).

2.1 Multiexciton Density Matrix Theory

Multiexciton dynamics for the case of weak exciton–vibrational coupling are best formulated in the framework of a density matrix theory reduced to the electronic (multiexciton) DOF. The reduced density operator reads

$$\hat{\rho}(t) = \text{tr}_{\text{vib}}\{\hat{W}(t)\} \,. \tag{13}$$

The trace refers to the vibrational DOF and $\hat{W}(t)$ denotes the nonequilibrium statistical operator of the complete multiexciton vibrational system. A well established approach is represented by the perturbative account for the exciton–vibrational coupling in the equation of motion for $\hat{\rho}(t)$ (cf., e.g., [2]). From a projection superoperator perspective such calculations are based on

$$\mathcal{P}... = \hat{r}_0^{(\text{eq})}\text{tr}\{...\} , \tag{14}$$

where $\hat{r}_0^{(\text{eq})}$ denotes the vibrational equilibrium statistical operator of the unexcited complex. The mentioned weak-coupling limit results in the following equation of motion

$$\frac{\partial}{\partial t}\hat{\rho}(t) = -\frac{i}{\hbar}\left(H_{\text{mx}}, \hat{\rho}(t)\right)_- + \left(\frac{\partial \hat{\rho}(t)}{\partial t}\right)_{\text{diss}} , \tag{15}$$

with H_{mx} being the multi-exciton Hamiltonian introduced in Sect. 1.1. The dissipative part takes the form

$$\begin{aligned}\left(\frac{\partial \hat{\rho}(t)}{\partial t}\right)_{\text{diss}} = -\sum_{u,v} \int_0^{t-t_0} \mathrm{d}\tau \Big[& C_{uv}(\tau)\left(\hat{\Pi}_u, \mathrm{e}^{-\mathrm{i}H_{\text{mx}}\tau/\hbar}\hat{\Pi}_v \hat{\rho}(t-\tau)\mathrm{e}^{\mathrm{i}H_{\text{mx}}\tau/\hbar}\right)_- \\ & -C_{vu}(-\tau)\left(\hat{\Pi}_u, \mathrm{e}^{-\mathrm{i}H_{\text{mx}}\tau/\hbar}\hat{\rho}(t-\tau)\hat{\Pi}_v \mathrm{e}^{\mathrm{i}H_{\text{mx}}\tau/\hbar}\right)_- \Big] . \end{aligned} \tag{16}$$

This formula uses $\hbar\sum_u \hat{V}_u(X)\hat{\Pi}_u$ for $H^{(N)}_{\text{ex-vib}}$, where $\hat{\Pi}_u$ equals $|\alpha\rangle\langle\beta|$ as well as $|\tilde{\alpha}\rangle\langle\tilde{\beta}|$ (i.e., u either abbreviates $(\alpha\beta)$ or $(\tilde{\alpha}\tilde{\beta})$). The $\hat{V}_u(X)$ are the parts depending on the vibrational coordinates according to $\hat{V}_{\alpha\beta} = \langle\alpha|H^{(1)}_{\text{ex-vib}}|\beta\rangle$ and $\hat{V}_{\tilde{\alpha}\tilde{\beta}} = \langle\tilde{\alpha}|H^{(2)}_{\text{ex-vib}}|\tilde{\beta}\rangle$ (for the simplest version cf. (11)). They determine the correlation functions $C_{uv}(\tau)$ which are defined with respect to the vibrational equilibrium of the unexcited complex.

Introducing multiexciton matrix elements of $\hat{\rho}(t)$ results in the various elements of the multiexciton density matrix, e.g., ρ_0, $\rho_{\alpha 0}$, $(\rho_{0,\alpha})$, $\rho_{\alpha\beta}$, $\rho_{\tilde{\alpha}0}$, $(\rho_{0\tilde{\alpha}})$, $\rho_{\tilde{\alpha}\beta}$, $(\rho_{\beta\tilde{\alpha}})$, and $\rho_{\tilde{\alpha}\tilde{\beta}}$. This density matrix approach with multiexciton-vibrational coupling included in a second-order perturbational treatment is well-know in dissipative quantum dynamics and often named multilevel Redfield-theory [2]. The density matrix equations obtained from (15) include four-index memory kernels which follow from (16). All the multi-exciton density matrix elements have to be calculated simultaneously when studying the femtosecond photoinduced dynamics [17, 18].

The most simple treatment of the dissipative multi-exciton dynamics is based on the neglect of memory effects and the application of the so-called secular approximation (for a detailed justification cf. e.g., [2,4]). As an example we present the equation of motion for the single-exciton density matrix $\rho_{\alpha\beta}$. Here, the possible coupling among diagonal and off-diagonal elements via the dissipative part does not take place (cf., e.g., [23], transitions into the two–exciton manifold have been neglected for simplicity):

$$\frac{\partial}{\partial t}\rho_{\alpha\beta} = -\mathrm{i}\Omega_{\alpha\beta}\rho_{\alpha\beta} + \delta_{\alpha\beta}\sum_{\gamma}(-k_{\alpha\to\gamma}\rho_{\alpha\alpha} + k_{\gamma\to\alpha}\rho_{\gamma\gamma}) \tag{17}$$

$$-(1-\delta_{\alpha\beta})(\gamma_\alpha + \gamma_\beta)\rho_{\alpha\beta} + \frac{i}{\hbar}\mathbf{E}(t)(\mathbf{d}_\alpha\rho_{0\beta} - \mathbf{d}^*_\beta\rho_{\alpha 0})\,. \tag{18}$$

The $\Omega_{\alpha\beta} = \Omega_\alpha - \Omega_\beta$ are transition frequencies following from the (single) exciton energies $\hbar\Omega_\alpha$, and the $\mathbf{d}_\alpha$ denote the transition dipole elements into exciton states $|\alpha\rangle$. Neglecting any vibrational modulation of the J_{mn} the exciton relaxation rates read

$$k_{\alpha\to\beta} = 2\pi\Omega^2_{\alpha\beta}(1+n(\Omega_{\alpha\beta}))\sum_m |C_\alpha(m)C_\beta(m)|^2[J_m(\Omega_{\alpha\beta}) - J_m(-\Omega_{\alpha\beta})]\,. \tag{19}$$

The $J_m = \sum_\xi g^2_m(\xi)\delta(\omega-\omega_\xi)$ denote the spectral densities caused by the exciton–vibrational coupling. The dephasing rates γ_α follow from the exciton relaxation rates as $\sum_\beta k_{\alpha\to\beta}/2$ if so-called pure dephasing contributions are neglected.

2.2 Simulation of Linear and Nonlinear Spectra

All developments in the field of Frenkel excitons found an immediate application to that part of photosynthetic research which concentrates on what is known as the early events of photosynthesis (excitation energy transfer and charge separation taking place on a ps and subpicosecond time region, for a review on somewhat older work see [8]). Multiexciton models like those explained in Sect. 2.1 (cf. also Fig. 2) are in the focus of interest when doing ultrafast spectroscopic experiments at antenna complexes. And the failure of a complete quantum chemical determination of all multiexciton states of a given antenna system (and all couplings to vibrational DOF) made the use of more simple models unavoidable. When using such a multiexciton model in most cases a complete knowledge of all parameters entering the model is not achievable. Then, a specification via the fit of measured spectra becomes necessary.

Linear Absorbance of the PS1 Antennae

As a particular example for such a fit of spectra we shortly comment on respective calculations for the PS1 (photosystem 1) core antenna system (in contrast to the LH2 shown in Fig. 3 the PS1 complex which is found in cyanobacteria comprises 96 chlorophyll molecules and includes the reaction center [23, 27]). Although the spatial structure of the PS1 is known with a 2.6 Å resolution [27] an exciton model like that derived from (5) in the foregoing sections cannot be build up completely. This is caused by the fact that the excitation energy E_{eg} of each chlorophyll is slightly changed by its specific protein environment. Fitting the linear absorption of the PS1 (in the Q_y-excitation region), however, allows one to complete the model (Figures 4 and 5).

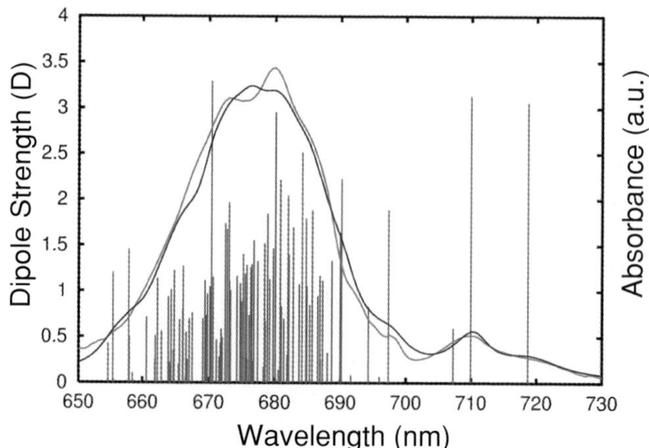

Fig. 4. Absorption spectrum of the PS1 antenna complex at 4 K. *Thin dotted line:* measured spectrum, *full line:* spectrum calculated according to (20) (for parameters see [28]), *vertical line:* exciton transition dipole moments $|\mathbf{d}_\alpha|$ at the excitonic energies $\hbar\Omega_\alpha$ (the respective line broadening is shown in Fig. 5)

In line with the presented density matrix theory which is suitable for weak exciton–vibrational coupling the absorbance follows as (for more details see [4, 23])

$$A(\omega) \sim \sum_\alpha |\mathbf{d}_\alpha|^2 \frac{\gamma_\alpha}{(\omega - \Omega_\alpha)^2 + \gamma_\alpha^2} \,. \tag{20}$$

The expression includes all (single) exciton energies $\hbar\Omega_\alpha$, the transition dipole elements $\mathbf{d}_\alpha$ and the dephasing rates (line-broadening) γ_α. All quantities have been determined in [23] by applying an evolutionary search algorithm. (For a discussion of the important influence of static structural and energetic disorder we refer to [23]).

Subpicosecond Transient Absorption of the LH2 Antennae

Since a number of excitation energy transfer processes in photosynthetic antenna systems takes place on a subpicosecond time–scale, pump probe spectroscopy is used to elucidate details of the dynamics. (Once respective data are available they are used, for example, to understand the optimization of the antennae by evolution to carry out excitation energy transfer efficiently and lossless). Within pump probe spectroscopy the pump pulse (with field-strength $\mathbf{E}_{\mathrm{pu}}$) excites the system and the probe pulse (with field-strength $\mathbf{E}_{\mathrm{pr}}$) probes the resulting excited state dynamics (see also Fig. 1). The probe pulse transient absorption spectrum (TAS) A_{pr} decomposed with respect to temporal and spectral contributions is used for an analysis.

Usually A_{pr} is deduced by calculating the third-order response function. The latter determines the polarization of the molecular complex proportional

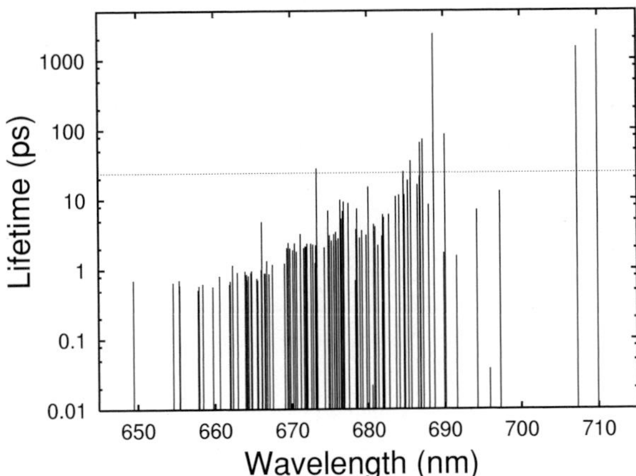

Fig. 5. Inverse dephasing rates $2/\sum_\beta k_{\alpha\beta}$ at T = 4 K vs. wavelength for all PS1 exciton levels shown in Fig. 4. (A reasonable value of the inverse pure dephasing rate is given by the *dashed line*)

to the dipole operator expectation value $< \hat{\mu}(t) >$ at the third power of the overall external field $\mathbf{E} = \mathbf{E}_{\text{pu}} + \mathbf{E}_{\text{pr}}$ (see, e.g., [1]).

Within the described multiexciton density matrix theory the polarization is obtained from $\text{tr}\{\hat{\rho}(t;\mathbf{E})\hat{\mu}\}$ with the multiexciton density operator depending in any order on the external field. Such a nonperturbative dependency on $\mathbf{E}$ simply follows from the solution of the field-driven multiexciton density matrix equations. To arrive at A_{pr} the respective part of the overall polarization has to be deduced (for details see, e.g., [18]).

Figure 6 displays respective results for the LH2 of purple bacteria (cf. Fig. 3). The lower panel nicely demonstrates the reproduction of experimental data (cf. [29]), whereas the upper panels show the internal multiexciton dynamics of the antenna by drawing the absolute values of all elements of the multiexciton density matrix (up to the two-exciton manifold) at different times. There are parts in the figures corresponding to identical manifold-numbers 0, 1, 2 on the horizontal and vertical axes. Those display the ground-state density matrix ρ_{00} as well as all elements of the single-exciton and two-exciton density matrices vs. energy. In the remaining parts, off-diagonal density matrix elements are shown determining transition polarizations or so-called coherences (for example, the combination (1,2) of manifold numbers corresponds to the elements of $\rho_{\alpha\tilde{\beta}}$).

If the pump pulse reaches its maximum (left upper panel) off-diagonal density matrix elements become large. But with increasing time dephasing results in a decay of these intra- and intermanifold off-diagonal elements and only diagonal elements survive forming the two-exciton distribution $P_{\tilde{\alpha}} = \rho_{\tilde{\alpha}\tilde{\alpha}}$ and the single-exciton distribution $P_\alpha = \rho_{\alpha\alpha}$. For comparison, the lower panel

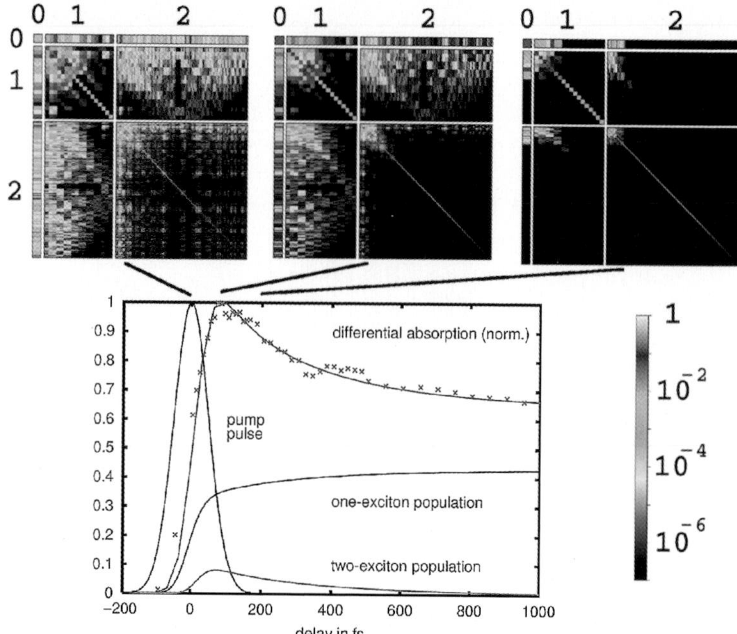

Fig. 6. Transient differential absorption and related multiexciton dynamics for the LH2 antenna complex shown in Fig. 3. Upper panels: absolute values of the multi-exciton density matrix (for gray code see lower part of the figure) at the pump pulse maximum as well as after 100 fs and after 200 fs (shown are all density matrix elements ordered with increasing energy from the left to the right as well as from the top to bottom, the numbers 0, 1, 2 indicate the different multiexciton manifolds). Lower panel: transient absorption vs. delay time between pump and probe pulse together with the overall single and two-exciton population. Shown is also the envelope of the 100 fs pump pulse (for experimental data which are displayed by crosses cf. [29])

displays the overall two-exciton population $P_2 = \sum_{\tilde{\alpha}} P_{\tilde{\alpha}}$ and the single-exciton counterpart $P_1 = \sum_{\alpha} P_{\alpha}$. With increasing time P_{α} relaxes into a thermal equilibrium distribution, whereas $P_{\tilde{\alpha}}$ vanishes. This latter effect is caused by the inclusion of exciton–exciton annihilation which represents a particular two-exciton decay channel (for more details see [2, 17, 18]).

3 Electronic Frenkel-Excitons: Beyond Weak Exciton–Vibrational Coupling

To go beyond the limit of weak exciton–vibrational coupling mainly two approaches have been described in literature. Both, however, exclusively concentrate on the dynamics within the single exciton manifold, and determine diagonal electronic (excitonic) matrix elements of the reduced density operator $\hat{\rho}(t)$, (13). One approach is based on the distribution function for a single molecular excitation $P_m(t) = \langle\phi_m|\hat{\rho}(t)|\phi_m\rangle$, whereas the other uses the (single) exciton distribution $P_\alpha(t) = \langle\alpha|\hat{\rho}(t)|\alpha\rangle$. The nonequilibrium quantum statistical background, however, is common to both descriptions. To get P_m or P_α the so-called Liouville space technique is applied ending up with generalized master equations for the particular distributions. The related memory kernels are the result of a complete perturbation expansion (see [30, 31] and the more recent presentation [2, 4, 21] for details).

3.1 Generalized Förster Theory

The first approach leading to the distribution $P_m(t)$ focuses on an expansion with respect to the Coulombic interchromophore coupling J_{mn} (cf. (5)) and is based on the following projection superoperator

$$\mathcal{P}\ldots = \sum_m \hat{r}_m^{(\mathrm{eq})} \hat{\Pi}_m \mathrm{tr}\{\hat{\Pi}_m \ldots\} \,. \tag{21}$$

The $\hat{\Pi}_m$ are given by $|\phi_m\rangle\langle\phi_m|$, and $\hat{r}_m^{(\mathrm{eq})}$ denotes the statistical operator for the vibrational equilibrium present if the mth molecule is excited. This excitation might be connected with an arbitrary displacement of the vibrational equilibrium configuration and thus the whole treatment is nonperturbative with respect to the coupling to the vibrational coordinates. The related rates follow from Fourier transformed memory kernels of the generalized master equations. They describe excitation energy transfer from molecule m to molecule n, and take the following form:

$$k_{m\to n} = -\mathrm{i}\mathrm{tr}\{\hat{\Pi}_n \mathcal{J} \tilde{\mathcal{G}}(\omega=0) \mathcal{J} \hat{r}_m^{(\mathrm{eq})} \hat{\Pi}_m\} \,. \tag{22}$$

The quantity $\mathcal{J}$ is the Liouville superoperator defined by the interchromophore interactions $\sim J_{mn}$ (cf. (5)), and $\tilde{\mathcal{G}}(\omega=0)$ denotes the Fourier–transformed Green's superoperator (but defined with $(1-\mathcal{P})\mathcal{J}$ instead of $\mathcal{J}$ alone [2, 4]). The lowest order rate expression (neglecting any vibrational coordinate dependence of the J_{mn})

$$k_{m\to n}^{(2)} = \frac{|J_{mn}|^2}{\hbar^2} \int dt\; \mathrm{tr}_{\mathrm{vib}}\{\hat{r}_m^{(\mathrm{eq})} \mathrm{e}^{\mathrm{i}H_m t/\hbar} \mathrm{e}^{-\mathrm{i}H_n t/\hbar}\} \tag{23}$$

reconstitutes the well-known Förster rate. Fourth-order rate expressions (resembling what is known as superexchange in electron transfer theory, see, e.g., [2]) have been investigated in [21, 32]. However, any experimental evidence for these generalizations could not be underlined so far.

3.2 Excitonic Potential Energy Surfaces

A second way of treating exciton–vibrational coupling beyond a perturbation expansion is based on the introduction of what might be called excitonic PES

$U_\alpha(Q) = \hbar\Omega_\alpha - \hbar\lambda_\alpha + \sum_\xi \hbar\omega_\xi(Q_\xi + 2g_{\alpha\alpha}(\xi))^2/4$ (cf. [4,33]). Here, the diagonal part of the exciton–vibrational coupling, (11), proportional to $g_{\alpha\alpha}(\xi)$ has been assumed to be large and has been included in the PES. The reorganization energy referring to a transition into such an excitonic PES of exciton level α reads $\hbar\lambda_\alpha = \sum_\xi \hbar\omega_\xi g^2_{\alpha\alpha}(\xi)$. The introduction of excitonic PESs corresponds to the following separtion of $\mathcal{H}_1$, (5):

$$\mathcal{H}_1 = \sum_{\alpha,\beta} \Big(\delta_{\alpha,\beta}(T_{\text{vib}} + U_\alpha) + (1 - \delta_{\alpha,\beta}) \sum_\xi \hbar\omega_\xi g_{\alpha\beta}(\xi) Q_\xi \Big) |\alpha\rangle\langle\beta| \,. \tag{24}$$

The assumed smallness of the $g_{\alpha\beta}(\xi)$ ($\alpha \neq \beta$) allows for a perturbational treatment. But the presence of the $g_{\alpha\alpha}(\xi)$ in the PES accounts for the dominant part of the exciton–vibrational coupling nonperturbatively (cf. [4,33–36]).

Within this scheme, but neglecting the off-diagonal parts of $g_{\alpha\beta}(\xi)$ the linear absorbance, for example, reads [4,33]:

$$A(\omega) \sim \sum_\alpha |d_\alpha|^2 \mathrm{e}^{-G_\alpha(0)} \int \mathrm{d}t \; \mathrm{e}^{\mathrm{i}(\omega - \Omega_\alpha - \lambda_\alpha) + G_\alpha(t)} \,. \tag{25}$$

The expression resembles the absorbance related to transitions between two independent states with harmonic PES. Besides the reorganization energies $\hbar\lambda_\alpha$ it includes the so-called lineshape functions $G_\alpha(t) = \sum_\xi g^2_{\alpha\alpha}(\xi)([1 + n(\omega_\xi)]\mathrm{e}^{-\mathrm{i}\omega_\xi t} + n(\omega_\xi)\mathrm{e}^{\mathrm{i}\omega_\xi t})$. The (25) clearly indicates that the electronic interchromophore coupling (resulting in the formation of exciton states) as well as the exciton–vibrational coupling both enter beyond any perturbation theory (a perturbational inclusion of the off-diagonal elements of the coupling matrix has been used in [34] to calculate the corresponding correction to the linear absorbance).

This treatment has been extended in [35] to calculate photon echo spectra of the photosynthetic antenna complex LH2 and pump probe spectra (in a doorway–window representation of the third-order response function). A derivation of rate equations for P_α is also included. They have been obtained in a similar way as those for P_m discussed in Sect. 3.1, but now with transition rates being of second order with respect to the $g_{\alpha\beta}(\xi)$ ($\alpha \neq \beta$). A recent application to fit transient absorbance of the LH2 can be found in [36]. There, it has been argued that such a treatment improves the spectra fit considerable. Unfortunately, the importance of the off-diagonal coupling constants has been not quantified.

4 Vibrational Frenkel-Excitons: Arbitrary Exciton–Vibrational Coupling

There exists a particular application of the Frenkel exciton concept to high-frequency molecular vibrations. It dates back to the seventies of the last century and concentrates on the study of vibrational excitons in α-helical polypeptides (see [37] for an overview as well as the Figs. 7 and 8). Since the localized high-frequency vibrations of the various amide groups forming the polypeptide chain are characterized by sufficiently large transition dipole moments the formation of Frenkel exciton states becomes possible.

The self-localization of these excitons and the formation of so-called *Davydov solitons* has been of main interest. But any univocal experimental confirmation has failed so far. Only recently some experimental indications on the formation of self-trapped excitons could be reported in [38]. These studies comprise subpicosecond infrared pump–probe measurements in the absorption range of the N–H amide group vibration of poly-γ-benzyl-L-glutamate helices. And they focused on two-exciton states. The concept of *vibrational* two-exciton states has been already introduced in the field of ultrafast infrared two-dimensional spectroscopy (cf. [16, 39]). Recently, the formation and self-trapping of two-exciton states in polypeptide chains has been also discussed theoretically in [40, 41].

It is believed that self-trapping follows from a sufficiently large change of the energy level scheme of the amide group high-frequency vibrations upon chain deformation. Therefore, the coupling to low-frequency vibrations of the chain (mainly longitudinal vibrations along the chain axis) should become large enough to suppress the quantum mechanical delocalization of the high-frequency vibrational quanta along the chain. To achieve a correct picture of self-trapping (self-localization) of multiexciton states, hence, it requires a complete quantum description of the coupled exciton–vibrational dynamics beyond any perturbation theory.

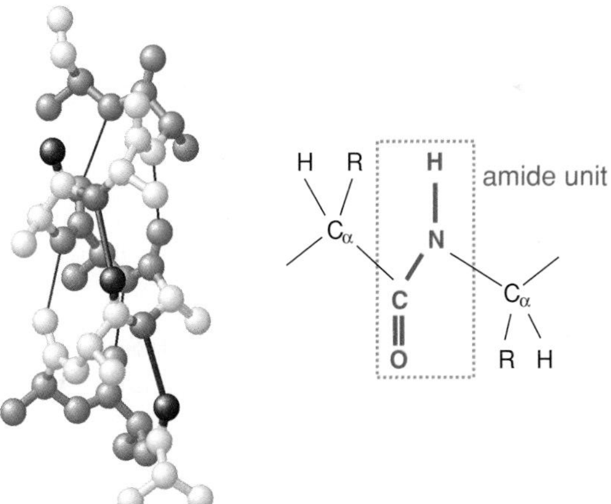

Fig. 7. Spatial structure of an α-helical polypeptide (the three lines of variable thickness indicate the sequence of hydrogen bridges connecting the amide units). In the right part the chemical structure of a single amide unit is shown (the so-called C_α carbon atoms bind residuals R which distinguish different amino acids by their chemical structure)

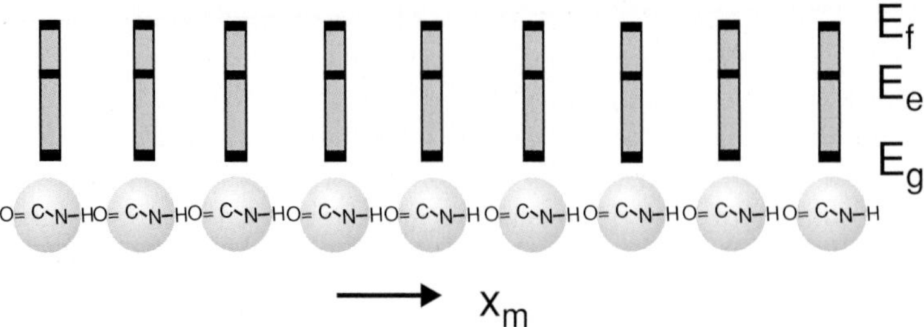

Fig. 8. Linear chain model of the sequence of hydrogen bridges connecting different amide units in an α-helical polypeptide (cf. Fig. 7). A three-level system is formed by the ground-state energy E_g, the energy of the first excited state E_e and of the first overtone energy E_f of a selected normal mode vibration of the amide unit. x_m is the one-dimensional displacement of the mth unit along the chain

It will be demonstrated in the following that the concept of multiexciton states introduced in Sect. 1.1 together with a proper treatment of the exciton–vibrational coupling is ready to describe self-trapping of vibrational multiexciton states and to compute related spectroscopic observables. Already recently it has been suggested by us in [20] that a (numerically) exact description of the exciton–vibrational coupling and thus of self-trapping becomes possible when applying the Multiconfiguration Time-Dependent Hartree (MCTDH) method [42,43] for a solution of the multidimensional vibrational Schrödinger equations. Calculating, additionally, the adiabatic multiexciton levels supports the understanding of vibrational exciton dynamics and related infrared spectra.

Such an analysis has been carried out in [20] by concentrating on the so-called amide I excitons and using a linear chain model of the α-helical peptide suggested, e.g., in [37,40]. Such a model incorporates a very selected number of vibrational DOF referring to the longitudinal displacement of the amide groups along the chain. We note here that self-localization has been also suspected for electronic Frenkel-excitons [5]. Moreover, some recent theoretical studies on adiabatic electronic excitons can be also found in [13,19].

4.1 Adiabatic Single and Two-Exciton States

If the reorganization of the vibrational coordinates upon multiexciton formation becomes large one may consider adiabatic states introduced in the (9) and (10) rather than the ordinary states (7) and (8). We will consider them for the N-H-amide group vibrations in a linear chain description of an α–helical polypeptide (cf. Figs. 7 and 8). These studies do not account for the helical structure of the polypeptide and neglect the coupling among different high-frequency amide group vibrations (cf. [44]). Nevertheless, the essence of the opposite action of exciton delocalization and self-trapping can be accounted for in the right way (see also Fig. 8).

What we have to expect is shown schematically in Fig. 1. Every single and two-exciton state forms a set of PESs $U_\alpha^{(1)}$ and $U_{\tilde{\alpha}}^{(2)}$, respectively. They are defined with respect to the longitudinal displacements x_m of the single amide group along the linear chain (additional local minima in the PESs are not shown for simplicity). Moreover, the states φ_a with $a = g, e, f$ of an individual excitable unit introduced in Sect. 1.1 correspond to the N-H-vibrational ground-state as well as to the first and second excited state, respectively. As a consequence the overall ground-state (1) and the single and double excited states (2) and (3), respectively, can be easily defined.

To carry out computations for N-H-vibrational excitons the respective coupling to the chain vibrations (of a chain with $N_{\rm au}$ amide units) has to be specified. As suggested in literature [45] the following potential can be taken

$$V(q, X) = \frac{1}{2}\chi \sum_{m=1}^{N_{\rm au}-1} q_m^2 [x_{m+1} - x_m] , \tag{26}$$

where q_m denotes the mth amide group N-H-vibrational coordinate. The matrix elements $\langle\phi_m|V(q,X)|\phi_m\rangle$ and $\langle\phi_{mn}|V(q,X)|\phi_{mn}\rangle$ together with the respective N-H-vibrational energy levels as well as the chain vibrational Hamiltonian define $\mathcal{H}_1$ and $\mathcal{H}_2$, (5) and (6), respectively. The inter-amide unit couplings J_{mn} and $J_{kl,mn}$ are used in the standard form of dipole dipole interaction (for parameters cf. [46]).

First, a diagonalization of the Hamiltonians of (5) and (6) taken at the equilibrium configuration X_0 of the peptide chain ground-state (at the absence of any N-H vibrational excitation) leads to the ordinary (unrelaxed) single and two-exciton states, (7) and (8), respectively. Related relaxed multiexciton levels are obtained in two steps. First, one carries out the calculation of the excited states for arbitrary values of X, i.e., one introduces adiabatic states, (9) and (10) with related adiabatic PES. And second, one searches for the minimum of every adiabatic PES. Figure 9 shows both energy values (relaxed and unrelaxed) for the single exciton states (upper panel) and the two-exciton states (lower panel). To characterize the localization of these states the participation ratios $\sum_m |C_\alpha^{(1)}(m; X_{\rm rel}^{(\alpha)})|^4$ and $\sum_{m,n} |C_{\tilde{\alpha}}^{(2)}(mn; X_{\rm rel}^{(\tilde{\alpha})})|^4$ at the respective relaxed chain configurations $X_{\rm rel}^{(\alpha)}$ and $X_{\rm rel}^{(\tilde{\alpha})}$ have been also drawn.

Since the potential, (26) is asymmetric the coupling to the chain vibrations is absent for the last amide unit ($m = N_{\rm au}$). This introduces in the uppermost part of the single as well as the two-exciton energy spectrum displayed in Fig. 9, a considerable shift. In the lower part of both spectra the reorganization energy reaches its maximum. This together with the shown large participation ratio indicates self-trapping of the excitations. In the single-exciton manifold only the lowest exciton state appears to be self-trapped whereas two-exciton states relax into a self-trapped configuration. It has been already shown by us in [20] that these computations represent an exploratory analysis, only.

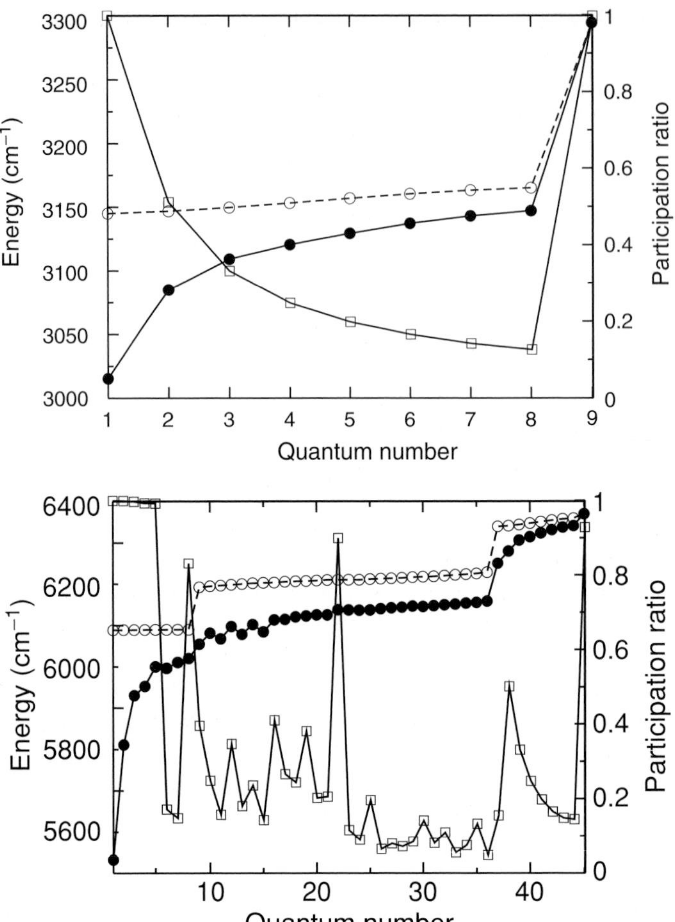

Fig. 9. Single- and two-exciton energies (caused by the coupling of amide unit N-H-vibrations and for a chain of 9 units). Upper panel: single-exciton energies vs. quantum number α (= 1,...,9). Lower panel: two-exciton energies vs. quantum number $\tilde{\alpha}$ (= 1,...,45, all shown energies are related to the minimum of the ground-state PES which value has been set equal to zero). *Open circles:* values of the PESs $U_{\alpha}^{(1)}$ and $U_{\tilde{\alpha}}^{(2)}$ at the chain equilibrium configuration X_0 in the ground-state. Full circles: values following after relaxation into the configuration $X_{\rm rel}^{(\alpha)}$ or $X_{\rm rel}^{(\tilde{\alpha})}$ corresponding to the minimum of $U_{\alpha}^{(1)}$ or $U_{\tilde{\alpha}}^{(2)}$, respectively, (the differences between the relaxed and unrelaxed energies define the reorganization energies $\hbar\lambda_{\alpha}$ and $\hbar\lambda_{\tilde{\alpha}}$). *Open squares*: participation ratio for all relaxed single and two exciton levels

The complete quantum description of the chain configurations lifts this obvious trapping discussed here and introduces quantum mechanical superpositions of self-trapped configurations (a fact which is used by default in variational descriptions of self-trapping [37]). Moreover, the correct energy spectrum in the region of the single- and two-exciton manifold appears as a mixture of the adiabatic levels (via nonadiabatic couplings) together with vibrational progressions caused by the chain vibrations. It is the advantage of the following considerations that all these effects can be accounted for when calculating the transient absorbance.

4.2 Exciton–Vibrational Quantum Dynamics

To arrive at a complete quantum description of the exciton–vibrational dynamics we introduce an expansion similar to that of the (9) and (10) but with the expansion coefficients $C_\alpha^{(1)}(m;X)$ and $C_{\tilde{\alpha}}^{(2)}(mn;X)$ now reinterpreted as time-dependent chain-vibrational wave functions $\psi_m(X,t)$ and $\psi_{mn}(X,t)$, respectively. Both sets of functions have to be supplemented by the wave function of the exciton ground state $\psi_0(X,t)$. The respective time-dependent Schrödinger equations are governed by the related matrix elements of the Hamiltonians $\mathcal{H}_0$, $\mathcal{H}_1$, and $\mathcal{H}_2$ introduced in the (4), (5), and (6), respectively.

These equations are solved in applying the MCTDH-method. It represents $\psi_m(x,t)$, for example, as a time-dependent superposition of time-dependent Hartree products (cf. [42]):

$$\psi_m(X,t) = \sum_{\zeta_1,\dots,\zeta_f} A^{(m)}(\zeta_1,\dots,\zeta_f;t) \prod_{j=1}^{f} \psi_{\zeta_j}^{(m)}(x_j,t) \,. \tag{27}$$

Within a single Hartree–product the index j counts the different vibrational coordinates x_j ($j=1,\dots,f$, with total number $f = N_{\rm au}-1$ in the present case). The $\psi_{\zeta_j}^{(m)}(x_j,t)$ are single chain-coordinate dependent wave functions. Their dependence on m indicates that they refer to the mth vibrational wave function in the single-exciton state expansion. Moreover, the particular index ζ_j indicates that $\psi_{\zeta_j}^{(m)}$ enters the Hartree product in the multiconfigurational ansatz with prefactor $A^{(m)}(\zeta_1,\dots,\zeta_f;t)$. The method appears as a modification of the standard basis-set expansion scheme by using time-dependent expansion functions which may be adapted to the actual state and thus can be drastically reduced in their overall number.

Calculations could be carried out up to chains with 9 peptide units (arriving at eight longitudinal chain coordinates). This leads to a computation of nine functions of the type $\psi_m(X,t)$ referring to the single-exciton manifold and 45 functions of the type $\psi_{mn}(X,t)$ referring to the two-exciton manifold. Related excitation energy dynamics in the single exciton manifold has been studied in [21] by drawing the local amide group population $P_m(t) = \int dX |\psi_m(X,t)|^2$ vs. time. In the following, the possibility of a rather exact computation of ψ_0, the ψ_m, and the ψ_{mn} is used to determine infrared transient absorption spectra like those measured in [38].

4.3 Transient Absorption Spectra

As already mentioned in Sect. 2.2 it represents a suitable experimental approach to measure the femtosecond transient absorption signal (TAS) when studying ultrafast molecular dynamics. In the present case it is of particular interest to study higher excited states which are characterized by a short life time. While the computation of the third-order response function becomes necessary in the general case of simulating the TAS (cf. also the discussion in Sect. 2.2) the description of a sequential pump–probe experiment is less sophisticated.

In the sequential pump–probe experiment which is of interest here [38] the pump and the probe pulse are well separated on a time-scale of single exciton relaxation. Therefore, it is suitable to start with the calculation of the probe–pulse response (the polarization P_{pr}) linear with respect to the probe pulse field E_{pr}. Although the related response function is defined in any order of the pump field it is not necessary to calculate this dependence. The sequential character of the experiment allows to assume the presence of a relaxed excited state when the probe pulse starts to act (of course, the considered time region needs to be below the life time of the excited state). Here, we take a relaxed mixed state of the exciton vibrational system which covers a somewhat depleted ground state with population w_0 and energy $\hbar\Omega_{\mathrm{rel}}^{(0)}$ and a relaxed state in the single exciton manifold with population $w_1 = 1 - w_0$ and energy $\hbar\Omega_{\mathrm{rel}}^{(1)}$. The wavefunctions of the respective pure states are written as $\psi_0^{(\mathrm{rel})}(X)|0\rangle$ and $\sum_m \psi_m^{(\mathrm{rel})}(X)|\phi_m\rangle$. They are computed via imaginary time propagation as the chain vibrational ground-state and the lowest chain single-exciton state.

Then, the differential TAS can be written as

$$\Delta A_{\mathrm{pr}}(\omega) \sim \mathrm{Im}\big(R_{\mathrm{pr}}^{(\mathrm{GB})}(\omega) + R_{\mathrm{pr}}^{(\mathrm{SE})}(\omega) + R_{\mathrm{pr}}^{(\mathrm{EA})}(\omega)\big)\,. \tag{28}$$

The expression includes Fourier transformed response functions $R^{(\mathrm{GB})}$, $R^{(\mathrm{SE})}$, and $R^{(\mathrm{EA})}$ referring to the ground state bleaching, the stimulated emission, and the excited state absorption signal, respectively. All functions are determined within a time-dependent formulation according to

$$\begin{aligned}
R_{\mathrm{pr}}^{(\mathrm{GB})}(t) &= \frac{i}{\hbar}\theta(t)(w_0 - 1)\mathrm{e}^{\mathrm{i}\Omega_{\mathrm{rel}}^{(0)}t}\int \mathrm{d}x \sum_m d_m^* \psi_0^{(\mathrm{rel})*}(x)\psi_m(x,t) \\
R_{\mathrm{pr}}^{(\mathrm{SE})}(t) &= -\frac{i}{\hbar}\theta(t)w_1\mathrm{e}^{-\mathrm{i}\Omega_{\mathrm{rel}}^{(1)}t}\int \mathrm{d}x \sum_m d_m^* \psi_m^{(\mathrm{rel})}(x)\psi_0^*(x,t) \\
R_{\mathrm{pr}}^{(\mathrm{EA})}(t) &= \frac{i}{\hbar}\theta(t)w_1\mathrm{e}^{\mathrm{i}\Omega_{\mathrm{rel}}^{(1)}t}\int \mathrm{d}x \sum_{m,n}\Big(\delta_{m,n}\tilde{d}_m^* \psi_m^{(\mathrm{rel})*}(x)\psi_{mm}(x,t) \\
&\quad + [1-\delta_{m,n}]\big(d_m^*\psi_n^{(\mathrm{rel})*}(x) + d_n^*\psi_m^{(\mathrm{rel})*}(x)\big)\psi_{mn}(x,t)\Big)\,.
\end{aligned} \tag{29}$$

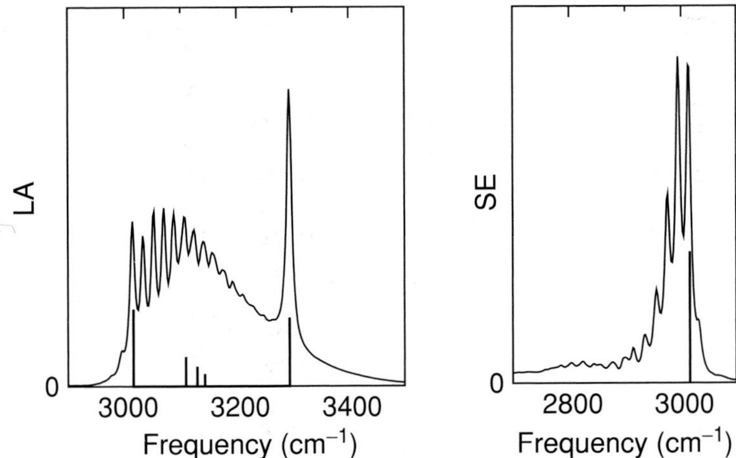

Fig. 10. Calculated linear absorption (left panel) and stimulated emission spectrum (right panel) of a sequential pump probe experiment in the spectral range of N-H-vibration of the amide units. The vertical lines give the position of the adiabatic single-exciton levels displayed in Fig. 9, and their length corresponds to squares of the respective transition dipole moments $\sim |\sum_m C_\alpha^{(1)}(m; X_{\mathrm{rel}}^{(\alpha)})|^2$

The local $\varphi_{\mathrm{g}} \to \varphi_{\mathrm{e}}$ and $\varphi_{\mathrm{e}} \to \varphi_f$ transition dipole moments are denoted as d_m and $\tilde{d}_m$, respectively. To obtain $\psi_m(x,t)$ in $R_{\mathrm{pr}}^{(\mathrm{GB})}$ a propagation within the single-exciton manifold becomes necessary using the initial condition $d_m^* \psi_0^{(\mathrm{rel})*}$. In a similar way $R_{\mathrm{pr}}^{(\mathrm{SE})}$ and $R_{\mathrm{pr}}^{(\mathrm{EA})}$ have to be calculated but now carrying out a propagation in the ground-state and in the two-exciton manifold, respectively.

To understand details of the spectra let us first compare $R_{\mathrm{pr}}^{(\mathrm{GB})}$ as shown in Fig. 10 with the single-exciton levels displayed in Fig. 9 (note the inclusion of levels with a sufficient large oscillator strength in Fig. 10). The spectrum (which is identical with the linear absorbance) is dominated by the lowest exciton level and a subsequent vibrational progression with some contributions of higher lying exciton levels. The contribution of the level which shift is caused by the chain end effect is also obvious. Changing to $R_{\mathrm{pr}}^{(\mathrm{EA})}$ in Fig. 11 a clear separation of the two lowest self-trapped two-exciton levels from the remaining levels can be found. At higher energies a number of delocalized two-exciton levels contribute. The resulting differential TAS comprizes all these contributions, in particular, it displays the signature of the self-trapped two-exciton states as observed in [38] (for more details see also [47]).

5 Concluding Remarks

An overview on picosecond and subpicosecond Frenkel exciton dynamics has been presented with particular emphasis on the description of exciton

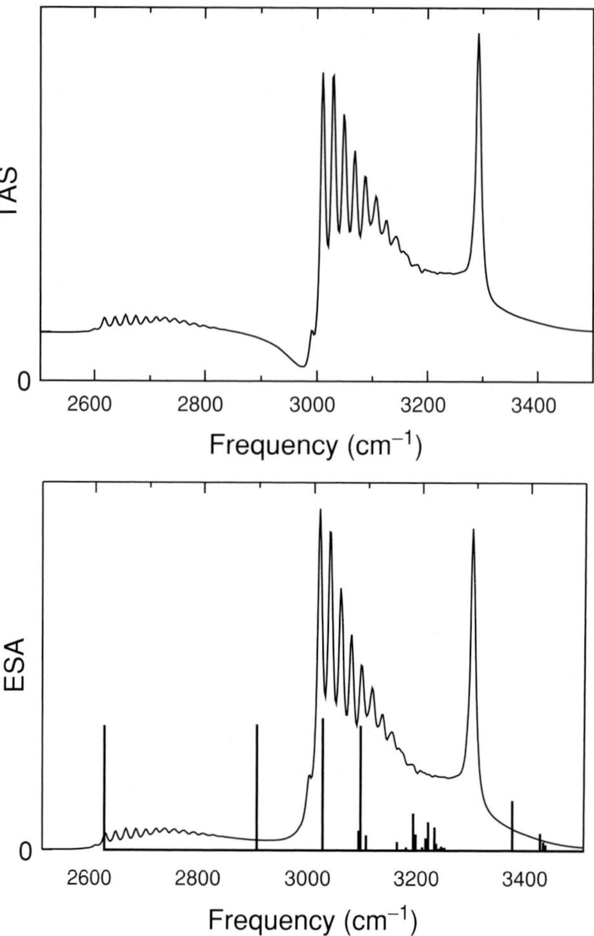

Fig. 11. Calculated differential transient absorption spectrum of a sequential pump probe experiment in the spectral range of N-H-vibration of the amide units (upper panel) The lower panel shows the related excited state absorption part. The vertical lines give the position of the adiabatic two-exciton levels (minus the energy of the lowest single-exciton level) of Fig. 9, and their length indicates the squares of the respective transition dipole moments (for details cf. [47])

vibrational coupling. Systems forming electronic as well as vibrational excitons have been considered on the basis of a common theoretical description (and with the restriction to weak static disorder).

The multiexciton density matrix theory presented in Sect. 2.1 has to be considered as standard in the field, with its advantages and disadvantage well understood. So, the method can be used as technique to obtain reference data for a proof if more sophisticated descriptions are necessary. The usefulness of this approach in simulating excitation energy dynamics in the rather ordered chromophore complexes of photosynthetic antenna systems has been demonstrated. The techniques presented in Sect. 3 may offer some improvements as shown in particular by the recent calculations of [36]. Nevertheless, at the moment further extended checks against experimental data are necessary.

The exact description of multiexciton vibrational dynamics as presented in Sect. 4 seems to be particularly suitable. However, it is restricted to a very limited number of modes (the agglomeration of modes as used in [48] to describe electron transfer may represent a possible way to overcome this restriction). Of course, another demand on theory would be an improvement of the used multiexciton models by quantum chemical calculations.

Finally, we note that it became also of interest to discuss femtosecond laser pulse control of exciton motion [24, 49]. In using appropriately tailored laser pulses of some 100 fs one may try to form particular multiexciton wave packets which lead, for example, to excitation energy localization at a single chromophore. Then, particular energy transfer pathways may be studied which are otherwise not accessible.

Acknowledgments

We acknowledge discussions with V. Sundström (Lund) and R. van Grondelle (Amsterdam). Our thanks are also to H.-D. Meyer (Heidelberg) and L. Wang (Berlin) for their assistance in using the MCTDH-package. Finally, we gratefully acknowledge the financial support of the Deutschen Forschungsgemeinschaft through Sfb 450 and project Ma 1356 – 8/1.

References

1. S. Mukamel, *Principles of Nonlinear Optical Spectroscopy*, (Oxford University Press, 1995)
2. V. May and O. Kühn: *Charge and Energy Transfer Dynamics in Molecular Systems* (Wiley, Berlin, 2000, second edition 2004)
3. H. van Amerongen, L. Valkunas, and R. van Grondelle: *Photosynthetic Excitons* (World Scientific, Singapore, 2000)
4. Th. Renger, V. May, and O. Kühn, Phys. Rep. **343** 137 (2001)
5. K. Timpmann, M. Rätsep, C. N. Hunter, and A. Freiberg, J. Phys. Chem. B **108**, 10581 (2004)
6. A. S. Davydov, *Theory of Molecular Excitons*, (Plenum, New York, 1962)
7. M. Kasha, in *Spectroscopy of the Excited State*, (Plenum, New York, 1976), pp. 337–351
8. R. M. Pearlstein, in *Excitons* (eds. E. I. Rashba and M. D. Sturge, North Holland, Amsterdam, 1982), p. 735
9. V. M. Agranovich and M. D. Galanin, in *Modern problems in condensed matter sciences*, (eds. V. M. Agranovich and A. A. Maradudin, North Holland, Amsterdam, 1982)

10. V. M. Kenkre and P. Reineker, in *Springer Tracts Mod. Phys.*, (Volume 94, Springer, Berlin Heidelberg New York, 1982)
11. T. Förster, Ann. Physik (Leipzig) **6**, 55 (1948)
12. D. L. Dexter, J. Chem. Phys **21**, 836 (1953)
13. T. Meier, Y. Zhao, V, Chernyak, and S. Mukamel, J. Chem. Phys. **107**, 3876 (1997)
14. T. Meier, V. Chernyak, and S. Mukamel, J. Phys. Chem. B **101**, 7332 (1997)
15. The storage of solar energy in energetically rich organic compounds represents the basis of life on earth and the related process is called photosynthesis. It starts with a primary charge separation in the photosynthetic *reaction center.* In most cases the initial excitation is supplied by *light-harvesting antennae*, which surround the reaction center to enlarge the cross-section for the capture of sunlight. There is a huge diversity of antenna complexes in bacteria and higher plants. For some of them the structure is known with an atomic resolution (for details see, e.g., [3, 4]). As it could be clarified in detail over the last three decades excitation energy transfer in photosynthetic systems takes place via Frenkel exciton mechanism
16. P. Hamm, M. Lim, and R. M. Hochstrasser, J. Phys. Chem. B **102** 6123 (1998)
17. B. Brüggemann and V. May, J. Chem. Phys. **118**, 746 (2003)
18. B. Brüggemann and V. May, J. Chem. Phys. **120**, 2325 (2004)
19. W. Beenken, M. Dahlbom, P. Kjellberg, and T. Pullerits, J. Chem. Phys. **117**, 5810 (2002)
20. D. Tsivlin and V. May, Chem. Phys. Lett. **408**, 360 (2005)
21. O. Kühn, V. Chernyak, and S. Mukamel, J. Chem. Phys. **105**, 8586 (1996)
22. There have been attempts to compute the electronic energy level structure for such a complex of chlorophyll molecules embedded into the carrier proteins (see references in [4, 18]). However, such quantum chemical approaches have been successful only in part since the consideration of the whole carrier proteins is beyond present day computational capabilities. Moreover, a direct computation of the coupling of excitons to intrachlorophyll vibrations and to those of the surrounding protein is also hopeless at the moment. Consequently, an approach has to be chosen which is based on additional assumptions mainly related to the energy level structure of the chlorophylls and their mutual interaction. Concrete parameter values are fixed by a comparison of this model with different experimental results. This just underlines the importance of the presented model
23. B. Brüggemann and K. Sznee, V. Novoderezhkin, R. van Grondelle, and V. May, J. Phys. Chem. B **108**, 13563 (2004)
24. B. Brüggemann and V. May, Gerald F. Small Festschrift, J. Phys. Chem. B **108**, 10529 (2004)
25. There does not exist a unique set of parameters for LH2 antennae. To offer an impression we quote the following values [18]: $d_{eg} = 6.32$ D and nearest–neighbor coupling $J_{m\,m\pm1} = 288 \ldots 322$ cm^{-1}. Transitions into the higher–excited state φ_f are described by the same value of the dipole operator as those into φ_e, and the related transition energy E_{fe} is 100 cm^{-1} larger than E_{eg}. The uniformly taken spectral density $J_m(\omega)$, (19) covers a prefactor $j_e = 1.5$ and different parts $\sim \omega^2 \exp(\omega/\omega_\nu)$ with the frequency constants ω_ν ranging from 10.5 cm^{-1} up to 350 cm^{-1}
26. To be complete we mention two approximations not indicated in the running text. First, thermal expectation values of the $H^{(N)}_{\text{ex-vib}}$ ($N = 1, 2$) to be incorporated into (15) have been neglected (cf. [2]). Furthermore, the possible external

field dependence of (16) also has not been taken into account (see, for example, D. Schirrmeister and V. May, Chem. Phys. Lett. **297**, 383 (1998),T. Mancal and V. May, Chem. Phys. **268**, 201 (2001))
27. P. Jordan, P. Fromme, H.T. Witt, O. Klukas, W. Saenger, and N. Krauß: Nature **909**, 411 (2001)
28. Used PS1 parameters are: mean transition energy into the Q_y–state: $E_{eg} =$ 14841 cm^{-1}, related dipole moment $d_{eg} = 6$ D, for the used spectral density see the remark in Fig. 3 and for the particular role of the special pair [23]
29. B. Brüggemann, J. L. Herek, V. Sundström, T. Pullerits, and V. May, J. Phys. Chem. B **105**, 11391 (2001)
30. M. Sparpaglione and S. Mukamel, J. Chem. Phys. **88**, 3263 (1988)
31. Y. Hu and S. Mukamel, J. Chem. Phys. **91**, 6973 (1989)
32. T. Kakitani, A. Kimura, H. Sumi, J. Phys. Chem. B **103**, 3720 (1999)
33. J. Schütze, B. Brüggemann, Th. Renger, and V. May, Chem. Phys. **275**, 333 (2002)
34. Th. Renger and R. A. Marcus, J. Phys. Chem. B **106**, 1809 (2002)
35. W. M. Zhang, T. Meier, V. Chernyak, and S. Mukamel, J. Chem. Phys. **108**, 7763 (1998)
36. V. I. Novoderezhkin, M. A. Palacios, H. van Amerongen, and R. van Grondelle, J. Phys. Chem. B **108**, 10363 (2004)
37. A. C. Scott, Phys. Rep. **217**, 1 (1992)
38. J. Edler, R. Pfister, V. Pouthier, C. Falvo, and P. Hamm, Phys. Rev. Lett. **93**, 106405 (2004)
39. S. Mukamel and R. M. Hochstrasser, *Special Issue on Multidimensional Spectroscopy*, Chem. Phys. **266** (2001)
40. V. Pouthier, Phys. Rev E **68**, 021909 (2003)
41. V. Pouthier and C. Falvo, Phys. Rev E **69**, 041906 (2004)
42. M. H. Beck, A. Jäckle, G. A. Worth, and H.-D. Meyer, Phys. Rep. **324**, 1 (2000)
43. G. A. Worth, M. H. Beck, A. Jäckle, and H.-D. Meyer, The MCTDH–Package, Version 8.3, University of Heidelberg, Heidelberg 2002 (see http://www.pci.uniheidelberg.de/tc/usr/mctdh/)
44. A. M. Moran, S.-M. Park, J. Dreyer, and S. Mukamel, J. Chem. Phys. **118**, 3651 (2003)
45. W. Förner, Phys. Rev. A **44**, 2694 (1991)
46. Parameters used in the simulations on infrared excitations in polypeptide are: $E_e - E_g = 3294$ cm^{-1}, $E_e - E_g - (E_f - E_e) = 120$ cm^{-1}, $J = 5.0$ cm^{-1}, $W =$ 13 N/m, $M = 92\ m_{\rm p}$, and $\chi = 300$ pN. ($m_{\rm p}$ denotes the proton mass)
47. D. Tsivlin, H.–D. Meyer, and V. May, J. Chem. Phys. **124**, 134907 (2006)
48. M. Thoss, this book.
49. B. Brüggemann and V. May, Chem. Phys. Lett. 400, 573 (2004).

Exciton and Charge-Transfer Dynamics in Polymer Semiconductors

Eric R. Bittner and John Glen S. Ramon

Summary. Organic semiconducting polymers are currently of broad interest as potential low-cost materials for photovoltaic and light-emitting display applications. We will give an overview of our work in developing a consistent quantum dynamical picture of the excited state dynamics underlying the photophysics. We will also focus upon the quantum relaxation and reorganization dynamics that occur upon photoexcitation of a couple of type II donor–acceptor polymer heterojunction systems. Our results stress the significance of vibrational relaxation in the state-to-state relaxation and the impact of curve crossing between charge-transfer and excitonic states. Furthermore, while a tightly bound charge-transfer state (exciplex) remain the lowest excited state, we show that the regeneration of the optically active lowest excitonic state in TFB:F8BT is possible via the existence of a steady-state involving the bulk charge-transfer state. Finally, we will discuss ramifications of these results to recent experimental studied and the fabrication of efficient polymer LED and photovoltaics.

1 Introduction

Over the past three decades, there has been an explosion of interest in developing semiconducting materials based upon π-conjugated organic polymers. Conducting polymers are generally lighter in weight, more flexible, and less expensive to synthesize and fabricate than their inorganic counterparts which are typically based upon copper or silicon. Such material properties are desirable for applications such as smart windows, electronic paper, and flexible flat screen displays. It has even been speculated that conductive polymers may play a significant role in the development of quantum and molecular computing.

Almost all organic solids and polymers are insulators. However, when the electronic states of the constituent molecules are extended over a significant length scale, as in the case of π-conjugated states, electrons can move quite freely along the backbone of the molecules. The polycyclic aromatic polymers

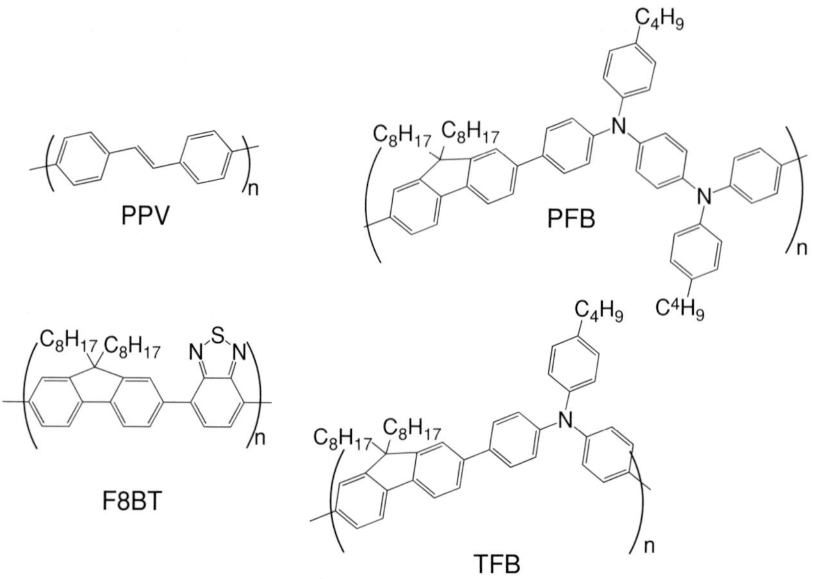

Fig. 1. Structures and common short-hand names of various conjugated polyphenylene derived semiconducting polymers that are of interest for fabricating luminescent devices

shown in Fig. 1 and phthalocyanine salt crystals are just some examples of these materials.

Typically, conjugated polymeric materials conduct electricity poorly compared to inorganic conductors. This is due to the intramolecular disorder intrinsic to a polymeric and glassy material. This disorder leads to trapping of polaronic charge carriers and hence a dramatic decrease in the carrier mobility. Recent work has focused upon improving the carrier mobility through either doping or through exploiting self-assembled systems and molecular crystals. In fact, recent observations of mobilities as high as 30 $\mathrm{cm^2\,V^{-1}\,s^{-1}}$ have been reported in rubrene [1] as well as several reports of high mobility in pentacene [2–8].

One of the earliest reported organic electronic devices was a voltage controlled switch fabricated from melanin (polyacetylene) by McGinness et al. [9] This original device is actually now in the Smithsonian Institution's collection of early electronic devices. These researchers also patented batteries and other devices made from organic semiconducting materials. Remarkably, even though this seminal work appeared in Science, the principal credit for the discovery and development of organic polymer semiconductors and "synthetic metals" goes to Heeger, MacDiarmid, and Shirakawa [10] who were jointly awarded the Nobel Prize in Chemistry in 2000. The high conductivity of doped polyacteylene, as well as a number of its semiconducting properties is largely explained by the simple one-dimensional lattice soliton model by

Su et al. [11–13] Finally for succinct history of the field of conducting polymers we, see Hush [14].

Organic semiconductors exhibit similar electronic properties as inorganic semiconductors. The highest occupied molecular orbitals (HOMOs) and the lowest unoccupied molecular orbitals (LUMOs) give rise to separate hole and electron conduction bands and a band gap. In organic semiconductors, these are π-type molecular orbitals. As with inorganic amorphous semiconductors, localized states due to disorder, tunneling, mobility gaps and phonon-assisted hopping all contribute to the conduction and mobility of charge carriers in the materials. Unlike inorganic materials, the electronic states of organic semiconductors can be easily modified by chemical modification of the polymer and through the addition of side-chains to the polymer backbone. Such chemical modifications can also be used to tune the mechanical and material properties while preserving desirable electronic properties. Furthermore, the quasi-one-dimensional nature of the π-states means that the density of electronic states is largely determined by the persistence length of the π-conjugation. Hence, polymer morphology will have a significant impact on the electronic density of states. Defects in the chain due to torsions, chemical impurities, and so on limit the persistence length of the π orbitals to the extent that one can consider conjugated polymer molecules to be a linked sequence of isolated quasi-one-dimensional states [15–19].

In light of the novel material and semiconducting properties of organic semiconductors, there have been significant advances in fabricating optical-electronic devices such as light-emitting diodes and photovoltaic cells based upon polymeric materials. Since OLED displays do not require backlighting, they are well suited for mobile applications such as cell phones, digital cameras, and flat-screen displays. According to data compiled by the Society for Information Display, the world-wide market for organic light emitting diodes in 2004 was approximately \$480 million. By 2008, that figure is estimated be anywhere between \$3 and \$8 billion.

In fact one of the economic driving forces behind the development of this technology is the quest for energy efficient light sources. In the US alone, six quadrillion BTU's energy per year is required to provide lighting, this is nearly 20% of all the energy used in buildings. Incandescent bulbs, which typically operate at $15\,\mathrm{lm\,W^{-1}}$, turn about 90% of that energy into heat and fluorescent bulbs, at $60–100\,\mathrm{lm\,W^{-1}}$, are a bit better in converting 70% of their energy into light. As of recently, there have been reports of very bright organic based white LEDS with efficiencies as high as nearly $60\,\mathrm{lm\,W^{-1}}$ [20] This efficiency, along with their relatively inexpensive fabrication and ability to be cast from solution over large surface areas make it highly likely that OLED based lighting technologies will soon become common place.

Organic LED devices are typically layered structures with luminescent media sandwiched between cathode and anode materials which are selected such that their Fermi energies roughly match the conduction and valence bands of the luminescent material. Often the semiconducting media itself consists of a hole transport layer and an electron transport layer engineered to facilitate the rapid diffusion of the injected carriers away from their image charges on the cathode or anode. These carriers are best described as polarons since electron–phonon coupling produces significant lattice reorganization about the carriers. Finally, a third, luminescent layer can be sandwiched between the transport layers. In this layer, the electron and hole polarons interact and combine to produce excitons. The individual spins of the electrons and holes are uncorrelated and only singlet excitons are radiatively coupled the ground electronic state. In the absence of singlet–triplet coupling, this places a theoretical upper-limit or 1:4 or 25% on the overall efficiency of an LED device and it has been long debated whether or not the efficiency of organic LED devices is in fact limited by this theoretical upper-limit.

The electronic properties of these materials are derived from the delocalized π orbitals found in conjugated polymers. The π electron system is primarily an intramolecular network extending along the polymer chain. For a linear chain, the valence and conduction π and π^* bands are typically 1–3 eV wide compared to the intermolecular bandwidth (due to π-stacking) of about 0.1 eV for well ordered materials. Thus, intrachain charge transport is extremely efficient; however, interchain transport typically limits the charge mobility for the usual size range of devices. The polymer backbone is held together through a σ bonding network. These bonds are considerably stronger than the π bonds and keep the molecule intact even following photoexcitation. Hence, we can consider the electronic dynamics as taking place within the π band and treat the localized σ bonds as skeletal framework.

Since the dielectric constant of organic semiconductors is relatively low, screening between charges is relatively weak. At a given radius, r_c, thermal fluctuation will be insufficient to break apart an electron/hole pair,

$$kT = \frac{e^2}{\epsilon r_c}$$

at 300 K, this radius is approximately 20 nm, which is on the order of a few molecular lengths. If we consider the electron/hole pair to be a hydrogenic-type system with effective masses equal to the free electron mass and dielectric constant of 3, the resulting binding energy is about 0.75 eV with an effective Bohr radius of 0.3 nm, which effectively confines the exciton to a single molecular unit. Finally, if we consider the electron/hole pair to be a pair of bound Fermions, exchange energy resulting from the antisymmetrization of the electron/hole wave function splits the spin-singlet and spin-triplet excitons by about 0.5–0.7 eV with the spin-triplet lying lower in energy than the singlet. While both species are relatively localized, singlets typically span about 10nm in well ordered materials while triplets are much more localized. In absence of spin–orbit coupling, emission from the triplet states is forbidden. Hence, triplet formation in electron/hole capture can dramatically limit the efficiency of a light-emitting diode device, although strong theoretical and experimental evidence indicates that singlet formation can be enhanced in long-chain polymers.

Experiments by various groups suggest that in long-chain conjugated polymer systems, the singlet exciton population can be greatly enhanced and that efficiencies as high as 60–80% can be easily achieved in PPV type systems [21]. On the other hand, in small oligomers, the theoretical upper limit appears to hold true. These initial experiments were then followed by a remarkable set of observations by Wohlgenannt and Vardeny [22] that indicate that the singlet to triplet ratio, $r > 1$ for wide range of conjugated polymer systems and that r scales universally with the polaron energy – which itself scales inversely with the persistence length of the π-conjugation

$$r \propto 1/n. \tag{1}$$

The electro-luminescent efficiency, ϕ, is proportional to the actual singlet population and is related to r via

$$\phi = r/(r+3).$$

Various mechanisms favoring the formation of singlets have been proposed for both interchain and intrachain e–h collisions. Using Fermi's golden rule, Shuai, Bredas and coworkers [23–25] indicate that the S cross-section for interchain recombination can be higher than the triplet one due to bond-charge correlations. Wohlgenannt et al. [26] employ a similar model of two parallel polyene chains. Both of these works neglect vibronic and relaxation effects. In simulating the intrachain collision of opposite polarons, Kobrak and Bittner [27–29] show that formation of singlets are enhanced by the near-resonance with the free e–h pair. The result reflects the fact that spin-exchange renders the triplet more tightly bound than the singlet and hence more electronic energy must be dissipated by the phonons in the formation of the former. The energy-conservation constraints in spin-dependent e–h recombination have been analyzed by Burin and Ratner [30] in an essential-state model. The authors point out that nonradiative processes (internal conversion, intersystem crossing) must entail C=C stretching vibrons since these modes couple most strongly to $\pi \rightarrow \pi^*$ excitations. Tandon et al. [31] suggest that irrespective of the recombination process, interchain or intrachain, the *direct transition* to form singlets should always be easier than triplets due to its smaller binding energy relative to the triplet. A comprehensive review of detailing the experiments and theory of this effect was presented by Wohlgenannt et al. [32]. By and large, recent theoretical models point towards the role of multiphonon relaxation and the scaling of the singlet/triplet splitting with chain length as dominant factors in determining this enhancement [33–36].

If we assume that the electron/hole capture proceeds via a series of microstates one can show that the ratio of the singlet to triplet capture cross-sections, r scale with the ratio of the exciton binding energies [37]

$$r \propto \frac{\epsilon_B^T}{\epsilon_B^S}.$$

If we take the singlet–triplet energy difference to be equal to twice the electron–hole exchange energy, $\Delta E_{\mathrm{ST}} = 2K$, and expand K in terms of the inverse conjugation length

$$K = K_\infty + K^{(1)}/n + \cdots ,$$

where $2K_\infty$ is the singlet–triplet splitting of an infinitely long polymer chain, one obtains

$$r \propto 1 + nK_\infty/\epsilon_B^S + \cdots$$

Since both $K_\infty > 0$, $r > 1$. Moreover, if we take $\epsilon_{\mathrm{B}}^{\mathrm{S}} \propto 1/n$, we obtain a simple and universal scaling law for the singlet–triplet capture ratio, $r \propto n$. This "universal" scaling law for r was reported by Wohlgennant et al. [22]. What is even more surprising, is that the the same scaling law (i.e., slope and intercept for $r = an + c$) describes nearly all organic conjugated polymer systems. Hence, it appears that a common set of electronic interaction parameters is transferable between a wide range of organic conjugated polymer systems.

Another general consequence of localized electronic states in molecular semiconductors is their effect on the molecule itself. Promoting an electron from a π bonding orbital to a π^* antibonding orbital decreases the bond order over several carbon–carbon bonds. This leads to a significant rearrangement of the bond-lengths to accommodate the changes in the electronic structure. By and large, for polymers containing phenyl rings, it is the C=C bond stretching modes and much lower frequency phenylene torsional modes that play significant roles in the lattice reorganization following optical excitation. This is evidenced in the strong vibronic features observed in the absorption and emission spectra of these materials.

Finally, one can fabricate devices using blends of semiconducting polymers which phase segregate. For example, the phenylene backbone in F8BT is very planar molecule facilitating very delocalized π-states. On the other hand, TFB and PFB are very globular polymers due to the triamide group in the chain. Consequently, phase segregation occurs due to more favorable π-stacking interactions between F8BT chains than between F8BT and TFB or PFB. Moreover, the electronic states in TFB and PFB are punctuated by the triamides. This difference in electronic states results in a band offset between the two semiconducting phases. When we place the materials in contact with each other, a p–n heterojunction forms.

One can think of the HOMO and LUMO energy levels of a given polymer as corresponding to the top and bottom of the valance and conduction bands, respectively. For the polymers under consideration herein, the relative band edges are shown in Fig. 2. In Type II heterojunction materials, the energy bands of the two materials are off set by ΔE. If the exciton binding energy $\varepsilon_{\mathrm{B}} > \Delta E$, excitonic states will the lowest lying excited state species, resulting in a luminescent material with the majority of the photons originating from the side with the lowest optical gap. Since the majority of the charge carriers are consumed by photon production, very little photocurrent will be observed.

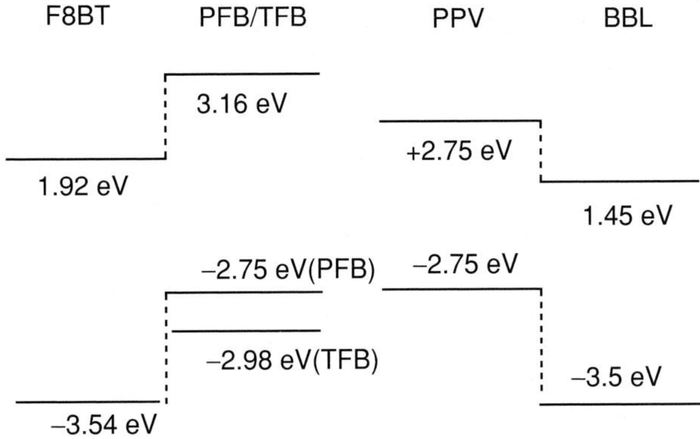

Fig. 2. Relative placement of the HOMO and LUMO levels for various conjugated polymers

On the other hand, charge transfer states across the interface will be energetically favored if $\varepsilon_{\mathrm{B}} < \Delta E$. Here, any exciton formed will rapidly decay in to a charge-separated state with the electron and hole localizing on either side of the junction. This will result in very little luminescence but high photocurrent. Consequently, heterojunctions of PPV and BBL which have a large band offset relative to the exciton binding energy are excellent candidate materials for organic polymer solar cells [38, 39].

Heterojunctions composed of TFB:F8BT and PFB:F8BT lie much closer to the exciton stabilization threshold as seen by comparing the relative band offsets in Fig. 2. Notice that the offset for TFB:F8BT is only slightly larger than 0.5 eV, which is approximately the exciton binding energy where as in PFB:F8BT the offset is >0.5 eV. Since such blends lie close to the stabilization threshold, they are excellent candidates for studying the relation between the energetics and the kinetics of exciton fission.

A comprehensive overview of the all the experimental and theoretical development in this field is well beyond the scope of a single chapter or single review article. Indeed, very good topical reviews exist and the reader is steered towards the Handbook of Conducting Polymers [40] for general overview, as well as *Organic Light-Emitting Devices: A Survey* [41] and *Conjugated Polymers: The Novel Science and Technology of Highly Conducting and Nonlinear Optically Active Materials* edited by Bredas and Silbey [42].

In this paper we present an overview of our recent work in developing a dynamical model for electronic relaxation processes in molecular semiconductors. We start with a brief primer on the excited states of molecular semiconductors and develop concepts from solid-state physics that are important in understanding molecular semiconductors. We then provide details of a model we have developed over the past few years which captures much of the salient physics for the photophysics of molecular semiconductors with nondegenerate ground states, such as PPV, F8BT, and related polymers. We then present an overview of our recent theoretical work aimed at understanding and modeling the state-to-state photophysical pathways in blended heterojunction materials [43,44].

2 Two-Band Configuration Interaction Model

Our basic description is derived starting from a model for the on-chain electronic excitations of a single conjugated polymer chain [36,45–47]. This model accounts for the coupling of excitations within the π-orbitals of a conjugated polymer to the lattice phonons using localized valence and conduction band Wannier functions ($|\overline{h}\rangle$ and $|p\rangle$) to describe the π orbitals and two optical phonon branches to describe the bond stretches and torsions of the the polymer skeleton

$$
\begin{aligned}
H = & \sum_{\mathbf{mn}} (F^{\circ}_{\mathbf{mn}} + V_{\mathbf{mn}}) A^{\dagger}_{\mathbf{m}} A_{\mathbf{n}} \\
& + \sum_{\mathbf{nm}i\mu} \left(\frac{\partial F^{\circ}_{\mathbf{nm}}}{\partial q_{i\mu}} \right) A^{\dagger}_{\mathbf{n}} A_{\mathbf{m}} q_{i\mu} \\
& + \sum_{i\mu} \omega^2_{\mu} (q^2_{i\mu} + \lambda_{\mu} q_{i\mu} q_{i+1,\mu}) + p^2_{i\mu},
\end{aligned}
\tag{2}
$$

where $A^{\dagger}_{\mathbf{n}}$ and $A_{\mathbf{n}}$ are operators that act upon the ground electronic state $|0\rangle$ to create and destroy electron/hole configurations $|n\rangle = |\overline{h}p\rangle$ with positive hole in the valence band Wannier function localized at h and an electron in the conduction band Wannier function p. Finally, $q_{i\mu}$ and $p_{i\mu}$ correspond to lattice distortions and momentum components in the ith site and μth optical phonon branch.

Wannier functions are essentially spatially localized basis functions that can be derived from the band-structure of an extended system. Quantities such as the exchange interaction and Coulomb interaction can be easily computed within the atomic orbital basis; however, there are many known difficulties in computing these within the crystal momentum representation. Because of this, is is desirable to develop a set of orthonormal spatially localized functions that can be characterized by a band index and a lattice site vector, R_{μ}. These are the Wannier functions, which we shall denote by $a_n(r - R_{\mu})$ and define in terms of the Bloch functions

$$
a_n(r - R_{\mu}) = \frac{\Omega^{1/2}}{(2\pi)^{d/2}} \int \mathrm{e}^{-\mathrm{i}kR_{\mu}} \psi_{nk}(r) \mathrm{d}k. \tag{3}
$$

The integral is over the Brillouin zone with volume $V = (2\pi)^d/\Omega$ and Ω is the volume of the unit cell (with d dimensions). A given Wannier function is defined for each band and for each unit cell. If the unit cell happens to contain multiple atoms, the Wannier function may be delocalized over multiple atoms. The functions are orthogonal and complete.

The Wannier functions are not energy eigenfunctions of the Hamiltonian. They are, however, linear combinations of the Bloch functions with different wave vectors and therefore different energies. For a perfect crystal, the matrix elements of H in terms of the Wannier functions are given by

$$\begin{aligned}\int a_l^*(r-R_\nu)H_o a_n(r-R_\mu)\mathrm{d}r &= \frac{\Omega}{(2\pi)^d}\int \mathrm{e}^{\mathrm{i}(qR_\nu - kR_\mu)}\psi_{lk}(r)H_o\psi_{nk}(r)\mathrm{d}r\,\mathrm{d}q\,\mathrm{d}k \\ &= \mathcal{E}_n(R_\nu - R_\mu)\delta_{nl},\end{aligned} \tag{4}$$

where

$$\mathcal{E}_n(R_\nu - R_\mu) = \frac{\Omega}{(2\pi)^d}\int \mathrm{e}^{\mathrm{i}k(R_\nu - R_\mu)}E_n(k)\,\mathrm{d}k.$$

Consequently, the Hamiltonian matrix elements in the Wannier representation are related to the Fourier components of the band structure, $E_n(k)$. Therefore, given a band structure, we can derive the Wannier functions and the single particle matrix elements, $F^{\circ}_{\mathbf{mn}}$.

The single-particle terms, $F^{\circ}_{\mathbf{mn}}$, are derived at the ground-state equilibrium configuration, $q_\mu = 0$, from the Fourier components f_r and $\overline{f}_r$ of the band energies in pseudomomentum space

$$\begin{aligned}F_{\mathbf{mn}} &= \delta_{\overline{mn}}\langle m|f|n\rangle - \delta_{mn}\langle\overline{m}|\overline{f}|\overline{n}\rangle \\ &= \delta_{\overline{mn}}f_{m-n} - \delta_{mn}\overline{f}_{\overline{m}-\overline{n}}.\end{aligned} \tag{5}$$

Here, f_{mn} and $\overline{f}_{mn}$ are the localized energy levels and transfer integrals for conduction-band electrons and valence-band holes. At the ground-state equilibrium geometry, $q_\mu = 0$, these terms can be computed as Fourier components of the one-particle energies in the Brillouin zone. For example, for the conduction band

$$f_{mn} = f_{m-n} = \frac{1}{B_z}\int_{B_z}\varepsilon_k \mathrm{e}^{\mathrm{i}k(m-n)}\,\mathrm{d}k, \tag{6}$$

where k is the pseudomomentum for a 1-dimensional lattice with unit period. For the case of cosine-shaped bands

$$\epsilon(k) = f_o + 2f_1\cos(k),$$

the site-energies are given by the center f_o and the transfer integral between adjacent Wannier functions is given by f_1. The band-structure and corresponding Wannier functions for the valence and conduction bands for PPV are shown in Fig. 3 [35, 45, 46]. For the intrachain terms, we use the hopping terms and site energies derived for isolated polymer chains of a given species, $t_{i,||}$, where our notation denotes the parallel hopping term for the ith chain ($i = 1, 2$). For PPV and similar conjugated polymer species, these are approximately 0.5 eV for both valence and conduction π bands.

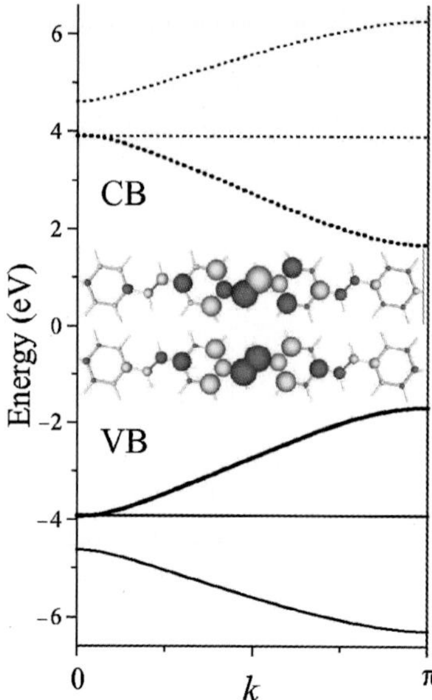

Fig. 3. Computed band-structure and vinylene-centered Wannier functions for PPV

The two-particle interactions are spin-dependent with

$$V^{\mathrm{T}}_{mn} = -\langle m\overline{n}||n\overline{m}\rangle \tag{7}$$

$$V^{\mathrm{S}}_{mn} = V^{\mathrm{T}}_{mn} + 2\langle m\overline{n}||\overline{m}n\rangle \tag{8}$$

for triplet and singlet combinations respectively with

$$\langle m\overline{n}||i\overline{j}\rangle = \int \mathrm{d}1 \int \mathrm{d}2\phi^*_m(1)\phi^*_{\overline{n}}(2)v(12)\phi_i(1)\phi_{\overline{j}}(2)$$

With the exception of geminate WFs, orbital overlap is small such that the two-body interactions are limited to Coulomb, $J(r)$ and exchange integrals, $K(r)$ reflecting e–h attraction and spin-exchange coupling nongeminate configurations. and transition dipole–dipole integrals $D(r)$ coupling only geminate singlet electron–hole pairs. Table 1 gives a listing of the electron–hole integrals and their parameters we have determined for PPV and similar poly-phenylene based conjugated chains. We have found that these are quite transferable amongst this class of conjugated polymers and allow us to focus upon modeling similar poly-phenylene chains through variation of the Wannier function band-centers (i.e., site energies) and band-widths (i.e., intrachain hopping integrals).

Since we will be dealing with interchain couplings, we make the following set of assumptions. First, the single-particle coupling between chains is expected to be small compared to the intramolecular coupling. For this, we assume that the perpendicular hopping integral $t_\perp = 0.01\,\mathrm{eV}$. This is consistent with LDA calculations performed by Vogl and Campbell [48] and with

Table 1. Electron/hole integrals for poly-phenylene-type polymer chains

Term	Functional form	Parameters
Direct Coloumb	$J(r) = J_o/(1 + r/r_o)$	$J_o = 3.092\,\mathrm{eV}$
		$r_o = 0.6840a$
Exchange	$K(r) = K_o\,\mathrm{e}^{-r/r_o}$	$K_o = 1.0573\,\mathrm{eV}$
		$r_o = 0.4743a$
Dipole–dipole	$D(r) = D_o(r/r_o)^{-3}$	$D_o = -0.03209\,\mathrm{eV}$
		$r_o = 1.0a$

Note: a = unit lattice spacing

the $t_\perp \approx 0.15 f_1$ estimate used in an earlier study of interchain excitons by Yu et al. [49] Furthermore, we assume that the $J(r)$, $K(r)$, and $D(r)$ two-particle interactions depend only upon the linear distance between two sites, as in the intrachain case. Since these are expected to be weak given that the interchain separation, d, is taken to be somewhat greater than the intermonomer separation.

Finally, the most important assumption that we make is that the site energies for the electrons and holes for the various chemical species can be determined by comparing the relative HOMO and LUMO energies to PPV. These are listed in Table 2. For example, the HOMO energy for PPV (as determined by its ionization potential) is $-5.1\,\mathrm{eV}$. For BBL, this energy is $-5.9\,\mathrm{eV}$. Thus, we assume that the $\overline{f}_o$ for a hole on a BBL chain is $0.8\,\mathrm{eV}$ lower than $\overline{f}_o$ for PPV at $-3.55\,\mathrm{eV}$. Likewise for the conduction band. The LUMO energy of PPV is $-2.7\,\mathrm{eV}$ and that of BBL is $-4.0\,\mathrm{eV}$. Thus, we shift the band center of the BBL chain $1.3\,\mathrm{eV}$ lower than then PPV conduction band center to $1.45\,\mathrm{eV}$. For the F8BT, TFB, and PFB chains, we adopt a similar scheme as discussed below. The site energies and transfer integrals used throughout are indicated in Table 2. We believe our model produces a reasonable estimate of the band offsets in the PN-junctions formed at the interface between these semiconducting polymers.

Table 2. Band centers and reported HOMO and LUMO levels for various polymer species

Molecule	ε_e	ε_h	HOMO	LUMO
PPV	2.75 eV	−2.75 eV	−5.1 eV[a]	−2.7 eV[a]
BBL	1.45	−3.55	−5.9[a]	−4.0[a]
F8BT	1.92 (2.42,1.42)	−3.54 (−3.04,−4.04)	−5.89[b]	−3.53[b]
PFB	3.16 (3.36,2.96)	−2.75 (−2.55,−2.95)	−5.1[b]	−2.29[b]
TFB	3.15 (3.35,2.95)	−2.98 (−2.78,-3.18)	−5.33[b]	−2.30[b]

Parenthesis indicate the modulation of the intramolecular valence and conduction band site energies

[a] See [53]

[b] See [54]

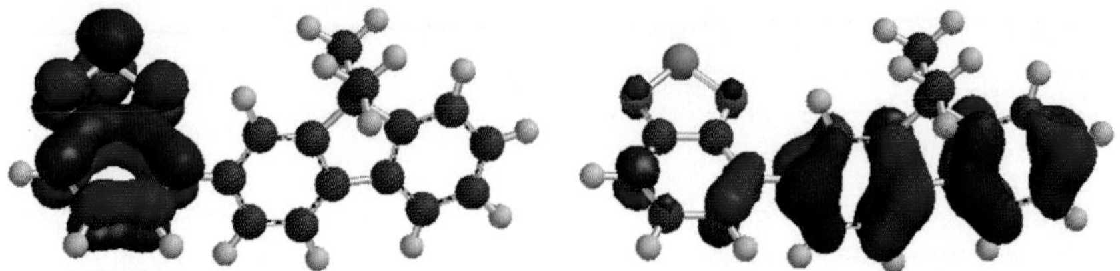

Fig. 4. Semiempirical (PM3) LUMO (*left*) and HOMO (*right*) orbitals for FBT monomer

A uniform site model for F8BT, TFB, and PFB, may be a gross simplification of the physical systems. For example, recent semi-empirical CI calculations by Jespersen and coworkers [50] indicate that the lowest energy singlet excited state of F8BT consists of alternating positive and negative regions corresponding to the electron localized on the benzothiadiazole units and the hole localized on the fluorene units. These are consistent with a previous study by Cornil et al. [51] which places the LUMO on the benzothiadiazole units. Cornil et al. [51] also report the HOMO and LUMO levels of the isolated fluorene and benzothiadiazole [51] indicating a $\Delta = 1.56\,\text{eV}$ offset between the fluorene and benzothiadiazole LUMO levels and a 0.66 eV offset between the fluorene and benzothiadiazole HOMO energy levels. Similarly, PM3 level calculations at the optimized geometry indicate an LUMO offset of 1.48 eV and a HOMO offset of 0.8 eV. The HOMO and LUMO orbitals for FBT (where we replaced the octyl side chains in F8BT with methyl groups) are shown in Fig. 4. This clearly indicates the localization of the HOMO and LUMO wave functions on the copolymer units.

We can include this alternation into our model by modulating the site energies of the F8BT chain [46]. Thus, in F8BT we include a 0.5 eV modulation of both the valence and conduction band site energies relative to the band center. Table 2. Hence, the fluorene site energies are at 2.42 and −3.04 eV for the conduction and valence band while the benzothiadiazole site energies are 1.42 and −4.04 eV. This results in a shift in the excitation energy to 0.28 eV relative to the unmodulated polymer and a 0.09 eV increase in the exciton binding energy. Furthermore, the absorption spectrum consists of two distinct peaks at 2.14 eV (563 nm) and 4.4 eV (281 nm) which are more or less on par with the 2.77 eV (448 nm) $S_o \rightarrow S_1$ and the 4.16 eV (298 nm) $S_o \rightarrow S_9$ transitions computed by Jespersen et al. [50] and observed at 2.66 and 3.63 eV by Stevens et al. [52].

For the case of parallel chains, we assume that there is no direct mechanical coupling between the chains. Consequently, each polymer chain is assumed to posses its own ensemble of phonon normal modes localized on the given chain and that there are no interchain phonon–phonon couplings. Moreover, since we have assumed bilinear coupling between on-site displacement coordinates $q_{i\mu}$ and hard-wall boundary conditions, the phonon normal mode frequencies for each mode ξ are given by

$$\omega_{\xi\mu}^2 = \omega_\mu^2 + 2\lambda_\mu \cos\left(\frac{\xi\pi}{2(N+1)}\right), \tag{9}$$

where N is the number of lattice sites in a given chain, ω_μ the band center for the μ-phonon band, λ the coupling, and $\xi = 1, \ldots, N$. In what follows, we shall condense our notation and adopt a generic ξ to denote both normal mode and band.

Finally, an important component in our model is the coupling between the electronic and lattice degrees of freedom. These we introduce via a linear coupling term of the form

$$\left(\frac{\partial f_{mn}}{\partial q_{i\mu}}\right)_o = \frac{S_\mu}{2}(2\hbar\omega_\mu^3)^{1/2}(\delta_{mi} + \delta_{ni}), \tag{10}$$

where S_μ is the Huang–Rhys factor which can be obtained from the vibronic features in the experimental photoemission spectrum. The Huang–Rhys factor, S is related to the intensity of the 0–n vibronic transition

$$I_{0-n} = \frac{e^{-S} S^n}{n!}. \tag{11}$$

For the case of conjugated polymers such as PPV and similar poly-phenylene vinylene species, the emission spectra largely consists of a series of well-resolved vibronic features corresponding to the C=C stretching modes in the phenylene rings with typical Huang–Rhys factors of $S = 0.6$ and a broad-featureless background attributed to either low frequency ring torsions (in the case of phenylene–vinylene polymers) or other low frequency modes with weak coupling to the electronic states with $S_\mu \approx 4$. On the other hand, the photoluminescence spectra of F8 shows a series of well resolved vibronic peaks with an energy separation of about 1,600 cm^{-1} [52,55]. Analysis of the Huang–Rhys factors of F8 in crystalline β phase and glassy states indicates a relatively low overall Huang–Rhys factor of $S = 0.6$ [55] which indicates that there is relatively little geometric relaxation following the transition from the excited to the ground state in these systems. This value of $S = 0.6$ is also in reasonable agreement with values estimated by Guha et al. [56] for ladder-type poly-*para*-phenylene and $S = 1.2$ for *para*-hexaphenyl. The modes which would couple a more planar excited state to a nonplanar ground state involve torsions between phenylene rings. These low frequency modes occur around 70 cm^{-1} and can not be spectroscopically resolved [55]. Based upon these observations, it seems reasonable from the standpoint of model building that a two phonon branches, one with $\omega = 1{,}600$ cm^{-1} and $S = 0.6$ and the other with $\omega = 70$ cm^{-1} and $S = 4$, provide a transferable set of electron/phonon couplings suitable for the conjugated polymers considered in this work.

Upon transforming H into the diabatic representation by diagonalizing the electronic terms at $q_{i\mu} = 0$, we obtain a series of vertical excited states

$|a_\circ\rangle$ with energies, ε_a° and normal modes, Q_ξ with frequencies, ω_ξ. (We will assume that the sum over ξ spans all phonon branches.)

$$H = \sum_a \varepsilon_a^\circ |a_\circ\rangle\langle a_\circ| + \sum_{ab\xi} g_{ab\xi}^\circ q_\xi (|a_\circ\rangle\langle b_\circ| + |b_\circ\rangle\langle a_\circ|)$$
$$+ \frac{1}{2}\sum_\xi (\omega_\xi^2 Q_\xi^2 + P_\xi^2). \tag{12}$$

The adiabatic or relaxed states can be determined then by iteratively minimizing $\varepsilon_a(Q_\xi) = \langle a|H|a\rangle$ according to the self-consistent equations

$$\frac{\mathrm{d}\varepsilon_a(Q_\xi)}{\mathrm{d}Q_\xi} = g_{aa\xi} + \omega_\xi^2 Q_\xi = 0. \tag{13}$$

Thus, each diabatic potential surface for the nuclear lattice motion is given by

$$\varepsilon_a(Q_\xi) = \varepsilon_a + \frac{1}{2}\sum_\xi \omega_\xi^2 (Q_\xi - Q_\xi^{(a)})^2. \tag{14}$$

These are shown schematically as S_a and S_b in Fig. 5 with Q_ξ being a collective normal mode coordinate. On can also view this figure as a slice through an N-dimensional coordinate space along normal coordinate Q_ξ. In this figure, ϵ_a° and ϵ_b° are the vertical energies taken at the ground-state equilibrium geometry $Q_\xi = 0$. The adiabatic energies, taken at the equilibrium geometry of each

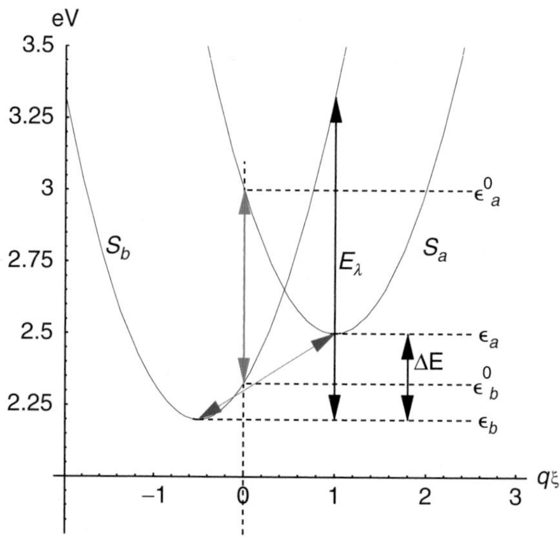

Fig. 5. Schematic representation of excited state Diabatic potentials obtained within our approach. The ground state configuration is taken as $q_\xi = 0$ with vertical excitation energies atϵ_a° and ϵ_b° and adiabatic (minimum) energies at ϵ_a and ϵ_b

excited state are denoted as ϵ_a and ϵ_b. While our model accounts for the distortions in the lattice due to electron/phonon coupling, we do not account for any adiabatic change in the phonon force constants within the excited states. Lastly, the electronic coupling between diabatic curves is given by g_{ab} which we compute at the ground-state geometry (g°_{ab}). We assume that both the diagonal g°_{aa} and off-diagonal g°_{ab} terms can be derived from the spectroscopic Huang–Rhys parameters.

The advantages of our approach is that it allows us to easily consider the singly excited states of relatively large conjugated polymer systems. Our model is built from both ab initio and experimental considerations and can in fact reproduce most of the salient features of the vibronic absorption and emission spectra for these systems. The model is limited in that we cannot include specific chemical configurational information about the polymers other than their conjugation length and gross topology. For isolated single chains, the model is rigorous. For multiple chains, our interchain parameterization does not stand on such firm ground since technically the Wannier functions are derived from a quasi-one-dimensional band structure. Nonetheless, our model and results provide a starting point for predicting and interpreting the complex photophysical processes within these systems. We next move on to describing the state-to-state interconversion proceses that occur following both photo- and electro-excitation.

3 State-to-State Relaxation Dynamics

The electronic levels in our model are coupled to the lattice phonons as well as the radiation field. Consequently, relaxation from a given electronic state can occur via state to state interconversion via phonon excitation or absorption or fluorescent decay to the S_0 ground state. For triplet excitations, only phonon transitions are allowed. For the singlets, fluorescence occurs primarily from the lowest S_n state independent of how the excitation was prepared. This certainly holds true for conjugated polymers in which both electroluminescence and photoluminescence originates from the same $S_1 = S_{xt}$ state. This implies that internal conversion dynamics are fast relative to the fluorescence lifetime.

Coupling the electronic relaxation dynamics to the vibrational dynamics is a formidable task. An exact quantum mechanical description of this is currently well beyond the state of the art of current computational methods. One can, however, compute the state-to-state rate constants using Fermi's golden rule and arrive at a reasonable picture.

If we assume that the vibrational bath described by $H_{\rm ph}$ remains at its ground-state geometry, then the state-to-state transition rates are easily given by Fermi's golden rule:

$$k^\circ_{ab} = \pi \sum_\xi \frac{g^2_{ab}}{\hbar\omega_\xi}(1 + n(\omega_{ab}))(\Gamma(\omega_\xi - \omega_{ab}) - \Gamma(\omega_\xi + \omega_{ab})). \qquad (15)$$

where $n(\omega)$ is the Bose–Einstein population for the phonons, Γ is a Lorentzian broadening in which the width is inversely proportional to the phonon lifetime used to smooth the otherwise discrete phonon spectrum, $\omega_{ab} = (\epsilon_a^o - \epsilon_b^o)/\hbar$. In order for a transition to occur, there must be a phonon of commensurate energy to accommodate the energy transfer. The coupling term, $g_{ab\xi}$, is the diabatic coupling in the diabatic Hamiltonian given in (12).

This static model is fine so long as either the nuclear relaxation has little effect on the state to state rate constant or if the electronic transitions occur on a time-scale which is short compared to the nuclear motion. However, if lattice reorganization does play a significant role, then we need to consider the explicit nuclear dynamics when computing the state-to-state rates. If we assume that vibrational relaxation within a given diabatic state is rapid compared to the interstate transition rate, we can consider the transitions as occurring between displaced harmonic wells

$$k_{ab} = \frac{2\pi}{\hbar} |V_{ab}|^2 \mathcal{F}, \tag{16}$$

where V_{ab} is the coupling between electronic states a and b and

$$\mathcal{F} = \mathcal{F}(E_{ab}) = \sum_{\nu_a} \sum_{\nu_b} P_{th}(\varepsilon_a(\nu_a)) |\langle \nu_a | \nu_b \rangle|^2 \delta(\epsilon_a(\nu_a) - \epsilon(\nu_b) + \Delta E_{ab}) \tag{17}$$

is the thermally averaged Franck–Condon weighted density of nuclear vibrational states. Here, ν_a and ν_b denote the vibronic states, P_{th} is the Boltzmann distribution over the initial states, $\epsilon_a(\nu_a)$ and $\epsilon(\nu_b)$ are the corresponding energies, and ΔE_{ab} is the electronic energy gap between a and b. In the classical limit, $\mathcal{F}$ becomes

$$\mathcal{F}(E_{ab}) = \frac{1}{\sqrt{4\pi E_\lambda k_B T}} \exp\left(-\frac{(E_\lambda + \Delta E_{ab})^2}{4 E_\lambda k_B T} \right), \tag{18}$$

where E_λ is the reorganization energy as sketched in Fig. 5.

Each of these terms can be easily computed from the diabatic Hamiltonian in (12). The diabatic coupling matrix element between the adiabatically relaxed excited states, $|V_{ab}|^2$, requires some care since we are considering transitions between eigenstates of different Hamitonians (corresponding to different nuclear geometries). Since the vertical $Q_\xi = 0$ states provide a common basis, $|a^\circ\rangle$, we can write

$$V_{ab} = \sum_{a^\circ b^\circ} \langle a | a^\circ \rangle g^\circ_{ab} \langle b^\circ | b \rangle, \tag{19}$$

where g°_{ab} is the diabatic matrix element computed at the equilibrium geometry of the ground-state.

Once we have the rate constants computed, it is a simple matter to integrate the Pauli master equation for the state populations

$$\dot{P}_a(t) = \sum_b (k_{ba}P_b - k_{ab}P_a) - k_a^{\mathrm{rad}} P_a, \tag{20}$$

where k_a^{rad} is the radiative decay rate of state a

$$k_a^{\mathrm{rad}} = \frac{|\mu_{a0}|^2}{6\epsilon_o \hbar^2}(1 + n(\omega_{a0}))\frac{\hbar\omega_{a0}^3}{2\pi c^3}, \tag{21}$$

where μ_{a0} are the transition dipoles of the excited singlets. These we can compute directly from the Wannier functions or empirically from the photoluminescence decay rates for a given system. Photon mediated transitions between excited states are highly unlikely due to the ω_{ab}^3 density factor of the optical field. In essence, so long as the nonequilibrium vibrational dynamics is not a decisive factor, we can use these equations to trace the relaxation of an electronic photo- or charge-transfer excitation from its creation to its decay including photon outflow measured as luminescence.

4 Exciton Regeneration Dynamics

Donor–acceptor heterojunctions composed of blends of TFB with F8BT and PFB with F8BT phase segregate to form domains of more or less pure donor and pure acceptor. Even though the polymers appear to be chemically quite similar, the presence of the triphenyl amine groups in TFB and PFB cause the polymer chain to be folded up much like a carpenter's rule. F8BT, on the other hand, is very rod-like with a radius of gyration being more or less equivalent to the length of a give oligomer. Molecular dynamics simulations of these materials by our group indicate that segregation occurs because of this difference in morphology and that the interface between the domains is characterized by regions of locally ordered π-stacking when F8BT rod-like chains come into contact with more globular PFB or TFB chains.

As discussed earlier, TFB:F8BT and PFB:F8BT sit on either side of the exciton destabilization threshold. In TFB:F8BT, the band offset is less than the exciton binding energy and these materials exhibit excellent LED performance. On the other hand, devices fabricated from PFB:F8BT where the exciton binding energy is less than the offset, are very poor LEDs but hold considerable promise for photovoltic devices. In both of these systems, the lowest energy state is assumed to be an interchain exciplex as evidenced by a red-shifted emission about 50–80 ns after the initial photoexcitation [57]. In the case of TFB:F8BT, the shift is reported to be 140 ± 20 meV and in PFB:F8BT the shift is 360 ± 30 meV relative to the exciton emission, which originates from the F8BT phase. Bearing this in mind, we systematically varied the separation distance between the cofacial chains from $r = 2a$–$5a$ (where $a =$ unit lattice constant) and set $t_\perp = 0.01$ in order to tune the Coulomb and exchange coupling between the chains and calibrate our parameterization.

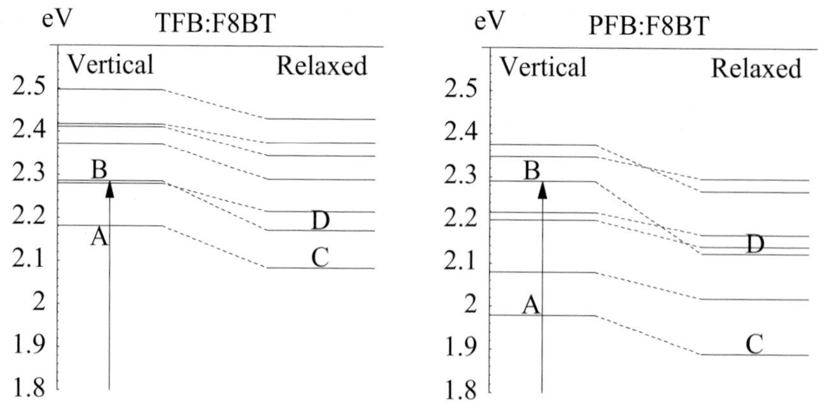

Fig. 6. Vertical and relaxed energies of the lowest lying states in the TFB:F8BT and PFB:F8BT heterojunctions. In each, A and C refer to the interchain exciplex state and B and D refer to the predominantly intrachain F8BT exciton state

For large interchain separations, the exciton remains localized on the F8BT chain in both cases. As the chains come into contact, dipole–dipole and direct Coulomb couplings become significant and we begin to see the effect of exciton destabilization. For TFB:F8BT, we select and interchain separation of $r = 2.8a$ giving a 104 meV splitting between the vertical exciton and the vertical exciplex and 87.4 meV for the adiabatic states. For PFB:F8BT, we chose $r = 3a$ giving a vertical exciton–exciplex gap of 310 meV and an adiabatic gap of 233 meV. In both TFB:F8BT and PFB:F8BT, the separation produce interchain exciplex states as the lowest excitations. with energies reasonably close to the experimental shifts.

Figure 4 compares the vertical and adiabatic energy levels in the TFB:F8BT and PFB:F8BT chains and Figs. 7 and 8 show the vertical and relaxed exciton and charge-separated states for the two systems. Here, sites 1–10 correspond to the TFB or PFB chains and 11–20 correspond to the F8BT chain. The energy levels labeled in Fig. 4 correspond to the states plotted in Figs. 7 and 8. We shall refer to states A and B as the vertical exciplex and vertical exciton and to states C and D as the adiabatic exciplex and adiabatic exciton respectively. Roughly speaking, a pure exciplex state will have the charges completely separated between the chains and will contain no geminate electron/hole configurations. Likewise, strictly speaking, a pure excitonic state will be localized to a single chain and have only geminate electron/hole configurations. Since site energies for the the F8BT chain are modulated to reflect to internal charge-separation in the F8BT copolymer as discussed above, we take our "exciton" to be the lowest energy state that is localized predominantly along the diagonal in the F8BT "quadrant".

In the TFB:F8BT junction, the lowest excited state is the exciplex for both the vertical and adiabatic lattice configurations with the hole on the

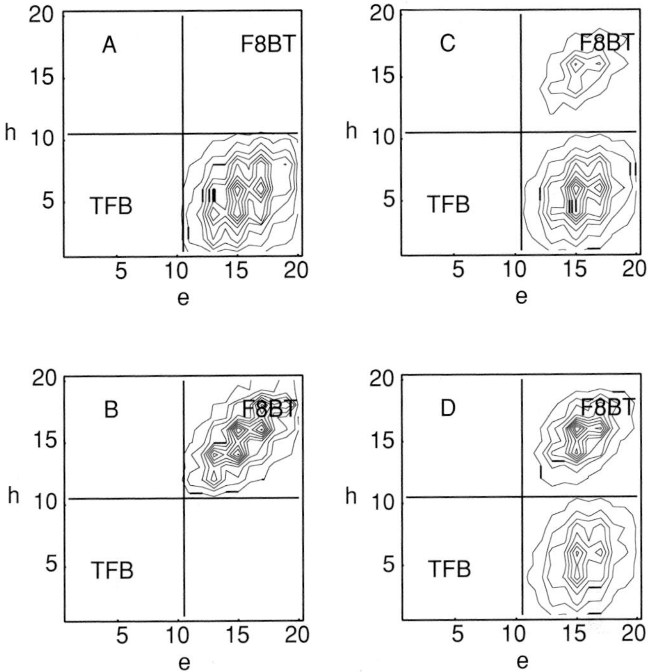

Fig. 7. Excited state electron/hole densities for TFB:F8BT heterojunction. The electron/hole coordinate axes are such that sites 1–10 correspond to TFB sites and 11–20 correspond to F8BT sites. Note the weak mixing between the interchain charge-separated states and the F8BT exciton in each of these plots

TFB and the electron on the F8BT (Fig. 7a,c). In the vertical case, there appears to be very little coupling between intrachain and interchain configurations. However, in the adiabatic cases there is considerable mixing between intra- and interchain configurations. First, this gives the adiabatic exciplex an increased transition dipole moment to the ground state. Second, the fact that the adiabatic exciton and exciplex states are only 87 meV apart means that at 300 K, about 4% of the total excited state population will be in the adiabatic exciton.

For the PFB:F8BT heterojunction, the band offset is greater than the exciton binding energy and sits squarely on the other side of the stabilization threshold. Here the lowest energy excited state (Figs. 8a,b) is the interchain charge-separated state with the electron residing on the F8BT (sites 11–20 in the density plots in Fig. 8) and the hole on the PFB (sites 1–10). The lowest energy exciton is almost identical to the exciton in the TFB:F8BT case. Remarkably, the relaxed exciton (Fig. 8D) shows slightly more interchain charge-transfer character than the vertical exciton (Fig. 8C). While the system readily absorbs at 2.3 eV creating a localized exciton on the F8BT,

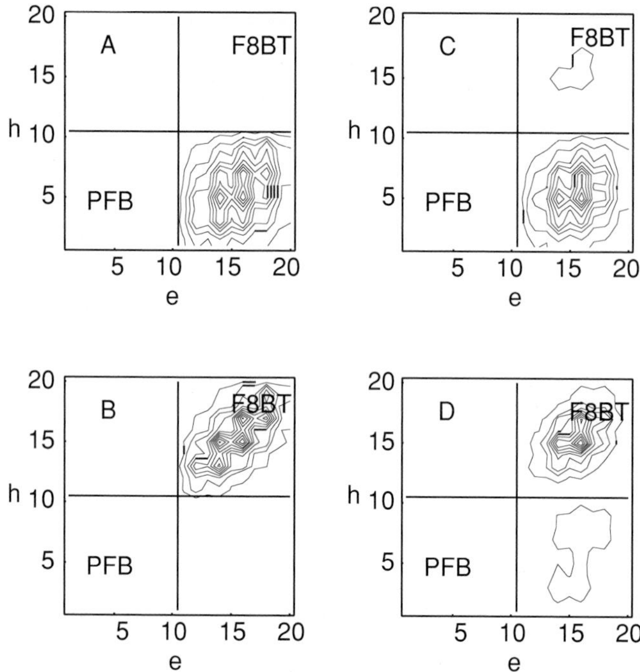

Fig. 8. Excited state electron/hole densities for PFB:F8BT heterojunction. The axes are as in previous figures except that sites 1–10 correspond to PFB sites and sites 11–20 to F8BT sites

luminescencc is entirely quenched since all population within the excited states is readily transfer to the lower-lying interchain charge-separated states with vanishing transition moments to the ground state.

In calculating the state-to-state interconversion rates for TFB:F8BT, we note two major differences between the static and the adiabatic Marcus–Hush approaches (see Fig. 9). First is the sparsity of the latter with transitions being limited to states with smaller energy differences. This leads to a relaxation dynamics that is more intertwined with the DOS. Second is the relatively faster rates calculated in the latter leading to interconversion lifetimes in the femtosecond (fs) to a couple of picosecond (ps) regime as opposed to hundreds of ps in the former. The same general difference is observed for PFB:F8BT (Not shown). These marked differences in the distribution of rates and their range of magnitudes are brought about by the introduction of the reorganization energy as a parameter in the rates calculation to complement the energy differences between the states. It provides a way to incorporate lattice distortions in the semiclassical limit into the relaxation dynamics. While this is not fully dynamical in its account of the lattice distortions, it improves upon the static approximation previously employed.

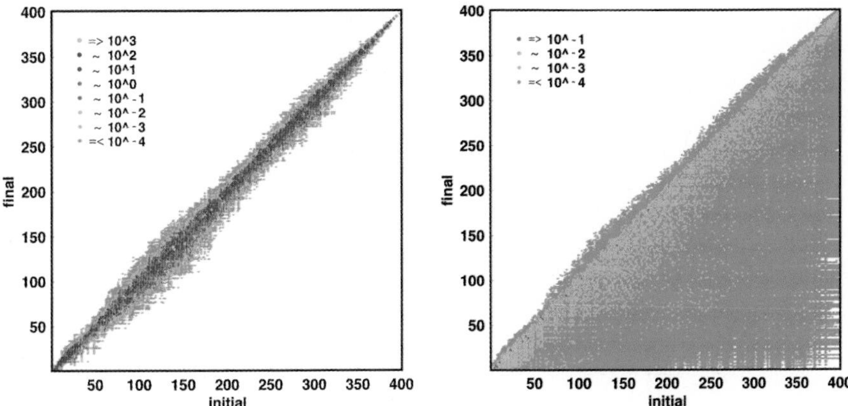

Fig. 9. (*Color online*) TFB–F8BT internal conversion rates distribution at 290 K. Rates are in ps^{-1}. Note the sparsity and relatively faster Marcus rates compared to the diabatic rates

The photoexcitation of heterojunction systems is simulated by populating a higher-lying excitonic state. Figure 10 shows the time-evolved populations of the lowest charge-transfer (CT) and excitonic (XT) vertical and relaxed states, respectively, in photoexcited TFB:F8BT and PFB:F8BT. We see that the relaxation to the lowest CT state is faster in TFB:F8BT than in PFB:F8BT. Furthermore, the relaxation from the XT state to the CT state occurs faster in the former. This is despite the XT state being formed faster in the latter for both cases. This is manifested more in the Marcus–Hush approach shown in Fig. 10 where despite reaching a maximum population of 0.86 in 250 fs as opposed to just 0.40 in 500 fs, the XT→CT interconversion is practically done in 2 ps in TFB:F8BT compared to 10 ps in PFB:F8BT. In addition, we note that the XT state reaches a steady-state population in TFB:F8BT whereas it goes to zero in PFB:F8BT. This small but nonzero population of the XT state is consistent with the distributed thermal population of states of 0.022 at 290 K owing to the fact that this XT state is 95 meV higher in energy relative to the lowest CT state [43].

Interestingly, while the overall XT→CT interconversion occurs in just a couple of ps in both heterojunction systems, a closer look into the rates reveal that this relaxation does not occur directly. Rather, it involves the next lowest CT state. Figure 11 show the relevant interconversions between the three lowest states of both systems: the lowest CT state(CT1), the next lowest CT state(CT2), and the lowest XT state(XT). It is worth noting that the considerable mixing between the intra-chain and interchain configurations of the former compared to those of the latter. In TFB:F8BT, the direct XT→CT1 transition ($\sim 10^{-3}$ ps^{-1}) is at least 3 orders of magnitude slower than the corresponding XT→CT2→CT1 transition route ($>1\,ps^{-1}$), the indirect route being consistent with the evolution data (Fig. 10). Thus, the

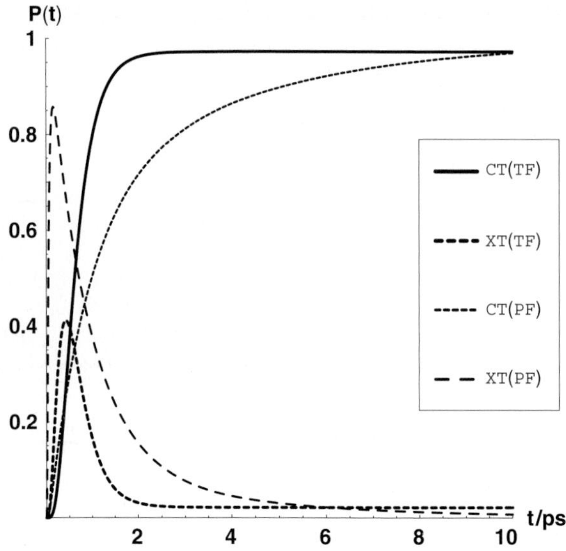

Fig. 10. Time-evolved populations of the lowest CT (*solid lines*) and XT (*dashed lines*) relaxed states of TFB:F8BT(TF) and PFB:F8BT(PF) in the Marcus–Hush approach at 290 K

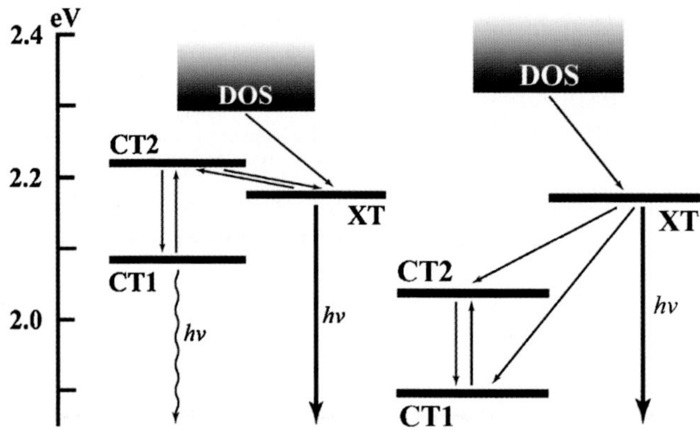

Fig. 11. Relevant Marcus–Hush interconversion rates for the three lowest states of (*left*) TFB:F8BT and (*right*) PFB:F8BT. In both cases, relaxation proceeds from the density of states (DOS) to the lowest excitonic state(XT) (offset to the right relative to the CT states for clarity of relaxation route) before relaxing to the lowest charge-transfer state(CT1). CT1 proceeds to equilibrium with the next higher CT state(CT2). In PFB:F8BT, CT2 has a lower energy than XT where as in TFB:F8BT, it has a higher energy. Also shown are the radiative rates emanating from the XT states which are strongly coupled to ground state, S_0, of both systems and the TFB:F8BT CT1 state which is just weakly coupled to S_0

XT→CT conversion occurs via the CT2 state and not directly. The reverse transitions for both routes are slower but have the same order of magnitude difference. This CT1→CT2→XT transition ($\sim 10^{-1}$ ps^{-1}) effectively presents a regeneration pathway for the XT. This leads to an XT state population that is always at equilibrium with the CT1 state.

In PFB:F8BT the XT→CT1 and XT→CT2 conversion occur at relatively the same rate ($\sim 10^{-1}$ ps^{-1}) while their reverse transitions are at least 2 orders of magnitude slower. Consequently, XT is not regenerated. The role played by CT2 as a bridge state is apparently relative to whether it has a slightly higher or lower energy than the XT as has been accounted by Morteani et al. [57,58]. Spontaneous transition rates are typically faster when going from a higher to a lower energy state than the reverse according to detailed balance. Here, CT1 is the exciplex state which exhibit sizable mixing with the bulk CT state (CT2). When CT2 has a higher energy than XT, such as in TFB:F8BT, a fraction of the population in CT2 converts to XT. If it has a lower energy relative to the XT state such as in PFB:F8BT, this regeneration of the XT, practically, does not occur.

To see the effect of temperature, the interconversion rates were calculated at 230, 290, and 340 K. Figure 12 shows how the interconversions among the three lowest states (CT1, CT2, and XT) of TFB:F8BT, as illustrated in

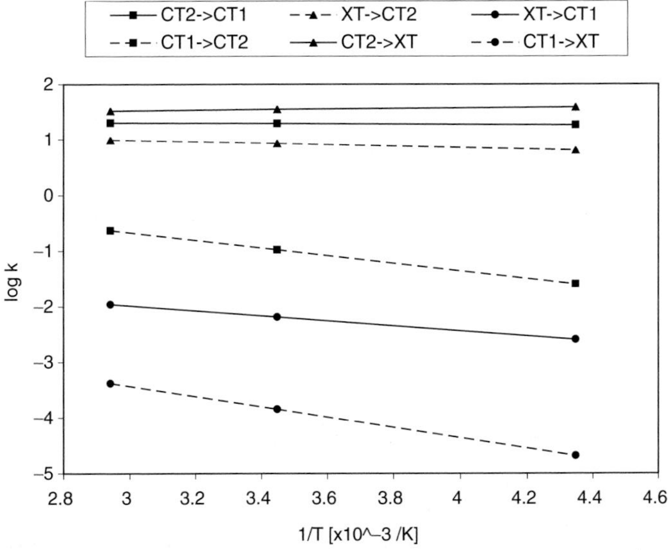

Fig. 12. Interconversion rates between the three lowest states of TFB:F8BT as a function of temperature (230, 290, and 340 K). Plot is given as log k vs. $1/T$. Transitions to lower energy states are given as *solid lines* while their reverse are given as *dashed lines*. The CT2↔CT1, CT2↔XT and XT↔CT1 are plotted as squares, triangles, and circles, respectively. All transition rates increase directly with temperature except the CT2→XT conversion which decreases as temperature increases

Fig. 11, vary with temperature. This dependence is given in an Arrhenius plot of log k vs. $1/T$ and gives a linear plot for each transition having a slope associated with the activation energy, E_{act}, for that particular transition. This activation energy has the expression

$$E_{\mathrm{act}} = \frac{(\Delta E - E_\lambda)^2}{4E_\lambda}. \tag{22}$$

Transitions to lower energy states are given as solid lines while those going to higher energy states are given as dashed lines. Curiously, although XT→CT1 is exothermic compared to XT→CT2 which is endothermic, the latter is a more favorable transition. This has to do with the fact that XT→CT1 has an activation energy almost three times greater than that of XT→CT2. As alluded to above, this is a consequence of the former being in the inverted region while the latter being in the normal region. In the inverted region, the larger ΔE is, the larger E_{act} as opposed to the more familiar normal region where E_{act} decreases as ΔE increases. Having stated this, however, we note that in the former, due to maximal overlap between the vibrational modes of the two states, transitions may be possible via tunneling processes. Overall, in TFB:F8BT, we see an increase in the fraction of the total excited state population in XT as temperature increases. At 230 K only 0.81% is in XT while at 290 and 340 K, 2.16% and 3.67% is in XT, respectively.

Finally, we note that all transition rates increase with temperature except for the CT2→XT in TFB:F8BT and XT→CT2 in PFB:F8BT which decrease with temperature. Such a trend, while not uncommon in chemical reactions, are though to be indicative of a more complicated transition mechanism as noted by Porter [59]. We surmise this to be due to the coupling between the low frequency vibrational modes of the initial state with the high frequency vibrational modes of the final state as in the case of an early transition state in reactive scattering.

5 Discussion

In this paper, we gave an overview of our recent work in developing a theoretical understanding interfacial excitonic dynamics in a complex material system. The results herein corroborate well with the experimental results on these systems. In particular, following either charge injection or photoexcitation, the system rapidly relaxes to form the interchain charge-separated species. In the experimental data, this occurs within the first 10 or so ps for the bulk material. Our calculations of a single pair of cofacial chains puts the exciplex formation at about 1 ps. Likewise, the experimental time-resolved emission indicates the regeneration occurs on a much longer time-scale with most of the time-integrated emission coming from regenerated excitons. This too, is shown in our calculations as evidenced by the slow thermal repopulation of the XT state in the TFB:F8BT system. Since this state has a significant transition dipole to the ground state, population transferred to this state can either decay back to the CT state via thermal fluctuations or decay to the ground state via the emission of a photon. Since this secondary emission is dependent upon the thermal population of the XT state at any given time, the efficiency of this process shows a strong dependency upon the temperature of the system.

The exciton model we present herein certainly lacks the molecular level of details so desired by materials chemists. However, it offers a tractable way of building from molecular considerations the salient physical interactions that give rise to the dynamics in the excited states of these extended systems. In building this model we make a number of key assumptions. First, and perhaps foremost, that the excited states are well described via bands of π orbitals and that from these bands we can construct localized Wannier functions. Hand in hand with this assumption is that within the general class of oligo-phenylene derived polymers, configuration interaction matrix elements, hopping integrals, electron/phonon couplings, and phonon spectra are transferable from one system to another. This is a fairly dangerous approximation since it discounts important contributions from heteroatoms, side-chains, and chain morphology. However, given that a single oligomeric chain of F8BT with 10 repeat units has well over 300 atoms, such potentially dire approximations are necessary in order to extract the important features of these very extended systems.

Second, we make the assumption that the explicit vibrational dynamics can be integrated out of the equations of motion for the electronic states. This is probably not too extreme of an assumption so long as we can assume that the phonons remain thermalized over the course of the electronic relaxation. However, looking back at the level correlation diagrams, crossings between diabatic states are present in this system and hence conical intersections between electronic states may play an important role. Finally, we discount the effects of electronic coherence. This, too, may have a profound impact upon the final state-to-state rate constants since it is well recognized that even a small amount of quantum coherence between states leads to a dramatic increase in the transition rate. Fortunately, many of the papers presented in this proceedings address these assumptions. Approaches, such as the MCTDH method presented by Thoss, the DFT based nonadiabatic molecular dynamics approach (NAMD) developed by Prezhdo, for example, are important strides towards achieving a molecular level understanding of complex photophysical processes in light-emitting and light-harvesting materials.

Acknowledgments

This work was sponsored in part by the National Science Foundation and by the Robert A. Welch Foundation. The authors also thank the organizers for putting together a highly stimulating conference in a wonderful location.

References

1. V. Podzorov, E. Menard, A. Borissov, V. Kiryukhin, J. A. Rogers, and M. E. Gershenson. Intrinsic charge transport on the surface of organic semiconductors. *Physical Review Letters*, 93(8):086602, 2004
2. W. Warta, L. B. Stehle, and N. Karl. pentacene. *Applied Physics A*, 36:163, 1985
3. Y. X. Xu, H. S. Byun, and C. K. Song. Organic integrated circuits based on pentacene tfts using pvp gate insulator, volume 5632, pages 332–340. SPIE, 2005
4. O. Ostroverkhova, D. G. Cooke, S. Shcherbyna, R. F. Egerton, F. A. Hegmann, R. R. Tykwinski, and J. E. Anthony. Bandlike transport in pentacene and functionalized pentacene thin films revealed by subpicosecond transient photoconductivity measurements. *Physical Review B (Condensed Matter and Materials Physics)*, 71(3):035204, 2005
5. H. L. Kwok. Investigation into the modeling of field-effect mobility in disordered organic semiconductors, volume 5838, pages 85–94. SPIE, 2005
6. J. Lee, D. K. Hwang, J.-M. Choi, K. Lee, J. H. Kim, S. Im, J. H. Park, and E. Kim. Flexible semitransparent pentacene thin-film transistors with polymer dielectric layers and nio[sub x] electrodes. *Applied Physics Letters*, 87(2):023504, 2005
7. Y. Kato, S. Iba, R. Teramoto, T. Sekitani, T. Someya, H. Kawaguchi, and T. Sakurai. High mobility of pentacene field-effect transistors with polyimide gate dielectric layers. *Applied Physics Letters*, 84(19):3789–3791, 2004
8. M. Ahles, R. Schmechel, and H. von Seggern. n-Type organic field-effect transistor based on interface-doped pentacene. *Applied Physics Letters*, 85(19):4499–4501, 2004
9. J. McGinness, Corry, and Proctor. *Science*, 183:853, 1974
10. C. K. Chiang, C. R. Fincher, Jr., Y. W. Park, A. J. Heeger, H. Shirakawa, E. J. Louis, S. C. Gau, and Alan G. MacDiarmid. Electrical conductivity in doped polyacetylene. *Physical Review Letters*, 39(17):1098–1101, 1977
11. Rudolf Peierls. *Surprises in theoretical physics*. Princeton University Press (Princeton, NJ), 1979
12. W. P. Su, J. R. Schrieffer, and A. J. Heeger. Solitons in polyacetylene. *Physical Review Letters*, 42(25):1698–1701, 1979
13. W. P. Su, J. R. Schrieffer, and A. J. Heeger. Soliton excitations in polyacetylene. *Physical Review B (Condensed Matter)*, 22(4):2099–2111, 1980
14. N. S. Hush. An overview of the first half-century of molecular electronics. *Ann. N.Y. Acad. Sci.*, 1006:120, 2003
15. J. Cornil, D. Beljonne, C. M. Heller, I. H. Campbell, B. K. Laurich, D. L. Smith, D. D. C. Bradley, K. Müllen, and J. L. Brédas. Photoluminescence spectra of oligo-paraphenyllenevinylenes: a joint theoretical and experimental characterization. *Chemical Physics Letters*, 278(1-3):139–145, 1997
16. G. C. Claudio and Eric R. Bittner. Random growth statistics of long-chain single molecule poly-(p-phenylene vinylene). *The Journal of Chemical Physics*, 115(20):9585–9593, 2001
17. G. Mao, J. E. Fischer, F. E. Karasz, and M. J. Winokur. Nonplanarity and ring torsion in poly(p-phenylene vinylene). a neutron-diffraction study. *The Journal of Chemical Physics*, 98(1):712–716, 1993

18. T. W. Hagler, K. Pakbaz, K. F. Voss, and A. J. Heeger. Enhanced order and electronic delocalization in conjugated polymers oriented by gel processing in polyethylene. *Physical Review B (Condensed Matter)*, 44(16):8652–8666, 1991
19. K. Pichler, D. A. Halliday, D. D. C. Bradley, P. L. Burn, R. H. Friend, and A. B. Holmes. Optical spectroscopy of highly ordered poly(p-phenylene vinylene). *Journal of Physics: Condensed Matter*, 5:7155–7172, 1993
20. R. F. Service. Organic leds look forward to a bright, white future. *Science*, 310:1762, 2005
21. R. H. Friend, R. W. Gymer, A. B. Holmes, J. H. Burroughes, R. N. Marks, C. Taliani, D. D. C. Bradley, D. A. dos Santos, J. L. Bredas, M. Lögdlund, and W. R. Salaneck. Electroluminescence in conjugated polymers. *Nature*, 397:121–128, 1999
22. M. Wohlgenannt, X. M. Jiang, Z. V. Vardeny, and R. A. J. Janssen. Conjugation-length dependence of spin-dependent exciton formation rates in pi-conjugated oligomers and polymers. *Physical Review Letters*, 88(19):197401, 2002
23. D. Beljonne, Z. Shuai, A. Ye, and J.-L. Bredas. Charge-recombination processes in oligomer- and polymer-based light-emitting diodes: A molecular picture. *Journal of the Society for Information Display*, 13(5):419–427, 2005
24. A. Ye, Z. Shuai, and J. L. Bredas. Coupled-cluster approach for studying the singlet and triplet exciton formation rates in conjugated polymer led's. *Physical Review B (Condensed Matter and Materials Physics)*, 65(4):045208, 2002
25. Z. Shuai, D. Beljonne, R. J. Silbey, and J. L. Bredas. Singlet and triplet exciton formation rates in conjugated polymer light-emitting diodes. *Physical Review Letters*, 84(1):131–134, 2000
26. M. Wohlgenannt, K. Tandon, S. Mazumdar, S. Ramasesha, and Z. V. Vardeny. correction: Formation cross-sections of singlet and triplet excitons in pi-conjugated polymers. *Nature*, 411(6837):617–617, 2001
27. M. N. Kobrak and E. R. Bittner. Quantum molecular dynamics study of polaron recombination in conjugated polymers. *Physical Review B (Condensed Matter and Materials Physics)*, 62(17):11473–11486, 2000
28. M. N. Kobrak and E. R. Bittner. A dynamic model for exciton self-trapping in conjugated polymers. i. theory. *The Journal of Chemical Physics*, 112(12):5399–5409, 2000
29. M. N. Kobrak and E. R. Bittner. A quantum molecular dynamics study of exciton self-trapping in conjugated polymers: Temperature dependence and spectroscopy. *The Journal of Chemical Physics*, 112(17):7684–7692, 2000
30. A. L. Burin and M. A. Ratner. Spin effects on the luminescence yield of organic light emitting diodes. *The Journal of Chemical Physics*, 109(14):6092–6102, 1998
31. K. Tandon, S. Ramasesha, and S. Mazumdar. Electron correlation effects in electron–hole recombination in organic light-emitting diodes. *Physical Review B (Condensed Matter and Materials Physics)*, 67(4):045109, 2003
32. M. Wohlgenannt, C. Yang, and Z. V. Vardeny. Spin-dependent delayed luminescence from nongeminate pairs of polarons in pi-conjugated polymers. *Physical Review B (Condensed Matter and Materials Physics)*, 66(24):241201, 2002
33. M. Wohlgenannt and O. Mermer. Single-step multiphonon emission model of spin-dependent exciton formation in organic semiconductors. *Physical Review B (Condensed Matter and Materials Physics)*, 71(16):165111, 2005
34. W. Barford. Theory of singlet exciton yield in light-emitting polymers. *Physical Review B (Condensed Matter and Materials Physics)*, 70(20):205204, 2004

35. S. Karabunarliev and E. R. Bittner. Dissipative dynamics of spin-dependent electron–hole capture in conjugated polymers. *The Journal of Chemical Physics*, 119(7):3988–3995, 2003
36. S. Karabunarliev and E. R. Bittner. Spin-dependent electron–hole capture kinetics in luminescent conjugated polymers. *Physical Review Letters*, 90(5):057402, 2003
37. E. R. Bittner and S. Karabunarliev. Energy relaxation dynamics and universal scaling laws in organic light-emitting diodes. *International Journal of Quantum Chemistry*, 95(4–5):521–531, 2003
38. M. M. Alam and S. A. Jenekhe. Efficient solar cells from layered nanostructures of donor and acceptor conjugated polymers. *Chemistry of Materials*, 16(23):4647–4656, 2004
39. S. A. Jenekhe and S. Yi. Efficient photovoltaic cells from semiconducting polymer heterojunctions. *Applied Physics Letters*, 77(17):2635–2637, 2000
40. T. A. Skotheim. *Handbook of Conducting Polymers, Third Edition.* CRC, Boca Raton, 2006
41. J. Shinar. *Organic Light-Emitting Devices: A Survey.* Springer, Berlin Heidelberg New York, 2004
42. J. L. Bredas and R. Silbey. *Conjugated Polymers: The Novel Science and Technology of Highly Conducting and Nonlinear Optically Active Materials.* Springer, Berlin Heidelberg New York, 1991
43. E. R. Bittner, J. G. S. Ramon, and S. Karabunarliev. Exciton dissociation dynamics in model donor–acceptor polymer heterojunctions. i. energetics and spectra. *The Journal of Chemical Physics*, 122(21):214719, 2005
44. J. G. S. Ramon and E. R. Bittner. Exciton dissociation dynamics in model donor–acceptor polymer heterojunctions. ii. kinetics and photophysical pathways. *The Journal of Physical Chemistry B* (in submission)
45. S. Karabunarliev and E. R. Bittner. Polaron–excitons and electron–vibrational band shapes in conjugated polymers. *The Journal of Chemical Physics*, 118(9):4291–4296, 2003
46. S. Karabunarliev and E. R. Bittner. Electroluminescence yield in donor–acceptor copolymers and diblock polymers: A comparative theoretical study. *The Journal of Chemical Physics*, 108(29):10219–10225, 2004
47. P. Karadakov, J.-L. Calais, and J. Delhalle. A localized-basis monoexcited configuration interaction technique for extended systems. *The Journal of Chemical Physics*, 94(12):8520–8528, 1991
48. P. Vogl and D. K. Campbell. First-principles calculations of the three-dimensional structure and intrinsic defects in trans-polyacetylene. *Physical Review B (Condensed Matter)*, 41(18):12797–12817, 1990
49. Z. G. Yu, M. W. Wu, X. S. Rao, X. Sun, and A. R. Bishop. Excitons in two coupled conjugated polymer chains. *Journal of Physics: Condensed Matter*, 8(45):8847–8857, 1996
50. K. G. Jespersen, W. J. D. Beenken, Y. Zaushitsyn, A. Yartsev, M. Andersson, T. Pullerits, and V. Sundstrom. The electronic states of polyfluorene copolymers with alternating donor–acceptor units. *The Journal of Chemical Physics*, 121(24):12613–12617, 2004
51. J. Cornil, I. Gueli, A. Dkhissi, J. C. Sancho-Garcia, E. Hennebicq, J. P. Calbert, V. Lemaur, D. Beljonne, and J. L. Bredas. Electronic and optical properties of polyfluorene and fluorene-based copolymers: A quantum-chemical characterization. *The Journal of Chemical Physics*, 118(14):6615–6623, 2003

52. M. A. Stevens, C. Silva, D. M. Russell, and R. H. Friend. Exciton dissociation mechanisms in the polymeric semiconductors poly(9,9-dioctylfluorene) and poly(9,9-dioctylfluorene-*co*-benzothiadiazole). *Physical Review B (Condensed Matter and Materials Physics)*, 63(16):165213, 2001
53. M. M. Alam and S. A. Jenekhe, *Chem. Mater.* 16:4647, 2004
54. A. C. Morteani, P. Sreearunothai, L. M. Herz, R. H. Friend, and C. Silva, *Phys. Rev. Lett.* 92:247402, 2004
55. A. L. T. Khan, P. Sreearunothai, L. M. Herz, M. J. Banach, and A. Kohler. Morphology-dependent energy transfer within polyfluorene thin films. *Physical Review B (Condensed Matter and Materials Physics)*, 69(8):085201, 2004
56. S. Guha, J. D. Rice, Y. T. Yau, C. M. Martin, M. Chandrasekhar, H. R. Chandrasekhar, R. Guentner, P. S. de Freitas, and U. Scherf. *Physical Review B*, 67:125204, 2003
57. A. C. Morteani, A. S. Dhoot, J.-S. Kim, C. Silva, N. C. Greenham, C. Murphy, E. Moons, S. Ciná, J. H. Burroughes, and R. H. Friend. Barrier-free electron–hole capture in polymer blend heterojunction light-emitting diodes. *Advanced Materials*, 15(20):1708, 2003
58. A. C. Morteani, P. Sreearunothai, L. M. Herz, R. H. Friend, and C. Silva. Exciton regeneration at polymeric semiconductor heterojunctions. *Physical Review Letters*, 92(24):247402, 2004
59. G. Porter. Flash photolysis and some of its applications. *Nobel Lecture*, pages 241–263, 1967

Dynamics of Resonant Electron Transfer in the Interaction Between an Atom and a Metallic Surface

J.P. Gauyacq and A.G. Borisov

Summary. Resonant Charge Transfer (RCT) between an atom and a metal surface corresponds to a one-electron energy-conserving transition between a discrete atomic level and the continuum of metallic states. In a static system (fixed atom-surface distance), RCT can be efficiently described by attributing a width, inverse of a finite lifetime, to the atomic level. The RCT rate is then given by the atomic level width. The use of the same description, based on an adiabatic approximation, is not always valid in a collisional context, when the atom moves with respect to the surface. We review some recent results obtained on this problem using a wave-packet propagation approach to describe the dynamics of RCT. The nonadiabatic character of RCT is illustrated on three different situations. (1) For a free-electron metal surface, the adiabatic approximation is found to hold. (2) For more realistic metal surface descriptions, the presence of a projected band gap is found to deeply influence the static RCT. However, significant non-adiabatic transitions can appear even at moderate velocities, which wash out the effect of the metal electronic band structure. (3) In the case of metal surfaces partly covered with adsorbates, the possibility of electronic transitions between three objects (the atom, the adsorbate, and the substrate) deeply affects the RCT, leading to various dynamical behaviors, very different from the predictions of the adiabatic approximation.

1 Introduction

Electron transfer between an ion (atom, molecule) and a metal surface determines the charge state of the species scattered or sputtered from the surface during a heavy particle impact on the surface. It is thus important, for e.g., negative ion beam production techniques [1] and for various surface analysis methods such as SIMS [2, 3] (secondary ion mass spectrometry), LEIS [4, 5] (low energy ion spectroscopy), or MDS [6–8] (metastable atom de-excitation spectroscopy) as well as for the determination of charge equilibrium between gas and surfaces. In addition, charge transfer between an atom (molecule) and a surface often play a key role as a step in surface reaction processes: transient states formed by electron transfer between adsorbates and the surface or between a projectile and the surface are often invoked as intermediates in surface reaction mechanisms [9, 10]. Owing to its fundamental and practical interests, charge transfer in ion-surface interaction has received a lot of attention, that have been reviewed at a few places [11–14].

The electron transfer process between an atomic particle and a metal surface corresponds to a discrete state-continuum transition and presents a few specific characteristics that are linked to the very different nature of the states involved in the two collision partners: atomic levels are discrete states localized in a finite region of space around the atom center, whereas metallic states form a continuum of states delocalized over the entire crystal. If the atomic level is degenerate with a continuum of metallic states, a one-electron resonant (energy conserving) transition between the atomic level and the metallic states is possible. It is usually termed resonant charge transfer (RCT). Multi-electron transitions also exist. For example, if there is a vacancy on an inner orbital of the atom, a metal electron can be transferred on this orbital, the energy defect of the capture being balanced by the excitation of another metal electron (Auger process) or by a collective excitation of the metal electrons (plasmon-assisted charge transfer) [15–18]. In the present chapter, we discuss the RCT process with an emphasis on its dynamical characteristics in the course of an ion-surface collision. Since it is a one-electron process, it is usually considered to be the most efficient charge transfer process, when it is energetically possible.

As said earlier, the atom–metal surface charge transfer is associated to a discrete state-continuum transition. In a static system (fixed atom–surface distance), such transitions are associated to an exponential decrease of the discrete state population with time (Fermi golden rule) [19]. The evolution from the discrete state to the continuum is irreversible. The interaction with the continuum of metallic states results in a finite width of the atomic levels, equal to the inverse of their lifetimes. Let us consider a time-dependent system, such as an atom–surface collision described in a semiclassical approach with a classical motion of the atom centers. It is very tempting to keep the same kind of description for the RCT as for the static system (fixed atom–surface distance), i.e., to assume that at each time along the trajectory, the transition rate between the discrete state and the continuum can be described by the width of the atomic level at the corresponding position, i.e., obtained in a static calculation with a fixed atom–surface distance. Equivalently, the atomic level is associated to a complex potential, the imaginary part of which gives the decay rate of the state. This adiabatic approximation is often referred to as the local complex potential approximation. Various theoretical approaches have been developed to go beyond this simple adiabatic approximation for ion–surface collisions, including in particular many body aspects [20–30]. In this chapter, we review some recent results on the dynamics of the RCT process obtained with a wave-packet propagation approach (WPP). Three cases are presented in detail (1) the case of a metal described in the free-electron model, where the adiabatic approximation is found to hold, (2) the case of a metal surface with a projected band gap in its electronic structure, which can strongly modify the RCT, and lead to important nonadiabatic effects, and (3) the case of a metal surface with adsorbates on it, where the existence of transitions between three objects (the projectile, the adsorbate, and the metal) leads to a variety of dynamical behaviors.

2 Wave-Packet Propagation (WPP) Treatment of the Charge Transfer Process

A useful theoretical framework for the treatment of the one-electron RCT process is to consider the time evolution of the active electron in the compound potential created by the projectile and the surface. We solve this problem with a wave packet propagation (WPP) approach (see e.g., [31, 32] for details on the WPP application to RCT). The active electron is described by a three-dimensional wave-packet $\Psi(\boldsymbol{r},t)$ defined on a grid of points and the time evolution of $\Psi(\boldsymbol{r},t)$ is given by the time dependent Schrodinger equation:

$$\begin{aligned} i\frac{\mathrm{d}\Psi(\boldsymbol{r},t)}{\mathrm{d}t} &= H\,\Psi(\boldsymbol{r},t) \;=\; (T+\,V)\,\Psi(\boldsymbol{r},t) \\ &= (T\,+\,V_{\mathrm{e-Surf}}\,+\,V_{\mathrm{e-Atom}}\,+\,\Delta\,V_{\mathrm{Surf}})\,\Psi(\boldsymbol{r},t), \end{aligned} \tag{1}$$

where T is the electron kinetic energy operator and V, the potential felt by the active electron. V is given by the sum of three terms: $V_{\mathrm{e-Surf}}$, the electron interaction with the metal surface, $V_{\mathrm{e-Atom}}$, the electron interaction with the atomic projectile core and $\Delta\,V_{\mathrm{Surf}}$, the change in the electron–surface interaction induced by the presence of the projectile core. The three potential terms are usually represented with model or pseudopotentials.

The $V_{\mathrm{e-Atom}}$ potentials are taken from earlier atomic physics studies. As an example, in the applications later with alkali atoms, the electron interaction with positive alkali ion cores are taken as the ℓ-dependent pseudopotentials from Bardsley [33], transformed using the Kleynman-Bylander procedure [34], allowing an efficient handling in the WPP propagation scheme. In order to study the effect of the target metal band structure on the RCT, we have used two different $V_{\mathrm{e-Surf}}$ terms representing two different physical situations: a free-electron metal and a metal with a projected band gap perpendicular to the surface. Free-electron metals are described with the local analytical potentials derived by Jennings et al. [35] from DFT slab calculations. This potential is constant inside the metal and joins an image potential outside the metal. Metal targets with a projected band gap are described with the model potentials by Chulkov et al. [36]. These potentials only depend of z, the electron coordinate normal to the surface and are invariant by translation parallel to the surface. Inside the metal, the potential oscillates with the lattice frequency opening a gap for the electron motion perpendicular to the surface. This oscillating potential smoothly joins an image potential outside the surface. These model potentials accurately reproduce the characteristics of the electronic band structure perpendicular to the surface [36]: energy position of the surface projected band gap, energies of the image states and of the surface states (or resonances). As shown later, these are the important features influencing the RCT process. The ΔV_{Surf} term corresponds to the polarization of the surface electronic density by its interaction with the projectile core; it is mainly important in the case of charged projectile cores and is then taken as the electron interaction with the classical electrical image of the core.

With the earlier choice of potentials, the projectile-surface system is invariant by rotation around the z-axis, normal to the surface and going through the projectile center and m the projection of the electron momentum on the z-axis is a good quantum number. We thus used cylindrical coordinates (ρ, ϕ, z) and the ϕ dependence of $\Psi(\boldsymbol{r}, t)$ can be factored out following:

$$\Psi(\boldsymbol{r}, t) = \sum_m \Psi_m(\rho, z, t)\, \mathrm{e}^{\mathrm{i}m\phi}, \tag{2}$$

where $\Psi_m(\boldsymbol{r}, t)$ is given by:

$$i\frac{\partial \Psi_m(\rho, z, t)}{\partial t} = H_m\, \Psi_m = \left(-\frac{1}{2}\frac{\partial^2}{\partial z^2} - \frac{1}{2\rho}\frac{\partial}{\partial \rho}\frac{1}{\rho}\frac{\partial}{\partial \rho} + \frac{m^2}{2\rho^2} + V\right) \Psi_m. \tag{3}$$

The WPP approach consists in propagating the electronic wave packet from a well-chosen initial condition, $\Psi(\vec{r}, t = 0) = \Phi_0(\boldsymbol{r})$. Usually, $\Phi_0(\boldsymbol{r})$ is chosen equal to the wave function of one of the bound states of the free projectile. The time propagation of the electron wave function is performed using the time-stepping algorithm:

$$\Psi_m(\rho, z, t + \mathrm{d}t) = \mathrm{e}^{-\mathrm{i}H_m \mathrm{d}t}\, \Psi_m(\rho, z, t). \tag{4}$$

The split operator approximation [37] is then used to compute the action of the exponential operators involved in the $\mathrm{e}^{-\mathrm{i}H_m \mathrm{d}t}$ time propagator:

$$\mathrm{e}^{\mathrm{i}(A+B)\mathrm{d}t} = \mathrm{e}^{\mathrm{i}A\mathrm{d}t/2}\, \mathrm{e}^{\mathrm{i}B\mathrm{d}t}\, \mathrm{e}^{\mathrm{i}A\mathrm{d}t/2} + O(\mathrm{d}t^3). \tag{5}$$

This allows to use propagation schemes appropriate for each part in the Hamiltonian [31]: coordinate representation for the local potential terms, pseudospectral approach [38] with fast Fourier transform or finite differences with Cayley transform and variable change for the kinetic energy terms.

Two different kinds of calculations are performed and discussed later: static and dynamic. In the static calculations, the atom–surface distance is fixed. From the time propagation one obtains the survival amplitude, $A(t)$, of the wave packet in the initial state:

$$A(t) = \langle \Phi_0(\boldsymbol{r}) \mid \Psi(\boldsymbol{r}, t) \rangle\,. \tag{6}$$

The Laplace transform of the survival amplitude yields $n(\omega)$, the density of states of the system projected on the initial state, which presents peaks with a finite width at the position of the quasistationary states (resonances) of the problem. One thus obtains the energy and the width of the resonances of the system, i.e., of the projectile states perturbed by their interaction with the surface. Alternatively, one can adjust the time dependence of the $A(t)$ function to the sum of a few exponentials representing the main quasistationary states in the wave packet [31]. In this static calculation, the width of a given projectile state, called static width later, is equal to the RCT rate and to the inverse of the lifetime of the state.

In the dynamic calculation, the projectile is moving with respect to the surface along a classical trajectory, given by $Z(t)$, the projectile surface distance as a function of time. The active electron is then evolving in a time-dependent potential. The propagation is started for a large enough projectile–surface distance with the initial wave packet equal to a bound state of the free projectile. The survival amplitude, $A(t)$, and probability, $P_{WPP}(t) = |A(t)|^2$, directly correspond to the survival of the initial state for the physical situation of a collision. At this point, one must stress that, due to the interaction with the surface, one can expect the initial projectile state to mix with other states, e.g., to get polarized, so that the above survival probability is a priori different from the survival of the system in the quasistationary state localized on the projectile. This feature can be very important in the case of a strong mixing between atomic states induced by the surface or in the case of mixing between atomic and surface states, such as occurs when adsorbate localized states are present on the surface.

3 Resonant Charge Transfer with a Free-electron Metal

The simplest description for a metal electronic structure is given by a free-electron model where electrons move freely in a constant potential inside the metal. This situation is schematized in Fig. 1 which presents the total potential felt by the active electron. In the case of a negative ion projectile interacting with a metal surface illustrated later, it is the sum of the electron–metal and electron–atom interactions. This potential exhibits two potential wells, one inside the metal and one around the atom, separated by a potential barrier. RCT consists in transitions between the states localized around the atom and the continuum of metal states. It can also be seen as the tunneling of the active electron through the potential barrier separating the projectile and the surface. Figure 1 is only a cut of the potential. In the full 3D problem, the potential barrier is the thinnest along the normal to the surface that goes through the atom center, the z-axis, so that electron transfer preferentially occurs along this direction.

Because of the interaction with the surface, in a static picture, the atomic levels acquire a finite width, Γ. Γ gives the transfer rate of the electron, it is a function of Z, the atom–surface distance. If the atomic level is above the Fermi level of the metal, it is degenerated with the empty part of the metal

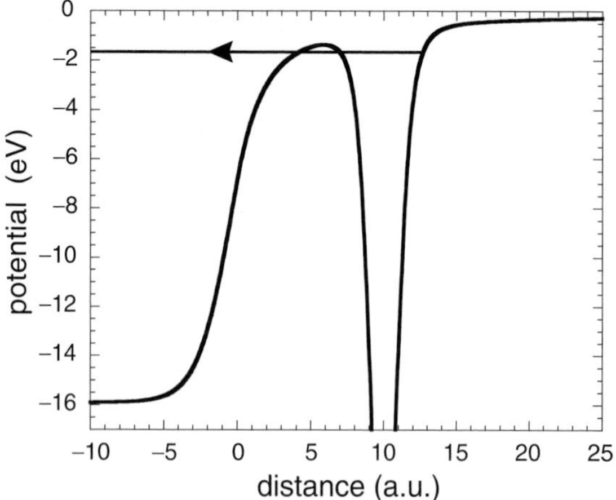

Fig. 1. Schematic picture of the potential felt by the electron active in the RCT between a negative ion and a free-electron metal surface. It is presented along the z-axis, that is normal to the surface and goes through the projectile center. The projectile is located at 10 a.u. from the surface and the metal surface is located at the origin of coordinates. The metal extends on the $z < 0$ side. The energy position of a negative ion state is indicated by the horizontal full line

conduction band (at 0 K) and the electron transfer occurs from the atom to the surface. If the atomic level is below the Fermi level it is degenerated with the occupied part of the metal conduction band and electron transfer to the metal is impossible. In that case, it is better to reformulate the problem in terms of vacancies, leading to the conclusion that a vacancy on the atom is transferred to the metal, i.e., that electron transfer occurs from the metal to the atom.

A few theoretical methods have been designed and applied to the determination of the energy and width of the atomic levels in a static situation (fixed projectile–surface distance) using model or pseudopotentials representations. Nonperturbative approaches basically look for the quasistationary states in the problem using different techniques: complex scaling [39], close-coupling scattering approach [40], stabilization [41,42], close-coupling [43,44], or wave packet propagation [32,45,46]. Energies and widths obtained in a static study can then be used to describe the RCT dynamics in a collision via an adiabatic assumption. It consists in assuming that the width of the atomic level computed in the static picture (fixed atom–surface distance), $\Gamma(Z)$, still gives the charge transfer rate when the atom moves with respect to the surface. If one also makes the assumption that the atom is following a classical trajectory, $Z(t)$, when approaching the surface, the evolution of the atomic state population, $P_{\mathrm{adia}}(t)$, can be described via a rate equation:

$$\frac{\mathrm{d}P_{\mathrm{adia}}}{\mathrm{d}t} = -\Gamma_{\mathrm{loss}}(Z(t))\, P_{\mathrm{adia}}(t) \;+\; \Gamma_{\mathrm{capture}}(Z(t))(1 - P_{\mathrm{adia}}(t)), \qquad (7)$$

where the time dependence of the capture and loss rates is given by $Z(t)$. Γ_{loss} and $\Gamma_{\mathrm{capture}}$ are the electron transfer rates. In the simplest situation, they are equal to the static width of the state, depending on the population of the metallic states degenerated with the atomic level. They can also include a statistical factor taking into account the different degeneracy of the different charge states [47]. Derivations of the rate equation have been presented using a semiclassical approximation [22, 48] or a high temperature limit [49]. They were all made through a broad band approximation, implicitly assuming the absence of structures in the continuum or of fast energy dependence of the various couplings in the continuum. The rate equation approach has been applied to a series of systems involving quasifree-electron metals such as Al, leading to predictions in quantitative agreement with experiments [50–53]. In the case of fast grazing angle collisions [14], the collision velocity perpendicular to the surface is very low and the adiabatic approximation (7) where the capture and decay rates incorporate the parallel velocity effect leads to a quantitative account of experimental results in a variety of collisional systems involving an Al metal target [50,51] which can reasonably well be described by a free-electron model. The so-called "parallel-velocity" effect is a consequence of the change of Galilean reference frame between the metal target and the projectile [54] that can strongly affect the RCT process in the case of fast collisions. Its treatment in the rate equation approach requires the computation of the partial electron transfer rates between the atomic levels and the different metallic states.

The WPP approach solves the dynamics of the problem exactly and can be used to test the validity of the adiabatic approximation in the free-electron metal case. The idea is to get the exact time dependence of the atomic level population, $P_{\mathrm{WPP}}(t)$, using the WPP approach,and then to extract from it an effective charge transfer rate, $G(Z)$ that can be directly compared to the static width $\Gamma(Z)$ obtained in the fixed atom calculation. This procedure directly tests the relevance of the adiabatic approximation for the charge transfer rate. Figure 2 presents such a comparison for the case of an H^- ion approaching an Al(111) surface at a normal velocity of 0.05 a.u. [55]. In the considered range of projectile–surface distances, the ion level is well above Fermi level and the electron transfer only occurs from the ion to the surface. The effective charge transfer rate is then defined by:

$$G(Z) = -\frac{1}{P_{\mathrm{WPP}}} \frac{\mathrm{d}P_{\mathrm{WPP}}}{\mathrm{d}t}. \tag{8}$$

Figure 2 shows that, the two rates, $\Gamma(Z)$ and $G(Z)$, perfectly agree over a very large range of projectile–surface distances, Z. It then fully confirms the validity of the adiabatic approximation, i.e., of the rate equation (7) in this case. This agreement covering a large collision velocity range is fully consistent with the success of the various theoretical studies based on the rate equation (7) in quantitatively accounting for experimental results.

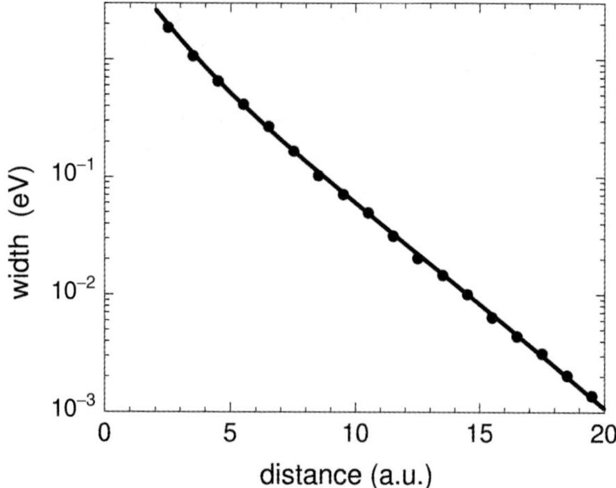

Fig. 2. Comparison of the effective width (*black circles*, see definition (8) in the text) with the width obtained in a static calculation (*full line*) for an H^- ion interacting with an Al(111) surface described in the free-electron model. The ion velocity is 0.05 a.u. The two widths are presented as functions of the ion–surface distance measured from the surface image reference plane

4 Effect of the Electronic Band Structure of the Metal Target

Free-electron metals being much idealized, one must wonder about the possible effects of the electronic band structure of the metal target. Indeed, the potential inside a metal is not constant, its periodicity according to the lattice structure leads to specificities in the electronic band structure. In a one-dimensional problem, a periodic potential leads to the existence of an energy gap in which propagation is impossible. In 3D, the periodicity leads to domains in the $(E, \hat{k})$ (energy, direction of the momentum) space where there is not any propagating state. The surface performs a cut through this structure and it can occur that propagation perpendicular to the surface is impossible for states in a certain energy range; such an energy gap is called a surface projected band gap. The impossibility of propagation in a surface projected band gap can lead to the existence of states localized in the surface area such as surface states or image states [56,57]. A projected band gap can be thought to deeply influence the RCT and even to be the most efficient feature of an electronic band structure to do so. Consider the electronic band structure of Cu(111) as described by the model potential from [36] and presented in Fig. 3. There is a band gap along the surface normal (vanishing $k_{//}$, electron momentum parallel to the surface) and the various states exhibit a parabolic dispersion as a function of $k_{//}$. As explained earlier, tunneling between the atom and the metal is much favored along the normal to the surface in the case of a free-electron metal surface and so, it mainly populates metal states that

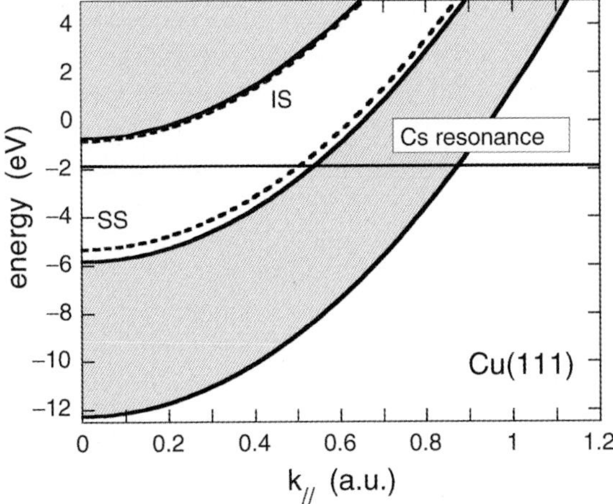

Fig. 3. Schematic picture of the electronic structure of the Cu(111) surface as described by Chulkov et al. potential [36]. The energy of the various states are presented as functions of $k_{//}$ the electron momentum parallel to the surface. The hatched areas correspond to 3D-propagating bulk states. The surface state and image state located inside the surface projected band gap are shown as *dashed lines*. The energy of a Cs atomic level interacting with the Cu(111) surface is shown by the *horizontal line*

are propagating along the surface normal, around $k_{//} = 0$. If the projectile level is located in front of a projected band gap, the only metal states available for resonant state transfer, i.e., the metal states with the same energy as the projectile state, correspond to a finite $k_{//}$. As seen in Fig. 3, these can be 3D-bulk states or a state in the surface state 2D-continuum with a given $k_{//}$. Thus, the $k_{//} \approx 0$ states that are the most active states in the RCT for a free-electron metal are not playing any role because of the projected band gap. In this case, one can then think that RCT should be deeply affected, and more specifically to be significantly weakened.

4.1 Static Systems: Alkali Adsorbates on Noble Metals

The projected band gap effect is illustrated on the example of excited states localized on alkali adsorbates on noble metals surfaces [58]. The (111) and (100) surfaces of noble metals exhibit surface projected band gaps in energy domains where atomic levels can lie. At low coverage, isolated alkalis adsorb as positive ions on metal surfaces [59–61]. Excited states corresponding to the transient capture of an electron around the adsorbed ion can be found in the energy range of the projected band gap. These states can be associated with atomic alkali states perturbed by their interaction with the surface [58]. Since, in the equilibrium situation, these states are not populated, they are usually studied by inverse photoemission or by two-photon photoemission.

In particular, time resolved two-photon photo-emission (TR-2PPE) allows the study in real time of the dynamics of the charge transfer between the adsorbate and the metal [62]. Theoretical studies of these systems using the WPP approach confirmed the very large effect of the projected band gap. The Cs/Cu(111) system exhibits the most spectacular effect [58]. At the adsorption distance, the RCT rate for the lowest lying state (termed "6s" even if it is much distorted by the interaction with the surface and is closer to a 6s–6p hybrid) is found to be equal to 7 meV, to be compared with 900 meV on a free-electron metal [63]. The projected band gap leads to a decrease of the electron transfer rate by two orders of magnitude, i.e., to a quasiblocking of the RCT in this case.

The band gap quasiblocking effect is illustrated in Fig. 4 which shows the wave packet associated to the "6s" excited state in the Cs/Cu(111) and in the Cs/free-electron metal systems. It presents the logarithm of the modulus of the electron wave function (electron density) of the transient excited state in cylindrical coordinates: the z-axis is normal to the surface and goes through the adsorbate center and ρ is the coordinate parallel to the surface. The Cs-metal surface distance is different in the two cases in order to have

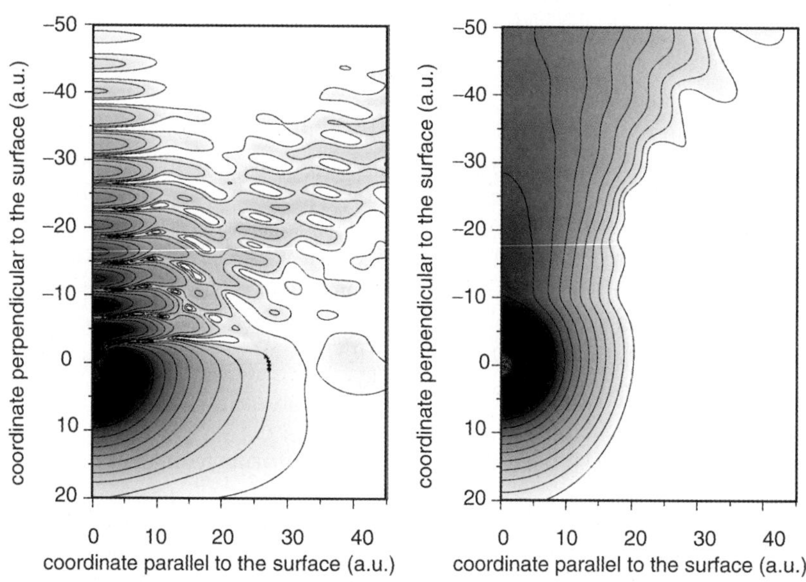

Fig. 4. Electron density of the resonant Cs(6s) states in front of a Cu(111) and a free-electron metal surface (left and right panel, resp.). The logarithm of the electron density is presented as a contour plot as a function of the coordinates perpendicular and parallel to the surface. The Cs atom is located at the origin of coordinates and the metal is on the negative coordinate side. In the left panel, Cs is at its adsorption distance, 3.5 a_0 from Cu(111) image reference plane. In the right panel, Cs is at 10. a_0 from the image reference plane. Large electron densities zones are in black, smaller electron densities are in gray with electron densities increasing with darkening gray

similar decay rates for the Cs transient state. One recognizes in both cases the distorted atomic wave function centered on the adsorbate center. In the free-electron case, the electron transfer appears as a strong electron flux along the surface normal. This flux is absent in the Cu(111) surface case and the outgoing electron flux, much weaker, goes into the metal at a finite angle from the surface normal. This is directly the signature of the projected band gap effect, which prevents RCT along the surface normal, requiring the electron to tunnel through a much thicker barrier. One can also notice a strong distortion of the electron cloud in the Cu(111) case, which corresponds to a short atom–surface distance. This polarization of the electronic cloud is induced by its interaction with the surface, it also contributes to the quasiblocking of the RCT process (see a discussion in [64]).

Usually, the RCT is thought to dominate the various possible electron transfer processes, however, when it is almost blocked, one should also consider multielectron effects i.e., electron transfer induced by inelastic interaction of the excited electron with the substrate electrons. Theoretical computation of the multielectron term [65] yields a transfer rate of 16.5 meV, leading to a total electron transfer rate of 23.5 meV, i.e., to a lifetime of the excited state of 28 fs. Experimental TR-2PPE studies also revealed very long lived states in the Cs/Cu(111) systems with lifetimes up to a few tens of fs [66–70], in excellent agreement with the theoretical predictions [63,65,71]. Similar results are found for other alkali/Cu(111), Cu(100) systems [66,68,72], the differences being associated with differences in the band gap locations or to differences between the alkali atoms [73]. So, in the static system (fixed atom–surface distance) a projected band gap has a very strong influence on the RCT, or in other words, the RCT process is quite sensitive to the electronic band structure of the metal.

4.2 Collisional Systems

As a first example, we can briefly mention the case of grazing angle collisions. In that case, the collision velocity vector makes a very small angle with the surface plane, so that the component of the velocity perpendicular to the surface remains small, even for very fast collisions. In such collisions (see e.g., a review in [14]), if we neglect corrugation parallel to the surface, the dynamics of the collision is governed by the small perpendicular velocity. The component parallel to the surface leads to the well-studied parallel velocity effect [14,54]. A detailed joint experimental-theoretical study of RCT in grazing angle collisions has been performed for electron capture by hydrogen atom and Li^+ ions in collisions on Cu(111) surfaces [74,75]. The theoretical part includes a static WPP study of the system, associated to an adiabatic approximation for treating the collision dynamics and the parallel velocity effect. In this low velocity system, the adiabatic approximation holds, as has been checked with the dynamical WPP approach. From the comparison between theory and experiment, it appears that the electronic band structure of Cu(111) plays an important role in the process. In particular electron transfer from and to the two-dimensional Cu(111) surface state is dominating the RCT and leads to results very different from the predictions made for a free-electron metal surface where only 3D-propagating bulk states are involved. In contrast, similar collisions on a Cu(110) surface, which do not exhibit a projected band gap, are very well reproduced by a free-electron modeling. So, in this case of collisions with the perpendicular collision energy typically in the eV range, the electronic band structure effect on RCT is present and deeply influencing and the adiabatic approach is efficient in accounting for experimental observations.

The situation is quite different if we consider higher collision velocities where nonadiabatic effects come into play. A first theoretical analysis of these effects was reported in the case of H^- ions interacting with a Cu(111) surface [31, 76]. In this system, the H^- ion level is in front of the projected band gap of the surface, and similarly to the cases discussed in the preceding section, this leads to a decrease of the static RCT rate compared to the free-electron case [31,76]. However, in the dynamical situation, it was found that very quickly as the collision velocity is increased, the dynamics of the RCT cannot be represented by the adiabatic approximation (7) anymore. Dynamical WPP calculations were performed for an H^- ion approaching the surface at constant velocity and the effective width was extracted from the decay of the H^- ion population, following (8). The effective width is presented in Fig. 5 as a function of the ion–surface distance for various collision velocities.

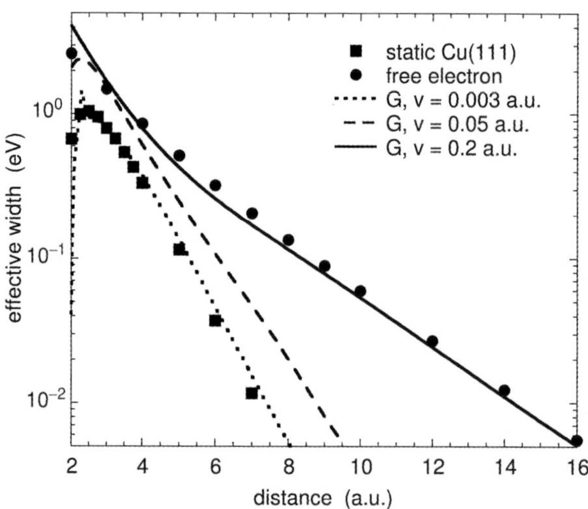

Fig. 5. Effective width for an H^- ion approaching a Cu(111) surface at different collision velocities: $v = 0.2$ a.u. (*full line*), 0.05 a.u. (*dashed line*), and 0.003 a.u. (*dotted line*). It is compared to the static width obtained in front of a Cu(111) surface (*full squares*) and of a free-electron metal surface (*full circles*). All widths are presented as a function of the ion–surface distance, measured from the surface image reference plane

It is also compared with the static width obtained in a static WPP calculation on Cu(111) and on a free-electron metal. As a first remark, the effective width varies with the collision velocity, bringing evidence of a nonadiabatic behavior. It appears that for the lowest velocity ($v = 0.003$ a.u) the effective width nicely agrees with the static width for Cu(111). However, for a 0.2 a.u. velocity, the effective width is quite different and is practically equal to the static width for a free-electron metal. For intermediate velocities, the effective width varies with the velocity in between the two limits given by the static width for Cu(111) and for a free-electron metal. Thus on Cu(111) the adiabatic approximation is only valid for very low collision velocities. As the collision velocity is increased, nonadiabatic transitions appear that tend to make the charge transfer on Cu(111) identical to that on a free-electron metal. This last feature can be linked to the time-dependence of the RCT. Indeed, the specificities of an electronic band structure are consequences of the periodic structure of the crystal lattice, i.e., they come from interference of waves scattered by the different lattice sites. This interference needs some time to set in and so does the band structure effect on the RCT.

Figure 6 shows the time dependence of the survival probability of an H^- ion at a fixed distance ($Z = 6\,a_0$) from a Cu(111) surface [31]. Two results are shown: for a free-electron metal and for the model Cu(111) surface (WPP approach). For late times, in both cases, the decay of the population is exponential with two very different time constants. The decay at late times on Cu(111) is much slower; indeed, the H^- ion level is inside the Cu(111) projected band gap and similarly to the Cs case discussed later, RCT is partly blocked. The situation is quite different at very short time. Below 30 a.u., the decay on Cu(111) is identical to that on the free-electron metal, it is followed

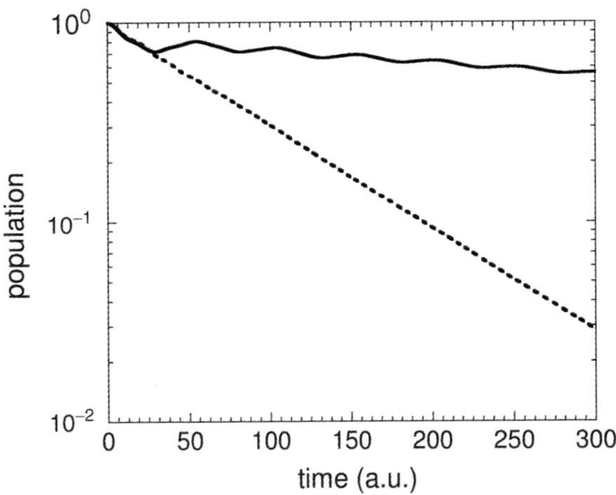

Fig. 6. Decay of the population of an H^- ion held at a fixed distance ($Z = 6\,a_0$) from a Cu surface. *Dashed line*: free-electron metal surface and *full line*: Cu(111) surface

by a transition region with oscillations in $P(t)$, before reaching the slow decay region at late times. This change of behavior corresponds to the onset of the effect of the band structure. For very small times, the electron wave-packet is localized around the ion and does not feel the periodic potential inside the metal. The electron then tunnels through the potential barrier separating the ion and the metal, this step is the same for Cu(111) and for the free-electron metal. After tunneling, the electron wave-packet enters the metal and is partly reflected by the various atomic planes, i.e., by the modulation of the $V_{\mathrm{e-Surf}}$ potential. All these reflections and the ensuing interference build up the blocking of the electron propagation along the surface normal and after a while, result in the drop of the electron transfer rate. It thus appears that the effect of the band structure on the RCT needs time to appear and that on very short time scales, the RCT behaves as on a free-electron metal. The critical time scale is of the order of a few back and forth travels of the electron between (111) reflecting planes and the ion. As a consequence, in a collision, if the effective collision time is shorter than this critical time, the RCT behavior will be similar to the one on a free-electron metal.

Experimentally, the energy variation of electron transfer in the H^-–Ag(111) collisional system has been interpreted as due to this short time effect [77]. More recently, the effect of nonadiabatic transitions in the RCT has been further studied in a joint experimental–theoretical study devoted to the neutralization of Li^+ ions by collision on Ag(100) surfaces [78]. Since the energies of the excited states of the Li projectile are too high compared to the Fermi energy of Ag(100), neutralization of Li^+ ions is dominated by electron capture into the Li(2s) ground state. The results of the static study of the Li(2s)–Ag(100) system are presented in Fig. 7. It presents the energy and the width of the Li(2s) level as a function of the Li-surface distance, in two cases: Ag(100) and a free-electron model. The level which correlates at infinite projectile–surface distance to the 2s atomic orbital is labeled "2s," although it is much mixed with other states by its interaction with the surface. Ag(100) exhibits a projected band gap, between −2.83 eV and +2.21 eV with respect to vacuum and a complete series of image states is present. It also exhibits a surface resonance located below the gap at −3.13 eV. In the free-electron model, the Li(2s) level energy steadily increases as the projectile approaches the surface, following the image charge potential variation. The presence of the surface resonance qualitatively influences the static picture in the case of Ag(100). An extra state splits off the Ag(100) surface resonance and mixes with the 2s state leading to an avoided crossing structure (see [32, 79–81] for a discussion of similar extra states). At large Z, the Li(2s) level is very close to the free-electron case and as Z decreases, it exhibits an avoided crossing with a state initially localized very close to the surface resonance. At small Z the two states have interchanged their character and the Li(2s) character is transferred to the higher level. As for the level width, it appears that except at very small Z, the Ag(100) and free-electron metal results are very similar;

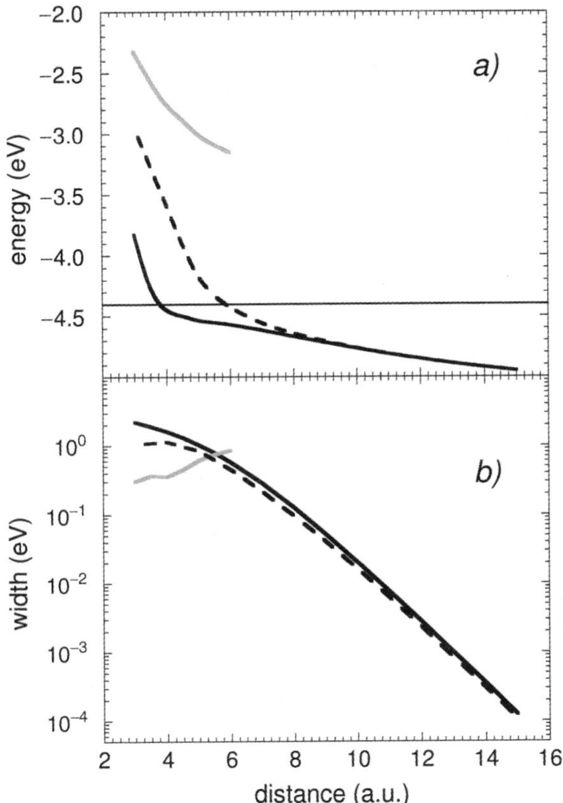

Fig. 7. Energy (part a) and width (part b) of the various states involved in Li^+ neutralization on a Ag surface, as a function of the Li-surface distance measured from the surface image reference plane. Energies (measured with respect to vacuum) and widths are obtained in WPP calculations performed for a fixed projectile–surface distance. Free-electron metal surface: *dashed line*. Ag(100) surface: state correlated at infinity with the Li(2s) level (*full black line*) and state splitting off at infinity from the Ag(100) surface resonance (*full gray line*). The Ag(100) Fermi energy is indicated by the thin horizontal *full line*

indeed, in both cases, RCT along the surface normal is possible leading to a large electron transfer rate at small Z.

In an adiabatic view, the avoided crossing in the Ag(100) case could be thought to deeply influence Li^+ neutralization. Indeed, neutralization by resonant electron capture only occurs at distances large enough for the Li(2s) level to be below the surface Fermi level. As seen in Fig. 7, the presence of the avoided crossing significantly widens the Z-region where Li^+ can capture an electron, in particular at small Z where the electron capture rate is large. So from the static picture, one would expect a much larger neutralization probability on Ag(100) than on a free-electron metal. This expectation is confirmed in Fig. 8 which shows experimental and theoretical results for the Li^+ neutralization as a function of the collision energy (scattered particles along the surface normal). As a first result, one can see that, in the entire investigated energy range, a large neutralization probability, over 90%, is obtained within the adiabatic approximation using the static Ag(100) results (Fig. 7). It is much larger than the corresponding result obtained with the adiabatic approximation, using the free-electron static results.

The result obtained via a dynamical WPP treatment for a Ag(100) surface is also shown in Fig. 8: it is quite different from the adiabatic Ag(100) result and it is much closer to the free-electron result. So, in this system, very important nonadiabatic effects are present and they tend to remove the effect of the electronic band structure, i.e., to make Ag(100) behave as a free-electron metal surface. One can also notice that at large velocities, the dynamical WPP result is very close to the free-electron result, whereas at the smallest investigated velocity, it is midway between the free-electron result and the adiabatic result, possibly indicating an onset of the Ag(100) band structure effect at small velocity. As for the experimental result, it lies close to the free-electron and to the dynamical-WPP results, confirming the earlier conclusions as well as the validity of the present WPP dynamical approach.

The results on the Li–Ag(100) system can be interpreted as the influence of nonadiabatic transitions increasing as the collision velocity goes up

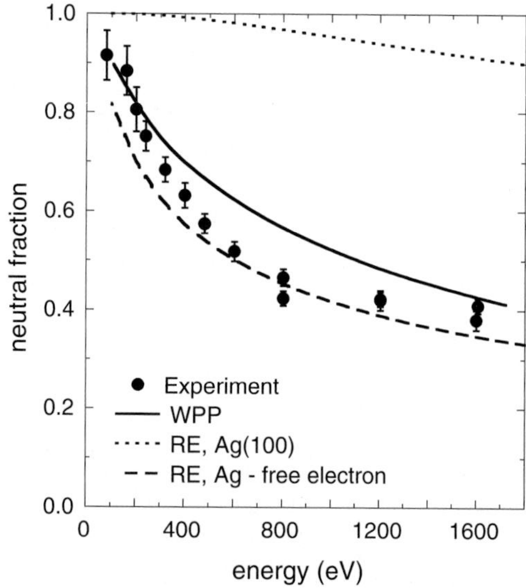

Fig. 8. Neutral fraction of Li particles reflected from a Ag(100) surface as a function of their energy. Experimental results from Canario et al. [78]: *black dots*. Theoretical results obtained in a rate equation approach (adiabatic approximation, (7)): *short dashed line* (Ag(100)) and *long dashed line* (free-electron metal). Dynamical WPP results for Ag(100): *full line*

(the experiments are in the 0.02–0.1 a.u. range). Equivalently, one can interpret it as the system behaving diabatically in the avoided crossing seen in Fig. 7, however, without a clear understanding of what is the diabatic character of the state crossing through the avoided crossing. Alternatively, following the earlier discussion on the H^-–Cu(111) system, one can say that in the collision energy range investigated in Fig. 8, the collision is too fast for an effect of the Ag(100) band structure to show up and Ag(100) behaves as a free-electron metal, at least from an RCT point of view.

4.3 Charge Transfer on a Metal Surface with Adsorbates

When adsorbates are present on a metal surface, they influence the electron transfer processes in collisions via nonlocal and local effects (see a review in [82]). The nonlocal effect is due to the change of the surface work-function induced by the presence of adsorbates. The surface work function change modifies the relative position of atomic and Fermi level and consequently the direction of the RCT. Local effects of the adsorbates on RCT arise because of changes in the electrostatic potential and in the electronic structure in the immediate vicinity of the adsorbates. In particular, quasistationary states such as the long-lived states discussed in Sect. 4.1 may be localized on adsorbates, bringing a three-body aspect into the RCT. In this case, the electron involved in the charge transfer can make transitions between the projectile, the adsorbate or the metal. This three-body aspect deeply influences the RCT process and is possibly associated to nonadiabatic transitions.

The local effects of adsorbates on the RCT are illustrated on the example of an H^- ion interacting with a Li adsorbate on an Al surface [83]. Figure 9 shows the energies and widths of the various states for a hydrogen projectile at a fixed position on the normal to the surface that goes through the adsorbate center (this geometry maximizes the local effects). At large distances, one recognizes the H^- ion state with its energy decreasing as the ion approaches the surface, due to electrostatic interactions with the adsorbate-surface system. Its width increases exponentially as the ion approaches the surface due to the increasing overlap between projectile and metal states. The state localized on the adsorbate has a different behavior: its properties, energy and width, are roughly independent of the projectile position when the latter is far away. As the ion approaches the surface, the energies of the projectile and adsorbate localized quasistationary states come close together and the two quasistationary states exhibit an avoided crossing in the complex energy plane. At small distances, it is the lowest energy state that exhibits the H^- ion characteristics with an energy close to the electrostatic prediction. One can also notice that the energy and width of an H^- ion approaching a clean Al surface are quite different (Fig. 9) confirming the importance of the local perturbation induced by the Li adsorbate.

The electron loss by an H^- ion approaching a Li adsorbate on an Al surface in the back-scattering geometry has been studied in the dynamical WPP approach [83] associated with a classical trajectory of the hydrogen projectile.

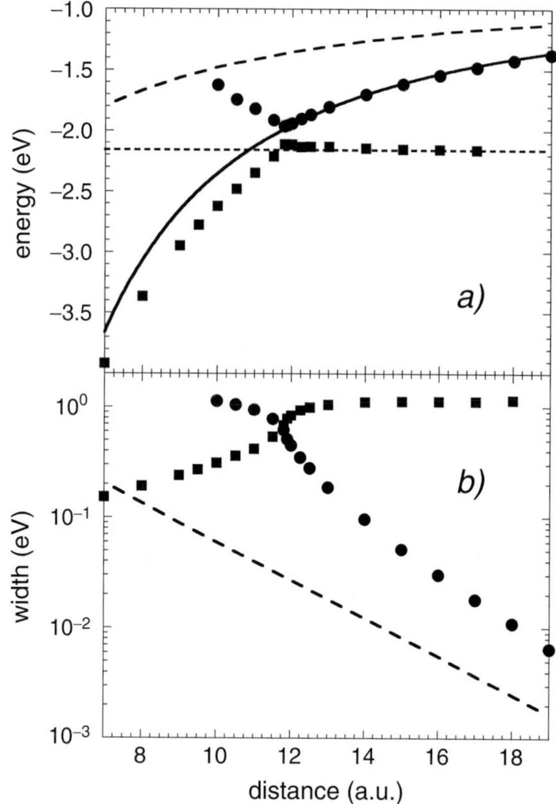

Fig. 9. Energy (part a) and width (part b) of the various quasistationary states in the case of an H^- ion interacting with a Li/Al system. The H^- ion is located on the normal to the Al surface that goes through the Li adsorbate center (back-scattering geometry). Energies (with respect to vacuum) and widths are presented as functions of the projectile distance from the Al image reference plane. Two quasistationary states are present that correlate at infinity to the H^- ion state (*full circles*) and to the quasistationary state localized on the Li adsorbate in the Li/Al system (*full squares*). *Full* and *short dashed lines*: electrostatic predictions for the H^- ion state and the Li localized state (these are obtained as the energy at infinity plus the electrostatic potential at the center of the atom, H or Li). *Long dashed lines*: energy and width of the H^- ion state interacting with a clean Al surface

A straight line trajectory with a constant velocity along the surface normal going through the adsorbate center was chosen for this study aiming at the characterization of the system dynamics. The dynamical behavior of the RCT has been studied by computing the effective decay rate of the H^- ion, similarly to the case discussed in Sect. 3, and by computing the energy spectrum of the electron transferred into the metal [83]. Both methods yield the same conclusion: strong nonadiabatic transitions occur in the avoided crossing in the investigated velocity range (0.005–0.04 a.u.). When the projectile–surface distance decreases, the system initially in the upper adiabatic state crosses through the avoided crossing and goes into the lower state and no sign of adiabatic behavior can be seen, even at the lowest velocity. Because of the finite width of both these states, this evolution is associated to a decrease of the total probability. One can notice that the investigated velocities are very low, too low even for the use of a straight line trajectory. In a more realistic approach, the projectile would first accelerate when approaching the surface, however, that can only strengthen the present result of the absence of adiabatic behavior at low energy. The dynamics of this system then appears to be never of the adiabatic kind and to be dominated by non-adiabatic transitions.

One can relate the discussion of the H^-–Li/Al system dynamics to the above discussion of the band structure effect. The lifetime of the adsorbate-localized state is very short, meaning that the interaction between the adsorbate localized level and the metal is strong. So it is reasonable to consider that the three-body system (adsorbate + projectile + metal surface) is in fact behaving as two coupled subsystems: projectile and (adsorbate + surface). The presence of the adsorbate then generates a broad structure in the metal state continuum with which the projectile level is interacting. In this picture, to first order, the energy of the projectile level follows the electrostatic prediction and does not exhibit any avoided crossing. The result of the dynamical study can be expressed as the system behaving as if the quasistationary state localized on the adsorbate were absent, i.e., as if it were completely incorporated into the continuum. In this sense, it can be compared with the disappearance of band structure effects in a collision, here it is the structure associated to the quasistationary state that is not seen in the collision.

5 Conclusions

We have summarized some recent theoretical results on the dynamics of the RCT in the course of an atom–surface collision. RCT is a one-electron, energy-conserving transition between a discrete atomic state and the continuum of metallic states. In a static system, a bound state-continuum interaction results in a finite lifetime of the discrete state and usually in an exponential decrease of the discrete state population with time. However, in the course of a collision, the dynamics of the RCT can be deeply modified by the existence of nonadiabatic transitions that qualitatively alter the RCT features. The use of a wave packet propagation approach allows the direct treatment of the RCT dynamics via the study of the time evolution of an electron in a time-dependent potential. This allows to characterize the main features of the dynamics of the atom–metal electron transfer.

On a free-electron metal, a quantitative account of the dynamics of the RCT in the course of a collision can be obtained with an adiabatic approximation that describes the atomic state population evolution by a classical rate equation, in which the transition rates are equal to those obtained in a static situation (fixed atom–surface distance).

The situation is quite different if structures are present in the continuum (these can come from the electronic band structure of the metal or from local perturbations induced by adsorbates on the surface). In the static situation, structures in the metal continuum such as a surface projected band gap or states localized on adsorbates can efficiently modify the RCT rate; as a striking example, Cu(111) projected band gap partially blocks the RCT in the Cs/Cu(111) system. However, these modifications in the static RCT characteristics do not always survive in a dynamical context. For fast enough collisions, the specificity of the metal surface disappears and the RCT dynamics is practically identical to that on a free-electron metal. This feature is attributed to a short time effect: if the collision is fast, the electron active in the RCT does not have time to probe in detail the target electronic structure and electron transfer has the same characteristics as on a structure-less metal.

Other types of systems also revealed strong nonadiabatic transitions that qualitatively modify the dynamics of atom–metal surface transitions. One can further mention two examples:

1. In the case of Cs adsorbates on Cu(111), the long-lived quasistationary state localized on the adsorbate almost behaves as a true bound state when a projectile hits the Cs adsorbate. The three-body system (projectile–adsorbate–metal) can then be considered as partly decoupled in two subsystems (projectile–adsorbate) and (metal). As a consequence, electron transfer in this system resembles much an atom–atom charge transfer and in particular, it loses its irreversible character allowing for Stuckelberg oscillations due to back and forth transitions between the adsorbate and the projectile to appear [84].
2. In the case of a thin metal film as the target, quantization of the electron motion in the direction perpendicular to the film surface results in static RCT rates quite different from those on a semi-infinite metal; in particular, the RCT rate exhibits a sharp saw-tooth behavior as a function of the atom–surface distance [85–88]. However, in a collision context, for fast enough collisions, the electron does not have enough time to travel back and forth between the two film surfaces, i.e., to feel the finite film thickness and the RCT dynamics is identical to that on an a semi-infinite metal [88].

Finally, one can stress that an adiabatic approximation can be very tempting for the treatment of charge transfer in atom–surface collisions. Discrete state-continuum transitions can be very heavy to treat exactly in a collisional context and reducing the effect of the continuum on the discrete state to a lifetime or to a local complex potential is a very appealing approximation. However, the above examples, all pertaining to the case of structured continua, show that nonadiabatic effects already appear at moderate collision velocities or even for all velocities in certain cases, making the adiabatic approximation inoperative in these systems. The breakdown of the adiabatic approximation is not uncommon in molecular reactive processes where, often, interactions between several potential energy surfaces strongly influence the dynamics. In the present case, we showed that a similar situation arises with electronic continua.

References

1. J. Ishikawa, in *Handbook of ion sources*, Ed. B. Wolf, (CRC Press, Boca Raton, 1995) p. 289
2. E. Taglauer in *Surface Analysis- The principal techniques*, Ed. J.C. Vickerman (Wiley, Chichester, 1997) p. 457
3. *Secondary Ion Mass Spectrometry*, Eds. G. Gillen, R. Lareau, J. Bennett and F. Stevie (Wiley, NY, 1998)
4. H. Niehus, W. Heiland and E. Taglauer, Surf. Sci. Rep. 17 (1993) 213
5. *Low Energy Ion-Surface Interactions*, Ed. J.W. Rabalais (Wiley, NY, 1994)
6. H. Conrad, G. Ertl, J. Köppers, S.W. Wang, K. Gerard and H. Haberland Phys. Rev. Lett. 42 (1979) 1082
7. F. Boszo, J.T. Yates, J. Arias, H. Metiu and R.M. Martin, J. Chem. Phys. 78 (1983) 4256
8. V. Kempter Comm. At. Mol. Phys. 34 (1998) 11
9. R.E. Palmer Prog. Surf. Sci. 41 (1992) 51
10. H. Guo, P. Saalfranck and T. Seidemann Prog. Surf. Sci. 62 (1999) 239
11. J. Los and J.J.C. Geerlings, Phys. Rep. 190 (1990)
12. *Low energy Ion-Surface Interactions*, Ed. J.W. Rabalais (Wiley, NY 1994)
13. J. Burgdörfer in *Review of fundamental processes and applications of atoms and ions* Ed. C.D. Lin, World Scientific, Singapore (1993) 517
14. H. Winter, Phys. Rep. 367 (2002) 387
15. H.D. Hagstrum Phys. Rev. 96 (1954) 325, 336
16. N. Lorente and R. Monreal Phys. Rev. B53 (1996) 9622; Surf. Sci. 370 (1997) 324
17. R.A. Baragiola and C.A. Dukes Phys. Rev. Lett. 76 (1996) 2547
18. M.A. Cazalilla, N. Lorente, R. Diez-Muino, J.P. Gauyacq, D. Teillet-Billy and P.M. Echenique Phys. Rev. B 58 (1998) 13991
19. C. Cohen-Tannoudji, B. Diu and F. Laloë *Mécanique quantique* (Hermann, Paris, 1977)
20. J. Burgdorfer, E. Kupfer and H. Gabriel Phys. Rev. A 35 (1987) 4963
21. K.W. Sulston, A.T. Amos and S.G. Davison Phys. Rev. B 37 (1988) 9121
22. D.C. Langreth and P. Nordlander Phys. Rev. B 43 (1991) 2541
23. E.C. Goldberg and F. Flores Phys. Rev. B 45 (1992) 8657
24. J.B. Marston, D.R. Andersson, E.R. Behringer, B.H. Cooper, C.A. Di Rubio, G.A. Kimmel and C. Richardson Phys. Rev. B 48 (1993) 7809
25. H. Shao, P. Nordlander and D.C. Langreth Phys. Rev. Lett. 77 (1996) 948, Phys. Rev. B 52 (1995) 2988
26. E.A. Garcia, E.C. Goldberg and M.C.G. Passeggi Surf. Sci. 325 (1995) 311
27. E.R. Behringer, D.R. Andersson, B.H. Cooper and J.B. Marston Phys. Rev. B 54 (1996) 14765, 14780
28. J. Merino, N. Lorente, M.Y. Gusev, F. Flores, M. Maazouz, L. Guillemot and V.A. Esaulov Phys. Rev. B 57 (1998) 1947

29. G. Katz, Y. Zeiri and R. Kosloff Surf. Sci. 425 (1999) 1
30. C.M. Dutta and P. Nordlander Prog. Surf. Sci. 67 (2001) 155
31. A.G. Borisov, A.K. Kazansky and J.P. Gauyacq Phys. Rev. B 59 (1999) 10935
32. A.G. Borisov, J.P. Gauyacq and S.V. Shabanov Surf. Sci. 487 (2001) 243
33. J.N. Bardsley, in Case studies in At. Phys. 4 (1974) 299
34. L. Kleinman and D.M. Bylander, Phys. Rev. Lett. 48 (1982) 1425
35. P.J. Jennings, R.O. Jones and M. Weinert, Phys. Rev. B 37 (1988) 6113
36. E.V. Chulkov, V.M. Silkin and P.M. Echenique Surf. Sci. 437 (1999) 330
37. M.D. Fleit, J.A. Fleck J. Chem. Phys. 78 (1983) 301
38. R. Kosloff J. Phys. Chem. 92, (1988) 2087
39. P. Nordlander and J.C. Tully Phys. Rev. Lett. 61 (1988) 990
40. D. Teillet-Billy and J.P. Gauyacq, Surf. Sci. 239 (1990) 343
41. F. Martin and M.F. Politis Surf. Sci. 356 (1996) 247
42. S.A. Deutscher, X. Yang and J. Burgdorfer Phys. Rev. A 55 (1997) 466
43. P. Kürpick, U. Thumm and U. Wille Nucl. Inst. Methods B 125 (1997) 273
44. B. Bahrim and U. Thumm Surf. Sci. 451 (2000) 1
45. V.A. Ermoshin and A.K. Kazansky Phys. Lett. A 218 (1996) 99
46. H. Chakraborty, T. Niederhausen and U. Thumm Phys. Rev. A 69 (2004) 052901
47. R. Zimny Surf. Sci. 233 (1990) 333
48. J.J.C. Geerlings, J. Los, J.P. Gauyacq and N.M. Temme Surf. Sci. 172 (1986) 257
49. R. Brako and D.M. Newns Surf. Sci. 108 (1978) 277
50. A.G. Borisov, D. Teillet-Billy and J.P. Gauyacq Phys. Rev. Lett. 68 (1992) 2842
51. A.G. Borisov, D. Teillet-Billy, J.P. Gauyacq, H. Winter and G. Dierkes Phys. Rev. B 54 (1996) 17166
52. S.B. Hill, C.B. Haich, Z. Zhou, P. Nordlander and F.B. Dunning Phys. Rev. Lett. 85 (2000) 5444
53. M. Maazouz, A.G. Borisov, V.A. Esaulov, J.P. Gauyacq, L. Guillemot, S. Lacombe and D. Teillet-Billy Phys. Rev. B 55 (1997) 13869
54. J.N.M. Van Wunnick, R. Brako, K. Makoshi and D.M. Newns Surf. Sci. 126 (1983) 618
55. J. Sjakste Thèse de doctorat, Orsay, 2004
56. P.M. Echenique and J.B. Pendry J. Phys. C 11 (1978) 2065
57. M.C. Desjonquères and D. Spanjaard, Concepts in Surface Science, Springer-Verlag Series in Surface Science 40 (1993)
58. J.P. Gauyacq, A.G. Borisov, M. Bauer, in *Time-resolved Photoemission from Solids: Principles and Applications* Eds. M. Aeschlimann and M. Wolf (Springer, Berlin Heidelberg New York), 2005
59. R.W. Gurney Phys. Rev. 47 (1935) 479
60. M. Scheffler, Ch. Doste, A. Fleszar, F. Maca, G. Wachutka and G. Barzel Phys. B 172 (1991) 143
61. M. Scheffler and C. Stampfl in *Electronic Structure, Handbook of Surface Science*, Eds. K. Horn and M. Scheffler (Elsevier Science, Amsterdam, 2000), Vol. 2, p. 285
62. H. Petek and S. Ogawa Prog. Surf. Sci. 56 (1997) 239
63. A.G. Borisov, A.K. Kazansky and J.P. Gauyacq Surf. Sci. 430 (1999) 165
64. J.P. Gauyacq, A.G. Borisov, G. Raşeev and A.K. Kazansky Faraday Discuss. 117 (2000) 15
65. A.G. Borisov, J.P. Gauyacq, A.K. Kazansky, E.V. Chulkov, V.M. Silkin and P.M. Echenique, Phys. Rev. Lett. 86 (2001) 488

66. M. Bauer, S. Pawlik and M. Aeschlimann Phys. Rev. B 55 (1997) 10040
67. S. Ogawa, H. Nagano and H. Petek, Phys. Rev. Lett. 82 (1999) 1931
68. M. Bauer, S. Pawlik and M. Aeschlimann Phys. Rev. B 60 (1999) 5016
69. H. Petek, M.J. Weida, H. Nagano and S. Ogawa Science 288 (2000) 1402
70. M. Bauer, M. Wessendorf, D. Hoffmann, C. Wiemann, A. Mönnich and M. Aeschlimann Appl. Phys. A 80, 987 (2005)
71. J.P. Gauyacq and A.K. Kazansky Phys. Rev. B 72 (2005) 045418
72. H. Petek, N. Nagano, M.J. Weida and S. Ogawa J. Phys. Chem. B 105 (2001) 6767
73. A.G. Borisov, J.P. Gauyacq, E.V. Chulkov, V.M. Silkin and P.M. Echenique Phys. Rev. B 65 (2002) 235434
74. T. Hecht, H. Winter, A.G. Borisov, J.P. Gauyacq and A.K. Kazansky Phys. Rev. Lett. 84 (2000) 251
75. T. Hecht, H. Winter, A.G. Borisov, J.P. Gauyacq and A.K. Kazansky Faraday Discuss. 117 (2000) 27
76. A.G. Borisov, A.K. Kazansky and J.P. Gauyacq Phys. Rev. Lett. 80 (1998) 1996
77. L. Guillemot and V.A. Esaulov Phys. Rev. Lett. 82 (1999) 4552
78. A.R. Canario, A.G. Borisov, J.P. Gauyacq and V.A. Esaulov Phys. Rev. B 71 (2005) 0121401(R)
79. A.G. Borisov, A.K. Kazansky and J.P. Gauyacq Phys. Rev. B 65 (2002) 205414
80. F.E. Olsson, M. Persson, A.G. Borisov, J.P. Gauyacq, J. Lagoute and S. Fölsch Phys. Rev. Lett. 93 (2004) 206803
81. L. Limot, E. Pehlke, J. Kröger and R. Berndt Phys. Rev. Lett. 94 (2005) 036805
82. J.P. Gauyacq and A.G. Borisov, J. Phys. C 10 (1998) 6585
83. J. Sjakste, A.G. Borisov and J.P. Gauyacq Nucl. Inst. Methods B 203 (2003) 49
84. J. Sjakste, A.G. Borisov and J.P. Gauyacq Phys. Rev. Lett. 92 (2004) 156101
85. A.G. Borisov and H. Winter, Z. Phys. D 37 (1996) 253
86. A.G. Borisov and H. Winter, Nucl. Inst. Methods B 115 (1996) 142
87. U. Thumm, P. Kürpick and U. Wille Phys. Rev. B 61 (2000) 3067
88. E. Yu. Usman, I.F. Urazgil'din, A.G. Borisov and J.P. Gauyacq Phys. Rev. B 64 (2001) 205405

I.2 From Multidimensional Dynamics to Dissipative Phenomena

Nonadiabatic Multimode Dynamics at Symmetry-Allowed Conical Intersections

H. Köppel

Summary. Conical intersections of potential energy surfaces have emerged as paradigms for nonadiabatic excited state processes and correspondingly complex nuclear dynamics. In this contribution a particular quantum dynamical approach is surveyed which has been developed and used in our groups over the years to describe molecular electronic spectra and ultrafast internal conversion processes in such situations. Particular attention is paid to the existence of a symmetry element in many cases; this allows one to formally diagonalize the electronic Hamiltonian, although at the expense of introducing a nonlocal potential. This can be viewed as an operator formulation of a block-diagonal structure of the secular matrix for the different irreducible representations existing in these cases. An application of the formalism is given to singlet excited states of furan.

1 Introduction

Vibronic coupling, i.e., the interaction of different electronic states through the nuclear motion, is of paramount importance for spectroscopy, collision processes, photochemistry, etc., and quite general for electronically excited state processes of even small polyatomic molecules. One of its most important consequences is the violation of the Born–Oppenheimer, or adiabatic, approximation [1] whereby the nuclear motion no longer proceeds on a single potential energy surface but rather on several surfaces simultaneously. Nonadiabatic coupling effects are of singular strength at degeneracies of these surfaces, in particular at conical intersections, which have emerged in recent years as paradigms of nonadiabatic excited-state dynamics in quite different fields [2,3].

In our groups (Heidelberg and Munich) we have developed over the past decades simple, but also efficient and rather flexible methods to deal with the nuclear dynamics in such systems, based on the so-called multimode vibronic coupling approach [4–7]. This approach relies on the well-established concept of diabatic electronic states [8–11], where the singularities of the adiabatic electronic wavefunctions at the intersection are removed by a suitable orthogonal transformation and the off-diagonal, or coupling, elements arise from the potential rather than kinetic energy (at least to a sufficiently good approximation). The potential coupling terms can be expanded in a Taylor series, and the truncation after the first (or second) order gives the linear (or quadratic) multimode vibronic coupling scheme. The resulting model potential energy surfaces turn out to be sufficiently flexible to cover a variety of interesting phenomena and be applicable to different molecular systems [4–7]. To generalize the approach, it has been suggested more recently that it be applied *only* to the adiabatic-to-diabatic (ATD) mixing angle [12–14]. This leads directly to the concept of regularized diabatic states [12–14], see also below. The resulting enormous increase in flexibility renders this concept applicable also to photochemical problems, at least in principle. To present both approaches in comparison, and give a representative current application, is a main objective of the present article.

Most of our applications of the above formalism to date are characterized by the existence of a symmetry element by which the interacting electronic states differ. This implies that the "original" symmetry has to be lowered in order for an interaction to become possible: there is a high-symmetry subspace in which the potential energy surfaces cross freely, and the associated conical intersection is thus termed "symmetry-allowed." In accord with this symmetry it is only nontotally symmetric modes that couple the states to first order, while totally symmetric modes provide for first-order intrastate couplings [4]. The vibronic secular matrix then block-diagonalizes according to the different irreducible representations of the interacting states [4]. In the present contribution we draw particular attention to this fact and show that it can be cast in an elegant operator formulation. This is basically independent of the aforementioned approximations and only a consequence of symmetry. The treatment formally diagonalizes the electronic potential energy matrix (in the diabatic representation), although at the expense of introducing nonlocal potential energy terms. The symmetry-adapted treatment is presented in its generality (for a two-state problem) in the next-but-one Sect. 3, following an exposition of the multimode vibronic coupling approach in the next Sect. 2. Important aspects of the numerical implementation are presented in Sect. 4, while an illustrative example (singlet excited states of furan) follows in Sect. 5. Concluding remarks as well as a short summary are provided in Sect. 6.

2 Vibronic Hamiltonians

2.1 The Linear Vibronic Coupling Approach

Throughout this work we utilize the concept of a diabatic electronic basis [8, 9, 11], where the interaction between the different states is described by a potential energy matrix $\mathbf{W}$, containing off-diagonal elements, while the nuclear kinetic energy operator T_N is taken to be diagonal to a sufficiently good approximation. The pertinent Hamiltonian $\mathcal{H}$ can then generally be written as

$$\mathcal{H} = T_N \mathbf{1} + \mathbf{W}, \tag{1}$$

where $\mathbf{1}$ denotes the unit matrix in electronic function space.

Owing to the smoothness of the diabatic states, the matrix elements of $\mathbf{W}$ can be expanded in a Taylor series in the nuclear displacement coordinates $\mathbf{Q} = (Q_1, Q_2, \ldots, Q_f)$. Taking the expansion to be around the origin $\mathbf{Q} = \mathbf{0}$ we can write these matrix elements as follows [4]:

$$W_{nn} = V_0(\mathbf{Q}) + E_n + \sum_i \kappa_{\mathrm{i}}^{(n)} Q_i + \sum_{i,j} \gamma_{ij}^{(n)} Q_i Q_j + \ldots, \tag{2a}$$

$$W_{nm} = \sum_{\mathrm{i}} \lambda_i^{(nm)} Q_i + \sum_{ij} \eta_{ij}^{(nm)} Q_i Q_j + \ldots \qquad (n \neq m). \tag{2b}$$

In (2), $V_0(\mathbf{Q})$ represents some "unperturbed" potential energy term which is often identified with that of the electronic ground state and treated in the harmonic approximation. The E_n denote vertical excitation (or ionization) energies, the quantities $\kappa_{\mathrm{i}}^{(n)}$ and $\lambda_{\mathrm{i}}^{(nm)}$ are first-order coupling constants (intra- and interstate, respectively) while the parameters $\gamma_{ij}^{(n)}$ and $\eta_{ij}^{(nm)}$ represent second-order coupling constants. In the linear vibronic coupling (LVC) approach the latter terms in (2) are neglected.

In the frequent case of different spatial symmetries of the interacting states there is a useful symmetry selection rule limiting the number of relevant vibrational modes. Denoting the irreducible representations of the electronic states n and m by Γ_n and Γ_m, respectively, and that of the vibrational mode by Γ_Q, we have [4]

$$\Gamma_n \times \Gamma_Q \times \Gamma_m \supset \Gamma_A, \tag{3}$$

where Γ_A denotes the totally symmetric representation of the point group in question. For two nondegenerate states of different spatial symmetry (3) implies that – in first order – only totally symmetric modes appear in the diagonal elements W_{nn}, while suitable nontotally symmetric modes enter the off-diagonal elements W_{nm} $(n \neq m)$ of (2). This will also apply to all subsequent examples mentioned below.

The LVC approach embodied in (2) has been applied for a long time to analyze the vibrational structure of electronic spectra and time-dependent (electronic and vibrational) dynamics of vibronically coupled systems (see [4–7] and references therein). Strong nonadiabatic coupling effects associated with conical intersections of potential energy surfaces could be unequivocally established in this way. We refer to this literature for a survey of these examples and also for a further discussion of the meaning and implications of the various terms in the Hamiltonian (2).

Over time, the strict LVC approach has been extended by including selected, or even all, second-order terms in (2). In some cases their effect on the spectrum turned out to be surprisingly large [15]. Fitting the LVC spectra to a result obtained with the inclusion of second-order terms implies effective LVC coupling constants which incorporate some of the higher-order effects. The use of second-order coupling terms may reduce the need for parameter adjustment when using ab initio calculated coupling constants to reproduce an experimental spectrum [16]. The ab initio determination of the coupling constants is relatively easy (with or without second-order terms) since no multidimensional grid and only a small number of energy points per mode are required owing to the model assumptions underlying (2).

Finally we point out the close relation of the general Hamiltonian (2) and model Hamiltonians frequently used in Jahn–Teller (JT) theory [17,18]. There, an analogous Taylor series is used, but many interrelations between the various coupling terms exist due to symmetry. Formally, these JT Hamiltonians are thus a special case of (2) and are recovered by imposing these restrictions. They have been successfully used to analyze and interpret even high-resolution molecular JT spectra [19]. Similar applications and extensions have been made to cover also couplings to nearby nondegenerate electronic states [18].

2.2 The Concept of Regularized Diabatic States

The concept of regularized diabatic states [12, 13] can be understood as a generalization of the "conventional" LVC approach (as outlined earlier) by applying it to the ATD mixing angle only. For illustrative purposes consider the case of two potential energy surfaces $V_1(\mathbf{Q})$ and $V_2(\mathbf{Q})$ intersecting at a point $Q_{\mathrm{g}} = Q_{\mathrm{u}} = 0$ in two-dimensional nuclear coordinate space. Let their behavior near the origin be described by $E_0 + \kappa Q_{\mathrm{g}} \pm \delta\kappa Q_{\mathrm{g}}$ along a symmetry-preserving coordinate Q_{g} (no interaction between the states) and by $E_0 \pm \lambda Q_{\mathrm{u}}$ along a symmetry lowering coordinate Q_{u} (inducing an interaction). The corresponding LVC Hamiltonian (2) can be written as follows:

$$\mathcal{H} = (T_N + V_0 + E_0 + \kappa Q_{\mathrm{g}})\mathbf{1} + \begin{pmatrix} \delta\kappa\, Q_{\mathrm{g}} & \lambda Q_{\mathrm{u}} \\ \lambda Q_{\mathrm{u}} & -\delta\kappa\, Q_{\mathrm{g}} \end{pmatrix} \tag{4a}$$

$$= H_0\mathbf{1} + \mathbf{W}^{(1)}. \tag{4b}$$

The corresponding ATD angle $\alpha(\mathbf{Q})$ is defined through the eigenvector relation

$$\mathbf{S}^{\dagger}(\mathcal{H} - T_N\mathbf{1})\mathbf{S} = \begin{pmatrix} V_1^{(1)} & 0 \\ 0 & V_2^{(1)} \end{pmatrix}, \tag{5a}$$

$$\mathbf{S} = \begin{pmatrix} \cos\alpha & \sin\alpha \\ -\sin\alpha & \cos\alpha \end{pmatrix}, \tag{5b}$$

where $V_1^{(1)}$ and $V_2^{(1)}$ are the adiabatic potential energy surfaces in first order, inherent to the LVC model Hamiltonian (4). For convenience the coordinate-dependence of the various quantities is suppressed in (5) and also below. The concept of regularized diabatic states consists in applying the LVC mixing angle α of (5) to the *general* adiabatic potential surfaces V_1 and V_2. After some elementary algebra this leads to the following result [12, 13]:

$$\mathcal{H}_{\text{reg}} = \left(T_N + \frac{V_1 + V_2}{2}\right)\mathbf{1} + \frac{V_1 - V_2}{V_1^{(1)} - V_2^{(1)}} \begin{pmatrix} \delta\kappa\, Q_{\text{g}} & \lambda Q_{\text{u}} \\ \lambda Q_{\text{u}} & -\delta\kappa\, Q_{\text{g}} \end{pmatrix}. \tag{6}$$

This expression is seen to reduce to the usual LVC result close to the intersection (when $V_1 \to V_1^{(1)}$ and $V_2 \to V_2^{(1)}$). On the other hand, for configurations far away from it (when the adiabatic approximation is valid), the general surfaces V_1 and V_2 are recovered, because the eigenvalues of the coupling matrix in (6) cancel the denominator of the preceding ratio (of potential energy differences). It thus interpolates “smoothly” between the two limits.

A theoretical justification of this procedure is obtained by noting that the linear terms of (4) determine the *singular* part of the full derivative couplings [12, 13] (corresponding to the full surfaces V_1 and V_2) near the intersection at $Q_g = Q_u = 0$. Thus, within the concept of regularized diabatic states, (6), the singular derivative couplings are eliminated, which motivates the nomenclature adopted. Note that all derivative couplings cannot be eliminated in the general case [10, 11, 20]. Thus, the concept of regularized diabatic states constitutes a natural extension of the usual adiabatic, or Born–Oppenheimer, approximation to intersecting electronic surfaces: all the singular couplings are eliminated and the others are neglected (group Born–Oppenheimer approximation).

The above scheme has been generalized to cover seams of symmetry-allowed conical intersections, where likewise all information needed for the construction is obtained from the potential energy surfaces alone [13]. The singular derivative couplings can thus be removed for the whole symmetry-allowed portion of the seam [13]. The same has been achieved recently for an accidental intersection (i.e., without any symmetry) in a two-dimensional nuclear coordinate space [14]. In the most general case, however, more information, e.g., from the derivative couplings, is needed [21]. We mention in passing that virtually all other schemes, proposed in the literature for constructing approximately diabatic states, rely on information on the adiabatic electronic wavefunctions [11]. Finally we point out that the concept of regularized diabatic states has been tested numerically for a number of different symmetry-allowed [13] and Jahn–Teller [12], as well as accidental [14], conical intersections, and very good agreement on dynamical quantities with appropriate reference data has been obtained. The neglect of nonsingular coupling terms apparently constitutes a very good approximation in these cases, as is also expected from general reasoning [20, 22]. Applications of the scheme have been reported for H_3 [23, 24] and NO_2 [25], and are ongoing for C_2H_2 [21, 26].

3 Symmetry-Adapted Formulation of the Hamiltonian

We now turn to the question of a symmetry-adapted formulation of the vibronic coupling Hamiltonian. A formulation will be achieved which covers not only the above, but even more general cases, namely that of any Abelian symmetry with a single type of relevant nontotally symmetric coupling mode. It is inspired by an earlier related development of Fulton and Gouterman [27]. We start from the general diabatic matrix representation of the Hamiltonian, (1). For the case of two interacting electronic states, on which we focus here, the potential energy matrix can be written explicitly (in an obvious notation) as follows:

$$\mathcal{H} = (T_N + \bar{W})\mathbf{1} + \begin{pmatrix} \Delta W & W_{12} \\ W_{12} & -\Delta W \end{pmatrix} . \tag{7}$$

Consider now the case of a symmetry element by which the interacting states differ, one being of g (gerade) the other of u (ungerade) symmetry. Consequently, the off-diagonal element W_{12} must also be antisymmetric with respect to that symmetry element in order to allow for an interaction between the states (whereas the diagonal elements will be symmetric). Let us denote a representative symmetric displacement coordinate by Q_{g}, the antisymmetric one by Q_{u} (without loss of generality these are taken to be dimensionless normal coordinates of some suitable harmonic oscillator). Then the matrix elements of $\mathbf{W}$ can be written as the following Taylor series in Q_{u}, guaranteeing the aforementioned symmetry requirements:

$$\begin{aligned} \bar{\mathrm{W}} &= \sum \bar{w}^{(m)}(Q_{\mathrm{g}})\, Q_{\mathrm{u}}^{2m} , \\ \Delta W &= \sum \Delta w^{(m)}(Q_{\mathrm{g}})\, Q_{\mathrm{u}}^{2m} , \\ W_{12} &= \sum w_{12}^{(m)}(Q_{\mathrm{g}})\, Q_{\mathrm{u}}^{2m+1} . \end{aligned} \tag{8}$$

The various expansion coefficients may depend on the symmetric coordinate Q_{g}. This formulation is immediately extended to several symmetric modes, while remarks for the case of several antisymmetric modes are provided below.

Next we introduce the following operator, acting in the space of the antisymmetric normal coordinate Q_{u}:

$$G = \mathrm{e}^{\mathrm{i}\pi b^{\dagger} b} \tag{9}$$

with the usual creation and annihilation operators

$$\begin{aligned} b &= \frac{1}{\sqrt{2}} \left(Q_{\mathrm{u}} + \frac{\partial}{\partial Q_{\mathrm{u}}} \right) , \\ b^{\dagger} &= \frac{1}{\sqrt{2}} \left(Q_{\mathrm{u}} - \frac{\partial}{\partial Q_{\mathrm{u}}} \right) , \end{aligned} \tag{10}$$

the subscript u being suppressed at the l.h.s. for simplicity of notation.

The operator G is easily verified to be hermitean as well as unitary, $G^2 = 1$. In the Hilbert space of harmonic oscillator eigenfunctions $|n\rangle$ its matrix representation is diagonal with elements $+1$ and -1 according to whether the eigenfunctions are symmetric or antisymmetric, respectively, under the reflection operation $Q_\mathrm{u} \mapsto -Q_\mathrm{u}$:

$$\langle n|G|n'\rangle = \delta_{nn'}(-)^n \tag{11}$$

Furthermore, also the following relations are easily found to hold:

$$\begin{aligned} GQ_\mathrm{u}G &= -Q_\mathrm{u}, \\ GP_\mathrm{u}G &= -P_\mathrm{u} \end{aligned} \tag{12}$$

with P_u being the momentum conjugate to Q_u. The above should make it apparent that the operator G represents nothing but the symmetry operation $Q_\mathrm{u} \mapsto -Q_\mathrm{u}$ (say σ, to have a reflection in mind) in the vibrational space of the mode Q_u. From the expansions of (8) also the following relations become clear immediately:

$$\begin{aligned} G\bar{W}G &= \bar{W}, \\ G\Delta WG &= \Delta W, \\ GW_{12}G &= -W_{12}. \end{aligned} \tag{13}$$

After these preparatory steps we now introduce our basic transformation U to achieve the desired symmetry-adapted representation:

$$U = \frac{1}{2}\begin{pmatrix} 1+G & 1-G \\ 1-G & 1+G \end{pmatrix}. \tag{14}$$

The matrix operator U is again found to be hermitean as well as unitary. Utilizing the above relations (12) and (13), the transformed Hamiltonian matrix can be re-written, after some elementary manipulations, as follows:

$$\begin{aligned} U\mathcal{H}U &= (T_N + \bar{W})\mathbf{1} + \frac{1}{4}\begin{pmatrix} 1+G & 1-G \\ 1-G & 1+G \end{pmatrix}\begin{pmatrix} \Delta W & W_{12} \\ W_{12} & -\Delta W \end{pmatrix}\begin{pmatrix} 1+G & 1-G \\ 1-G & 1+G \end{pmatrix} \\ &= (T_N + \bar{W})\mathbf{1} + \begin{pmatrix} W_{12} + G\Delta W & 0 \\ 0 & W_{12} - G\Delta W \end{pmatrix}. \end{aligned} \tag{15}$$

Equation (15) looks surprising at first glance, since a formal diagonalization of the Hamiltonian matrix has been achieved: the coupling elements W_{12} appear now in the diagonal of the matrix in (15), while the diagonal elements ΔW are multiplied with the reflection operator G (with which they commute, see (13)).

The result of (15) represents an operator formulation of a symmetry-adapted block-diagonalization of the matrix Hamiltonian (7). While this will also become apparent below from the matrix representation of $\mathcal{H}$ in the vibrational space of the mode Q_u, we demonstrate this here more formally from an analysis of the symmetry operator itself. In the original basis of (1) the symmetry operator $\mathcal{S}$ extends the operator G as follows:

$$\mathcal{S} = \begin{pmatrix} G & 0 \\ 0 & -G \end{pmatrix} . \tag{16}$$

Indeed, $\mathcal{S}$ can be shown to commute with $\mathcal{H}$ and $\mathbf{W}$ of (1), and to have opposite eigenvalues for the two underlying electronic states with even and odd quanta of the mode Q_{u}, respectively, (owing to the different signs of G in the two diagonal elements of $\mathcal{S}$). In the transformed basis corresponding to (14) it is straightforward to show

$$\begin{aligned} USU &= \frac{1}{4}\begin{pmatrix} 1+G & 1-G \\ 1-G & 1+G \end{pmatrix}\begin{pmatrix} G & 0 \\ 0 & -G \end{pmatrix}\begin{pmatrix} 1+G & 1-G \\ 1-G & 1+G \end{pmatrix} \\ &= \begin{pmatrix} 1 & 0 \\ 0 & -1 \end{pmatrix} . \end{aligned} \tag{17}$$

This shows explicitly that the two diagonal elements of the transformed Hamiltonian of (15) correspond to the different eigenvalues $+1$ and -1 of the transformed symmetry operator $\mathcal{S}$.

Equation (15) and the subsequent developments represent the main result of this section and the major methodological result of this paper. It generalizes and extends the earlier results of Fulton and Gouterman [27] by a more transparent formulation (giving the unitary transformation (14) instead of a projection operator formalism, and also explicit expressions such as (9) for the operator G). As stated above, further generalization to several totally symmetric modes (g-modes) is trivial, since it amounts merely to a multidimensional argument of the various functions $w^{(m)}$ in (8) without affecting any of the (anti)commutation relations leading to (15). Somewhat less evident, but also simple, is the generalization to several coupling modes (u-modes) $Q_{\mathrm{u},j} (j = 1, 2, ..)$ according to the substitution

$$W_{12} \rightarrow \sum w_{12}^{(j)} (Q_{\mathrm{g}})\, Q_{\mathrm{u},j} , \tag{18}$$

where I have confined myself to the linear coupling terms for simplicity of notation. This situation is dealt with by the analogous substitution

$$G \rightarrow \prod \mathrm{e}^{\mathrm{i}\pi b_j^\dagger b_j} , \tag{19}$$

where the creation and annihilation operators $b_j^\dagger$ and b_j refer to the mode $Q_{\mathrm{u},j}$, in an obvious notation. It can be verified quite easily that the key relations (13) still hold for the substituted quantities of (18, 19), the latter still being unitary and Hermitean. Thus, all conditions are met to recover (15) also in this more general case. The same can be shown to hold when higher-order terms w.r.t. the u-modes are included in (18): since the sum of all exponents of these modes has to be odd due to symmetry, the same (anti)commutation relations remain valid also there, when using the substituted operator according to (19), and the block-diagonalized Hamiltonian (15) is recovered by the analogous transformation as in (14).

Finally, the most general Abelian case would be to allow for different types of nontotally symmetric modes such that their product, or multiple products thereof, have the correct symmetry behavior. While similar considerations as above are possible also there [28], a systematic exposition of this situation is beyond the scope of this work. On the other hand, for a single (type of) coupling mode and a special case of the LVC model, similar developments as above have already been worked out for the non-Abelian case, that is, for point groups and a vibronic coupling problem with degenerate electronic states and vibrational modes [29].

4 Numerical Implementation

4.1 General

Numerical applications of the above formalism have focused to date on photoinduced dynamics, where the vibronic coupling is operative in the final electronic-state manifold reached by the photoexcitation or -detachment process [5–7]. The initial electronic state is typically not part of the interacting system, and used to define the reference potential V_0 in (2). Together with the kinetic energy operator T_N this is often described in the harmonic approximation. The spectral intensity distribution of the photoexcitation spectrum is treated by Fermi's golden rule, either in the time-independent or in the time-dependent framework. Since both approaches are well established in the literature, the formalism is not repeated here. Suffice it to say that within the time-independent framework we employ the Lanczos algorithm, which is very well suited for our purposes since it converges fastest on the quantities of interest, that is, individual vibronic lines for low energies and the spectral envelope for medium and high vibronic energies [6]. In effect, the time-consuming step in either of the two approaches consists in the matrix–vector multiplication.

Two different variants of representing the state vector can be distinguished. In the LVC approach a basis set expansion is usually employed, relying on multidimensional harmonic oscillator wave functions as defined by the reference potential V_0. Within the concept of regularized diabatic states we are dealing with general functional forms of the potentials and coupling elements (see (6)) which can be conveniently treated by grid (FFT and DVR) methods. The integration of the time-dependent Schrödinger equation is usually achieved through standard methods (like the short iterative Lanczos scheme [30]). Within the LVC treatment, for genuinely multimode problems we are also relying on the multiconfiguration time-dependent Hartree (MCTDH) method [31,32]. This wave packet propagation method uses optimized time-dependent one-particle basis functions and thus arrives at a very compact representation of the state vector, at the expense of more complicated equations of motion. Owing to the structural simplicity of the LVC Hamiltonian matrix, (2), it is nevertheless particularly efficient for this purpose.

4.2 Symmetry Adaptation

Of special interest for the present work is the numerical implementaton of the symmetry-adapted formulation, (15), of the vibronic Hamiltonian. This is quite straighforward within the basis set approach, where the operator G is given by a diagonal matrix, with elements $+1$ and -1, see (11). Thus, the sign change of G with the vibrational quantum number n of the u-mode implies a switching of the diagonal elements $\bar{W} \pm G\Delta W$ between the interacting states: n odd amounts to (say) the first state, while n even amounts to the second state. This is precisely the result of a (numerical) block diagonalization of the vibronic secular matrix, see Fig. 1 of [4]. Two submatrices of this type result, with the notion "first" and "second" state being interchanged in the two matrices. Another evidence lies in the matrix representation of the operator U, (14). This is just the unit matrix for n even and the (1,2) transposition matrix for n odd, according to

$$\langle 2m|U|2m\rangle = \begin{pmatrix} 1 & 0 \\ 0 & 1 \end{pmatrix}, \tag{20}$$

$$\langle 2m+1|U|2m+1\rangle = \begin{pmatrix} 0 & 1 \\ 1 & 0 \end{pmatrix}. \tag{21}$$

With this somewhat shorthand notation adopted, the vibrational integration is meant to be performed for each (electronic) matrix element of U, (14), separately. The switching expressed by (20, 21) represents the well-known instance of vibronic coupling theory, namely, that even quanta of the (non-totally symmetric) coupling mode in one state combine with odd quanta in the other state to form the vibronic eigenstates, because of the same vibronic symmetry. Both reasonings confirm that the formal diagonalization achieved in (15) amounts to an operator formulation of the symmetry-adapted block diagonalization of the vibronic Hamiltonian.

We now turn to grid methods (see, e.g., [33, 34]). Here, a transformation from a suitable basis set to a grid is employed, on which the position operator Q_u for the nontotally symmetric mode is diagonal. The coordinates at the grid points are obtained, for example, by diagonalizing the position operator (here Q_u) in a suitable basis, such as harmonic oscillator wavefunctions [33, 34]. By transforming the operator G, (11) with the same transformation matrix, but also by direct geometric reasoning, it can be seen that the matrix elements of G on the grid read as

$$\langle Q_n|G|Q_m\rangle = \delta_{Q_m,-Q_n}. \tag{22}$$

This means, as emphasized before, that G represents a reflection operation at the origin $Q_{\mathrm{u}} = 0$, thus connecting only grid points of equal modulus and opposite sign. In the frequent case that the grid points are arranged symmetrically around the origin as in the following diagonal matrix

$$\mathbf{Q} = \begin{pmatrix} -Q_l & .. & 0 & 0 & 0 & 0 & ... & 0 \\ .. & .. & & & & & ... & ... \\ 0 & & -Q_2 & & & & & 0 \\ 0 & & & -Q_1 & & & & 0 \\ 0 & & & & Q_1 & & & 0 \\ 0 & & & & & Q_2 & & 0 \\ .. & .. & & & & & .... & ... \\ 0 & .. & 0 & 0 & 0 & 0 & ... & Q_l \end{pmatrix} \tag{23}$$

the grid representation of G takes an appearance similar to the unit matrix, but with an arrangement of nonzero entries which is orthogonal to the diagonal:

$$\mathbf{G} = \begin{pmatrix} 0 & ... & 0 & 0 & 0 & 0 & ... & 1 \\ ... & ... & & & & & ... & ... \\ 0 & & 0 & & & & 1 & 0 \\ 0 & & & 0 & 1 & & & 0 \\ 0 & & & 1 & 0 & & & 0 \\ 0 & & 1 & & & & 0 & 0 \\ ... & ... & & & & & ... & ... \\ 1 & ... & 0 & 0 & 0 & 0 & ... & 0 \end{pmatrix} . \tag{24}$$

The important point to note is that, due to the Kronecker-δ in (22) there is just one nonzero element of $\mathbf{G}$ per line and column in the grid representation, (24). Therefore, although G is a nonlocal operator, its implementation in grid-based methods does not introduce an extra computational effort compared to the (local) potential itself. Full advantage of the symmetry blocking, leading to half the length of the state vector to be dealt with, can thus be taken care of also there. This is expected to be of considerable help in the treatment of general coupled potential energy surfaces, as appear, for example, within the concept of regularized diabatic states.

5 Application to Singlet-Excited Furan

5.1 Model System and Potential Energy Surfaces

In this section the general concepts are illustrated by an application to singlet excited states of furan in the energy range 5.6–6.8 eV. Furan is a prototype heteroaromatic molecule [35] and has at least five singlet electronic states in this energy range [36–38]. The lowest four are considered in the dynamical treatment simultaneously, along with 13 nonseparable vibrational modes.

This multistate multimode treatment has been developed earlier [39,40]. Here we review some of the key results obtained there and augment them by a new representation of the potential energy surfaces as well as new population dynamics.

The calculations rely on the LVC approach, (2) and thus constitute a model type of treatment: The underlying potential energy surfaces are those of the LVC model, and the singular derivative couplings are eliminated only within the accuracy to which these surfaces reproduce the full ones near the conical intersections (remarks on this will be provided later). The pertinent molecular point group (for the ground state equilibrium geometry) is C_{2v}, and the electronic states in question transform according to the A_2, B_2, A_1, and B_1 irreducible representations. Utilizing the symmetry selection rule, (3), one readily arrives at the following Hamiltonian matrix for this interacting manifold ($H_0 = T_N + V_0$):

$$\mathcal{H} = H_0\mathbf{1} + \begin{pmatrix} E_{A_2} + \sum k_s^1 Q_s & \sum \lambda_s^{12} Q_s & \sum \lambda_s^{13} Q_s & \sum \lambda_s^{14} Q_s \\ \sum \lambda_s^{12} Q_s & E_{B_2} + \sum k_s^2 Q_s & \sum \lambda_s^{23} Q_s & \sum \lambda_s^{24} Q_s \\ \sum \lambda_s^{13} Q_s & \sum \lambda_s^{23} Q_s & E_{A_1} + \sum k_s^3 Q_s & \sum \lambda_s^{34} Q_s \\ \sum \lambda_s^{14} Q_s & \sum \lambda_s^{24} Q_s & \sum \lambda_s^{34} Q_s & E_{B_1} + \sum k_s^4 Q_s \end{pmatrix} . \quad (25)$$

Here the vertical excitation energies and the totally symmetric vibrational modes enter in the diagonal elements, in an obvious nomenclature. The summation index s is used throughout in (25), although the modes appearing in the various matrix elements are different according to (3). All symmetries of the vibrational modes come into play in the off-diagonal elements; the modes transform according to the various irreducible representations as follows:

$$\Gamma_{\text{vib}} = 8A_1 \oplus 3A_2 \oplus 3B_1 \oplus 7B_2 . \quad (26)$$

Extensive equation-of-motion coupled-cluster (EOM-CCSD) calculations have been undertaken, in combination with an augmented cc-pVDZ basis set to determine the parameters entering (25). For technical details of the calculation I refer to earlier work [38] As a result, 13 of the 21 vibrational modes of furan are found to be excited significantly in the system. The linear coupling constants of the relevant totally symmetric modes, as well as the vertical excitation energies resulting from the EOM-CCSD treatment, are collected in Table 1. Also included in the table are the corresponding harmonic vibrational frequencies which enter the zero-order (harmonic oscillator) Hamiltonian $T_N + V_0$ of (1, 2) and which are used in the subsequent calculations. The original literature is referred to for the other parameters [38,39]. Suffice it to say that the only modes neglected are the in-plane hydrogen stretching modes and (basically) the out-of-plane hydrogen bending modes (there are two of them in each of the four symmetry species of C_{2v}).

Table 1. Selected parameter values (all in eV) used in the dynamical calculations on furan. The excitation energies and coupling constants are determined by the EOM-CCSD method, the harmonic vibrational frequencies by the MP2 method [38, 39]

mode	ω	$^1A_2$	$^1B_2$	$^1A_1$	$^1B_1$
ν_3	0.1885	0.155	0.192	0.252	0.169
ν_4	0.1773	0.130	0.213	0.107	0.129
ν_5	0.1443	0.002	0.063	0.199	0.006
ν_6	0.1384	0.107	0.067	−0.017	0.075
ν_7	0.1265	0.044	0.105	0.032	0.027
ν_8	0.1085	0.056	0.071	−0.016	0.048
E	—	6.01	6.44	6.72	6.78

The close energetic proximity of the electronic states, and the rather large number of active vibrational modes, leads to low-energy surface crossings between most pairs of the potential energy surfaces. While there is no interaction for totally symmetric distortions due to symmetry (that is, within the C_{2v} molecular point group), the interaction becomes possible upon suitable nontotally symmetric distortions. There is thus a series of seams of symmetry-allowed conical intersections, according to the nomenclature adopted in the introduction. Rather than specifying the energetic minima on these seams numerically (as was done already before [40]), I find the schematic drawing of Fig. 1 illustrative. This shows three cuts in the multidimensional coordinate space (of the totally symmetric vibrations) designed so as to minimize the energy of the crossing seam for three pairs of potential energy surfaces. The cuts are straight lines, characterized by expressions that have been derived earlier [4]. The pairs of states are specified in each of the panels, and each minimum-energy crossing is emphasized by the circle surrounding it. The three lowest singlet excited states, being of A_2, B_2, and A_1 symmetry are indeed all interconnected through conical intersections in an energy range close to the vertical excitation energies (see Table 1) and thus relevant to the absorption spectrum (see later).

As is evident from the above, there is a whole set of seams of symmetry-allowed conical intersections between the various potential energy surfaces. These arise not only in the totally symmetric vibrational subspace, but also in lower symmetry. Consider, for example, the A_1 and B_2 potential energy surfaces which interact and repel each other upon distortion along a B_2 vibrational mode [39]. In the resulting C_s point group, the state correlating with B_2 still has a different symmetry than that correlating with the A_2 lowest excited state. The corresponding potential energy surfaces cross freely, but they interact upon further distortion along a B_1 vibrational mode. The situation is illustrated in Fig. 2. This figure displays nicely a symmetric pair of intersections that arise because of the double minimum shape of the upper, B_2 state surface (due to the repulsion with the still higher, A_1 state surface, not shown

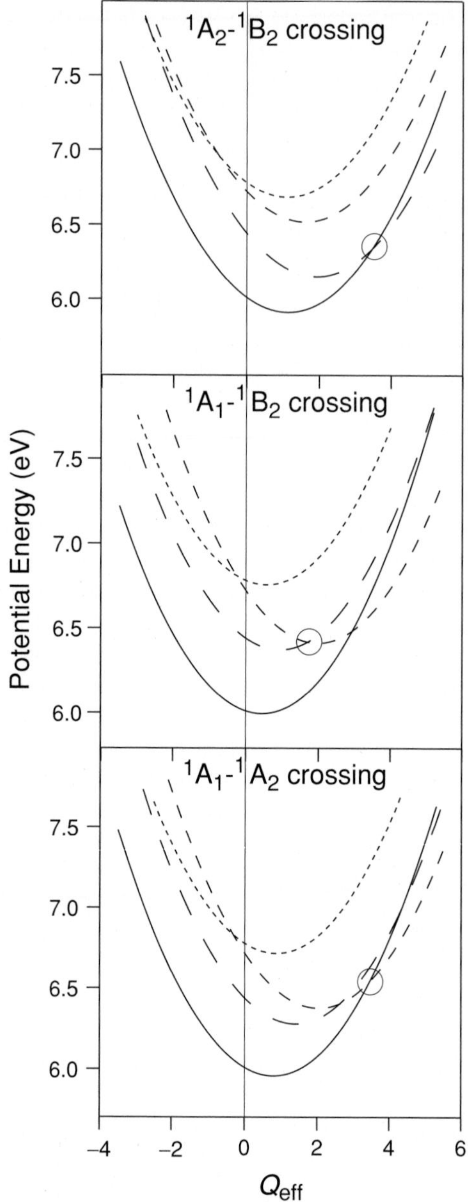

Fig. 1. Representative cuts through the potential energy surfaces of singlet-excited furan, chosen so as to minimize the energies of the crossing seams for the pairs of surfaces specified in the three panels. In the center of the Franck–Condon zone, the energetic ordering of the states is as follows: $^1A_2$ (*full lines*), $^1B_2$ (*long dashed lines*), $^1A_1$ (*short dashed lines*), $^1B_1$ (*dotted lines*)

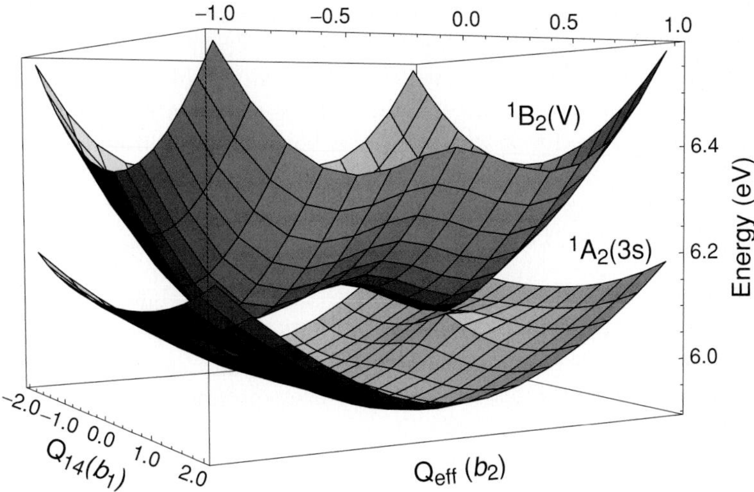

Fig. 2. Perspective drawing of the $^1A_2$ and $^1B_2$ state potential energy surfaces of electronically excited furan in the space of a B_1 and an effective B_2 vibrational mode (the latter are denoted by b_1 and b_2 in the figure). For more details see text

in the drawing). The intersections are arranged symmetrically around the origin $Q(b_2) = 0$ because of the symmetric shape of the potential energy along this coordinate; they are also symmetry-allowed for the reason stated above. They may be called "twin intersections," and are considered an instructive example of coupling between more than two electronic states and involving more than one type of nontotally symmetric vibrational mode. A rather complex nuclear dynamics can be expected to prevail in such a situation.

The accuracy of the LVC model underlying these potential energy surfaces has been checked by comparing its predictions on the stationary points with the results of a full geometry optimization (using, of course, the same ab initio method of calculation) [39]. The bond lengths and bond angles have thus been found to agree within ~0.01 Å and 1–2°, respectively, while the corresponding potential energy data differ by no more than 0.03 eV for all four states (with very few exceptions). Thus, the LVC model allows for a very reliable description of singlet-excited furan and can thus be used with confidence for the treatment of the nuclear dynamics.

5.2 Photoabsorption Spectrum and Population Dynamics

The LVC description established in Sect. 5.1 has been used for a variety of dynamical studies on the system following photoexcitation [39, 40]. The general computational framework employed is as described in Sect. 4.1, and more technical details can be found in the earlier work. The symmetry adaptation has been generally employed in the Lanczos calculations, and implicitly also in the MCTDH calculations for the wavepacket propagation (although not in the explicit formulation developed above). The Lanczos scheme proved feasible for most of the calculations, except for those with all 13 vibrational modes and four electronic states: these latter computations rely on a underlying basis of $\sim 2.10^{12}$ harmonic oscillator wavefunctions, being reduced in number to $\sim 10^6$ time-dependent single-particle functions by virtue of the MCTDH contraction effect. Apparently, the contraction effect is crucial to render the calculations numerically tractable. The results presented below have all been obtained in this way.

In Fig. 3 results for the photoabsorption spectrum are presented and compared to the experimental recording of Palmer et al. [36]. The overall agreement achieved is considered very satisfactory, although not quantitative. Nevertheless, all essential features observed are reproduced by the calculation. The energy scale of the lower panel is an absolute one, and the excitation energies thus deviate from experiment by less than 0.2 eV. Here it should be stated that a single quantity has actually been slightly adjusted, namely, the

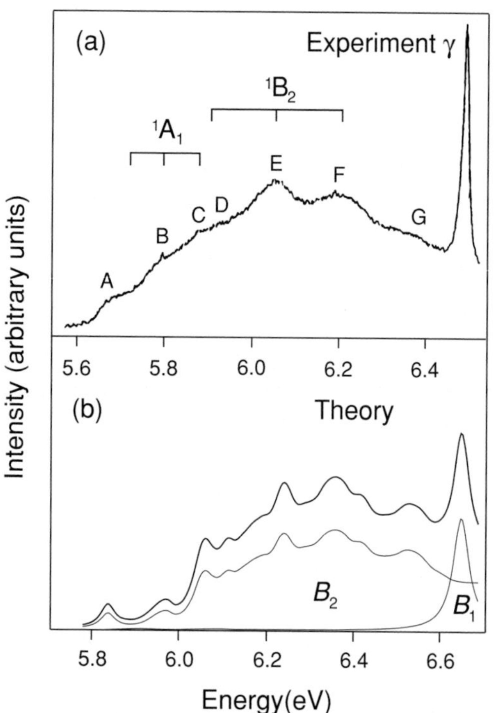

Fig. 3. Comparison of experimental [36] (upper panel) and theoretical (lower panel) photoabsorption spectrum of furan in the energy range 5.6–6.7 eV. In the lower panel, the spectral intensity is decomposed into the contributions from transitions to the ${}^1B_2$ and ${}^1B_1$ final states

vertical excitation energy of the B_1 state been increased by 0.25 eV relative to the EOM-CCSD result. This adjusted datum is given in Table 1 and has been used in all calculations reported here. All other parameters, however, are pure ab initio results, and also the vibrational frequencies have not been scaled in any way.

A closer analysis of the calculated absorption spectrum by various reduced dimensionality calculations [39] reveals the following key features. As also shown by the additional curves in Fig. 3b, the spectral intensity is largely due to the ${}^1B_2$ electronic state and, to a smaller extent, also to the ${}^1B_1$ state of furan. The transition to the ${}^1A_1$ state has almost vanishingly small oscillator strength, while that to the ${}^1A_2$ state is dipole forbidden in the C_{2v} point group [38]. Nevertheless, the lowest energy range in the spectrum is below the minimum of the ${}^1B_2$ state, and this part of the spectrum is characterized by an excitation of odd quanta of B_1 vibrational modes in the A_2 state, i.e., an effect of intensity borrowing from the ${}^1B_2$ state. For higher energies, above ~6.2 eV, all the various conical intersections, discussed above, come into play and render the nuclear motion completely nonadiabatic. The vibronic line structure is correspondingly highly complex and leads to the diffuse appearance of the spectral envelopes in Fig. 3. The latter is thus not an artifact of the finite propagation time (of 200 fs), amounting to a limited resolution, but represents a characteristic spectral feature for such a final state electronic manifold [4, 6]. Only the relatively sharp B_1 spectral peak at ~6.7 eV seems less affected by the nonadiabaticity. This is found to correspond to a Rydberg excited state of furan, as also the lowest-energy A_2 state. The unambiguous interpretation of these spectral features in terms of the Rydberg states in question represents an improved assignment of the experimental recordings of this system.

We now turn to the population dynamics (internal conversion processes) corresponding to these spectral bands. Figure 4 presents such results for a vertical transition to the ${}^1B_2$ (upper panel) and ${}^1A_1$ (lower panel) excited states of furan. As stated before, the transition to the ${}^1B_2$ state carries most of the oscillator strength, and therefore this result is most directly related to experiment. Figure 4a highlights a typical feature, namely an ultrafast internal conversion process on conically intersecting potential energy surfaces, proceeding on the time scale of typically 10–20 fs. This is of the order of a characteristic (totally symmetric) vibrational period, which means that the transition to the lower surface is virtually complete after a single encounter of the wave packet at the intersection [3, 5, 6]. As the figure shows, most of the population interconverts directly to the A_2 ground state and relatively little to the higher excited A_1 state. This is natural on energetic grounds and in view of the different densities of vibrational states. It explains why the photochemistry of this prototype heterocyclic molecule takes place on the A_2 potential energy surface (as assumed in the literature [41]), although the transition to this state is dipole forbidden. The $B_2 \rightarrow A_2$ internal conversion process is so fast that it precedes all other primary photochemical events.

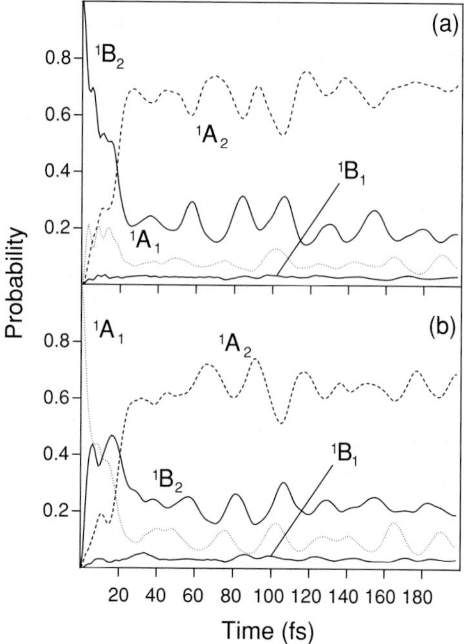

Fig. 4. Time-dependent electronic populations of singlet-excited furan for optical transitions to the $^1B_2$ state (*upper panel*) and $^1A_1$ state (*lower panel*). The excitation is broad-band, that is, the initial wave packet is located at the respective potential energy surface in the centre of the Franck–Condon zone

Figure 4b presents analogous results for the transition to the higher excited $^1A_1$ electronic state. While less important from an experimental point of view, the electronic populations shown there serve to illustrate genuine multistate features of internal conversion dynamics: In view of the existence of several electronic states lower in energy than that prepared initially, there is a stepwise transition to lower-energy states, first the next-lower $^1B_2$ state, then the $^1A_2$ lowest excited state. Only little population is transferred to the $^1B_1$ state, being still higher in energy than the $^1A_1$ state. Again, all processes proceed on the same ultrafast time scale as before. The curves of Fig. 4 represent benchmark results for highly complex, multistate nonadiabatic dynamics, i.e., involving more than 2 or 3 coupled potential energy surfaces. We mention that the $^1B_1$ state, when excited initially, is found to undergo slower decay (on a time scale of ~100 fs) owing to its weaker coupling to the other states.

6 Conclusions

In this contribution, I have surveyed salient features of a specific quantum dynamical approach to study the nonadiabatic nuclear motion on conically intersecting potential energy surfaces. This approach has been established in the literature over an extended time period, and complex structures in many electronic spectra as well as ultrafast internal conversion processes been treated successfully [3–6]. Two different lines of approach can be distinguished: for larger systems, with more than 3–6 relevant degrees of freedom, the LVC approach in its original formulation [4] is still the method of choice, although its applicability has to be explicitly checked for an individual example. However, the computational efficiency renders it most attractive, besides its conceptual simplicity: 10–20 nonseparable degrees of freedom can be included almost routinely in this way, especially with efficient wave-packet propagation techniques like the MCTDH method [16, 32]. In more recent work, emphasis has been shifted to include several (3–5) coupled electronic states in the analysis, with a correspondingly richer variety of phenomena and effects, thus highlighting even more the complexity of the nonadiabatic nuclear motion. The example presented here, the singlet excited states of furan, belongs to this category. For another system with degenerate electronic states and vibrational modes, I refer to the benzene radical cation [42, 43].

For smaller systems, where a more accurate description is possible and desired, the concept of regularized diabatic states offers a relatively simple alternative, where the LVC Hamiltonian is applied only to the adiabatic-to-diabatic mixing angle [13, 14]. This enables the treatment of general potential energy surfaces, with the same computational effort as for uncoupled potential energy surfaces (putting aside here the effort for an ab initio energy point itself).

The emphasis in this article, as in most of our applications of this formalism to date, was on symmetry-allowed conical intersections, where an interaction between the different electronic states becomes possible only by a suitable asymmetric distortion, leading to a lowering of the molecular symmetry. It has been worked out above that this allows for a formal diagonalization of the vibronic coupling Hamiltonian, rendering it electronically diagonal, but at the expense of introducing a nonlocal potential. The latter involves essentially the reflection operator in the coordinate space of the symmetry-lowering, i.e., coupling mode. This can be viewed as an operator formulation of the block-diagonalization (symmetry adaptation) of the vibronic secular matrix. Although within basis set methods this has been used before in the literature [4], the current analysis leads to a simple computational scheme also for grid methods, where the secular matrix is likewise reduced in size by a factor of two, without other disadvantages involved. Thus, full benefit can be taken of the symmetry blocking also for general coupled potential energy surfaces. While the above always refers to a quantum treatment of the nuclear motion, the same formulation of (15) can be used also within direct dynamics approaches, at least in principle. The tractability in numerical applications remains to be seen. Nevertheless, these developments are hoped to be of use in further studies of the complex nuclear dynamics at symmetry-allowed conical intersections.

Acknowledgments

The author is grateful to E. Gromov for a fruitful collaboration on singlet-excited furan, and for help with preparing the figures. He is indebted to L.S. Cederbaum, W. Domcke and S. Mahapatra for a long-term collaboration on the vibronic coupling problem. This work has been supported financially by the Deutsche Forschungsgemeinschaft.

References

1. M. Born and K. Huang. Dynamical theory of crystal lattices – App. VIII. Oxford University Press, 1954
2. Y. Haas, M. Klessinger, and S. Zilberg, editors. *Chem. Phys.*, volume 259. Elsevier, 2000
3. W. Domcke, D. Yarkony, and H. Köppel, editors. Conical intersections: Electronic structure, dynamics and spectroscopy. World Scientific, Singapore, 2004
4. H. Köppel, W. Domcke, and L. S. Cederbaum. *Adv. Chem. Phys.*, 57:59, 1984
5. W. Domcke and G. Stock. *Adv. Chem. Phys.*, 100:1, 1997
6. H. Köppel and W. Domcke. Encyclopedia in computational chemistry. page 3166. Wiley, New York, 1998
7. H. Köppel, W. Domcke, and L. S. Cederbaum. Conical intersections: Electronic structure, dynamics and spectroscopy. In Domcke et al. [3], page 323
8. W. Lichten. *Phys. Rev.*, 131:229, 1963
9. F. T. Smith. *Phys. Rev.*, 179:111, 1969
10. M. Baer. *J. Phys. Chem.*, 35:112, 1975
11. T. Pacher, L. S. Cederbaum, and H. Köppel. *Adv. Chem. Phys.*, 84:293, 1993
12. A. Thiel and H. Köppel. *J. Chem. Phys.*, 110:9371, 1999
13. H. Köppel, J. Gronki, and S. Mahapatra. *J. Chem. Phys.*, 115:2377, 2001
14. H. Köppel and B. Schubert. *Mol. Phys.*, 104:1069, 2006
15. S. Mahapatra, G. A. Worth, H.-D. Meyer, L. S. Cederbaum, and H. Kóppel. *J. Phys. Chem. A*, 105:5567, 2001
16. G. A. Worth, H.-D. Meyer, and L. S. Cederbaum. Conical intersections: Electronic structure, dynamics and spectroscopy. In Domcke et al. [3], page 583
17. I. B. Bersuker. *Chem. Rev.*, 101:1067, 2001
18. H. Köppel. Conical intersections: Electronic structure, dynamics and spectroscopy. In Domcke et al. [3], page 429
19. T. A. Barckholtz and T. A. Miller. *Int. Rev. Phys. Chem.*, 17:435, 1998
20. C. A. Mead and D. G. Truhlar. *J. Chem. Phys.*, 77:6090, 1982
21. H. Köppel. *Faraday Discuss.*, 127:35, 2004
22. B. K. Kendrick, C. A. Mead, and D. G. Truhlar. *Chem. Phys.*, 277:31, 2002
23. S. Mahapatra and H. Köppel. *J. Chem. Phys.*, 109:1721, 1998
24. B. J. Rao, S. Mahapatra, H. Köppel, and M. Jungen. *J. Chem. Phys.*, 123:134325, 2005
25. S. Mahapatra, H. Köppel, L. S. Cederbaum, P. Stampfuß, and W. Wenzel. *Chem. Phys.*, 259:211, 2000
26. B. Schubert, H. Köppel, and H. Lischka. *J. Chem. Phys.*, 122:184312, 2005
27. R. L. Fulton and M. Gouterman. *J. Chem. Phys.*, 35:1059, 1961

28. H. Köppel. unpublished results
29. H. Köppel, L. S. Cederbaum, and W. Domcke. *J. Chem. Phys.*, 89:2023, 1988
30. C. Leforestier et al. *J. Comput. Phys.*, 94:59, 1991
31. U. Manthe, H.-D. Meyer, and L. S. Cederbaum. *J. Chem. Phys.*, 97:3199, 1992
32. M. H. Beck, A. Jäckle, G. A. Worth, and H.-D. Meyer. *Phys. Rep.*, 324:1, 2000
33. D. O. Harris, G. G. Engerholm, and W. D. Gwinn. *J. Chem. Phys.*, 43:1515, 1965
34. J. C. Light, I. P. Hamilton, and J. V. Lill. *J. Chem. Phys.*, 82:1400, 1985
35. R. A. Jones, E. C. Taylor, and A. Weissberger, editors. The chemistry of heterocyclic compounds, volume 48. Wiley, New York, 1992
36. M. H. Palmer, I. C. Walker, C. C. Ballard, and M. F. Guest. *Chem. Phys.*, 192:111, 1995
37. O. Christiansen and P. Jørgensen. *J. Am. Chem. Soc.*, 120:3423, 1998
38. E. V. Gromov, A. B. Trofimov, J. Schirmer, N. M. Vitkovskaya, and H. Köppel. *J. Chem. Phys.*, 119:737, 2003
39. E. Gromov, A. Trofimov, N. Vitkovskaya, H. Köppel, J. Schirmer, H.-D. Meyer, and L. S. Cederbaum. *J. Chem. Phys.*, 121:4585, 2004
40. H. Köppel, E. Gromov, and A. Trofimov. *Chem. Phys.*, 304:35, 2004
41. M. D'Auria. *J. Org. Chem.*, 65:2494, 2000
42. H. Köppel, M. Döscher, I. Bâldea, and H.-D. Meyer. *J. Chem. Phys.*, 117:2657, 2002
43. I. Bâldea and H. Köppel. *J. Chem. Phys.*, 124:064101, 2006

Non-Markovian Dynamics at a Conical Intersection: Ultrafast Excited-State Processes in the Presence of an Environment

I. Burghardt

Summary. A high-dimensional environment coupled to a conical intersection can substantially influence the excited-state decay as well as the ensuing dephasing and relaxation processes. We use a reduced dynamics approach, via cumulant expansion techniques, to show that two phases can be distinguished in the system–environment dynamics: (a) an initial, short time scale on which the environment's effects are coherent ("inertial"), and are entirely determined by *three effective environmental modes* as recently introduced in [Cederbaum, Gindensperger, Burghardt, Phys. Rev. Lett. **94**, 113003 (2005)]; (b) a longer time scale, on which dissipative effects set in, due to the coupling between the effective modes and the (many) residual bath modes. The short-time effects can play a key role in the ultrafast nonadiabatic events at the conical intersection. The overall picture corresponds to a "Brownian oscillator" type dynamics, and is generally non-Markovian. An example is given for a 22-dimensional model system related to the D_1–D_0 conical intersection in the butatriene cation; for this system, explicit quantum dynamical calculations are feasible using the Multi-Configuration Time-Dependent Hartree (MCTDH) technique.

1 Introduction

Conical intersections are ubiquitous occurrences in the excited states of polyatomic systems, signalling an extreme breakdown of the Born–Oppenheimer approximation [1–5]. Due to their particular, double cone topology, they provide highly efficient photochemical decay mechanisms. The decay at a conical intersection is typically ultrafast, with a characteristic time scale of femtoseconds to picoseconds.

While conical intersections have been characterized in detail for many isolated gas phase species over the past decades, the effects of an environment on a conical intersection – in an intramolecular situation, solvent, or even in highly complex systems like proteins – have more recently become a topic of intense interest. Dynamical aspects relating to high-dimensional model environments and dissipative effects have been considered, e.g., in [6–8]. The explicit inclusion of a cluster, solvent or protein environment has been addressed in the hybrid quantum mechanical/molecular mechanical studies of [9–11] as well as the model studies of [12, 13]. As shown by these studies, the conical intersection topology is indeed extremely sensitive to environment-induced perturbations: these can shift the location of the conical intersection, or could even remove the degeneracy altogether. The importance of environmental effects is underscored by recent photochemical experiments, ranging from high-dimensional intramolecular situations [14,15] to the photochemistry of biological chromophores like retinal [16, 17] and the chromophores of the photoactive yellow protein [18] and the green fluorescent protein [11].

Of key importance is the influence of the environment on the characteristic, ultrafast time scale of the excited-state decay. If a large number of environmental modes couple to the conical intersection, one may ask whether (a) characteristic cumulative effects arise, and (b) if so, whether these effects are essentially of dissipative character, or whether they exhibit a coherent, "inertial" component which would typically arise on the shortest time scale available to the system. In this latter case, the dynamics would be of non-Markovian character.[1] The present discussion will show that cumulative effects can indeed be identified, and that two phases can be distinguished in the dynamical evolution of the system–environment supermolecular system: (a) an initial, short time scale on which the environment's effects are *entirely* coherent, or non-dissipative; (b) a longer time scale, on which dissipative effects set in and become dominant. This perspective will be developed in the framework of a suitable system–bath theory approach, in terms of a cumulant expansion of the subsystem propagator [19–22].

The present analysis is closely connected to our recent work [23–25], where we have shown for a multi-dimensional environment which couples to a conical intersection that *three collective environmental modes* can be identified which capture the short-time dynamics exactly. These modes result from an orthogonal coordinate transformation of the original N-mode system. The transformation in question can be considered to generalize the construction of an effective "cluster" mode for Jahn–Teller situations in solids, by O'Brien and others [26–29]: Here, a single effective mode was shown to carry all information on the width and asymmetry of the spectral envelope. For general conical intersection situations, three modes are necessary to describe the initial decay dynamics exactly [23–25].

Since the effective modes are in turn coupled to a set of (many) residual bath modes, the overall picture corresponds to a "Brownian oscillator" type dynamics, as recognized early on by Toyozawa and Inoue [30] and by Kubo and collaborators [31]. The dissipative effects exerted by the high-dimensional residual bath act with a *delay*, since the environment's influence during the

[1] Typically, non-Markovian behavior arises if the characteristic system vs. bath time scales are not well separated, and bath correlation times are long [19–22]. Such effects can acquire a predominant role if the observed processes are fast, and occur on the characteristic system/bath time scale.

earliest time scale is determined exclusively by the effective modes. Due to the fact that the decay at the conical intersection can be extremely rapid, the short-time, "inertial" effects determined by the effective modes can be of crucial importance.

A reduced dynamics analysis via a cumulant expansion of the subsystem propagator, to be detailed later, shows that the first few moments of the propagator are reproduced exactly if the overall bath is replaced by the effective modes, in the absence of the residual modes. This proves that the short-time behavior is entirely determined by the truncated "effective-mode bath", thereby confirming and extending our results of [23–25], which were based on a moment analysis for the wavepacket autocorrelation function. The present approach provides a systematic route for including mixed states and thermal effects, and for developing approximate treatments for intermediate and long time scales.

For illustration, we consider an example relating to the intramolecular dynamics at the D_1–D_0 conical intersection in the butatriene cation [23–25]. Here, a 2-mode subsystem, which provides an appropriate zeroth-order description of the conical intersection, is coupled to a finite-dimensional, intramolecular 20-mode bath (at $T = 0$). We use this system to illustrate the main aspects of our analysis, and to establish the connection between the characteristic quantities of the overall system (which remains in a pure state) and the subsystem (which evolves into a mixed state). For the system under consideration, a direct calculation including all bath modes can be carried out using efficient quantum propagation methods, in particular the multiconfiguration time-dependent Hartree (MCTDH) technique [32–35].

The remainder of the chapter is organized as follows. In Sect. 2, we review the construction of effective modes at a conical intersection. Sect. 3 addresses a reduced dynamics formulation, in conjunction with a moment (cumulant) expansion of the subsystem propagator. Sect. 4 gives a discussion of an alternative system–bath partitioning scheme, Sect. 5 addresses an example relating to the multidimensional intramolecular situation mentioned earlier, and Sect. 6 concludes.

2 Multi-Mode System–Bath Hamiltonian at a Conical Intersection

In the following, we consider a model Hamiltonian describing multi-mode processes at a conical intersection. We distinguish a "system" part which contains an electronic (two-level) subsystem, along with a certain number of nuclear modes which couple strongly to the electronic subsystem. The "bath" part is composed of a – potentially very large – number of nuclear modes which also couple to the electronic system. For certain nuclear geometries in the combined system and bath nuclear coordinates, a degeneracy arises, corresponding to a conical intersection point (or, in higher dimensions, a seam or $(N-2)$-dimensional intersection space) [1–5]. In general, we will assume that the system part by itself features a conical intersection. However, the analysis also includes situations where a conical intersection is generated by the interaction with the environment, along with the limiting case where all nuclear modes are part of the bath subspace (see the discussion of Sect. 5).

2.1 System–Bath Perspective

In accordance with the above, we consider a system–bath partitioning,

$$\hat{H} = \hat{H}_{\mathrm{S}} + \hat{H}_{\mathrm{SB}} + \hat{H}_{\mathrm{B}} \tag{1}$$

with the system part [23–25]

$$\hat{H}_{\mathrm{S}} = \hat{V}_{\Delta} + \sum_{i=1}^{N_{\mathrm{S}}} \left[\frac{\omega_{\mathrm{S},i}}{2} \left(\hat{p}_{\mathrm{S},i}^2 + \hat{x}_{\mathrm{S},i}^2 \right) + \hat{V}_{\mathrm{S},i}(\hat{x}_{\mathrm{S},i}) \right], \tag{2}$$

where $\hat{V}_{\Delta} = -\Delta\,\hat{\sigma}_z$ gives the electronic splitting, with $\hat{\sigma}_z = |1\rangle\langle 1| - |2\rangle\langle 2|$ the operator representation of the Pauli matrix, and $\hat{p}_i = (\hbar/i)\,\partial/\partial x_i$. The potential part $\hat{V}_{\mathrm{S},i}$ represents the coupling of the ith mode to the electronic subsystem and is of the form,

$$\hat{V}_{\mathrm{S},i}(\hat{x}_{\mathrm{S},i}) = \hat{v}_1(\hat{x}_{\mathrm{S},i})\,\hat{1} + \hat{v}_z(\hat{x}_{\mathrm{S},i})\,\hat{\sigma}_z + \hat{v}_x(\hat{x}_{\mathrm{S},i})\,\hat{\sigma}_x . \tag{3}$$

This form of the potential, in conjunction with the diagonal form of the kinetic energy, corresponds to a so-called (quasi-)diabatic representation [1, 4, 5, 36].

A particular instance is given by a linearized form at the conical intersection, i.e., the so-called linear vibronic coupling (LVC) model [1, 4, 5, 36],[2]

$$\hat{V}_{\mathrm{S},i}(\hat{x}_{\mathrm{S},i}) = \kappa_{\mathrm{S},i}^{(+)}\,\hat{x}_{\mathrm{S},i}\,\hat{1} + \kappa_{\mathrm{S},i}^{(-)}\,\hat{x}_{\mathrm{S},i}\,\hat{\sigma}_z + \lambda_{\mathrm{S},i}\,\hat{x}_{\mathrm{S},i}\,\hat{\sigma}_x . \tag{4}$$

In general, the ith nuclear mode can couple both to $\hat{\sigma}_z$ (diagonally) and to $\hat{\sigma}_x = |1\rangle\langle 2| + |2\rangle\langle 1|$ (off-diagonally). If the system is characterized by symmetry – i.e., in the case of so-called symmetry-allowed conical intersections [37–40] – the modes which couple diagonally (tuning modes) are distinct from those which couple off-diagonally (coupling modes). The basic, two-dimensional conical intersection topology is represented by the combination of one coupling mode and one tuning mode.

Further, the bath Hamiltonian $\hat{H}_{\mathrm{B}}$ represents the zeroth-order Hamiltonian for a – potentially large – number of environmental modes,

[2] By a linear expansion around the conical intersection, the LVC model accounts for the so-called removable part of the nonadiabatic coupling [37–40]. This model can be augmented so as to yield a correct, global representation of the adiabatic surfaces away from the conical intersection geometry [38].

$$\hat{H}_{\mathrm{B}} = \sum_{i=1}^{N_{\mathrm{B}}} \left[\frac{\omega_{\mathrm{B},i}}{2} \left(\hat{p}_{\mathrm{B},i}^2 + \hat{x}_{\mathrm{B},i}^2 \right) \right] \tag{5}$$

Finally, the system–bath interaction is given in terms of the electronic–nuclear interaction, which is of the same form as the linear vibronic coupling potential of (4),

$$\hat{H}_{\mathrm{SB}} = \sum_{i=1}^{N_{\mathrm{B}}} \left[\kappa_{\mathrm{B},i}^{(+)} \, \hat{x}_{\mathrm{B},i} \, \hat{1} + \kappa_{\mathrm{B},i}^{(-)} \, \hat{x}_{\mathrm{B},i} \hat{\sigma}_z + \lambda_{\mathrm{B},i} \, \hat{x}_{\mathrm{B},i} \, \hat{\sigma}_x \right] . \tag{6}$$

Note that there is no direct coupling between the N_S system nuclear modes and the N_{B} bath nuclear modes, but the coupling acts entirely via the electronic subsystem.[3]

While an analysis of the system–bath dynamics could be undertaken for the present form (1)–(6) of the Hamiltonian, we choose in the following a different approach, by first introducing a coordinate transformation in the bath subspace [23–25]. This transformation combines the effect of the (many) bath modes which couple to the electronic subsystem into few – actually no more than three – effective modes. The transformation is detailed in the following.

2.2 Effective-mode Transformation in the Bath Subspace

Following the analysis of [23–25], we note that the bath modes produce cumulative effects by their coupling to the electronic two-level system. Thus, the interaction Hamiltonian equation (6) can be formally re-written in terms of a set of *three collective bath modes* $(\hat{X}_{\mathrm{B},+}, \hat{X}_{\mathrm{B},-}, \hat{X}_{\mathrm{B},\Lambda})$,

$$\hat{H}_{\mathrm{SB}} = \hat{X}_{\mathrm{B},+} \, \hat{1} + \hat{X}_{\mathrm{B},-} \, \hat{\sigma}_z + \hat{X}_{\mathrm{B},\Lambda} \, \hat{\sigma}_x \tag{7}$$

defined as

$$\begin{aligned} \hat{X}_{\mathrm{B},+} &= \sum_{i=1}^{N_{\mathrm{B}}} \kappa_{\mathrm{B},i}^{(+)} \, \hat{x}_{\mathrm{B},i} \, , \\ \hat{X}_{\mathrm{B},-} &= \sum_{i=1}^{N_{\mathrm{B}}} \kappa_{\mathrm{B},i}^{(-)} \, \hat{x}_{\mathrm{B},i} \, , \\ \hat{X}_{\mathrm{B},\Lambda} &= \sum_{i=1}^{N_{\mathrm{B}}} \lambda_{\mathrm{B},i} \, \hat{x}_{\mathrm{B},i} \, , \end{aligned} \tag{8}$$

[3] The interaction Hamiltonian can be understood to correspond to a generalized spin-boson model, as pointed out in [41]. The conventional spin-boson Hamiltonian only includes a system–bath interaction term proportional to $\hat{\sigma}_z$, while the coupling term proportional to $\hat{\sigma}_x$ is coordinate-independent [22, 42].

which reflect the collective shift ($\hat{X}_{\mathrm{B},+}$), tuning ($\hat{X}_{\mathrm{B},-}$), and coupling ($\hat{X}_{\mathrm{B},\Lambda}$) effects induced by the bath. These modes are, however, not orthogonal, and are not of direct relevance for dynamical considerations.[4]

In [23–25], we have introduced three *orthogonal* effective coordinates ($\hat{X}_{\mathrm{B},1}, \hat{X}_{\mathrm{B},2}, \hat{X}_{\mathrm{B},3}$) which will turn out to play a crucial role for the system–bath dynamics on short time scales. These coordinates can be related to the ($\hat{X}_{\mathrm{B},+}, \hat{X}_{\mathrm{B},-}, \hat{X}_{\mathrm{B},\Lambda}$) by an orthogonalizing transformation [24],

$$\begin{pmatrix} \hat{X}_{\mathrm{B},1} \\ \hat{X}_{\mathrm{B},2} \\ \hat{X}_{\mathrm{B},3} \end{pmatrix} = \mathbf{W}^{-1} \begin{pmatrix} \hat{X}_{\mathrm{B},+} \\ \hat{X}_{\mathrm{B},-} \\ \hat{X}_{\mathrm{B},\Lambda} \end{pmatrix} \tag{9}$$

with $\mathbf{W}^{-1}$ the inverse of the transformation matrix

$$\mathbf{W} = \begin{pmatrix} \frac{1}{2} K_1 \bar{\kappa}_{\mathrm{B}}^{(+)} & \frac{1}{2} K_2 \bar{\kappa}_{\mathrm{B}}^{(-)} & 0 \\ \frac{1}{2} K_1 \bar{\kappa}_{\mathrm{B}}^{(-)} & \frac{1}{2} K_2 \bar{\kappa}_{\mathrm{B}}^{(+)} & 0 \\ \bar{\lambda}_{\mathrm{B}} \Lambda_1 & \bar{\lambda}_{\mathrm{B}} \Lambda_2 & \bar{\lambda}_{\mathrm{B}} \Lambda_3 \end{pmatrix} \tag{10}$$

with the parameters $\bar{\lambda}_{\mathrm{B}} = \left[\sum_i (\lambda_{\mathrm{B},i})^2\right]^{1/2}$ and $\bar{\kappa}_{\mathrm{B}}^{(\pm)} = (\bar{\kappa}_{\mathrm{B}}^{(1)} \pm \bar{\kappa}_{\mathrm{B}}^{(2)})$, where $\bar{\kappa}_{\mathrm{B}}^{(1,2)} = \left[\sum_i (\kappa_{\mathrm{B},i}^{(+)} \pm \kappa_{\mathrm{B},i}^{(-)})^2\right]^{1/2}$; $K_{1,2}$ are normalization constants and the Λ_i are defined as in [25].

The interaction Hamiltonian of (6) and (7) reads as follows in terms of the new, orthogonal coordinates [23–25]:

$$\begin{aligned} \hat{H}_{\mathrm{SB}} = & \frac{1}{2} (K_1 \bar{\kappa}_{\mathrm{B}}^{(+)} \hat{X}_{\mathrm{B},1} + K_2 \bar{\kappa}_{\mathrm{B}}^{(-)} \hat{X}_{\mathrm{B},2}) \, \hat{1} + \frac{1}{2} (K_1 \bar{\kappa}_{\mathrm{B}}^{(-)} \hat{X}_{\mathrm{B},1} + K_2 \bar{\kappa}_{\mathrm{B}}^{(+)} \hat{X}_{\mathrm{B},2}) \, \hat{\sigma}_z \\ & + \bar{\lambda}_{\mathrm{B}} (\Lambda_1 \hat{X}_{\mathrm{B},1} + \Lambda_2 \hat{X}_{\mathrm{B},2} + \Lambda_3 \hat{X}_{\mathrm{B},3}) \, \hat{\sigma}_x . \end{aligned} \tag{11}$$

The effective modes ($\hat{X}_{\mathrm{B},1}, \hat{X}_{\mathrm{B},2}, \hat{X}_{\mathrm{B},3}$) are the first three members of a set of N_{B} new coordinates which are generated by an overall unitary transformation from the original coordinates $\{\hat{x}_{\mathrm{B},i}\}$, $\hat{\boldsymbol{X}}_{\mathrm{B}} = \boldsymbol{T}^{-1} \hat{\boldsymbol{x}}_{\mathrm{B}}$ [23, 25]. This overall transformation yields the 3-mode interaction Hamiltonian (11) and the bath Hamiltonian in the form[5]

---

[4] The modes ($\hat{X}_{\mathrm{B},+}, \hat{X}_{\mathrm{B},-}, \hat{X}_{\mathrm{B},\Lambda}$) have a direct significance, though, for topological aspects. In the case where all nuclear modes are included in the bath subspace, the two modes ($\hat{X}_{\mathrm{B},-}, \hat{X}_{\mathrm{B},\Lambda}$) span the branching plane, where the degeneracy is lifted [43, 44]. The third coordinate $\hat{X}_{\mathrm{B},+}$ acts as a shift, or "seam coordinate", along which the degeneracy is preserved.

[5] Note that the transformation leaves the components of the Hamiltonian physically unchanged; we therefore keep the symbols $\hat{H}_{\mathrm{SB}}$ and $\hat{H}_{\mathrm{B}}$.

$$\hat{H}_{\rm B} = \sum_{i=1}^{N_{\rm B}} \frac{\Omega_{{\rm B},i}}{2}(\hat{P}_{{\rm B},i}^2 + \hat{X}_{{\rm B},i}^2)\hat{1} + \sum_{i,j=1,j>i}^{N_{\rm B}} d_{ij}\left(\hat{P}_{{\rm B},i}\hat{P}_{{\rm B},j} + \hat{X}_{{\rm B},i}\hat{X}_{{\rm B},j}\right)\hat{1}. \quad (12)$$

As a result of the transformation, bilinear couplings in the coordinates and momenta now occur within the bath subspace. Importantly, the $(N_{\rm B} - 3)$ "residual" bath modes do not couple to the electronic subsystem, but couple instead to the three effective modes $(\hat{X}_{{\rm B},1}, \hat{X}_{{\rm B},2}, \hat{X}_{{\rm B},3})$.

We note for completeness that the definition of the effective modes is not unique. The coordinates $(\hat{X}_{{\rm B},1}, \hat{X}_{{\rm B},2}, \hat{X}_{{\rm B},3})$ are a member of a *manifold* of coordinate triples which are interrelated by orthogonal transformations [24]. Two choices are of particular relevance: (1) First, a definition of the new coordinates which eliminates the bilinear couplings d_{ij} within the effective-mode subspace, and creates a diagonal form of the kinetic energy in that subspace [25]. (2) Second, a definition leading to *topology-adapted* vectors, two of which lie in the branching plane [24]. This choice further connects to the adiabatic $(\mathbf{g}, \mathbf{h}, \mathbf{s})$ vectors discussed by Yarkony [3,45].

The effective-mode transformation is conceptually related to early work by Toyozawa and Inoue [30] on the identification of an "interaction mode" in Jahn–Teller systems, and further, to work by O'Brien and others [26–29] on the construction of a "cluster mode". Our recent results reported in [23–25] represent a generalization beyond the Jahn–Teller case, to generic conical intersection situations described by the LVC Hamiltonian (6), which requires consideration of *three* effective modes.

2.3 Hierarchical Description of the Bath, and an Effective Hamiltonian

With the new form of the Hamiltonian $\hat{H} = \hat{H}_{\rm S} + \hat{H}_{\rm SB} + \hat{H}_{\rm B}$, now using (11) for $\hat{H}_{\rm SB}$ and (12) for $\hat{H}_{\rm B}$, a hierarchy of modes has been introduced in the bath subspace: (a) the three effective modes, which are distinguished by their role in determining the system–bath interaction $\hat{H}_{\rm SB}$, and (b) the $(N_{\rm B} - 3)$ residual bath modes which couple in turn to the effective modes. The new bath Hamiltonian $\hat{H}_{\rm B}$ of (12) can thus be split as follows:

$$\hat{H}_{\rm B} = \hat{H}_{\rm B}^{\rm eff} + \hat{H}_{\rm B}^{\rm eff\text{-}res} + \hat{H}_{\rm B}^{\rm res} \quad (13)$$

with the effective (eff) 3-mode bath portion

$$\hat{H}_{\rm B}^{\rm eff} = \sum_{i=1}^{3} \frac{\Omega_{{\rm B},i}}{2}(\hat{P}_{{\rm B},i}^2 + \hat{X}_{{\rm B},i}^2)\hat{1} + \sum_{i,j=1,j>i}^{3} d_{ij}\left(\hat{P}_{{\rm B},i}\hat{P}_{{\rm B},j} + \hat{X}_{{\rm B},i}\hat{X}_{{\rm B},j}\right)\hat{1} \quad (14)$$

the effective–residual (eff–res) mode interaction

$$\hat{H}_{\rm B}^{\rm eff\text{-}res} = \sum_{i=1}^{3}\sum_{j=4}^{N_{\rm B}} d_{ij}\left(\hat{P}_{{\rm B},i}\hat{P}_{{\rm B},j} + \hat{X}_{{\rm B},i}\hat{X}_{{\rm B},j}\right)\hat{1} \quad (15)$$

and a definition analogous to (14) for the residual (res) Hamiltonian $\hat{H}_{\rm B}^{\rm res}$ comprising the $(N_{\rm B} - 3)$ residual bath modes.

Since the transformation leaves some freedom in determining the coupling constants d_{ij}, one may choose these couplings to vanish within the effective-mode and residual-mode spaces, while only the d_{ij}'s occurring in the interaction term (15) remain non-zero [23, 25].

Overall, the transformation leads to a "Brownian oscillator" type picture [46, 47], by which the effective $(\hat{X}_{\mathrm{B},1}, \hat{X}_{\mathrm{B},2}, \hat{X}_{\mathrm{B},3})$ modes are coupled to the residual $(\hat{X}_{\mathrm{B},4}, \ldots, \hat{X}_{\mathrm{B},N_\mathrm{B}})$ modes via $\hat{H}_\mathrm{B}^{\text{eff-res}}$. The hierarchical picture of the system–bath interaction is illustrated in Fig. 1.

Since the system–bath interaction $\hat{H}_{\mathrm{SB}}$ of (11) is entirely carried by the effective modes $(\hat{X}_{\mathrm{B},1}, \hat{X}_{\mathrm{B},2}, \hat{X}_{\mathrm{B},3})$, one may conjecture what is the effect of truncating the bath at the level of the 3-mode $\hat{H}_\mathrm{B}^{\mathrm{eff}}$ contribution contained in $\hat{H}_\mathrm{B}$, see (13). That is, consider replacing the overall Hamiltonian $\hat{H} = \hat{H}_\mathrm{S} + \hat{H}_{\mathrm{SB}} + \hat{H}_\mathrm{B}$ by the modified Hamiltonian $\hat{H}'$ [23–25]

$$\hat{H}' = \hat{H}_\mathrm{S} + \hat{H}_{\mathrm{SB}} + \hat{H}_\mathrm{B}^{\mathrm{eff}}, \tag{16}$$

where the N_B-mode bath space was approximated by three effective modes, i.e., the bath Hamiltonian $\hat{H}_\mathrm{B}$ of (13) was approximated by $\hat{H}_\mathrm{B}^{\mathrm{eff}}$.

When replacing the original $(N_\mathrm{S} + N_\mathrm{B})$-mode Hamiltonian $\hat{H}$ by the $(N_\mathrm{S}+3)$-mode effective Hamiltonian $\hat{H}'$ of (16), one would expect that (a) the short-time, "inertial" dynamics is correctly reproduced, while (b) the dynamics on an intermediate time scale, which is also determined by the coupling to the residual bath, is not very well reproduced. Thus, coherent artifacts are expected to appear since the multimode nature of the bath has been disregarded.

Yet, even at the level of the reduced 3-mode bath of (16), the analysis can be of interest, due to the key importance of the short-time dynamics in

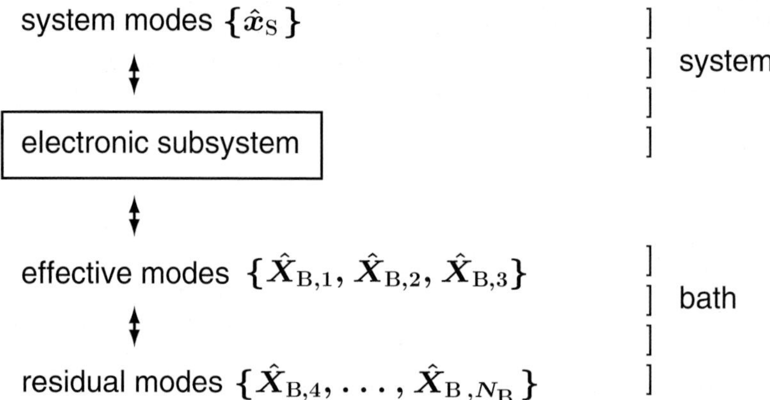

Fig. 1. Chain of interactions resulting from the transformation within the bath subspace. The effective modes $(\hat{X}_{\mathrm{B},1}, \hat{X}_{\mathrm{B},2}, \hat{X}_{\mathrm{B},3})$ couple directly to the electronic subsystem while the residual modes are in turn coupled to the effective modes. As a consequence of this hierarchical structure, the effective modes entirely determine the short-time, "inertial" components of the system–bath dynamics. The residual modes interact indirectly with the system, on intermediate and long time scales

the decay at the conical intersection. For intermediate and long time scales, the residual bath needs to be taken into account, at least in an approximate fashion.

In the following, the implications of the hierarchical system–environment interaction for the dynamical evolution are examined in detail, from a reduced dynamics perspective.

3 Reduced Propagator, Moments, and Short-time Dynamics

The goal of the present section is to show that the hierarchical structure of the transformed bath Hamiltonian translates to a hierarchy in the *dynamics* of the system–bath interaction. In particular, it is shown that the effective mode part of the bath accounts for all short-time effects which the actual bath exerts upon the system. The effects of the residual bath modes come into play on an intermediate time scale.

We will address these issues in the framework of a "reduced dynamics" formulation and, specifically, by referring to a moment, or cumulant expansion of the subsystem propagator [19,22,47,48], see (23) and (30) later. This formulation is particularly appropriate in view of addressing the influence of the environment on the short-time dynamics (in fact, more appropriate than the alternative master equation approaches [22,48]).

This perspective is connected to our previous analysis [23–25] in terms of a moment expansion of the wavepacket autocorrelation function $\langle\psi_{\mathrm{SB}}(t_0)|\psi_{\mathrm{SB}}(t)\rangle = \langle\psi_{\mathrm{SB}}(t_0)|\exp(-i\hat{H}t)|\psi_{\mathrm{SB}}(t_0)\rangle$. (Note that we use the convention $\hbar = 1$ here and in the following.) By this analysis, we have shown that the first four moments $\langle\psi_{\mathrm{SB}}(t_0)|\hat{H}^n|\psi_{\mathrm{SB}}(t_0)\rangle$, $n = 0,\ldots,3$, of the Hamiltonian are preserved if $\hat{H}$ is replaced by the effective $(N_{\mathrm{S}}+3)$-mode Hamiltonian $\hat{H}'$ of (16). As demonstrated later, a similar conclusion can be drawn from the moment expansion of the subsystem propagator.

3.1 Reduced Propagator

The evolution of the subsystem (comprising the relevant electronic and nuclear degrees of freedom) can be characterized by reduced equations of motion for the subsystem density operator $\hat{\rho}_{\mathrm{S}} = \mathrm{Tr}_{\mathrm{B}}\,\hat{\rho}_{\mathrm{SB}}$, where Tr_{B} is the trace operation with respect to the bath subspace,

$$\hat{\rho}_{\mathrm{S}}(t) = \hat{\hat{U}}_{\mathrm{sub}}(t,t_0)\hat{\rho}_{\mathrm{S}}(t_0). \tag{17}$$

The propagator $\hat{\hat{U}}_{\mathrm{sub}}$ is a superoperator (denoted by a double hat symbol) acting upon the subsystem density. $\hat{\hat{U}}_{\mathrm{sub}}$ in principle gives an exact representation of the dynamics. Its equation of motion can be constructed to be of local-in-time form or else of non-local in time form as detailed in Appendix A (while fully accounting for non-Markovian effects in both cases). In the Markovian limit, corresponding to a separation of system vs. bath time scales [19–22], the propagator is local in time and independent of the initial conditions.

Equation (17) can be derived from the evolution equation for the overall system under the Liouvillian $\hat{\hat{L}} = [\,\hat{H},\,\cdot\,]$, with the Hamiltonian (1), expressed in the original coordinates or else in the transformed coordinates. Given the unitary evolution of the overall density $\hat{\rho}_{\mathrm{SB}}$

$$\begin{aligned}\hat{\rho}_{\mathrm{SB}}(t) &= \hat{U}^{\dagger}(t,t_0)\,\hat{\rho}_{\mathrm{SB}}(t_0)\hat{U}(t,t_0)\\ &\equiv \hat{\hat{U}}(t,t_0)\,\hat{\rho}_{\mathrm{SB}}(t_0)\end{aligned} \tag{18}$$

with the propagator $\hat{U}(t,t_0) = \exp(-\mathrm{i}\hat{H}(t-t_0))$ and the associated Liouvillian propagator $\hat{\hat{U}}(t,t_0) = \exp(-\mathrm{i}\hat{\hat{L}}(t-t_0))$, the evolution for the subsystem density $\hat{\rho}_S(t)$ follows as:

$$\begin{aligned}\hat{\rho}_{\mathrm{S}}(t) &= \mathrm{Tr}_{\mathrm{B}}\left[\hat{\hat{U}}(t,t_0)\,\hat{\rho}_{\mathrm{SB}}(t_0)\right]\\ &= \hat{\hat{U}}_{\mathrm{S}}(t,t_0)\,\mathrm{Tr}_{\mathrm{B}}\left[\hat{\hat{U}}_{\mathrm{int}}(t,t_0)\,\hat{\rho}_{\mathrm{SB}}(t_0)\right]\end{aligned} \tag{19}$$

with $\hat{\hat{U}}_{\mathrm{S}}(t,t_0) = \exp(-\mathrm{i}\hat{\hat{L}}_{\mathrm{S}}(t-t_0))$, given $\hat{\hat{L}}_{\mathrm{S}} = [\,\hat{H}_{\mathrm{S}},\,\cdot\,]$, and where an interaction representation propagator was introduced as follows:

$$\begin{aligned}\hat{\hat{U}}_{\mathrm{int}}(t,t_0) &= \mathcal{T}\exp\left(-\mathrm{i}\int_{t_0}^{t}\mathrm{d}t'\,(\hat{\hat{L}}_{\mathrm{SB}}(t') + \hat{\hat{L}}_{\mathrm{B}})\right)\\ &= \hat{\hat{1}} + \sum_n(-\mathrm{i})^n\int_{t_0}^{t}\mathrm{d}\tau_1\int_{t_0}^{\tau_1}\mathrm{d}\tau_2\ldots\int_{t_0}^{\tau_{n-1}}\mathrm{d}\tau_n\,(\hat{\hat{L}}_{\mathrm{SB}}(\tau_1) + \hat{\hat{L}}_{\mathrm{B}})\\ &\quad\times(\hat{\hat{L}}_{\mathrm{SB}}(\tau_2) + \hat{\hat{L}}_{\mathrm{B}})\ldots(\hat{\hat{L}}_{\mathrm{SB}}(\tau_n) + \hat{\hat{L}}_{\mathrm{B}})\end{aligned} \tag{20}$$

with $\mathcal{T}$ the time-ordering operator. Note that $\hat{\hat{U}}_{\mathrm{int}}$ has been defined for the interaction Liouvillian $\hat{\hat{L}}_{\mathrm{SB}}(t) = \hat{\hat{U}}_0^{\mathrm{S}\dagger}(t,\tau)\hat{\hat{L}}_{\mathrm{SB}}\hat{\hat{U}}_0^{\mathrm{S}}(t,\tau)$, i.e., with respect to the *system's* zeroth-order propagator. This choice is in contrast to the usual interaction representation (defined with respect to $\hat{\hat{U}}_0^{\mathrm{S}}\hat{\hat{U}}_0^{\mathrm{B}}$), and is motivated by the fact that we propose to examine the influence of the bath on short time scales (or, equivalently, in a regime of long correlation times).

We consider an initial system–bath state which is uncorrelated,[6] $\hat{\rho}_{\mathrm{SB}}(t_0) = \hat{\rho}_{\mathrm{S}}(t_0)\otimes\hat{\rho}_{\mathrm{B}}(t_0)$ (see Appendix B for the explicit form of $\hat{\rho}_{\mathrm{SB}}(t_0)$), so that one obtains the following evolution equation in the system subspace:

$$\hat{\rho}_{\mathrm{S}}(t) = \hat{\hat{U}}_{\mathrm{S}}(t,t_0)\,\mathrm{Tr}_{\mathrm{B}}\left[\hat{\hat{U}}_{\mathrm{int}}(t,t_0)\,\hat{\rho}_{\mathrm{B}}(t_0)\right]\hat{\rho}_{\mathrm{S}}(t_0). \tag{21}$$

[6] An extension to correlated initial conditions is feasible, as described, e.g., in [21, 49].

Comparison with (17) yields an explicit expression for the subsystem propagator

$$\hat{\hat{U}}_{\mathrm{sub}}(t,t_0) = \mathrm{Tr}_{\mathrm{B}}\left[\hat{\hat{U}}(t,t_0)\,\hat{\rho}_{\mathrm{B}}(t_0)\right]$$
$$= \hat{\hat{U}}_{\mathrm{S}}(t,t_0)\,\mathrm{Tr}_{\mathrm{B}}\left[\hat{\hat{U}}_{\mathrm{int}}(t,t_0)\,\hat{\rho}_{\mathrm{B}}(t_0)\right]. \tag{22}$$

Equation (17) with the reduced propagator $\hat{\hat{U}}_{\mathrm{sub}}$ of (22) describes the subsystem evolution exactly. In the following, we consider a *moment expansion of* $\hat{\hat{U}}_{\mathrm{sub}}$ which allows one to envisage various approximation schemes.

3.2 Moment Expansion

From (22) and the definition of the interaction representation propagator $\hat{\hat{U}}_{\mathrm{int}}$, the following moment expansion for $\hat{\hat{U}}_{\mathrm{sub}}$ follows immediately:

$$\hat{\hat{U}}_{\mathrm{sub}}(t,t_0) = \hat{\hat{U}}_{\mathrm{S}}(t,t_0)\left(\hat{\hat{1}} + \sum_{n=1}^{\infty}\hat{\hat{M}}_n(t,t_0)\right) \tag{23}$$

with

$$\hat{\hat{M}}_n(t,t_0) = (-i)^n \int_{t_0}^{t} \mathrm{d}\tau_1 \int_{t_0}^{\tau_1} \mathrm{d}\tau_2 \ldots \int_{t_0}^{\tau_{n-1}} \mathrm{d}\tau_n\,\hat{\hat{m}}_n(\tau_1,\ldots,\tau_n) \tag{24}$$

Here, the moments $\hat{\hat{m}}_n$ are defined as

$$\hat{\hat{m}}_n(\tau_1,\ldots,\tau_n) = \mathrm{Tr}_{\mathrm{B}}\{(\hat{\hat{L}}_{\mathrm{SB}}(\tau_1) + \hat{\hat{L}}_{\mathrm{B}})\ldots(\hat{\hat{L}}_{\mathrm{SB}}(\tau_n) + \hat{\hat{L}}_{\mathrm{B}})\hat{\rho}_{\mathrm{B}}(t_0)\}. \tag{25}$$

The series equation (23) represents a chronologically ordered expansion of the propagator.

The moments of equations (24)–(25) are (super)operators acting on the subsystem density. In order to establish a connection to the (scalar) moments in the bath subspace, we note that the system–bath interaction of (6) or (11) can be expressed as a sum of products, $\hat{H}_{\mathrm{SB}} = \sum_k c_k \hat{h}_k^{\mathrm{S}} \otimes \hat{h}_k^{\mathrm{B}}$,

$$\hat{\hat{L}}_{\mathrm{SB}} = [\,\hat{H}_{\mathrm{SB}},\,\cdot\,]$$
$$= \sum_k c_k\,[\,\hat{h}_k^{\mathrm{S}}\hat{h}_k^{\mathrm{B}},\,\cdot\,], \tag{26}$$

where the electronic system operators $\hat{h}_k^{\mathrm{S}}$ correspond to the Pauli matrices $\{\,\hat{\sigma}_x,\hat{\sigma}_y,\hat{\sigma}_z\,\}$, while the bath operators $\hat{h}_k^{\mathrm{B}}$ relate to the nuclear bath degrees of freedom. In the following, we will formally include the bath Hamiltonian in this form, i.e., $\hat{H}_{\mathrm{B}} = \sum_k \hat{1}^{\mathrm{S}} \otimes \hat{h}_k^{\mathrm{B}}$.

Using (25) and (26), we obtain for the moment operators acting on the subsystem density

$$\hat{\hat{m}}_n(\tau_1, \ldots, \tau_n)\, \hat{\rho}_{\rm S}(t_0) = {\rm Tr}_{\rm B}\Big\{ (\hat{\hat{L}}_{\rm SB}(\tau_1) + \hat{\hat{L}}_B) \ldots$$

$$\ldots (\hat{\hat{L}}_{\rm SB}(\tau_n) + \hat{\hat{L}}_{\rm B})\, \hat{\rho}_{\rm B}(t_0) \Big\} \hat{\rho}_{\rm S}(t_0)$$

$$= {\rm Tr}_{\rm B}\Big\{ \Big[\sum_k c_k\, \hat{h}_k^{\rm S}(\tau_1) \hat{h}_k^{\rm B}, \ldots$$

$$\ldots \Big[\sum_{k'} c_{k'} \hat{h}_{k'}^{\rm S}(\tau_n) \hat{h}_{k'}^{\rm B}, \hat{\rho}_{\rm B}(t_0) \otimes \hat{\rho}_{\rm S}(t_0) \Big] \Big] \Big\}. \tag{27}$$

For example, the second-order contribution leads to

$$\hat{\hat{m}}_2\, \hat{\rho}_S(t_0) = \sum_k \sum_{k'} c_k c_{k'} \,[\, \hat{h}_k^{\rm S}(\tau_1), \hat{h}_{k'}^{\rm S}(\tau_2) \hat{\rho}_{\rm S}(t_0) \,]\, {\rm Tr}\Big\{ \hat{h}_k^{\rm B} \hat{h}_{k'}^{\rm B} \hat{\rho}_{\rm B}(t_0) \Big\}$$

$$- c_k c_{k'} \,[\, \hat{h}_k^{\rm S}(\tau_1), \hat{\rho}_{\rm S}(t_0)\, \hat{h}_{k'}^{\rm S}(\tau_2) \,]\, {\rm Tr}\Big\{ \hat{h}_{k'}^{\rm B} \hat{h}_k^{\rm B} \hat{\rho}_{\rm B}(t_0) \Big\} \tag{28}$$

which involves commutators of the system operators along with the bath moments $\mathcal{M}_{\rm B,2}^{(kk')} = {\rm Tr}\{\hat{h}_k^{\rm B} \hat{h}_{k'}^{\rm B} \hat{\rho}_{\rm B}(t_0)\} = \langle 0_{\rm B}|\, \hat{h}_k^{\rm B} \hat{h}_{k'}^{\rm B} |0_{\rm B}\rangle$. Here, $\hat{\rho}_{\rm B}(t_0) = |0_{\rm B}\rangle\langle 0_{\rm B}|$ was used, see Appendix B.

More generally, the bath moments $\mathcal{M}_{{\rm B},n}$ are the following – scalar – quantities derived from operators acting in the bath subspace only:

$$\mathcal{M}_{{\rm B},n}^{(k\ldots k')} = {\rm Tr}\Big\{ \hat{h}_k^{\rm B} \ldots \hat{h}_{k'}^{\rm B}\, \hat{\rho}_{\rm B}(t_0) \Big\} = \langle 0_{\rm B}| \hat{h}_k^{\rm B} \ldots \hat{h}_{k'}^{\rm B} |0_{\rm B}\rangle. \tag{29}$$

If approximations are sought for within the moment expansion formulation, a central criterion will thus be the faithful representation of – or, a good approximation of – the bath moments $\mathcal{M}_{{\rm B},n}$.

3.3 Moment Expansion: Cumulants

The basic moment expansion (23) is of limited usefulness, except for a perturbation development in the regime of very long correlation times [22]. One therefore turns to a resummation of the series equation (23) in terms of so-called cumulant expansions [19–22, 50]. In particular, a resummation can be carried out in such a way as to obtain the form

$$\hat{\hat{U}}_{\rm sub}(t, t_0) = \hat{\hat{U}}_{\rm S}\, \mathcal{T} \exp\Big[\sum_{n=1}^{\infty} (-i)^n \hat{\hat{K}}_n(t, t_0) \Big] \tag{30}$$

with

$$\hat{\hat{K}}_n(t,t_0) = \int_{t_0}^{t} \mathrm{d}\tau_1 \ldots \int_{t_0}^{\tau_{n-1}} \mathrm{d}\tau_n\, \hat{\hat{\theta}}_n(\tau_1,\ldots,\tau_n). \tag{31}$$

The expansion equation (30) is also referred to as a partially time-ordered series [19–22, 50].

Here, the first two cumulants are related as follows to the moments of (25):

$$\begin{aligned}\hat{\hat{\theta}}_1(\tau_1) &= \hat{\hat{m}}_1(\tau_1),\\ \hat{\hat{\theta}}_2(\tau_1,\tau_2) &= \hat{\hat{m}}_2(\tau_1,\tau_2) - \hat{\hat{m}}_1(\tau_1)\hat{\hat{m}}_1(\tau_2).\end{aligned} \tag{32}$$

The cumulants, or connected averages, vanish whenever an n-time average "decorrelates" and reduces to a product of low-order moments. This allows for a truncation of the series (30) according to different criteria than for the series (23).[7]

Different cumulant expansions can be defined, which exhibit different statistical properties. E.g., apart from the definition (30) for the partially ordered series, a so-called fully chronologically ordered series can be introduced [20,50]. The different types of expansions are associated with different generators in the equations of motion for $\hat{\hat{U}}_{\mathrm{sub}}$, see the discussion of Appendix A. With particular regard to the short-time properties, the partially ordered series (30) is distinct in that it yields a Gaussian distribution in the static limit [20]. We will consider this series further in the following discussion of the short-time evolution.

3.4 Short-Time Evolution

The moment expansions (23) and (30) can be carried out for the Hamiltonian equation (1) in the original coordinates or else in the transformed coordinates of Sect. 2.2. Since the coordinate sets $\{\hat{x}_{\mathrm{S},1},\ldots,\hat{x}_{\mathrm{S},N_{\mathrm{S}}},\hat{x}_{\mathrm{B},1},\ldots,\hat{x}_{\mathrm{B},N_{\mathrm{B}}}\}$ and $\{\hat{x}_{\mathrm{S},1},\ldots,\hat{x}_{\mathrm{S},N_{\mathrm{S}}},\hat{X}_{\mathrm{B},1},\ldots,\hat{X}_{\mathrm{B},N_{\mathrm{B}}}\}$ are related by an orthogonal transformation, the moments of the propagator remain unchanged as a result of the transformation.

The advantage of the new coordinate set lies in the approximations that the transformed Hamiltonian suggests. In particular, one would expect useful dynamical approximations to result from the $(N_{\mathrm{S}}+3)$-mode truncated Hamiltonian $\hat{H}'$ of (16), which accounts for the effective environmental modes while disregarding the residual modes.

Indeed one finds that the first three moments $\hat{\hat{m}}_n(\tau_1,\ldots,\tau_n)$, $n=1,\ldots,3$, of the expansion equations (23)–(25) are *unchanged* if the $(N_{\mathrm{S}}+N_{\mathrm{B}})$-mode

[7] In particular, the series equation (30) can be truncated at the second order if $\lambda\tau_{\mathrm{c}} \ll 1$, with λ the coupling strength associated with the system–bath interaction and τ_{c} the characteristic correlation time [22].

Hamiltonian $\hat{H} = \hat{H}_{\mathrm{S}} + \hat{H}_{\mathrm{SB}} + \hat{H}_{\mathrm{B}}$ is replaced with the $(N_{\mathrm{S}}+3)$-mode Hamiltonian $\hat{H}' = \hat{H}_{\mathrm{S}} + \hat{H}_{\mathrm{SB}} + \hat{H}_{\mathrm{B}}^{\mathrm{eff}}$ of (16) – or, equivalently, if the bath is replaced with the 3-mode effective bath portion, $\hat{H}_{\mathrm{B}} \to \hat{H}_{\mathrm{B}}^{\mathrm{eff}}$, see equation (13). Likewise, the corresponding orders of the cumulant expansion equation (30) are unchanged.

The invariance of the moments is due to the fact that the bath moments $\mathcal{M}_{\mathrm{B},n}$ of (29), for $n \leq 3$, only depend upon the effective modes and are therefore unchanged when replacing the N_{B}-mode bath with the 3-mode truncated bath. The higher-order moments, starting from the fourth order, also depend on the interaction between the effective modes and the residual modes. (For an explicit demonstration of the moment calculation, we refer to [24, 25].)

The fact that the first few moments are reproduced when replacing $\hat{H} \to \hat{H}'$ $(\hat{H}_{\mathrm{B}} \to \hat{H}_{\mathrm{B}}^{\mathrm{eff}})$ implies that the 3-mode truncated bath acts as a *surrogate bath on short time scales*. An effective propagator can be defined as follows:

$$\hat{\hat{U}}_{\mathrm{eff}}(t, t_0) = \exp\Big(-\mathrm{i}\hat{\hat{L}}_{\mathrm{eff}}(t - t_0)\Big) \tag{33}$$

with the $(N_{\mathrm{S}}+3)$-mode Liouvillian $\hat{\hat{L}}_{\mathrm{eff}} = [\,\hat{H}', \cdot\,]$, with $\hat{H}'$ of (16). The associated subsystem propagator reads as follows:

$$\begin{aligned} \hat{\hat{U}}_{\mathrm{sub}}^{\mathrm{eff}}(t, t_0) &= \mathrm{Tr}_{\mathrm{B}}\Big[\hat{\hat{U}}_{\mathrm{eff}}(t, t_0)\,\hat{\rho}_{\mathrm{B}}(t_0)\Big] \\ &= \hat{\hat{U}}_{\mathrm{S}}(t, t_0)\,\mathrm{Tr}_{\mathrm{B}}\Big[\hat{\hat{U}}_{\mathrm{int}}^{\mathrm{eff}}(t, t_0)\,\hat{\rho}_{\mathrm{B}}(t_0)\Big]. \end{aligned} \tag{34}$$

The propagator $\hat{\hat{U}}_{\mathrm{sub}}^{\mathrm{eff}}$ has the same short-time properties as the original propagator $\hat{\hat{U}}_{\mathrm{sub}}$, since its first few moments are identical.

In addition, by drawing on the previous cumulant expansion analysis, a *short-time propagator* can be constructed which again has the same first moments as both $\hat{\hat{U}}_{\mathrm{sub}}$ and $\hat{\hat{U}}_{\mathrm{sub}}^{\mathrm{eff}}$

$$\hat{\hat{U}}_{\mathrm{sub}}^{\mathrm{short}}(t, t_0) = \hat{\hat{U}}_{S}(t, t_0)\,\mathcal{T}\exp\left[\sum_{n=1}^{3}(-i)^n \hat{\hat{K}}_n(t, t_0)\right]. \tag{35}$$

Here, the cumulant expansion was truncated at the third order. This approximation is appropriate if the effects of the second and third cumulants are dominant, and the interaction with the bath brings about a rapid decay of correlations.[8] An example of this approximation is given in Fig. 2a, b of Sect. 5.

[8] In the frequency domain, the spectral features are expected to be very broad; this is illustrated, e.g., by the study of pressure broadening in [51].

3.5 Autocorrelation Functions

Apart from the reduced propagator which has been at the center of the discussion so far, quantities of interest include autocorrelation functions, as illustrated in Sect. 5. We consider, in particular, autocorrelation functions of the following type, which can be formulated for both the overall density $\hat{\rho}_{\rm SB}$ and the reduced density $\hat{\rho}_{\rm S}$,

$$
\begin{aligned}
\mathcal{C}(t, t_0) &= \left[{\rm Tr}\left\{ \hat{\rho}^{\dagger}_{\rm SB}(t_0) \hat{\hat{U}}(t,t_0)\, \hat{\rho}_{\rm SB}(t_0) \right\} \right]^{1/2} \\
&= \left[{\rm Tr_S}\left\{ \hat{\rho}^{\dagger}_{\rm S}(t_0)\, \hat{\hat{U}}{}^{\rm c}_{\rm sub}(t,t_0)\, \hat{\rho}_{\rm S}(t_0) \right\} \right]^{1/2} ,
\end{aligned}
\tag{36}
$$

where a separable initial condition $\hat{\rho}_{\rm SB}(t_0) = \hat{\rho}_{\rm S}(t_0) \otimes \hat{\rho}_{\rm B}(t_0)$ was again assumed (see Appendix B). The modified subsystem propagator $\hat{\hat{U}}{}^{\rm c}_{\rm sub}$ is given as

$$
\hat{\hat{U}}{}^{\rm c}_{\rm sub}(t,t_0) = \hat{\hat{U}}_{\rm S}(t,t_0)\, {\rm Tr_B}\left[\hat{\rho}_{\rm B}(t_0)\, \hat{\hat{U}}_{\rm int}(t,t_0)\, \hat{\rho}_{\rm B}(t_0) \right]
\tag{37}
$$

i.e., containing a projection onto $\hat{\rho}^{\dagger}_{\rm B}(t_0) = \hat{\rho}_{\rm B}(t_0)$ as compared with the definition equation (22) of $\hat{\hat{U}}_{\rm sub}(t,t_0)$.

A moment (cumulant) development can be carried out for $\hat{\hat{U}}{}^{\rm c}_{\rm sub}$ by complete analogy with the series expansions discussed earlier. The same observations thus hold as in the analysis of the earlier sections: In particular, the first few moments of the autocorrelation function are reproduced accurately by the 3-mode effective bath, i.e., when replacing $\hat{\hat{U}}{}^{\rm c}_{\rm sub} \to \hat{\hat{U}}{}^{\rm eff,c}_{\rm sub}$, see also (44)–(45).

If the overall system is prepared in a pure state $\hat{\rho}_{\rm SB} = |\psi_{\rm SB}\rangle\langle\psi_{\rm SB}|$ (and remains in a pure state), the expression (36) simplifies so as to yield the absolute value of the wavepacket autocorrelation function

$$
\begin{aligned}
\mathcal{C}(t, t_0)\Big|_{\hat{\rho}_{\rm SB}={\rm pure}} &= |\langle \psi_{\rm SB}(t_0) | \hat{U}(t,t_0) | \psi_{\rm SB}(t_0) \rangle| \\
&= |\langle \psi_{\rm SB}(t_0) | \psi_{\rm SB}(t) \rangle| .
\end{aligned}
\tag{38}
$$

The corresponding relation for the subsystem is again given by the second line of (36) since the subsystem state $\hat{\rho}_{\rm S}(t) = {\rm Tr_B}[\hat{\rho}_{\rm SB}(t)] = \hat{\hat{U}}_{\rm sub}(t,t_0)\hat{\rho}_{\rm S}(t_0)$ generally corresponds to a mixed state – even if $\hat{\rho}_{\rm SB}$ remains pure. This is a consequence of the system–bath correlations which are created at $t > t_0$ due to the system–bath interaction. The same is true for the modified subsystem state $\hat{\rho}^{\rm c}_{\rm S}(t) = \hat{\hat{U}}{}^{\rm c}_{\rm sub}(t,t_0)\hat{\rho}_{\rm S}(t_0)$ of (36). Therefore, even if the overall system remains in a pure state, it is necessary to consider the general mixed-state correlation function (36) if dynamical calculations are carried out in the system subspace.

The example discussed in Sect. 5 relates to such a pure-state situation for the overall system. To connect to our previous analysis of [23–25], the moment development based upon the pure-state expression (38) is briefly summarized in Appendix C.

4 Alternative system–bath partitioning

The earlier sections have shown that the representation of the bath in terms of primary, effective modes $(\hat{X}_{\rm B,1}, \hat{X}_{\rm B,2}, \hat{X}_{\rm B,3})$ and secondary, residual modes $(\hat{X}_{\rm B,4}, \ldots, \hat{X}_{{\rm B},N_{\rm B}})$ leads to a sequential picture of the system–bath interaction. Only the effective modes appear in the interaction with the electronic subsystem, cf. $\hat{H}_{\rm SB}$ of (11), while the residual modes impact indirectly upon the subsystem evolution, via their coupling to the effective modes. This chain-like interaction, illustrated in Fig. 1, entails a sequential-in-time dynamics of the system–environment interaction.

The impact of the effective modes in the *absence* of the residual modes is represented by the $(N_S + 3)$-mode effective Hamiltonian $\hat{H}'$ of (16), and is essentially of non-dissipative nature. The collective modes $(\hat{X}_{\rm B,1}, \hat{X}_{\rm B,2}, \hat{X}_{\rm B,3})$ carry the dynamical tuning, coupling, and shift effects exerted by the bath, which add to analogous effects generated by the system modes. These effects can lead, e.g., to a displacement of the conical intersection, changes in topology, and changes in the dynamics at the conical intersection, all of which can have a key influence on the passage through the conical intersection.

The three effective modes *entirely* determine the bath's response on the shortest time scales. This is proven by the moment (cumulant) expansion of the propagator $\hat{\hat{U}}_{\rm sub}$ of (23) and (30), whose first three moments (cumulants) are reproduced exactly when replacing $\hat{H} \to \hat{H}'$ of (16) (i.e., $\hat{H}_{\rm B} \to \hat{H}_{\rm B}^{\rm eff}$ in (13)). The $(N_{\rm S} + 3)$-mode propagator $\hat{\hat{U}}_{\rm sub}^{\rm eff}$ and the exact propagator $\hat{\hat{U}}_{\rm sub}$ have identical cumulant expansions up to the third order. Hence, $\hat{\hat{U}}_{\rm sub}^{\rm eff}$ acts as a *surrogate propagator on short time scales.* If all higher-order cumulants, beyond the third order, are disregarded, $\hat{\hat{U}}_{\rm sub}$ and $\hat{\hat{U}}_{\rm sub}^{\rm eff}$ can in turn be replaced by their short-time approximant $\hat{\hat{U}}_{\rm sub}^{\rm short}$ of (35).

Since the residual modes do *not* contribute to the first few moments, the multi-mode effects contained in the residual bath are "inactive" on the shortest time scales. Dissipation acts with a delay, setting in on an intermediate time scale. This is a characteristic instance of a non-Markovian dynamics.

This perspective suggests the possibility of a new partitioning of the overall Hamiltonian, by which the effective mode part of the bath becomes *part of a modified system Hamiltonian.* The new partitioning of the Hamiltonian reads

$$\hat{H} = \hat{H}_{\rm S}'' + \hat{H}_{\rm SB}'' + \hat{H}_{\rm B}'' \tag{39}$$

with the augmented subsystem part

$$\begin{aligned}\hat{H}''_{\mathrm{S}} &= \hat{H}' \\ &= \hat{H}_{\mathrm{S}} + \hat{H}_{\mathrm{SB}} + \hat{H}^{\mathrm{eff}}_{\mathrm{B}} , \end{aligned} \tag{40}$$

see (16), while the system–bath coupling part now represents the coupling between the three effective modes and the remaining $(N_{\mathrm{B}} - 3)$ bath modes,

$$\hat{H}''_{\mathrm{SB}} = \hat{H}^{\mathrm{eff\text{-}res}}_{\mathrm{B}} , \tag{41}$$

see (15), and the new bath corresponds to the residual modes

$$\hat{H}''_{\mathrm{B}} = \hat{H}^{\mathrm{res}}_{\mathrm{B}} . \tag{42}$$

Depending on the physical nature of the system, the residual bath can be amenable to a treatment by models for dissipation – within the Markovian limit, in the simplest case – or else to an explicit but approximate dynamical description. Since the dominant non-Markovian effects have been formally eliminated (by shifting the effective bath modes into the subsystem space), the formulation of a reduced dynamics is generally simpler for the modified system–bath problem (39).

The concept of modifying the system–bath partitioning according to (39) has been suggested previously by Kubo and collaborators [31], in conjunction with a Markovian description of a residual phonon bath in solids. A more recent application of the same concept is given in [52].

5 Vibronic-Coupling Dynamics for a System–bath Model

5.1 Model System

We illustrate the above development for the ultrafast, femtosecond scale decay dynamics in an intramolecular situation involving approximately 20 modes. The model system under consideration is closely related to the low-lying D_1–D_0 conical intersection in the butatriene cation. The system has been described in an early analysis [53,54] in terms of two strongly coupled modes; the predominant role of the latter has been confirmed by a recent comprehensive dynamical study involving all normal modes [55]. It is thus appropriate to consider the two strongly coupled modes – along with the electronic subsystem – as the "system" while the remaining modes act as an intramolecular "bath" (i.e., a finite-dimensional, zero-temperature bath). Alternatively, one could think of *all* nuclear modes as a bath coupled to the electronic subsystem. From this viewpoint, the conical intersection topology as such is entirely a feature of the environment.

In [23–25], we have explicitly constructed the decomposition into effective vs. residual modes for this finite-dimensional intramolecular model bath, using the transformation described in Sect. 2.2. The goal of the present discussion is to review our previous results in light of the reduced dynamics perspective developed above, i.e., in terms of the reduced propagators $\hat{\hat{U}}_{\rm sub}$, $\hat{\hat{U}}^{\rm eff}_{\rm sub}$, and $\hat{\hat{U}}^{\rm short}_{\rm sub}$ and the associated correlation functions of Sect. 3. We will mainly focus upon the short-time decay at the conical intersection, captured by the effective modes, rather than the intermediate and long-time effects exerted by the residual modes. Work in progress addresses various approximation schemes for the residual bath.

Indeed, the residual bath has rather specific properties for this system, which is intermediate between a low-dimensional intramolecular dynamics and a high-dimensional, dissipative situation. Nevertheless, even the low-dimensional bath under consideration acts so as to induce an effectively irreversible behavior of the relevant autocorrelation function,[9] as can be inferred from the decay of the "exact" 22-mode correlation functions of Fig. 2. This suggest that, even though the reduced dynamics framework developed above strictly applies only for the case $N_{\rm B} \rightarrow \infty$, useful approaches to the modeling of the residual bath can be derived even for finite-dimensional situations.

For the present system, quantum-dynamical calculations for all degrees of freedom, including the bath modes, are feasible using efficient quantum-dynamical techniques, in particular the multiconfiguration time-dependent Hartree (MCTDH) method [32–35]. We can thus relate quantities which are calculated explicitly for the overall system (which remains pure-state) to "reduced" quantites which characterize the subsystem that evolves into a mixed state for times $t > t_0$, see Sect. 3.5.

Since a detailed account of the model system has been given in [24, 56], only a brief summary of the main aspects is provided here. The system comprises 22 modes overall, two of which are assigned as "system" modes (unless the system part is restricted to the electronic subsystem). The remaining 20 "bath" modes are weakly coupled. The bath modes fall into three groups: (a) tuning modes, with $\kappa_i^{(+)}, \kappa_i^{(-)} \neq 0; \lambda_i = 0$, (b) coupling modes, with $\lambda_i \neq 0; \kappa_i^{(+)}, \kappa_i^{(-)} = 0$, and (c) non-symmetric modes, which do not conform to the symmetry of the molecular system. One may consider this model as a combination of an *intra*molecular bath (obeying the molecular symmetries) with an *inter*molecular, nonsymmetric bath [56]. The frequencies of the respective groups of bath modes are chosen to be randomly distributed within intervals that in part are close to the 2-mode subsystem frequencies (for the tuning modes), and in part are distributed over considerably lower frequencies (for the nonsymmetric modes). The model parameters are specified in [24, 56].

[9] Indeed, due to the short observation time scales relevant for the dynamical events at the conical intersection, even a comparatively low-dimensional bath would appear effectively irreversible. That is, the Poincaré recurrence time is always long as compared with the observation time scale.

5.2 Autocorrelation Functions and Spectra

In the following, we consider the autocorrelation functions introduced in Sect. 3.5, which directly reflect the properties of the – exact or approximate – propagators, defined with respect to the overall system or else the subsystem. We focus on the case (38) corresponding to an overall system that remains in a pure state. As a reference, the exact correlation function comprising all 22 modes is calculated

$$\mathcal{C}(t,t_0) = |\langle\psi_{\mathrm{SB}}(t_0)|\hat{U}(t,t_0)|\psi_{\mathrm{SB}}(t_0)\rangle| \tag{43}$$

with the propagator $\hat{U}(t,t_0) = \exp(-\mathrm{i}\hat{H}(t-t_0))$ for the Hamiltonian $\hat{H} = \hat{H}_{\mathrm{S}} + \hat{H}_{\mathrm{SB}} + \hat{H}_{\mathrm{B}}$ of (1). According to (36), $\mathcal{C}(t,t_0)$ also corresponds to the subsystem correlation function

$$\mathcal{C}(t,t_0) = \left[\mathrm{Tr}_{\mathrm{S}}\left\{\hat{\rho}_{\mathrm{S}}^{\dagger}(t_0)\ \hat{\hat{U}}_{\mathrm{sub}}^{\mathrm{c}}(t,t_0)\hat{\rho}_{\mathrm{S}}(t_0)\right\}\right]^{1/2}, \tag{44}$$

where $\hat{\rho}_{\mathrm{S}}^{\mathrm{c}}(t) = \hat{\hat{U}}_{\mathrm{sub}}^{\mathrm{c}}(t,t_0)\hat{\rho}_{\mathrm{S}}(t_0)$ is a *mixed* state for $t > t_0$. (Recall that the index c indicates that the subsystem propagator contains a projection onto the initial bath state, see (37).) $\mathcal{C}(t,t_0)$ can thus be calculated either from the time-evolving wavefunction $|\psi_{\mathrm{SB}}(t)\rangle$ (as in the present study), or else from the time-evolving reduced density $\hat{\rho}_{\mathrm{S}}^{\mathrm{c}}(t)$.

Following the discussion of Sect. 3.4 and Sect. 3.5, we now consider two types of approximate correlation functions. First, we address the effective-mode approximation of (34),

$$\begin{aligned}\mathcal{C}_{\mathrm{eff}}(t,t_0) &= |\langle\psi_{\mathrm{SB}}(t_0)|\hat{U}_{\mathrm{eff}}(t,t_0)|\psi_{\mathrm{SB}}(t_0)\rangle| \\ &= \left[\mathrm{Tr}_{\mathrm{S}}\left\{\hat{\rho}_{\mathrm{S}}^{\dagger}(t_0)\ \hat{\hat{U}}_{\mathrm{sub}}^{\mathrm{eff,c}}(t,t_0)\hat{\rho}_{\mathrm{S}}(t_0)\right\}\right]^{1/2}\end{aligned} \tag{45}$$

with the propagator $\hat{U}_{\mathrm{eff}} = \exp(-\mathrm{i}\hat{H}'(t-t_0))$ derived from the effective (N_S+3)-mode Hamiltonian $\hat{H}' = \hat{H}_{\mathrm{S}}+\hat{H}_{\mathrm{SB}}+\hat{H}_{\mathrm{B}}^{\mathrm{eff}}$ of (16). The subsystem propagator $\hat{\hat{U}}_{\mathrm{sub}}^{\mathrm{eff,c}}(t,t_0)$ is defined analogously to $\hat{\hat{U}}_{\mathrm{sub}}^{\mathrm{c}}(t,t_0)$, i.e., by including a projection onto $\hat{\rho}_{\mathrm{B}}(t_0)$ as compared with the definition equation (34) of $\hat{\hat{U}}_{\mathrm{sub}}^{\mathrm{eff}}(t,t_0)$.

Second, the short-time cumulant approximation equation (35) is considered,

$$\begin{aligned}\mathcal{C}_{\mathrm{short}}(t,t_0) &= |\langle\psi_{\mathrm{SB}}(t_0)|\psi_{\mathrm{SB}}(t)\rangle|\Big|_{\text{3rd order cumulant}} \\ &= \left[\mathrm{Tr}_{\mathrm{S}}\left\{\hat{\rho}_{\mathrm{S}}^{\dagger}(t_0)\ \hat{\hat{U}}_{\mathrm{sub}}^{\mathrm{short,c}}(t,t_0)\hat{\rho}_{\mathrm{S}}(t_0)\right\}\right]^{1/2}\end{aligned} \tag{46}$$

The first few cumulants defining the propagator $\hat{\hat{U}}_{\mathrm{sub}}^{\mathrm{short,c}}$ can be calculated either from the overall (N_S+N_B)-mode Hamiltonian, or else from the $(N_{\mathrm{S}}+3)$-mode effective Hamiltonian.

In Fig. 2, the respective correlation functions (43)–(46) are shown [24,25]. We consider the two types of system–bath partitioning that were mentioned earlier: (1) The system part $\hat{H}_S$ is restricted to the electronic two-level system, such that the effective-mode transformation is applied to *all* nuclear coordinates; $\hat{H}' = \hat{H}_S + \hat{H}_{SB} + \hat{H}_B^{eff}$ of (16) is thus a 3-mode Hamiltonian. (2) The system part $\hat{H}_S$ comprises the two most strongly coupled nuclear modes in addition to the electronic subsystem; $\hat{H}'$ is thus a five-mode Hamiltonian (including two system modes and three effective bath modes). Panels (a)–(d) of Fig. 2 refer to case (1) while panels (e) to (h) refer to case (2). All panels on the l.h.s. of the figure relate to the ground (D_0) state, while the panels on the r.h.s. relate to the excited (D_1) state. A reference calculation ("exact") for the overall 22-mode system is shown in all panels.

In panels (a) and (b), the initial decay of the autocorrelation function $\mathcal{C}(t, t_0)$ of (43)–(44), on a 20 fs time scale, is compared with the approximants $\mathcal{C}_{\text{eff}}(t, t_0)$ of (45) and $\mathcal{C}_{\text{short}}(t, t_0)$ of (46). The system–bath partitioning (1) is chosen, that assigns all nuclear modes to the "bath" subspace. Two features are noteworthy: (1) The initial, Gaussian decay is common to the three correlation functions, as predicted in Sect. 3.4,[10] and (2) the effective-mode approximation $\mathcal{C}_{\text{eff}}(t, t_0)$ remains very close to the exact correlation function for times which noticeably exceed the validity of the short-time approximant $\mathcal{C}_{\text{short}}(t, t_0)$. This indicates that while the first few moments of $\hat{\hat{U}}_{\text{sub}}(t, t_0)$ are reproduced exactly by $\hat{\hat{U}}_{\text{sub}}^{\text{eff}}(t, t_0)$, the effective propagator also provides a very good approximation for a certain number of moments beyond the third-order.

In panels (c) and (d), the same calculations are compared on an intermediate time scale, up to 60 fs. The figure illustrates that artificial recurrences tend to appear when using $\hat{H}'$, i.e., for $\mathcal{C}_{\text{eff}}(t, t_0)$, due to the fact that the residual bath modes are neglected. Clearly the effects of the residual bath need to be included, at least in an approximate fashion.

In panels (e) to (h), the focus is shifted to the system–bath partitioning (2) which includes the two most strongly coupled nuclear modes in the "system" part. Panels (e) and (f) illustrate that the artificial recurrences are much less pronounced for the combination of the 2-mode system and the 3-mode effective bath ("sys+eff (5)"). Indeed the result of the five-mode calculation remains close to the exact result for comparatively long times. In practice, it is therefore of importance to identify strongly coupled modes and include these in the "system" part.

For reference, panels (g) and (h) also show the correlation function $\mathcal{C}_{\text{sys}}(t, t_0) = |\langle \psi_S(t_0)|\hat{U}_S(t, t_0)|\psi_S(t_0)\rangle|$ for the 2-mode isolated system, in the

[10] Since only the even-order moments contribute to $\mathcal{C}_{\text{short}}(t, t_0)$, the decay of the short-time correlation function is purely Gaussian, i.e., is determined by the second moment.

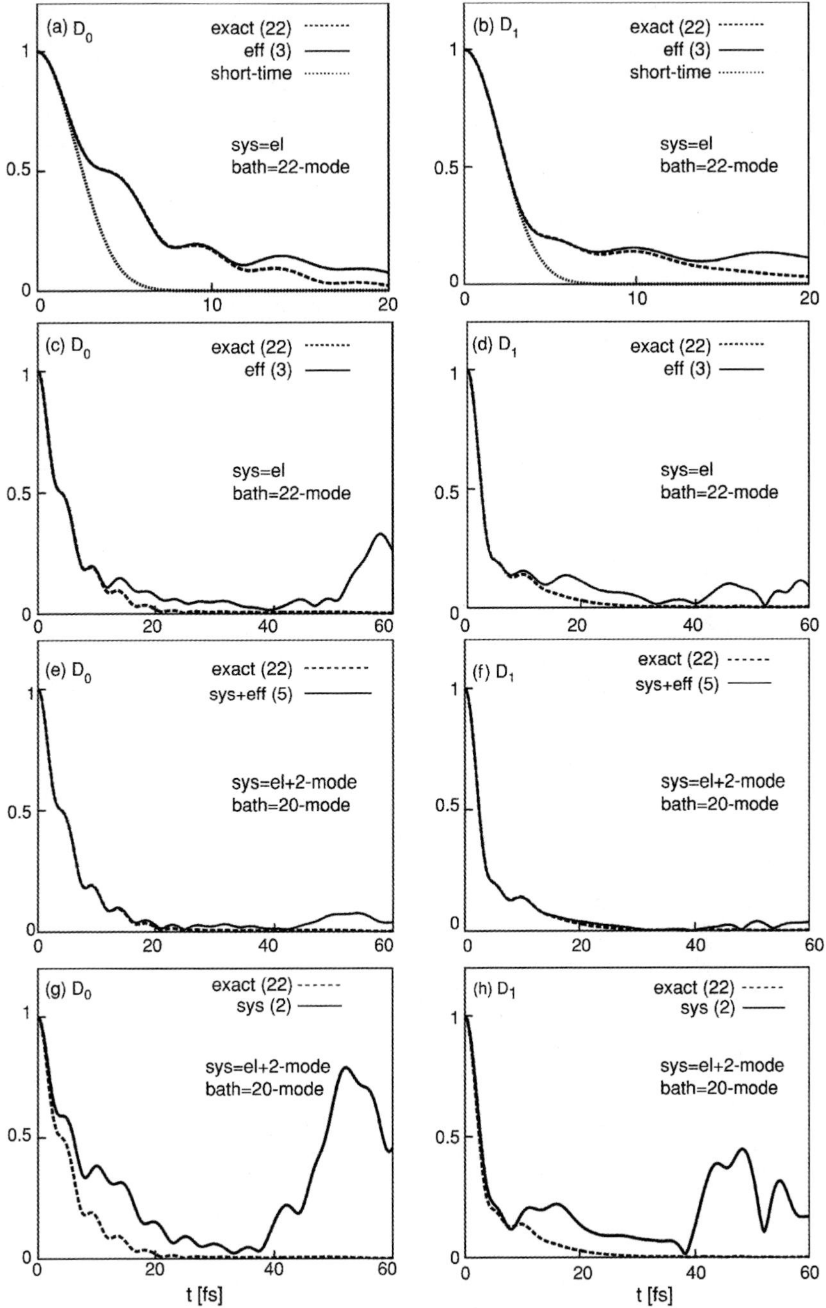

Fig. 2. Autocorrelation functions $\mathcal{C}(t, t_0)$, $\mathcal{C}_{\text{eff}}(t, t_0)$, and $\mathcal{C}_{\text{short}}(t, t_0)$ for the 2-state, 22-mode system–bath model discussed in Sect. 5 (data reproduced from [24]). All lhs (rhs) panels relate to the D_0 (D_1) state. See text for detailed explanations

absence of the bath modes. $\mathcal{C}_{\rm sys}(t,t_0)$ is clearly a less good approximation for the overall dynamics – both on short and intermediate time scales – than $\mathcal{C}_{\rm eff}(t,t_0)$ (or, for the shortest time scale, a less good approximation than $\mathcal{C}_{\rm short}(t,t_0)$).

Finally, Fig. 3 shows the associated spectrum, which represents a sum over the spectra obtained from the autocorrelation functions with initial conditions in one or the other diabatic state. Note the characteristic "interfering" band structure, which is a signature of the conical intersection [1, 53].

To summarize, the autocorrelation functions discussed above reflect the characteristic decay properties of the respective propagators, which can be defined either for the overall system or for the subsystem, see (43)-(46). While we can always refer back to the pure-state wavepacket autocorrelation functions in the present case, the connection to the subsystem correlation functions paves the way for general mixed-state situations. We expect that the main features of the dynamics carry over to situations which include thermal fluctuations.

The correlation functions confirm the role of the effective (N_S+3) mode propagator $\hat{\hat{U}}^{\rm eff}_{\rm sub}(t,t_0)$ (or $\hat{U}_{\rm short}(t,t_0)$) as a surrogate propagator on short time scales. The "inertial regime" which we define here as the time interval over which the effective mode approximation is valid, corresponds to the initial decay of the autocorrelation function, but can extend markedly beyond the

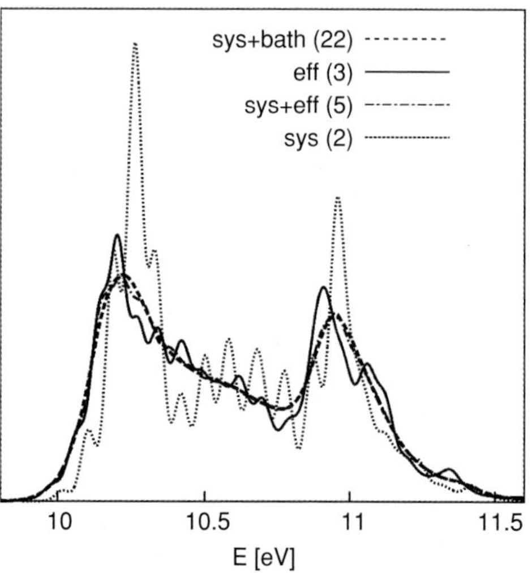

Fig. 3. Spectra obtained by Fourier transformation of the autocorrelation functions shown in Fig. 2, reproduced from [24]. The spectra represent superpositions of the spectra obtained from the autocorrelation functions for the individual diabatic states. The traces shown in the figure are defined in accordance with Fig. 2, and correspond to low-resolution spectra (with a resolution of about 40 meV, obtained by imposing a Gaussian damping with a decay constant of 40 fs)

very initial, Gaussian decay. Beyond the inertial time scale, multi-mode effects set in and induce dephasing and dissipation. These effects require the explicit calculation and/or approximate modeling of the residual bath.

6 Conclusions and Outlook

The purpose of the present analysis has been to develop a system–bath theory perspective describing the impact of a high-dimensional environment on the dynamics at a conical intersection. We have envisaged a scenario by which many environmental modes couple to the electronic subsystem, in addition to a limited number of strongly coupled "system" modes. We have addressed this situation within the LVC approximation for the environmental modes (see (6)), while no approximation is assumed a priori for the system modes.

A cornerstone of our analysis is the effective mode transformation which we have recently developed [23–25]. By an orthogonal coordinate transformation, *three* effective modes can be identified which carry all short-time effects of the environment on the evolution at the conical intersection. These modes correspond to the cumulative tuning, coupling, and shift effects exerted by the environment. They are in turn coupled to a residual bath composed of the remaining $(N_{\mathrm{B}} - 3)$ modes. This chain-like picture of interactions corresponds to a generalized Brownian oscillator model, as illustrated in Fig. 1.

An analysis by cumulant expansion techniques leads to the conclusion that the hierarchical structure of the transformed bath Hamiltonian translates to a hierarchy in the *dynamics* of the system–bath interaction. In particular, we conclude that (a) a short-time, "inertial" regime exists which is entirely determined by the three effective modes [23–25];[11] (b) the $(N_{\mathrm{B}} - 3)$ residual modes come into play on an intermediate time scale, *via* their coupling to the effective modes. A *separation of time scales within the bath* is thus observed, relating to the partitioning between the effective and residual modes. These conclusions have been obtained by consideration of the reduced propagator $\hat{\hat{U}}_{\mathrm{sub}}$ of (30), thus confirming our earlier analysis based upon a moment expansion of the wavepacket autocorrelation function [23, 25].

The present analysis has focused on the general formulation in terms of a subsystem propagator, and on the short-time limit determined by the "inertial" effects exerted by the effective modes. We have shown that the propagators $\hat{\hat{U}}_{\mathrm{sub}}$, $\hat{\hat{U}}^{\mathrm{eff}}_{\mathrm{sub}}$ of (34), and $\hat{\hat{U}}^{\mathrm{short}}_{\mathrm{sub}}$ of (35) have *identical cumulant expansions up to the third order*. The short-time propagator $\hat{\hat{U}}^{\mathrm{short}}_{\mathrm{sub}}$ can be identified as a limiting description in a regime where all higher-order cumulants vanish. However, the effective propagator $\hat{\hat{U}}^{\mathrm{eff}}_{\mathrm{sub}}$ tends to give a good approximation

[11] A related situation is encountered in describing the effects of a polar or polarizable solvent environment on a conical intersection situation. Here, a "solvent coordinate", or collective polarization mode, is introduced which also gives rise to pronounced inertial dynamical effects [12].

even at longer times, i.e., for higher-order moments. This is especially so if the "system" part comprises the most strongly coupled modes, as can be seen from the numerical results of Sect. 5.

Beyond the initial, inertial regime, the multi-mode, dissipative effects carried by the residual bath cannot strictly be neglected. These effects become dominant on longer time scales and include in particular, energy relaxation and dephasing phenomena which play a crucial role once the system has traversed the conical intersection. (See, e.g., [7, 8], for a detailed discussion of these aspects.) Future developments will address the systematic formulation of approximation schemes for the subsystem propagator on intermediate and long time scales.

As an alternative strategy, discussed in Sect. 4, the effective mode portion of the bath can be integrated into a modified system Hamiltonian, in view of the effective modes' coherent, "non-dissipative" effects. Depending on the physical nature of the system, one could envisage, e.g., a Markovian approximation scheme for the residual bath. This picture goes back to early work by Kubo and collaborators [31], in connection with the coupling to phonon modes in a solid.

Finally, in the vein of the example discussed in Sect. 5, one can resort to an explicit dynamical treatment of the combined system and bath dynamics, which is feasible either by the powerful multiconfigurational quantum dynamical techniques based upon the MCTDH method [32–35], or else by mixed quantum–classical techniques [44, 45, 47, 57]. Here, the reformulation of the Hamiltonian according to (11)–(12) may offer numerical advantages in the treatment of the residual bath modes. Among the multiconfigurational quantum approaches, several variants have been specifically designed for a hybrid system–bath dynamics, namely the self-consistent hybrid approach of [61], the multilayer formulation of [62], and the G-MCTDH method of [63,64] which involves a moving basis of Gaussian functions. In future work, we will report on the application of these methods, in conjunction with the reduced dynamics formulation reported here.

Acknowledgments

The author thanks Lorenz Cederbaum and Etienne Gindensperger for continued exchange on the topic of this chapter. Further, thanks are due to Casey Hynes for discussions on related issues. Financial support was granted by the Centre National de la Recherche Scientifique (CNRS), France.

Appendix A Equation of Motion for the Subsystem Propagator

In this appendix, we consider the equations of motion for the subsystem propagator $\hat{\hat{U}}_{\rm sub}$ of (17) [19–21,50]. Two types of equations can be formally derived, one of which is local in time

$$\frac{\partial}{\partial t}\hat{\hat{U}}_{\mathrm{sub}}(t,t_0) = \hat{\hat{\Gamma}}(t,t_0)\hat{\hat{U}}_{\mathrm{sub}}(t,t_0), \tag{A.1}$$

while the other is non-local in time

$$\frac{\partial}{\partial t}\hat{\hat{U}}_{\mathrm{sub}}(t,t_0) = \int_{t_0}^{t} \mathrm{d}t' \,\hat{\hat{\Xi}}(t,t';t_0)\hat{\hat{U}}_{\mathrm{sub}}(t',t_0). \tag{A.2}$$

In practice, perturbation series (using, in particular, the moment expansions discussed in Sect. 3) are applied to derive explicit equations for the generators $\hat{\hat{\Gamma}}$ and $\hat{\hat{\Xi}}$.

Specifically, the cumulant expansion (30) can be shown to obey the local-in-time equation (A.1), with the explicit form [19–21, 50]

$$\hat{\hat{\Gamma}}(t,t_0) = \sum_{n=1}^{\infty}(-1)^n \frac{\partial \hat{\hat{K}}_n(t,t_0)}{\partial t}. \tag{A.3}$$

A resummation of the cumulant expansion (30) in terms of a fully chronologically ordered series yields the non-local in time equation (A.2).

Both representations (A.1) and (A.2) are formally exact. Importantly, the local-in-time form of (A.1) does not imply any Markovian approximation. The non-local in time equation (A.2) is closely related to the Nakajima–Zwanzig equation [65–67], or generalized master equations [20, 22], which are usually derived by projection operator techniques.

In the Markovian limit, which implies the rapid decay of system–bath correlations, both (A.1) and (A.2) lead to equations of motion that are local in time. We now have a generator $\hat{\hat{\Gamma}}_{\mathrm{Markov}}$ which is independent of time (and independent of the initial condition at time t_0). Markovian equations are valid on a coarse-grained time scale, with $t - t_0 > \tau_{\mathrm{c}}$, with τ_{c} the characteristic system–bath correlation time, beyond which the bath is assumed to "forget" the initial correlations and approach a stationary state.

Appendix B Initial Conditions

In the context of the present discussion, the initial state of the overall system corresponds to the form [24, 25]

$$|\psi_{\mathrm{SB}}(t_0)\rangle = \tau_1\,|0\rangle_{\mathrm{vib}} \otimes |1\rangle + \tau_2\,|0\rangle_{\mathrm{vib}} \otimes |2\rangle, \tag{B.1}$$

where $|1\rangle$ and $|2\rangle$ denote the electronic states and

$$|0\rangle_{\mathrm{vib}} = |0_S\rangle \otimes |0_{\mathrm{B}}\rangle \tag{B.2}$$

is the non-interacting vibrational ground state, separable with respect to the system vs. bath modes.

The corresponding initial density operator is thus given as

$$\begin{aligned}\hat{\rho}_{\mathrm{SB}}(t_0) &= |\psi_{\mathrm{SB}}(t_0)\rangle\langle\psi_{\mathrm{SB}}(t_0)| \\ &= \sum_{ij} \tau_i \tau_j^* \, |i\rangle\langle j| \otimes |0_{\mathrm{S}}\rangle\langle 0_{\mathrm{S}}| \otimes |0_{\mathrm{B}}\rangle\langle 0_{\mathrm{B}}|\end{aligned} \tag{B.3}$$

representing a separable system–bath state,

$$\hat{\rho}_{\mathrm{SB}}(t_0) = \hat{\rho}_{\mathrm{S}}(t_0) \otimes \hat{\rho}_{\mathrm{B}}(t_0) \tag{B.4}$$

with $\hat{\rho}_{\mathrm{B}}(t_0) = |0_{\mathrm{B}}\rangle\langle 0_{\mathrm{B}}|$ and the reduced subsystem density at time t_0,

$$\hat{\rho}_{\mathrm{S}}(t_0) = \mathrm{Tr}_{\mathrm{B}}\{\hat{\rho}_{\mathrm{SB}}(t_0)\} = \sum_{ij} \tau_i \tau_j^* \, |i\rangle\langle j| \otimes |0_{\mathrm{S}}\rangle\langle 0_{\mathrm{S}}|, \tag{B.5}$$

where $\mathrm{Tr}\{\hat{\rho}_{\mathrm{B}}(t_0)\} = 1$ was used.

Appendix C Moment expansion of the pure-state autocorrelation function

In this appendix, we consider the moment expansion of pure-state autocorrelation functions, derived from the Hamiltonian analog of the Liouvillian propagator that was at the center of the discussion of Sect. 3. If the pure-state expression for the autocorrelation function, $\mathcal{C}(t,t_0) = |C(t,t_0)|$, see (38), with $C(t,t_0) = \langle\psi_{\mathrm{SB}}(t_0)|\psi_{\mathrm{SB}}(t)\rangle = \langle\psi_0|\hat{U}|\psi_0\rangle$, is taken as a starting point, one can introduce a moment expansion for the propagator $\hat{U}$ which is entirely analogous to (23) and (30), except that we now refer to a Hamiltonian (rather than Liouvillian) setting. Using the separable initial condition $|\psi_{\mathrm{SB}}(t_0)\rangle = |0_{\mathrm{S}}\rangle \otimes |0_{\mathrm{B}}\rangle$, we obtain

$$C(t,t_0) = \langle\psi_0|\hat{U}|\psi_0\rangle = \langle 0_{\mathrm{S}}|\hat{U}_{\mathrm{sub}}|0_{\mathrm{S}}\rangle \tag{C.1}$$

with the (non-Hamiltonian) subsystem propagator

$$\begin{aligned}\hat{U}_{\mathrm{sub}}(t,t_0) &= \hat{U}_{\mathrm{S}}(t,t_0)\, \langle 0_{\mathrm{B}}|\hat{U}_{\mathrm{int}}|0_{\mathrm{B}}\rangle \\ &= \hat{U}_{\mathrm{S}}(t,t_0)\left(\hat{1} + \sum_{n=1}^{\infty} \hat{M}_n(t,t_0)\right)\end{aligned} \tag{C.2}$$

and the moments of the interaction representation propagator $\hat{U}_{\mathrm{int}}$

$$\hat{M}_n(t,t_0) = (-i)^n \int_{t_0}^{t} \mathrm{d}\tau_1 \int_{t_0}^{\tau_1} \mathrm{d}\tau_2 \ldots \int_{t_0}^{\tau_{n-1}} \mathrm{d}\tau_n \, \hat{m}_n(\tau_1, \ldots, \tau_n) \tag{C.3}$$

with the moment operators

$$\hat{m}_n(\tau_1, \ldots, \tau_n) = \mathrm{Tr_B}\{(\hat{H}_{\mathrm{SB}}(\tau_1) + \hat{H}_{\mathrm{B}}) \ldots (\hat{H}_{\mathrm{SB}}(\tau_n) + \hat{H}_{\mathrm{B}})\hat{\rho}_{\mathrm{B}}(t_0)\}. \quad (\mathrm{C.4})$$

With the product form of the Hamiltonian, $\hat{H}_{\mathrm{SB}} = \sum_k c_k \hat{h}_k^{\mathrm{S}} \hat{h}_k^{\mathrm{B}}$, see (26), the moments (C.4) reduce to products of system operators and bath moments. The latter are again of the form (29). A similar conclusion holds for the associated cumulant expansion of $\hat{U}_{\mathrm{sub}}(t, t_0)$

$$\hat{U}_{\mathrm{sub}}(t, t_0) = \hat{U}_{\mathrm{S}}(t, t_0)\, \mathcal{T} \exp\left[\sum_n (-i)^n \hat{K}_n(t, t_0)\right]. \quad (\mathrm{C.5})$$

Alternatively, a moment expansion of the overall propagator $\hat{U}$ (including the system part) can be considered, without resorting to an interaction representation. This yields a direct moment expansion of $C(t, t_0)$,

$$\begin{aligned} C(t, t_0) &= 1 + \sum_{n=1}^{\infty} \tilde{M}_n(t, t_0) \\ &= 1 + \sum_{n=1}^{\infty} (-i(t - t_0))^n \frac{1}{n!} \langle \psi_{\mathrm{SB}}(t_0)| H^n |\psi_{\mathrm{SB}}(t_0)\rangle. \end{aligned} \quad (\mathrm{C.6})$$

By resummation of the series (C.6), the corresponding cumulant expansion is obtained. With separable initial conditions, the matrix elements $\langle \psi_{\mathrm{SB}}(t_0)| H^n |\psi_{\mathrm{SB}}(t_0)\rangle$ separate into system and bath contributions, where the bath matrix elements are again of the form (29). This latter perspective was used in [23–25].

References

1. H. Köppel, W. Domcke, and L. S. Cederbaum, Adv. Chem. Phys. **57**, 59 (1984)
2. D. R. Yarkony, Rev. Mod. Phys. **68**, 985 (1996)
3. D. R. Yarkony, Acc. Chem. Res. **31**, 511 (1998)
4. G. A. Worth and L. S. Cederbaum, Ann. Rev. Phys. Chem. **55**, 127 (2004)
5. H. Köppel, W. Domcke, and L. S. Cederbaum, 2004, in *Conical Intersections*, ed. W. Domcke, D. R. Yarkony, and H. Köppel, World Scientific, New Jersey, 2004, Vol. 15, p. 323
6. A. Raab, G. A. Worth, H. Meyer, and L. S. Cederbaum, J. Chem. Phys. **110**, 936 (1999)
7. A. Kühl and W. Domcke, J. Chem. Phys. **116**, 263 (2002)
8. D. Gelman, G. Katz, R. Kosloff, and M. A. Ratner, J. Chem. Phys. **123**, 134112 (2005)
9. G. Groenhof, M. Bouxin-Cademartory, B. Hess, S. P. de Visser, H. J. C. Berendsen, M. Olivucci, A. E. Mark and M. A. Robb, J. Am. Chem. Soc. **126**, 4228 (2004)

10. T. Andruniów, N. Ferré, and M. Olivucci, Proc. Natl. Acad. Sci. USA **101**, 17908 (2004)
11. A. Toniolo, S. Olsen, L. Manohar, and T. J. Martínez, Faraday Discuss. Chem. Soc. **127**, 149 (2004)
12. I. Burghardt, L. S. Cederbaum, and J. T. Hynes, Faraday Discuss. Chem. Soc. **127**, 395 (2004)
13. D. Laage, I. Burghardt, T. Sommerfeld, and J. T. Hynes, J. Phys. Chem. A **107**, 11271 (2003)
14. V. Blanchet, M. Z. Zgierski, T. Seideman, and A. Stolow, Nature **401**, 52 (1999)
15. O. Gessner, E. T. H. Chrysostom, A. M. D. Lee, D. M. Wardlaw, M. L. Ho, S. J. Lee, B. M. Cheng, M. Z. Zgierski, I. C. Chen, J. P. Shaffer, C. C. Haydena, and A. Stolow, Faraday Discuss. Chem. Soc. **127**, 193 (2004)
16. R. W. Schoenlein, L. A. Peteanu, R. A. Mathies, and C. V. Shank, Science **254**, 412 (1991)
17. R. W. Schoenlein, L. A. Peteanu, R. A. Mathies, and C. V. Shank, Science **266**, 422 (1994)
18. U. K. Genick, S. M. Soltis, P. Kuhn, I. L. Canestrelli, and E. D. Getzoff, Nature **392**, 206 (1998)
19. R. Kubo, J. Math. Phys. **4**, 174 (1963)
20. S. Mukamel, I. Oppenheim, and J. Ross, Phys. Rev. A **17**, 1988 (1978)
21. A. Royer, Phys. Rev. Lett. **77**, 3272 (1996)
22. N. G. van Kampen, *Stochastic Processes in Physics and Chemistry*, North-Holland, Amsterdam, (1992)
23. L. S. Cederbaum, E. Gindensperger, and I. Burghardt, Phys. Rev. Lett. **94**, 113003 (2005)
24. I. Burghardt, E. Gindensperger, and L. S. Cederbaum, Mol. Phys. **104**, 1081 (2006)
25. E. Gindensperger, I. Burghardt, and L. S. Cederbaum, J. Chem. Phys. **124**, 144103 (2006)
26. M. C. M. O'Brien, J. Phys. C **5**, 2045 (1971)
27. R. Englman and B. Halperin, Ann. Phys. **3**, 453 (1978)
28. E. Haller, L. S. Cederbaum, and W. Domcke, Mol. Phys. **41**, 1291 (1980)
29. L. S. Cederbaum, E. Haller, and W. Domcke, Solid State Commun. **35**, 879 (1980)
30. Y. Toyozawa and M. Inoue, J. Phys. Soc. Jpn. **21**, 1663 (1966)
31. T. Takagahara, E. Hanamura, and R. Kubo, J. Phys. Soc. Jpn. **44**, 728 (1978)
32. H.-D. Meyer, U. Manthe, and L. S. Cederbaum, Chem. Phys. Lett. **165**, 73 (1990)
33. U. Manthe, H.-D. Meyer, and L. S. Cederbaum, J. Chem. Phys. **97**, 3199 (1992)
34. M. H. Beck, A. Jäckle, G. A. Worth, and H.-D. Meyer, Phys. Rep. **324**, 1 (2000)
35. G. A. Worth et al., The MCTDH Package, Version 8.2, 2000,; H.-D. Meyer, The MCTDH Package, Version 8.3, 2002; see http://www.pci.uni-heidelberg.de/tc/usr/mctdh
36. H. Köppel and W. Domcke, 1998, in *Encyclopedia in Computational Chemistry*, Wiley, New York, p. 3166
37. T. Pacher, H. Köppel, and L. S. Cederbaum, Adv. Chem. Phys. **84**, 293 (1993)
38. H. Köppel, J. Gronki, and S. Mahapatra, J. Chem. Phys. **115**, 2377 (2001)
39. D. R. Yarkony, J. Chem. Phys. **112**, 2111 (2000)
40. B. K. Kendrick, C. A. Mead, and D. G. Truhlar, Chem. Phys. **277**, 31 (2002)

41. J. Gilmore and R. H. McKenzie, J. Phys.: Condens. Matter **17**, 1735 (2005)
42. U. Weiss, *Quantum Dissipative Systems*, World Scientific, Singapore, (1999)
43. J. Atchity, S. S. Xantheas, and K. Ruedenberg, J. Chem. Phys. **95**, 1862 (1991)
44. M. J. Paterson, M. J. Bearpark, M. A. Robb, L. Blancafort, and G. A. Worth, Phys. Chem. Chem. Phys. **7**, 2100 (2005)
45. D. R. Yarkony, Faraday Discuss. Chem. Soc. **127**, 325 (2004)
46. A. Garg, J. N. Onuchic, and V. Ambegaokar, J. Chem. Phys. **83**, 4491 (1985)
47. S. Mukamel, *Principles of Nonlinear Optical Spectroscopy*, Oxford University Press, New York/Oxford, (1995)
48. K. Lindenberg and B. J. West, *The Nonequilibrium Statistical Mechanics of Open and Closed Systems*, VCH Publishers, New York, (1990)
49. I. Burghardt, J. Chem. Phys. **114**, 89 (2001)
50. B. Yoon, J. M. Deutch, and J. H. Freed, J. Chem. Phys. **62**, 4687 (1975)
51. A. Royer, Phys. Rev. A **22**, 1625 (1980)
52. Y. Kayanuma and H. Nakayama, Phys. Rev. B **57**, 13099 (1998)
53. L. S. Cederbaum, W. Domcke, H. Köppel, and W. von Niessen, Chem. Phys. Lett. **26**, 169 (1977)
54. L. S. Cederbaum, H. Köppel, and W. Domcke, Int. J. Quant. Chem. **S15**, 251 (1981)
55. C. Cattarius, G. A. Worth, H. Meyer, and L. S. Cederbaum, J. Chem. Phys. **115**, 2088 (2001)
56. E. Gindensperger, I. Burghardt, and L. S. Cederbaum, J. Chem. Phys. **124**, 144104 (2006)
57. C. C. Martens and J.-Y. Fang, J. Chem. Phys. **106**, 4918 (1997)
58. R. Kapral and G. Ciccotti, J. Chem. Phys. **110**, 8919 (1999)
59. I. Horenko, M. Weiser, B. Schmidt, and C. Schütte, J. Chem. Phys. **120**, 8913 (2004)
60. E. Roman and C. C. Martens, J. Chem. Phys. **121**, 11572 (2004)
61. M. Thoss, H. Wang, and W. H. Miller, J. Chem. Phys. **115**, 2991 (2001)
62. H. Wang and M. Thoss, J. Chem. Phys. **119**, 1289 (2003)
63. I. Burghardt, H.-D. Meyer, and L. S. Cederbaum, J. Chem. Phys. **111**, 2927 (1999)
64. I. Burghardt, M. Nest, and G. A. Worth, J. Chem. Phys. **119**, 5364 (2003)
65. S. Nakajima, Prog. Theor. Phys. **20**, 948 (1958)
66. R. Zwanzig, J. Chem. Phys. **33**, 1338 (1960)
67. R. Zwanzig, Physica A **30**, 1109 (1964)

Density Matrix Treatment of Electronically Excited Molecular Systems: Applications to Gaseous and Adsorbate Dynamics

D.A. Micha, A. Leathers, and B. Thorndyke

Summary. The quantum mechanical density operator provides a consistent treatment of a many-atom system in contact with a physical environment, as needed to describe a complex molecular system undergoing a localized electronic excitation induced by interaction with light, or by atomic collisions. Treatments are presented where the degrees of freedom of the many-atom system are separated into quantal and classical-like ones, and the equation of motion of the density operators are derived by means of a partial Wigner transform. A computational procedure introduces approximations of short wavelengths in phase space, and effective potentials that guide trajectory bundles. The dynamics and spectra of electronically excited systems are treated introducing a basis set of many-electron states calculated in advance, or in terms of time-dependent molecular orbitals in a first principles approach to dynamics, and are used in applications on photodissociation of a diatomic and on collisional excitation in atomic collisions. Interactions with a medium are described by reduced density operators that satisfy equations of motion with dissipation and fluctuation terms. Both delayed and instantaneous dissipation are considered, and are involved in applications to femtosecond photodesorption and to vibrational relaxation of adsorbates.

1 Introduction

The quantum mechanical density operator provides a consistent treatment of a many-atom system in contact with a physical environment, as needed to describe a complex molecular system undergoing a localized electronic excitation induced by interaction with light, or by atomic collisions. The density operator (DOp) satisfies the Liouville–von Neumann (L–vN) equation, [1–3] which involves the Hamiltonian operator of the whole system and also accounts for thermodynamical constrains through its initial conditions. When the system of interest is only part of the whole, the treatment can be based on its reduced density operator (RDOp) . This satisfies a modified L–vN equation including dissipative rates and has been used in treatments of molecular spectra [4–8] and dynamics [8–10] in a medium.

The quantum mechanical calculation of spectral and dynamical properties is very demanding even for systems with a few (less than ten) atoms. A promising alternative approach is to separate the degrees of freedom of the many-atom system into quantal and classical-like ones, and to develop a consistent treatment of their interaction. References to this very active area of research, as they relate to electronic transitions in molecular systems, can be found in our recent publications. [11–13] Among several available methods, the one which introduces the Wigner transform [6, 14] is well suited for a quantum–classical formalism based on the density operator. Here we follow a treatment which introduces a partial Wigner transform (PWT) for molecular systems [15, 16].

The classification of degrees of freedom into quantal and classical ones is particularly useful in electronically excited molecular systems. The electronic motions must be treated in terms of quantum mechanics while the motion of nuclei, or atomic cores, can instead frequently be described as classical-like and given in terms of trajectories in phase space, starting from sets of initial conditions properly chosen to account for quantal distributions. The criterion here is that the associated de Broglie wavelengths should be short compared to distances over which interatomic potential energies change. The treatment can be done introducing a basis set of many-electron states calculated in advance, or in terms of time-dependent molecular orbitals (TDMOs) in a first principles approach to dynamics and spectra as we will show in the following applications on photodissociation of a diatomic and on collisional excitation in atomic collisions.

The equations for the RDOp contain terms describing energy dissipation and fluctuation effects as a locally excited molecular subsystem interacts with an extended medium. The total system can be partitioned into a primary (or p-) region to be treated in detail, interacting with a secondary (or s-) region treated only in terms of its statistical properties. Depending on the times scales of motions in both regions, the dissipative phenomena may occur with a delay described by a memory function, or it may happen instantaneously at each time. In some cases the instantaneous dissipation may further be independent of time, and is termed a Markovian dissipation. These cases will be discussed in the applications that follow, on the femtosecond photodesorption of adsorbates and on the vibrational relaxation of adsorbates.

2 Density Operator Treatment for Finite Systems

2.1 Quantum–Classical Treatment for Finite Systems

The state of a many-atom system is given by a density operator $\hat{\Gamma}(t)$ which satisfies the L–vN equation of motion,

$$i\hbar\partial\hat{\Gamma}/\partial t = \hat{H}(t)\hat{\Gamma}(t) - \hat{\Gamma}(t)\hat{H}(t), \tag{1}$$

where $\hat{H}(t)$ is the Hamiltonian operator of the whole system; the equation must be solved for the initial condition $\hat{\Gamma}(t_{\text{in}}) = \hat{\Gamma}_{\text{in}}$ and normalization $\text{tr}[\hat{\Gamma}(t)] = 1$. Expectation values of physical operators $\hat{A}$ of the whole system are obtained from the trace, $\langle A(t)\rangle = \text{tr}[\hat{\Gamma}(t)\hat{A}]/\text{tr}[\hat{\Gamma}(t)]$, which also depends on initial conditions. Introducing quantum variables (position and spin variables) $q = (q_1, \ldots, q_n)$ and quasiclassical variables (describing nearly classical motions in terms of trajectories) $Q = (Q_1, \ldots, Q_N)$, the density operator can be expanded in a partial coordinate representation using the set of states $\{|Q\rangle\}$, as

$$\hat{\Gamma}(t) = \int \mathrm{d}Q \int \mathrm{d}Q' \; |Q\rangle \hat{\Gamma}(Q, Q', t)\langle Q'|, \tag{2}$$

where the function $\hat{\Gamma}(Q, Q', t)$ is yet an operator in the quantal variables.

The PWT is obtained introducing the new coordinates $R = (Q + Q')/2$ and $S = Q - Q'$, in abbreviated notations, and the integral transform [14]

$$\hat{\Gamma}_{\text{W}}(P, R, t) = (2\pi\hbar)^{-N} \int \mathrm{d}^N S \; \exp(\mathrm{i}P \cdot S/\hbar)\langle R - S/2|\hat{\Gamma}|R + S/2\rangle, \tag{3}$$

where (P, R) are variables corresponding to momenta and position in a classical limit. The normalization of this density operator is obtained from a trace over quantum variables and an integral over P and R as $\text{tr}[\hat{\Gamma}_{\text{W}}] = \text{tr}_{\text{qu}}[\int \mathrm{d}R\mathrm{d}P \;\; \hat{\Gamma}_{\text{W}}(P, R, t)] = 1$. From this operator it is possible to obtain the quasiclassical phase density $\gamma(P, R, t) = \text{tr}_{\text{qu}}[\hat{\Gamma}_{\text{W}}(P, R, t)]$ and the quantal density operator $\hat{\Gamma}_{\text{qu}} = \int \mathrm{d}R\,\mathrm{d}P\; \hat{\Gamma}_{\text{W}}(P, R, t)$.

A partially Wigner transformed operator $\hat{A}_{\text{W}}$ is similarly defined by

$$\hat{A}_{\text{W}}(P, R) = \int \mathrm{d}^N S \; \exp(\mathrm{i}P \cdot S/\hbar)\langle R - S/2|\hat{A}|R + S/2\rangle \tag{4}$$

and physical properties are obtained from the trace as

$$\langle A\rangle = \text{tr}(\hat{\Gamma}_{\text{W}}\hat{A}_{\text{W}}) = \text{tr}_{\text{qu}}[\int \mathrm{d}R\,\mathrm{d}P\hat{\Gamma}_{\text{W}}(P, R, t)\hat{A}_{\text{W}}(P, R)]. \tag{5}$$

Quantities in the PWT are yet operators on the quantal variables, and their order must be preserved in products.

2.2 Coupled Quantal and Quasiclassical Variables

Taking the PWT of the L–vN equation gives the equation of motion for $\hat{\Gamma}_{\text{W}}$. This can be further developed for given Hamiltonians. Here we are interested in the dynamics of coupled quantum and quasiclassical variables that follows from a Hamiltonian operator

$$\hat{H} = H^{(\text{qu})}(\hat{p}, \hat{q}) + H^{(\text{cl})}(\hat{P}, \hat{Q}) + H^{(\text{cq})}(\hat{p}, \hat{q}, \hat{P}, \hat{Q}) \tag{6}$$

with terms corresponding to quantal and quasiclassical Hamiltonian functions of position and momentum operators $\{\hat{p}, \hat{q}\}$ and $\{\hat{P}, \hat{Q}\}$ plus their coupling energy $H^{(\mathrm{cq})}$, a function of all the variables . Their PWT give $H_{\mathrm{W}}^{(\mathrm{qu})} = H_{\mathrm{qu}}(\hat{p}, \hat{q})$, $H_{\mathrm{W}}^{(\mathrm{cl})} = H_{\mathrm{cl}}(P, R) = P^2/(2M) + V(R)$, and $H_{\mathrm{W}}^{(\mathrm{cq})} = H_{\mathrm{cq}}(\hat{p}, \hat{q}, P, R)$, so that $\hat{H}_{\mathrm{W}} = \hat{H}_{\mathrm{qu}} + H_{\mathrm{cl}} + \hat{H}_{\mathrm{cq}}$, to be replaced in the equation of motion. This leads to

$$\partial \hat{\Gamma}_{\mathrm{W}}/\partial t = (\mathrm{i}\hbar)^{-1}[\hat{H}_{\mathrm{qu}}, \hat{\Gamma}_{\mathrm{W}}(t)] + (\mathrm{i}\hbar)^{-1}[(H_{\mathrm{cl}} + \hat{H}_{\mathrm{cq}}) \exp(-\mathrm{i}\hbar \overleftrightarrow{\Lambda}/2) \hat{\Gamma}_{\mathrm{W}}(t) \\ - \hat{\Gamma}_{\mathrm{W}}(t) \exp(-\mathrm{i}\hbar \overleftrightarrow{\Lambda}/2)(H_{\mathrm{cl}} + \hat{H}_{\mathrm{cq}})] \quad (7)$$

in terms of the Moyal bidirectional operator

$$\overleftrightarrow{\Lambda} = \frac{\overleftarrow{\partial}}{\partial P} \cdot \frac{\overrightarrow{\partial}}{\partial R} - \frac{\overleftarrow{\partial}}{\partial R} \cdot \frac{\overrightarrow{\partial}}{\partial P}, \quad (8)$$

to be solved with initial conditions given by $\hat{\Gamma}_{\mathrm{W}}(P, R, t_{\mathrm{in}}) = \hat{\Gamma}_{\mathrm{W,in}}(P, R)$. The equation of motion for the quasiclassical phase density $\gamma(P, R, t)$ is found by taking the trace of this equation over quantal variables, and the equation of motion for $\hat{\Gamma}_{\mathrm{qu}}$ follows by instead integrating over R and P.

The initial conditions must be specified for both quantal and quasiclassical density functions. At an initial time t_{in}, the distribution of (P, R) values must be obtained from the PWT of initial conditions, so that the distribution is not simply classical. Initial distributions of (P, R) in phase space fall not only on classical allowed regions but also in regions around them. The correct calculation of expectation values requires a sum over all relevant points in phase space, and may be inaccurate if restricted to trajectories with purely classical initial conditions. For this reason we refer to (P, R) as quasiclassical or classical-like variables.

2.3 Expansion in Quantum States

It is convenient to deal with the quantum degrees of freedom introducing a basis set of N_B quantum states, parametrically dependent on the phase space variables (P, R). They can be arranged as a row matrix $|\mathbf{\Phi}(P, R)\rangle = [|\Phi_1(P, R)\rangle, |\Phi_2(P, R)\rangle, ...]$, taken here to be orthonormal, to obtain the matrix representation $\hat{\Gamma}_{\mathrm{W}} = |\mathbf{\Phi}\rangle \mathbf{\Gamma}_{\mathrm{W}} \langle \mathbf{\Phi}|$. Dropping in what follows the subindex W in the matrix, so that $\mathbf{\Gamma}_{\mathrm{W}} = \mathbf{\Gamma}$, the DM equation is of the form

$$\partial \mathbf{\Gamma}/\partial t = (\mathrm{i}\hbar)^{-1}[\mathbf{H}_{\mathrm{qu}}, \mathbf{\Gamma}(t)] + (\mathrm{i}\hbar)^{-1}[(H_{\mathrm{cl}}\mathbf{I} + \mathbf{H}_{\mathrm{cq}}) \overleftrightarrow{\mathbf{L}} \mathbf{\Gamma}(t) \\ - \mathbf{\Gamma}(t) \overleftrightarrow{\mathbf{L}} (H_{\mathrm{cl}}\mathbf{I} + \mathbf{H}_{\mathrm{cq}})], \quad (9)$$

where $\mathbf{I}$ is the identity matrix and $\overleftrightarrow{\mathbf{L}} = \langle \mathbf{\Phi}| \exp(-\mathrm{i}\hbar \overleftrightarrow{\Lambda}/2) |\mathbf{\Phi}\rangle$ is yet a bidirectional operator.

When a few quantum states are involved, the basis set can contain a small number of many-electron states. It can instead be a set of atomic orbitals in a treatment that introduces TDMOs as linear combinations of atomic orbitals, in a first principles description of the atomic dynamics. It can also be a set of vibrational states or of vibronic states when vibrational motions are included among the quantum variables. We will show in what follows examples for each of these choices.

3 The Semiclassical Limit

3.1 Coupled Quantum and Classical Equations

The formalism of the PWT provides an approach useful for approximations in applications to many-atom systems. In this subsection and the next one we describe a procedure based on two basic approximations which lead to coupled quantal–classical equations suitable for calculations [17]. Each approximation is of first-order in an expansion in a small parameter, so that its limitations can in principle be found by estimating the higher order terms.

In the first approximation, the PWT equations are given in a semiclassical limit, obtained to lowest order in $\hbar \overleftrightarrow{\Lambda}$, so that for two operators $\hat{A}$ and $\hat{B}$,

$$\hat{A}\exp(-\mathrm{i}\hbar\overleftrightarrow{\Lambda}/2)\hat{B} \simeq \hat{A}(1-\mathrm{i}\hbar\overleftrightarrow{\Lambda}/2)\hat{B}, \tag{10}$$

which is justified provided the operators are slowly varying functions of phase space variables (P, R). Further it follows that $-\hat{A}\overleftrightarrow{\Lambda}\hat{B} = \{\hat{A}, \hat{B}\}$ a Poisson bracket in a given order. This leads to

$$\begin{aligned}\frac{\partial \hat{\Gamma}_{\mathrm{W}}}{\partial t} &= (\mathrm{i}\hbar)^{-1}[\hat{H}_{\mathrm{qu}} + \hat{H}_{\mathrm{cq}}, \hat{\Gamma}_{\mathrm{W}}] + \{H_{\mathrm{cl}}, \hat{\Gamma}_{\mathrm{W}}\} \\ &+ \frac{1}{2}(\{\hat{H}_{\mathrm{cq}}, \hat{\Gamma}_{\mathrm{W}}\} - \{\hat{\Gamma}_{\mathrm{W}}, \hat{H}_{\mathrm{cq}}\}),\end{aligned} \tag{11}$$

where we find to the right a first term corresponding to quantal motion, followed by a term involving only classical motion, and finally a classical–quantum coupling term which cannot be expressed as a commutator of $\hat{\Gamma}_{\mathrm{W}}$ with a Hamiltonian and therefore describes a (*nonintuitive*) new term which does not appear in other treatments based only on physical considerations. This is a partial differential equation for a density operator in the quantum states and of first-order in the $2N+1$ variables (P, R, t), which must be solved starting with the initial value $\hat{\Gamma}_{\mathrm{W}}(P, R, t_{\mathrm{in}})$.

An equation for the quasiclassical phase space density $\gamma(P, R, t)$ follows from the trace over quantum variables, giving

$$\partial\gamma/\partial t - \{H_{\mathrm{cl}}, \gamma\} = \frac{1}{2}\mathrm{tr}_{\mathrm{qu}}(\{\hat{H}_{\mathrm{cq}}, \hat{\Gamma}_{\mathrm{W}}\} - \{\hat{\Gamma}_{\mathrm{W}}, \hat{H}_{\mathrm{cq}}\}) = \mathrm{tr}_{\mathrm{qu}}(\{\hat{H}_{\mathrm{cq}}, \hat{\Gamma}_{\mathrm{W}}\}) \tag{12}$$

written in terms of Poison brackets and showing that it it coupled to the equation for the density operator.

3.2 Trajectories from Effective Potentials and Forces

It is possible to further simplify the equations taking advantage of the quasi-classical nature of the P and R variables, by introducing effective potentials or forces to guide their motion through phase space, by the approximation

$$\{\hat{H}_{\text{cq}}, \hat{\Gamma}_{\text{W}}\} \simeq \{V', \hat{\Gamma}_{\text{W}}\} \tag{13}$$

with $V'(P,R,t)$ an effective potential function relating to the coupling Hamiltonian of quantal and classical variables. This leads to a new potential $\mathcal{V}(P,R,t) = V(R)+V'(P,R,t)$, and a new classical Hamiltonian $H'_{\text{cl}}(P,R,t) = H_{\text{cl}}(P,R) + V'(P,R,t)$, so that the equation for $\hat{\Gamma}_{\text{W}}$ becomes

$$\frac{\partial \hat{\Gamma}_{\text{W}}}{\partial t} = (\text{i}\hbar)^{-1}[\hat{H}_{\text{qu}} + \hat{H}_{\text{cq}}, \hat{\Gamma}_{\text{W}}(t)] + \{H'_{\text{cl}}, \hat{\Gamma}_{\text{W}}\}, \tag{14}$$

which takes the usual form found in the literature, with quantum plus classical terms to the right. This may be justified for example if the density operator varies slowly with classical positions and momenta. Possible choices for the effective potential are the *Ehrenfest potential* $\mathcal{V}(P,R,t) = V(R) + \text{tr}_{\text{qu}}[\hat{\Gamma}_{\text{W}}(P,R,t)\hat{H}_{\text{cq}}]/\gamma(P,R,t)$ or the *average path potential* $\mathcal{V}(P,R,t) = V(R) + H_{\text{cq}}[q_t(P,R), p_t(P,R), P, R]$ with $q_t(P,R) = \text{tr}_{\text{qu}}[\hat{\Gamma}_{\text{W}}(P,R,t)\hat{q}]$ and similarly for p_t, or the potential from the *effective (Hellmann–Feynman) force*

$$\frac{\partial \mathcal{V}(P,R,t)}{\partial R} = \frac{\partial V}{\partial R} + \text{tr}_{\text{qu}}[\hat{\Gamma}_{\text{W}}(P,R,t)\frac{\partial \hat{H}_{\text{cq}}}{\partial R}]/\gamma(P,R,t). \tag{15}$$

The same approximation can be made in the equations of motion for γ to obtain

$$\frac{\partial \gamma}{\partial t} = \{H'_{\text{cl}}, \gamma\}, \tag{16}$$

which is the usual equation of motion of the purely classical density of phase space.

A more accurate equation includes the quantum–classical operator coupling, after adding and subtracting the V' term, and reads

$$\begin{aligned}\frac{\partial \hat{\Gamma}_{\text{W}}}{\partial t} &= (\text{i}\hbar)^{-1}[\hat{H}_{\text{qu}} + \hat{H}_{\text{cq}}, \hat{\Gamma}_{\text{W}}] + \{H'_{\text{cl}}, \hat{\Gamma}_{\text{W}}\} \\ &\quad + \frac{1}{2}(\{\hat{H}_{\text{cq}} - V', \hat{\Gamma}_{\text{W}}\} - \{\hat{\Gamma}_{\text{W}}, \hat{H}_{\text{cq}} - V'\}),\end{aligned} \tag{17}$$

where the last term is a first-order quantum–classical coupling correction. This could be estimated to make sure that it is small in a given application and to insure that the previous equations would give accurate results. These operator equations become, after expansions in a basis of quantal states, sets of coupled equations for density matrix elements.

The time-evolution of the phase space density $\gamma(t)$ can be obtained from a bundle of trajectories generated by the effective potential at each initial condition. It is known from the Wigner transform or path integral descriptions of quantum dynamics that trajectories are in principle coherently coupled. The effective potentials of our quantum–classical description account for this indirectly through an average over the quantal density matrix. This provides a big advantage in numerical applications because each trajectory can be independently propagated from its initial condition. Other choices have been suggested instead of the effective potential $\mathcal{V}$, to account in more detail for the different dependence of each density matrix element with time, [15,16,18–21] and provide alternative couplings of quantal and classical variables.

Our equations for γ and $\hat{\Gamma}_{\mathrm{W}}$ must be solved simultaneously. To proceed, it is convenient to introduce the functions $R(t)$ and $P(t)$, solutions of the Hamiltonian equations

$$\frac{\mathrm{d}R}{\mathrm{d}t} = \frac{\partial H'_{\mathrm{cl}}}{\partial P}, \quad \frac{\mathrm{d}P}{\mathrm{d}t} = -\frac{\partial H'_{\mathrm{cl}}}{\partial R} \tag{18}$$

with $H'_{\mathrm{cl}} = P^2/(2M) + \mathcal{V}(P, R, t)$, and initial conditions $R_{\mathrm{in}} = R(t_{\mathrm{in}})$ and $P_{\mathrm{in}} = P(t_{\mathrm{in}})$. Introducing the total time derivative

$$\frac{\mathrm{d}\hat{\Gamma}_{\mathrm{W}}}{\mathrm{d}t} = \frac{\partial \hat{\Gamma}_{\mathrm{W}}}{\partial t} + \frac{\mathrm{d}R}{\mathrm{d}t} \cdot \frac{\partial \hat{\Gamma}_{\mathrm{W}}}{\partial R} + \frac{\mathrm{d}P}{\mathrm{d}t} \cdot \frac{\partial \hat{\Gamma}_{\mathrm{W}}}{\partial P} \tag{19}$$

and similarly for γ, we find that γ and $\hat{\Gamma}_{\mathrm{W}}$ depend on the parameters $\{R_{\mathrm{in}}, P_{\mathrm{in}}\}$. When the first-order coupling correction is neglected, they satisfy the simple equations

$$\mathrm{d}\hat{\Gamma}_{\mathrm{W}}/\mathrm{d}t = (\mathrm{i}\hbar)^{-1}[\hat{H}_{\mathrm{qu}} + \hat{H}_{\mathrm{cq}}(t), \hat{\Gamma}_{\mathrm{W}}(t)]\,, \quad \mathrm{d}\gamma/\mathrm{d}t = 0 \tag{20}$$

with total derivatives with respect to time, instead of the previous partial derivatives, and with the second equality indicating conservation of the phase-space density. In this way the many-atom description has been reduced to the simultaneous solution of the above equation for the quantal density operator coupled to the Hamiltonian equations for the classical variables, for given initial classical values. The first-order coupling can be added to the equation for $\hat{\Gamma}_{\mathrm{W}}$ if desired.

The procedure we have described allows for the numerical integration of individual trajectories for each set of initial conditions, which greatly simplifies calculations in applications. It would seem as if the trajectories would then be noninteracting, while we know that they should interact quantum mechanically. In fact, the trajectories are indirectly coupled in our treatment through the effective potential, which is constructed from the quantal terms in the shared hamiltonian operator and then evolves differently for each initial condition.

Expectation values of properties can be obtained from integrals over initial classical values, considering that the element of volume in phase space is independent of time so that $\mathrm{d}R\,\mathrm{d}P = \mathrm{d}R_{\mathrm{in}}\mathrm{d}P_{\mathrm{in}}$. With this, we find that

$$\langle A \rangle = \mathrm{tr}(\hat{\Gamma}_{\mathrm{W}}\hat{A}_{\mathrm{W}}) = \mathrm{tr}_{\mathrm{qu}}[\int \mathrm{d}R_{\mathrm{in}}\mathrm{d}P_{\mathrm{in}}\ \hat{\Gamma}_{\mathrm{W}}(P,R,t)\hat{A}_{\mathrm{W}}(P,R)], \tag{21}$$

where R and P are known functions of their initial values along the classical trajectories, and the integral can be constructed as the equations of motion are integrated over time for each initial condition.

4 Propagation of the Density Matrix

4.1 Propagation in a Local Interaction Picture

Oscillations in time of quantal states are usually much faster that those of the quasiclassical variables. Since both degrees of freedom are coupled, it is not efficient to solve their coupled differential equations by straightforward timestep methods. Instead it is necessary to introduce propagation procedures suitable for coupled equations with very different time scales: short for quantal states and long for quasiclassical motions. The following treatment parallels the formulation introduced in our previous review on this subject [11]. Our procedure introduces a unitary transformation at every interval of a time sequence, to create a local interaction picture for propagation over time.

As $P(t)$ and $R(t)$ change over time, basis functions $|\mathbf{\Phi}(P,R)\rangle$ generate matrix representations that vary over time, and the hamiltonian matrix takes the form $\mathbf{H} = \mathbf{H}_{\mathrm{qu}} + \mathbf{H}_{\mathrm{cq}} - \mathrm{i}\hbar\langle\mathbf{\Phi}|\mathrm{d}\mathbf{\Phi}/\mathrm{d}t\rangle$. At a given time, this matrix can be decomposed into a term $\mathbf{H}_0$ for fixed positions and zero velocities plus a term that depends on the instantaneous velocity and drives the system to its new phase space location. The hamiltonian $\mathbf{H}_0$ can be used to generate a local interaction picture to propagate the density matrix. The computational procedure starts with the matrix representation $\hat{\Gamma}_{\mathrm{W}} = |\mathbf{\Phi}\rangle\mathbf{\Gamma}\langle\mathbf{\Phi}|$, and the DM equation is of the form

$$\begin{aligned}\frac{\mathrm{d}\mathbf{\Gamma}}{\mathrm{d}t} &= (\mathrm{i}\hbar)^{-1}[\mathbf{H}_{\mathrm{qu}} + \mathbf{H}_{\mathrm{cq}} - \mathrm{i}\hbar\langle\mathbf{\Phi}|\frac{\mathrm{d}\mathbf{\Phi}}{\mathrm{d}t}\rangle, \mathbf{\Gamma}(t)] \\ &+\frac{1}{2}(\{\mathbf{H}_{\mathrm{cq}} - V'\mathbf{I}, \mathbf{\Gamma}(t)\} - \{\mathbf{\Gamma}(t), \mathbf{H}_{\mathrm{cq}} - V'\mathbf{I}\})\end{aligned} \tag{22}$$

with the full time derivative to the left. The last two terms can be made negligible in applications, with a suitable choice of the V' potential.

The coupled quantum–classical equations must be solved for the initial conditions at $t = t_{\mathrm{in}} : R_{\mathrm{in}} = R(t_{\mathrm{in}})$ and $P_{\mathrm{in}} = P(t_{\mathrm{in}})$, and $\mathbf{\Gamma}_{\mathrm{in}} = \mathbf{\Gamma}(t_{\mathrm{in}})$. In the time-propagation, the matrices and trajectory variables are assumed known at a time t_0; the density matrices are first obtained as they relax over the interval $t_0 \le t \le t_0 + \Delta t$ while keeping the quasiclassical variables fixed. They are the solutions of the equations

$$\mathrm{i}\hbar \mathrm{d}\mathbf{\Gamma}^0/\mathrm{d}t = \mathbf{H}_0 \mathbf{\Gamma}^0(t) - \mathbf{\Gamma}^0(t)\mathbf{H}_0, \tag{23}$$

which shows that the density matrix changes with time as it relaxes from its (nonstationary) value at t_0. The initial conditions in the interval are $\mathbf{\Gamma}^0(t_0) = \mathbf{\Gamma}_0$. Since the Hamiltonian matrix is now constant in time, these coupled equation are simple first-order differential equations with constant coefficients, and can be integrated by diagonalizing the matrix of coefficients. The results are sums of rapidly oscillating functions in time, reflecting the rapid quantal transitions.

In reality the quasiclassical variables are changing and one must account for the driving effect of their displacement due to the finite velocities within the interval $t_0 \le t \le t_1$. Provided this is small, and insofar as the quasiclassical motions are slower than the quantal ones, one can assume that the driving effect will only give corrections to the relaxing densities; this can be verified by shortening the time interval and repeating the calculations. The corrected densities are obtained writing

$$\mathbf{\Gamma}(t) = \mathbf{\Gamma}^0(t) + \mathbf{U}^0(t)\mathbf{\Gamma}'(t)\mathbf{U}^0(t)^\dagger, \quad \mathbf{U}^0(t) = \exp[-\tfrac{\mathrm{i}}{\hbar}\mathbf{H}_0(t - t_0)] \tag{24}$$

for the density matrix, where $\mathbf{U}^0$ defines a unitary transformation to a local interaction picture at each time t_0.

Replacing this in the L–vN equation, it is found that

$$\begin{aligned} \mathrm{i}\hbar\frac{\mathrm{d}\mathbf{\Gamma}'}{\mathrm{d}t} &= [\mathbf{V}, \mathbf{\Gamma}_0] + [\mathbf{V}, \mathbf{\Gamma}'], \\ \mathbf{V}(t) &= \mathbf{U}^0(t)[\mathbf{H}(t) - \mathbf{H}_0]\mathbf{U}^0(t)^\dagger. \end{aligned} \tag{25}$$

Here the matrix $\mathbf{V}$ contains the velocity dependent quasiclassical displacements within $\mathbf{H}(t)$ and therefore gives a driving effect. Formally, the solution for the density matrix correction is

$$\begin{aligned} \mathbf{\Gamma}'(t) &= \mathbf{\Delta}'(t) + (\mathrm{i}\hbar)^{-1}\int_{t_0}^{t} \mathrm{d}t' \, [\mathbf{V}(t'), \mathbf{\Gamma}'(t')], \\ \mathbf{\Delta}'(t) &= (\mathrm{i}\hbar)^{-1}\int_{t_0}^{t} \mathrm{d}t' \, [\mathbf{V}(t'), \mathbf{\Gamma}_0], \end{aligned} \tag{26}$$

where the driving term $\mathbf{\Delta}'$ can be obtained from a quadrature, and the second term can be made negligible by controlling the size of increments of t.

4.2 The Relax-and-Drive Computational Procedure

Straightforward stepwise integration of the coupled Hamiltonian and L–vN differential equations would be inefficient and possibly computationally inaccurate, because the fast quantal oscillations demand very small time steps, while the slow quasiclassical motions must be followed over long times, requiring many steps. An alternative is to separately do some of the integrations by quadratures. A simple and yet useful procedure employs the first-order correction $\mathbf{\Gamma}'(t) = \mathbf{\Delta}'(t)$ and an adaptive step size for the quadrature and propagation. The density matrix is approximated in each interval by

$$\mathbf{\Gamma}(t) = \mathbf{\Gamma}^0(t) + \mathbf{U}^0(t)\mathbf{\Delta}'(t)\mathbf{U}^0(t)^\dagger \tag{27}$$

with the first term describing relaxation and the second one giving the driving effect.

To advance from t_0 to $t_1 = t_0 + \Delta t$, the quasiclassical trajectory is first advanced to the time $t_{1/2} = t_0 + \Delta t/2$ and the relaxing density $\mathbf{\Gamma}^0(t)$ is calculated at this time; then the correction $\mathbf{\Delta}'(t_1)$ is obtained with the (easily improved) integrand approximation

$$\mathbf{V}(t) = \mathbf{U}^0(t)[\mathbf{H}(t_{1/2}) - \mathbf{H}(t_0)]\mathbf{U}^0(t)^\dagger, \tag{28}$$

which allows an analytical integration of each matrix element. This is finally followed by recalculation of the quasiclassical trajectory and full density matrix at time t_1. To ensure an accurate propagation, the step size Δt is varied to keep the density matrix correction within high and low tolerances in the interval, in accordance with $\varepsilon_{\mathrm{low}} \leq \parallel \mathbf{\Delta}'(t) \parallel / \parallel \mathbf{\Gamma}^0(t) \parallel \leq \varepsilon_{\mathrm{high}}$ and the normalization is checked. This leads to an efficient adaptation of the step size, so that for example in a collision it will start large, will then decrease, and later increase again after the interaction forces have disappeared. The propagation accuracy can also be verified by reversing the propagation direction in time.

This sequence, based on relaxing the density matrix for fixed nuclei and then correcting it to account for quasiclassical motions has been called the *relax-and-drive procedure*, and has been numerically implemented in several applications involving electronic rearrangement in atomic collisions [11].

5 Gaseous Dynamics

5.1 Photodissociation of NaI

A two-state model of the NaI molecule involves two diabatic potential curves and an interaction coupling them around their crossing. State $|1\rangle$ describes a covalent bonding between Na and I, while state $|2\rangle$ describes the ionic species Na^+ and I^-. The Hamiltonian elements are [22] $H_{11}(R) = A_1 \exp[-\beta_1(R - R_0)]$, $H_{22}(R) = [A_2 + (B_2/R)^8]\exp(-R/\rho) - 1/R - (\lambda^+ + \lambda^-)/2R^4 - C_2/R^6 - 2\lambda^+\lambda^-/R^7 + \Delta E_0$, and $H_{12}(R) = A_{12}\exp[-\beta_{12}(R - R_x)^2]$, with the model parameters given in [23]. This gives a potential well for the ionic state 2 coupled to a repulsive potential for the neutrals state 1 in the region of their crossing.

Although this is a simple special case, it provides a test of all the important features of our treatment. In particular, it tests the present use of effective forces because it involves a nearly bound motion in the ionic state 2 coupled to a nearly free motion in the neutrals state 1, with very different forces for each independent state. The quantum–classical approach with our effective potentials allows us to follow the populations and coherence of the ionic and neutral states for NaI, starting on its excited state after excitation by a femtosecond pulse, with a quantal distribution of initial conditions. Average distance and velocity changes can also be calculated to gain insight into the nature of the dissociation. The following calculations were done with the effective Hellmann–Feynman forces.

The computational procedure introduces the matrix representation $\hat{\Gamma}_{\mathrm{W}} = |\mathbf{\Phi}\rangle\mathbf{\Gamma}\langle\mathbf{\Phi}|$, in terms of the row matrix $|\mathbf{\Phi}\rangle = [|1,R\rangle, |2,R\rangle]$. The propagation of the density matrix over time requires integration of the sets of coupled differential equations for the quasiclassical trajectories and for the density matrix. Here we work with a diabatic electronic basis set for which $\langle\mathbf{\Phi}|\mathrm{d}\mathbf{\Phi}/\mathrm{d}R\rangle = 0$. The coupled equations are, $\mathrm{d}P/\mathrm{d}t = -\partial H'_{\mathrm{cl}}/\partial R$ and $\mathrm{d}R/\mathrm{d}t = \partial H'_{\mathrm{cl}}/\partial P$ as before, and

$$\mathrm{d}\mathbf{\Gamma}/\mathrm{d}t = (\mathrm{i}\hbar)^{-1}(\mathbf{H}\mathbf{\Gamma} - \mathbf{\Gamma}\mathbf{H}) \tag{29}$$

that must be solved for the initial conditions at $t = t_{\mathrm{in}} : R_{\mathrm{in}} = R(t_{\mathrm{in}})$ and $P_{\mathrm{in}} = P(t_{\mathrm{in}})$, and $\mathbf{\Gamma}_{\mathrm{in}} = \mathbf{\Gamma}(t_{\mathrm{in}})$. This has been done with the relax-and-drive procedure.

Populations and Coherence

We construct the initial DM from the lowest vibrational state of the harmonic well of the ionic potential. At $t = 0$, the wavefunction undergoes a sudden optical promotion to the neutral curve, so that the PWTDM becomes,

$$\Gamma_{11}(P,R) = \pi^{-1}\exp\{-[(R-R_0)/\sigma]^2 - \sigma^2(P-P_0)^2\}, \tag{30}$$

with $\Gamma_{12} = \Gamma_{21} = \Gamma_{22} = 0$.

We have defined three populations [23]: Ionic $\left(\eta_2 = \int_0^\infty \mathrm{d}R \int \mathrm{d}P\Gamma_{22}(R,P,t)\right)$, bound neutral $\left(\eta_1^b = \int_0^{R_x} \mathrm{d}R \int \mathrm{d}P\Gamma_{11}(R,P,t)\right)$ and free neutral $\left(\eta_1^f = \int_{R_x}^\infty \mathrm{d}R\,\mathrm{d}P\Gamma_{11}(R,P,t)\right)$, and introduce the coherence amplitude $\eta_{12} = \int_0^\infty \mathrm{d}R \int \mathrm{d}P\ \Gamma_{12}(R,P,t)$. These quantities evolve in time while coupled to the evolution of a grid of phase space points arising from the initial discretization of the P and R variables, which are then followed as they move over time. A total of 40 points along both P and R (for a total of 1,600 points) have been used in the following calculations.

Numerical results have been obtained solving our equations with the present quantum–classical propagation scheme, and also solving the full quantum coupled differential equations with the split-operator-fast-Fourier-transform (SO-FFT) method [24] to generate wavepacket solutions, for comparison to ascertain the accuracy of our procedure.

The ionic and covalent populations are displayed in Fig. 1. We see oscillations in the populations between ionic and covalent states, repeating approximately every 40,000 au, or about 1 ps. The results from the effective potential quantum–classical Liouville equation (EP-QCLE) is quantitatively similar to the exact results from the SO-FFT up to around 3 ps.

The quantum coherence, shown in Fig. 2, initially peaks through the first crossing, but it is substantially diminished through subsequent crossings. The EP-QCLE shows quantitatively similar results to the SO-FFT calculations.

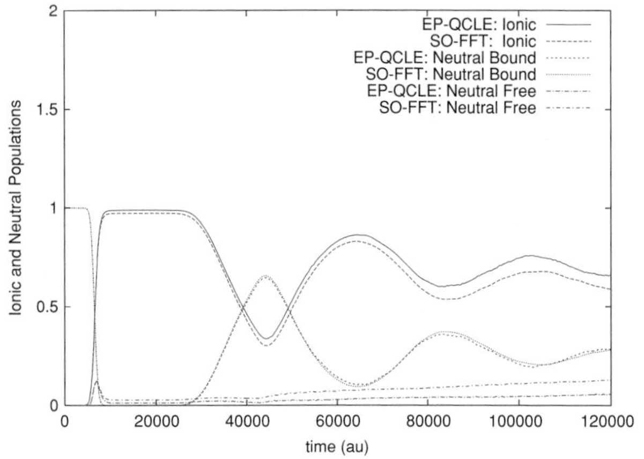

Fig. 1. Ionic and neutral populations over time, from [23]

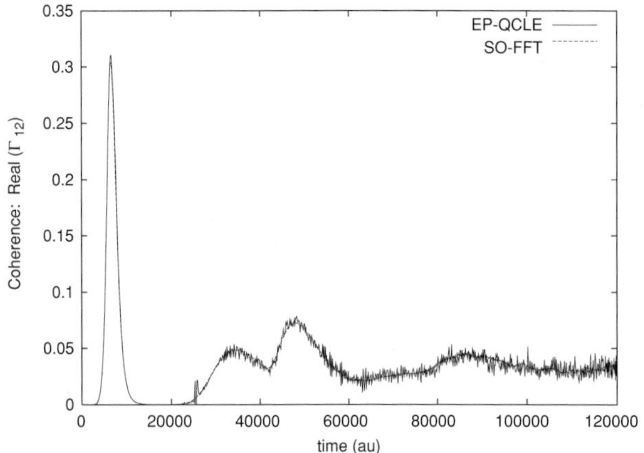

Fig. 2. Coherence as a function of time, from [23]

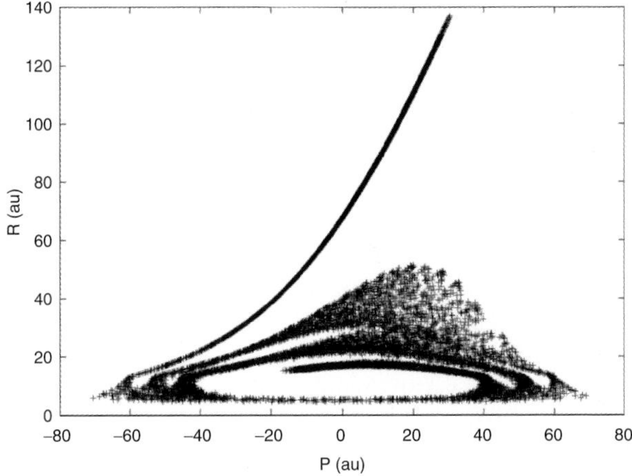

Fig. 3. Phase space grid at the end of the simulation, from [23]

The time evolution of the average value of the position, $\langle R \rangle$, and its dispersion ΔR have also been calculated [23] and show very good agreement with accurate results up to about 1.0 ps. At later times the average maintains qualitative accuracy up to 3 ps, while the dispersion ΔR starts to diverge from the accurate values. The dispersion is much more sensitive to the larger asymptotic populations in the SO-FFT simulation.

Phase Space Evolution

The deformation of the phase space grid has been followed from its initial rectangular shape to the distribution plotted in Fig. 3, and shows characteristics of both free and bound motions.

One set of grid points rapidly moves over time from the center of the grid, quickly straightening to reflect a negligible force on the points. These points represent the asymptotically free neutral components of the PWTDM. A second set of points circles around, gaining velocity and position, then turning. The formed ellipses are characteristic of the phase space of classical particles in a well. Therefore the quasiclassical motion of the PWTDM points under the Hellmann–Feynman force correctly show the features of motion on both ionic and covalent curves. This can also be seen in a sequence containing frames at each 4,000 au [23].

5.2 Collisional Excitation

The PWT of the density operator can be used to describe atomic and molecular collisions involving electronic excitations, by combining a time-dependent many-electron treatment with a quasiclassical description of the atomic motions in a first principles treatment. The terms first principles, ab initio and direct molecular dynamics equivalently refer to a class of methods for studying the dynamical motion of atoms while the electronic structure is generated "on the fly," and atomic forces are computed directly from the electronic structure of the system. A report along these lines [11] has covered treatments combining eikonal (or short wavelength) wavefunctions for the atomic motions and TDMOs for electrons. The present PWT of the DOp allows for a generalization where the short wavelength limit is applied to phase space. To illustrate the procedure, we present here results for the processes M(nl) + Ng $\rightarrow$ M($n'l'$) + Ng, where M is an alkali atom and Ng a noble gas atom, and in particular for Li($2s$) + He $\rightarrow$ Li($2p$) + He. In this case it is possible to obtain accurate results using atomic pseudopotentials, that reduce the many-electron problem to a single-electron case [25].

The electronic hamiltonian, writen in terms of the pseudopotentials $\hat{V}_{\rm el}^{\rm PP}$, using here atomic units with $\hbar = e = m_{\rm e} = 1$ and neglecting spin–orbit coupling, is

$$\hat{H}_{\rm el}^{\rm PP} = -\tfrac{1}{2}\nabla^2_{\boldsymbol{r}_{\rm A}} + \hat{V}_{\rm el}^{\rm PP}\,, \quad \hat{V}_{\rm el}^{\rm PP} = \hat{V}_{\rm A}(\boldsymbol{r}_A) + \hat{V}_{\rm AB}(\boldsymbol{r}_A, \boldsymbol{R}). \tag{31}$$

Here A refers to the alkali atom and B to the noble-gas atom and $\boldsymbol{R}$ is the relative position of the two centers. The atomic pseudopotential $\hat{V}_{\rm A}$ describes the interaction between the valence electron at $\boldsymbol{r}_{\rm A}$ and the center A. The term $\hat{V}_{\rm AB}$ contains the interaction between the electron and the Ng atom, electron–cores and core–core potentials. The PWT allows introduction of a classical-like hamiltonian for the nuclear motions of form

$$\mathcal{H}(\boldsymbol{P}, \boldsymbol{R}) = \frac{\boldsymbol{P}\cdot\boldsymbol{P}}{2M} + \mathcal{V}(\boldsymbol{P}, \boldsymbol{R}), \tag{32}$$

where $\boldsymbol{P}$ is the relative momentum, and the effective potential $\mathcal{V}$ can be written as $\mathcal{V} = {\rm tr}(\hat{\rho}\hat{H}_{\rm el}^{\rm PP})/tr(\hat{\rho})$, where $\hat{\rho} = |\psi\rangle\langle\psi|$ is the electronic density operator. The dynamics is carried out by solving the Hamilton equations

$$\mathrm{d}\boldsymbol{R}/\mathrm{dt} = \partial\mathcal{H}/\partial\boldsymbol{P}, \qquad \mathrm{d}\boldsymbol{P}/\mathrm{dt} = -\partial\mathcal{H}/\partial\boldsymbol{R} \tag{33}$$

coupled to the time-dependent differential equation for the density operator $\hat{\rho}(t)$

$$\mathrm{i}\partial\hat{\rho}/\partial t = \hat{H}_{\rm el}^{\rm PP}\hat{\rho} - \hat{\rho}\hat{H}_{\rm el}^{\rm PP}, \tag{34}$$

where the time derivative here is $(\partial/\partial t)_{\boldsymbol{r}} = (\partial/\partial t)_{\boldsymbol{r},\boldsymbol{R}} + (\mathrm{d}\boldsymbol{R}/\mathrm{dt})\cdot\partial/\partial\boldsymbol{R}$, so that it implicitly includes gradient couplings between electronic states.

The TDMO ψ can be expanded as a linear combination of the traveling atomic functions ξ_μ

$$\psi_i(\boldsymbol{r}, t) = \textstyle\sum_\mu \xi_\mu(\boldsymbol{r}, t) c_{i\mu}(t), \quad \xi_\mu(\boldsymbol{r}, t) = T_m(\boldsymbol{r}, t)\chi_\mu(\boldsymbol{r}), \tag{35}$$

where the c's are complex expansion coefficients, $\chi_\mu(\boldsymbol{r})$ is an atomic orbital centered at core position $\boldsymbol{R}_m(t)$ for the electron with position $\boldsymbol{r}$, and $T_m(\boldsymbol{r}, t)$ is an electron translation factor.

The density operator in the basis of traveling atomic orbitals is written as

$$\hat{\rho}(t) = \sum_{\mu\nu} \mid \xi_\mu \rangle P_{\mu\nu}(t) \langle \xi_\nu \mid = \mid \boldsymbol{\xi} \rangle \mathbf{P} \langle \boldsymbol{\xi} \mid, \tag{36}$$

where $\mathbf{P}$ is the density matrix, with matrix elements $P_{\mu\nu}(t) = \sum_{\mathrm{occ}\ i} c^*_{\mu i}(t)$ $c_{\nu i}(t)$, and the differential equation for the density matrix is then transformed into

$$\mathrm{i}\dot{\mathbf{P}} = \mathbf{W}\mathbf{P} - \mathbf{P}\mathbf{W}^\dagger \ , \quad \mathbf{W} = \mathbf{S}^{-1}\mathbf{H}_\mathrm{T}, \tag{37}$$

where $\mathbf{S} = \langle \boldsymbol{\xi} \mid \boldsymbol{\xi} \rangle$ is the atomic overlap matrix, and $\mathbf{H}_\mathrm{T}$ is the hamiltonian matrix in the traveling atomic basis.

Calculations require generating electron integrals, as described in [25], to construct the matrices in $\mathbf{W}$. The density matrix $\mathbf{P}$ can then be propagated with the relax-and-drive procedure mentioned before. Some results are presented in Fig. 4, [25] obtained with four different sets of atomic basis functions: Basis I (6s5p2d/4s4p2d); Basis II (6s5p3d/4s4p3d); Basis III (7s6p4d/5s5p4d); and Basis IV (9s9p5d/7s7p5d). Results for sets III and IV are indistinguishable in the graphics. The results with the largest basis set are in very good agreement with experiments over a wide range of laboratory collision energies (1.0–10.0 keV), and illustrate the importance of using large basis sets to obtain reliable results.

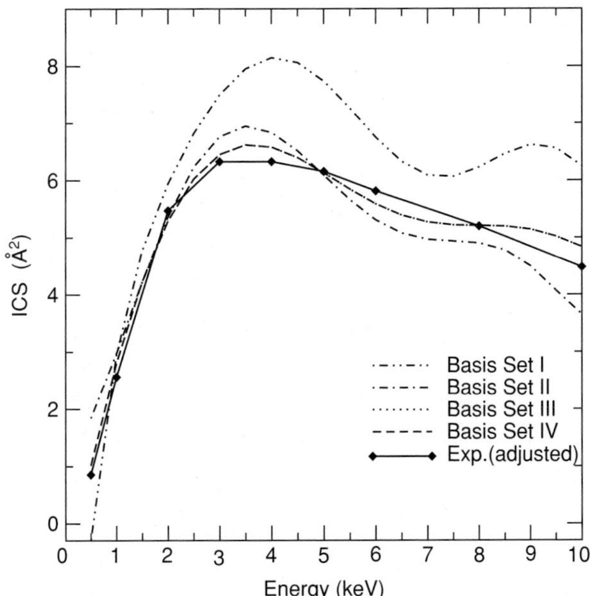

Fig. 4. Integral cross-sections for the excitation Li(2s) to Li(2p) in Li–He collisions (E_lab = 1.0–10.0 keV), comparing results with several basis set and experimental data, from [25]

6 Dissipative Dynamics in Extended Systems

6.1 Equation of Motion of the Reduced Density Operator

Localized dynamics in a complex molecular system, induced by light absorption or by collisions, is accompanied by energy dissipation into the medium and by effects of fluctuation forces. When a molecular subsystem of interest is strongly coupled to its surroundings, it is convenient to define a primary or p-region including the subsystem and neighboring atoms, and a remaining secondary or s-region. This is illustrated in Fig. 5, for energy dissipation between times t and t'. The dissipation is generally delayed and involves a memory term in the dissipative rate. In special cases the s-region undergoes an instantaneous dissipation at each time t, or an instantaneous and time-independent dissipation (a Markovian dissipation).

The extension of the previous treatment to a system in a medium starts from $\hat{\Gamma}(t)$ to derive the equation of motion of the reduced density operator for the p-region, (RDOp) $\hat{\rho}(t) = \mathrm{tr}_s[\hat{\Gamma}(t)]$, involving the trace over s-region variables. This equation includes a dissipative term, given by a Liouville superoperator $\hat{\mathcal{L}}_p^{(D)}$ that can be obtained from the interaction with an s-region. The PWT is applied only to the p-region so that the quantum and classical Hamiltonians of the previous section refer to the p-region. The s-region can be described in terms of its collective motions and a distribution of initial

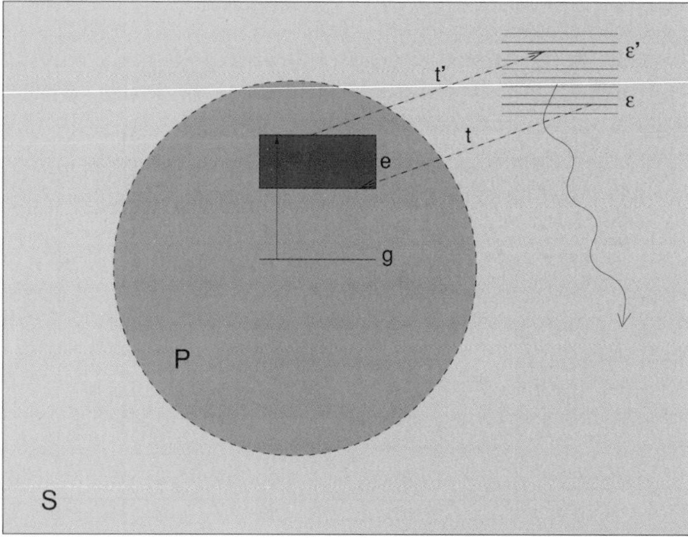

Fig. 5. Interaction of primary (p) and secondary (s) regions for transitions between p-states g and e, coupled to an s-region where dissipation of energy occurs between times t and t'

motion amplitudes [17,26] when the region is a surface or a crystal, and it can be described by means of stochastic dynamics or hydrodynamics when it is a fluid or dense gas.

The dissipative term in the L–vN equation has been derived in several ways. Three models of current interest are based on dissipative potentials [26–28], on dissipative rate operators [29, 30], and on a memory obtained to second-order in the p–s coupling, [8, 31, 32]. The treatments take simpler forms when the coupling of p- and s-regions is of the bilinear form $\hat{H}_{\mathrm{ps}} = \sum_j \hat{A}_{\mathrm{p}}^{(j)} \hat{B}_{\mathrm{s}}^{(j)}$, in which case the dissipation effects can be expressed in terms of time-correlation functions of the s-region.

Strong couplings must be expected between p- and s-regions when the latter is activated for example by light absorption or atomic collisions, or when there are chemical bonds between atoms in the p- and s-regions. A perturbative treatment of their interaction would not suffice; an alternative is to obtain approximate solutions to the L–vN equation assuming that they can be *factorized after averaging* over the distribution of initial s-region properties, as

$$\overline{\hat{\Gamma}}(t) = \hat{\rho}(t) \otimes \overline{\hat{\Gamma}^{(\mathrm{s})}}(t) \tag{38}$$

at all times, with $\overline{\hat{\Gamma}^{(\mathrm{s})}}$ describing the s-region and with normalizations $\mathrm{tr}_{\mathrm{p}}\hat{\rho} = 1$ and $\mathrm{tr}_{\mathrm{s}}\overline{\hat{\Gamma}^{(\mathrm{s})}} = 1$. If for example the s-region involves collective modes such as phonons or charge density waves, then the averaging is done over the distribution of initial values of mode amplitudes and phases. The above product form is more general than the usual Fano factorization [3], insofar the latter assumes that the secondary region can be described by a time-independent (and usually equilibrium) density operator. Our factorization allows for active media, as found in femtosecond pulse excitations of complex systems. This expression leads to coupled equations for p- and s-regions, which can be constructed to provide mean-field solutions, or more generally to give selfconsistently correlated states, as we next describe.

To derive an equation for $\hat{\rho}(t)$ including p–s correlations, we start from the full L–vN equation for $\hat{\Gamma}(t)$. Introducing Liouville superoperators shown as caligraphic symbols, such as $\hat{\mathcal{H}} = [\hat{H}, \bullet]$, to write

$$\mathrm{i}\hbar\partial\hat{\Gamma}/\partial t = \hat{\mathcal{H}}\hat{\Gamma}(t). \tag{39}$$

We will transform this into an integrodifferential form as has been done to display correlations in the s-region [4], summarizing the derivation first without an external field. Solving formally for the density operator, with a decomposition $\hat{H} = \hat{F} + \hat{H}_F$, where $\hat{F}$ is a convenient, possibly time-dependent, effective hamiltonian to be defined, we have

$$\hat{\Gamma}(t) = \hat{\mathcal{U}}_0(t)\hat{\Gamma}(0) + (\mathrm{i}\hbar)^{-1} \int_0^t \mathrm{d}t'\, \hat{\mathcal{U}}_0(t,t')\hat{\mathcal{H}}_F(t')\hat{\Gamma}(t'), \tag{40}$$

where $\hat{\mathcal{U}}_0(t,t') = \exp_T[-\mathrm{i}\int_{t'}^{t} \mathrm{d}t''\hat{\mathcal{F}}(t'')/\hbar]$, written as a time-ordered exponential. Replacing this integral form in the original L–vN equation gives

$$\mathrm{i}\hbar\partial\hat{\Gamma}/\partial t = \hat{\mathcal{F}}\hat{\Gamma}(t) - \hat{\mathcal{R}}(t)\hat{\Gamma}(0) + (\mathrm{i}\hbar)^{-1}\int_0^t \mathrm{d}t'\,\hat{\mathcal{M}}(t,t')\hat{\Gamma}(t'). \tag{41}$$

Here $\hat{\mathcal{R}}(t) = \hat{\mathcal{H}}_F\hat{\mathcal{U}}_0(t)$ is an energy fluctuation term and $\hat{\mathcal{M}}(t,t') = \hat{\mathcal{H}}_F\hat{\mathcal{U}}_0(t,t')$ $\hat{\mathcal{H}}_F = \hat{\mathcal{R}}(t)\hat{\mathcal{R}}(t')^{\dagger}$ is a dissipative memory term. Taking the trace over s-variables on both sides one obtains a generalized Langevin equation (or GLE) for $\hat{\rho}$.

To obtain equations for selfconsistently correlated $\hat{\rho}$ and $\overline{\hat{\Gamma}^{\mathrm{s}}}$, it is convenient to make the choice of effective Hamiltonian

$$\hat{F} = \hat{F}_{\mathrm{p}} + \hat{F}_{\mathrm{s}} - \langle\langle \hat{H}_{\mathrm{ps}}\rangle\rangle\ , \quad \hat{F}_{\mathrm{p}} = \hat{H}_{\mathrm{p}} + \hat{G}_{\mathrm{p}} \tag{42}$$

with $\hat{G}_{\mathrm{p}} = \mathrm{tr}_{\mathrm{s}}[\hat{H}_{\mathrm{ps}}\overline{\hat{\Gamma}^{(\mathrm{s})}}]$ and $\langle\langle \hat{H}_{\mathrm{ps}}\rangle\rangle = \mathrm{tr}_{\mathrm{ps}}[\hat{H}_{\mathrm{ps}}\hat{\rho}\overline{\hat{\Gamma}^{(\mathrm{s})}}]$, and similarly for the s-operators. This definition leads to $\hat{H}_F = \hat{H}_{\mathrm{ps}} - (\hat{G}_{\mathrm{p}} + \hat{G}_{\mathrm{s}}) + \langle\langle \hat{H}_{\mathrm{ps}}\rangle\rangle$, a residual coupling due to the nonfactorized correlation of motions in the p- and s-regions which averages to zero at all times. Using the factorized form of $\overline{\hat{\Gamma}}$ on the right hand side of the equation for $\hat{\rho}$, one obtains

$$\mathrm{i}\hbar\partial\hat{\rho}/\partial t = \hat{\mathcal{F}}_{\mathrm{p}}\hat{\rho}(t) - \hat{\mathcal{R}}_{\mathrm{p}}(t)\hat{\rho}(0) + (\mathrm{i}\hbar)^{-1}\int_0^t \mathrm{d}t'\,\hat{\mathcal{M}}_{\mathrm{p}}(t,t')\hat{\rho}(t'), \tag{43}$$

where $\hat{\mathcal{R}}_{\mathrm{p}} = \mathrm{tr}_{\mathrm{s}}[\hat{\mathcal{R}}\overline{\hat{\Gamma}^{(\mathrm{s})}}]$ and $\hat{\mathcal{M}}_{\mathrm{p}} = \mathrm{tr}_{\mathrm{s}}[\hat{\mathcal{M}}\overline{\hat{\Gamma}^{(\mathrm{s})}}]$ are expressions for fluctuation and dissipative terms from the p–s coupling. Dissipation in the s-region can be similarly treated, making the same stochastic medium assumptions to obtain the equation of motion for $\hat{\Gamma}^{(\mathrm{s})}(t)$.

An alternative procedure for the derivation of dissipative rates relies on projection operator techniques for the L–vN equation [8, 33–36]. The treatment is more general than the SCF one when the s-region is at equilibrium, but involve more complicated equations. For a bath at equilibrium, with DOp $\hat{\Gamma}^{(\mathrm{s})}_{\mathrm{eq}}$, a projection superoperator $\hat{P}_{\mathrm{eq}}$ is defined by $\hat{\Gamma}_{\mathrm{P}}(t) = \hat{\mathcal{P}}_{\mathrm{eq}}\hat{\Gamma}(t) = \hat{\Gamma}^{(\mathrm{p})}(t)\hat{\Gamma}^{(\mathrm{s})}_{\mathrm{eq}}/\mathrm{tr}_{\mathrm{s}}(\hat{\Gamma}^{(\mathrm{s})}_{\mathrm{eq}})$, where $\hat{\Gamma}^{(\mathrm{p})} = \mathrm{tr}_{\mathrm{s}}(\hat{\Gamma})$ is as before the reduced density operator of the p-region, and a complementary projection superoperator is defined by $\hat{\mathcal{Q}}_{\mathrm{eq}} = \hat{\mathcal{I}} - \hat{\mathcal{P}}_{\mathrm{eq}}$. The projected density operator $\hat{\Gamma}_{\mathrm{P}}$ has the factorized form of an SCF approximation, here for a medium at equilibrium, and $\hat{\Gamma}_{\mathrm{Q}}(t) = \hat{Q}_{\mathrm{eq}}\hat{\Gamma}(t)$ describes correlation corrections. Projecting the L–vN equation with both $\hat{\mathcal{P}}_{\mathrm{eq}}$ and $\hat{\mathcal{Q}}_{\mathrm{eq}}$, one finds coupled equations of motion for $\hat{\Gamma}_{\mathrm{P}}$ and $\hat{\Gamma}_{\mathrm{Q}}$, which can be formally solved to obtain an integrodifferential equation for $\hat{\Gamma}_{\mathrm{P}}$. It provides a generalization with a delayed dissipation term containing a memory superoperator.

When the bilinear coupling $\hat{H}_{ps} = \sum_j \hat{A}_p^{(j)} \hat{B}_s^{(j)}$ is weak, the memory can be approximated to second-order in the coupling, and is expressed in terms of time-correlation functions $C_s^{(jk)}(t) = \hbar^{-2} \langle \Delta \hat{B}_s^{(j)}(t) \Delta \hat{B}_s^{(k)}(0) \rangle$ of the s-region, where $\Delta \hat{B}_s^{(j)} = \hat{B}_s^{(j)} - \langle \hat{B}_s^{(j)} \rangle_s$. Taking the trace over s-variables then gives, for operators in the interaction picture generated by $\hat{H}_0 = \hat{H}_p + \hat{H}_s$, [8]

$$
\begin{aligned}
\partial \hat{\rho} / \partial t = (\mathrm{i}\hbar)^{-1} \sum_j \langle B_s^{(j)} \rangle [\hat{A}_p^{(j)}, \hat{\rho}(t)] \\
- \sum_{j,k} \int_0^t \mathrm{d}t' \, \{ C_s^{(jk)}(t-t') [\hat{A}_p^{(j)}(t), \hat{A}_p^{(k)}(t') \hat{\rho}(t')] \\
- C_s^{(kj)}(-t+t') [\hat{A}_p^{(j)}(t), \hat{\rho}(t') \hat{A}_p^{(k)}(t')]_- \},
\end{aligned}
\tag{44}
$$

which identifies a memory kernel superoperator $\hat{\mathcal{K}}(t, t')$ in this approximation. A variety of methods have been developed to integrate these equations of motion [37–41].

6.2 Instantaneous Dissipation

The equation for $\hat{\rho}$ is simplified when the s-region can be described as a stochastic medium where fluctuations relax rapidly toward mean values and the delay of the dissipative memory can be neglected. This can be done in the context of the selfconsistent factorization when (1) the fluctuation forces average to zero on the primary time scale, i.e., $\overline{\hat{\mathcal{R}}(t)\hat{\Gamma}(0)} = 0$; and (2) the memory kernel describes instantaneous dissipation, so that $\overline{\hat{\mathcal{M}}(t,t')\hat{\Gamma}(t')} = \delta(t-t')\hat{\mathcal{W}}(t)\hat{\Gamma}(t)$, giving a time-dependent dissipative potential superoperator. The equation for $\hat{\rho}(t)$ is then

$$
\partial \hat{\rho} / \partial t = (\mathrm{i}\hbar)^{-1} [\hat{F}_p, \hat{\rho}(t)] + \hat{\mathcal{L}}_p^{(D)} \hat{\rho}(t), \tag{45}
$$

where $\hat{\mathcal{L}}_p^{(D)} = -\hat{\mathcal{W}}_p(t)/(2\hbar)$ and $\hat{\mathcal{W}}_p(t) = \overline{\mathrm{tr}_s[\hat{\mathcal{W}}(t)\hat{\Gamma}^{(s)}(t)]}$ is an instantaneous dissipative potential quadratic in $\hat{H}_F$. This dissipative potential superoperator depends generally on the time t, but in some cases it can be assumed to be independent of t, giving a Markovian approximation. The dissipative term cannot be written as a commutator of the RDOp with a Hamiltonian, and therefore it is necessary to solve the differential equation directly for the RDOp.

A popular choice for the Markovian dissipative superoperator follows from the so-called Lindblad-type expression, [29,30] which amounts in our notation to

$$
\hat{\mathcal{L}}_p^{(D)} \hat{\rho}(t) = \sum_L \left\{ \hat{C}_p^{(L)} \hat{\rho}(t) \hat{C}_p^{(L)\dagger} - \left[\hat{C}_p^{(L)\dagger} \hat{C}_p^{(L)}, \hat{\rho}(t) \right]_+ \Big/ 2 \right\}, \tag{46}
$$

where the $\hat{C}_p^{(L)}$ are operators in the p-region constructed from information about relaxation and decoherence times in the s-region. This form maintains complete positivity, and also leads to an RDOp $\hat{\rho}(t)$ of constant norm.

The operators $\hat{C}_{\mathrm{p}}^{(L)}$ can be constructed as combinations of position and momentum operators in the p-region, or from empirical transition rates $k_{\alpha' \leftarrow \alpha}$ between orthonormal eigenstates $\Phi_{\alpha}^{\mathrm{p}}$ and $\Phi_{\alpha'}^{p}$ of $\hat{F}_p$ [42]. The index L then refers to a given transition $\alpha \rightarrow \alpha'$, and the corresponding operator is $\hat{C}_{\mathrm{p}}^{(L)} = \sqrt{k_{\alpha' \leftarrow \alpha}} |\Phi_{\alpha'}^{\mathrm{p}}\rangle\langle\Phi_{\alpha}^{\mathrm{p}}|$.

The PWT of the resulting equation of motion can be obtained expressing a product $[\hat{A}\hat{B}\hat{C}]_{\mathrm{W}}$ in terms of each operator PWT, and keeping the first-order in the operator $\hbar \overleftrightarrow{\Lambda}$. The resulting equation of motion for $\hat{\rho}_{\mathrm{W}}$ also leads to a new equation for the phase space density γ, after taking the trace over quantum variables.

Alternatively, the equation of motion can be obtained from the Hamiltonian $\hat{F}_{\mathrm{pW}}$, using its eigenstates $|\Phi_I^{\mathrm{p}}(P, R)\rangle$ for mixed quantum–classical states I at each phase space point (P, R) to construct the operators $\hat{C}_{\mathrm{pW}}^{(L)}(P, R)$ of the Lindblad expression, with semiempirical rates $k_{J \leftarrow I}(P, R)$. The resulting matrix equation is

$$\frac{\mathrm{d}\boldsymbol{\rho}}{\mathrm{d}t} = (\mathrm{i}\hbar)^{-1}[\mathbf{F}_{\mathrm{qu}} + \mathbf{F}_{\mathrm{cq}} - \mathrm{i}\hbar\langle\boldsymbol{\Phi}|\frac{\mathrm{d}\boldsymbol{\Phi}}{\mathrm{d}t}\rangle, \boldsymbol{\rho}] \\ -(1/2)\sum_L \{[\mathbf{C}^{(L)\dagger}\mathbf{C}^{(L)}, \boldsymbol{\rho}]_+ - 2\mathbf{C}^{(L)\dagger}\boldsymbol{\rho}\mathbf{C}^{(L)}\}, \quad (47)$$

where $\mathbf{C}^{(L)} = [\sqrt{k_{J \leftarrow I}}]$ is an $N_B \times N_B$ matrix. The DM depends on the initial conditions in phase space, and it can be obtained as before on a grid, now constructed in the p-region phase space. The matrix equation is equivalent to a set of coupled linear equations for functions of time, which must be simultaneously integrated with the classical density in phase space, $\gamma(t)$. The propagation of the DM, which generally changes rapidly over time compared to $\gamma(t)$, can again be done with our *relax-and-drive* procedure. This advances time from t_0 to t_1 by first generating a relaxing $\boldsymbol{\rho}^0(t)$ from $\mathbf{F}(t_0)$ and then correcting it by quadratures to account for the driving term $\Delta\mathbf{F}(t) = \mathbf{F}(t) - \mathbf{F}(t_0)$. Similar equations can be derived for the time evolution of the RDOp and RDM in the s-region.

7 Adsorbate Dynamics

7.1 Photodesorption

The main steps in the femtosecond photodesorption of CO from Cu(001) are excitation by the substrate, followed by energy transfer to the adsorbate region and break-up of the Cu–C bond [43]. The desorption dynamics is fast compared with vibrational motions in the substrate metal, so that only electronic excitation and de-excitation of its electrons must be considered. The steps are as follows.

$$\mathrm{CO}(\mathbf{v})/\mathrm{Cu}(001) \xrightarrow{\text{light}} \mathrm{CO}(\mathbf{v})/\mathrm{Cu}(001)^* \text{ (light absorption)}$$

$$\mathrm{CO}(\mathbf{v})/\mathrm{Cu}(001)^* \rightarrow \mathrm{CO}(\mathbf{v}') + \mathrm{Cu}(001)^* \text{ (break-up)}$$

corresponding to an indirect photodesorption, where **v** indicates the collection of vibrational quantum numbers for the normal modes of the adsorbate. The modes with the lowest excitation energy, and most likely to be excited during desorption, are the so-called frustrated translation and frustrated rotations [44]. The position of the center of mass of the CO above the surface is called Z, the frustrated translation coordinate parallel to the surface is x, and the frustrated rotation angles are (θ, ϕ), as shown in Fig. 6, in a cluster model CO/Cu_6 for the adsorbate site. The potential energy surfaces and distance dependent transition dipoles were calculated from the electronic structure of CO/Cu_n clusters and were parametrized for calculations of the dynamics [28].

In the application that follows, an external electric field pulse $\mathcal{E}(t)$ lasting femtoseconds, first excites the s-region and leads to a density operator $\hat{\Gamma}^{\mathrm{s}} = \hat{\Gamma}^{\mathrm{s}}_0 + \hat{\Gamma}^{\mathrm{s}}_l$ where the second term results from the response of the s-region to the field. This then shows as an indirect excitation of the p-region, through the SCF potential $\hat{G}_{\mathrm{p}} = \hat{G}^0_{\mathrm{p}} + \hat{G}^l_{\mathrm{p}}$. The second term here is expressed as the field coupled to an effective p-dipole operator which can be parametrized from experiment, or alternatively it is written as the coupling of the p-dipole operator to an effective field in the p-region, as has been recently derived from a theory of the nonlinear response of the s-region to a pulse of light [45].

In our model, the transfer of energy from the substrate metal to the adsorbate region is mediated by the dipole–dipole interaction

$$\hat{H}_{\mathrm{ps}} = \int \mathrm{d}^3 r_{\mathrm{s}} \, \frac{\hat{\boldsymbol{D}}_{\mathrm{p}}(\boldsymbol{r}_{\mathrm{p}}) \cdot \hat{\boldsymbol{P}}_{\mathrm{s}}(\boldsymbol{r}_{\mathrm{s}}) - 3[\hat{\boldsymbol{D}}_{\mathrm{p}}(\boldsymbol{r}_{\mathrm{p}}) \cdot \boldsymbol{n}_{\mathrm{p}}]\hat{\boldsymbol{P}}_{\mathrm{s}}(\boldsymbol{r}_{\mathrm{s}}) \cdot \boldsymbol{n}_{\mathrm{s}}}{|\boldsymbol{r}_{\mathrm{s}} - \boldsymbol{r}_{\mathrm{p}}|^3} \tag{48}$$

from which the SCF potential $\hat{G}_{\mathrm{p}}$, a dissipative potential $\hat{W}_{\mathrm{p}}$ and dissipative rates in the s-region can be derived. Here $\hat{\boldsymbol{D}}_{\mathrm{p}}$ is the dipole operator of the p-region, $\hat{\mathbf{P}}_{\mathrm{s}}$ is the dipole operator per unit volume in the s-region, and $\boldsymbol{n}_i = \boldsymbol{r}_i / r_i$, $i = \mathrm{p}, \mathrm{s}$, denotes a unit vector in the p- or s-region. This simplifies for

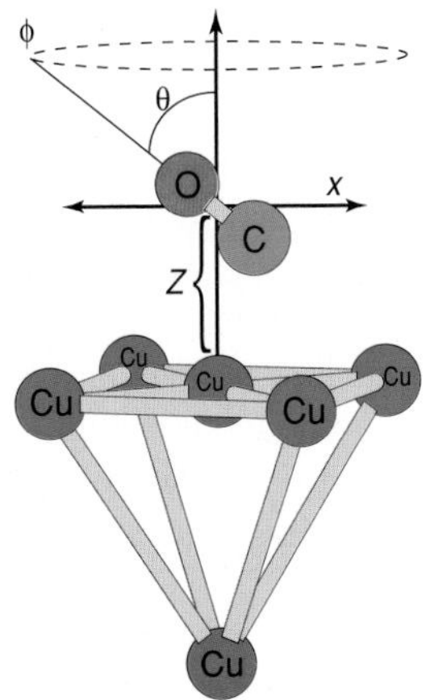

Fig. 6. The CO/Cu_6 cluster model of CO/Cu(001) for the adsorbate region

an electric field of long wavelength polarized parallel to the surface, to give for the SCF potential [45]

$$\hat{G}_{\mathrm{p}}(\boldsymbol{r}_{\mathrm{p}},t)=\hat{D}_{\mathrm{p}}(z_{\mathrm{p}})D_{\mathrm{s}}(t;Z_{\mathrm{s}})|Z_{\mathrm{s}}-z_{\mathrm{p}}|^{-3}, \tag{49}$$

where $D_{\mathrm{s}}(t;Z_{\mathrm{s}})$ is the average substrate dipole induced by the applied field inside the metal at distance Z_{s}.

In our previous work [27,28,45], we have implemented the dissipative potential approach in a computationally convenient way, starting instead with a total density operator expressed in terms of density amplitudes $\Psi_\mu(t)$ with statistical weights w_μ, as $\hat{\Gamma}=\sum_\mu w_\mu|\Psi_\mu\rangle\langle\Psi_\mu|$. An average over initial conditions in the s-region is assumed to give factorized weights $w_\mu = w^p_\alpha w^s_\beta$ and amplitudes $\overline{\Psi_\mu(t)} = \Psi^p_\alpha(t)\Psi^s_\beta(t)$, used to construct as above an integrodifferential equation for the p-region amplitude. The p-density operator is

$$\hat{\rho}(t)=\sum_\alpha w_\alpha|\Psi^p_\alpha\rangle\langle\Psi^p_\alpha| \tag{50}$$

and the assumptions of instantaneous dissipation give then p-amplitude equations

$$\mathrm{i}\hbar\frac{\partial}{\partial t}|\psi^p_\alpha\rangle=(\hat{F}_{\mathrm{p}}-\mathrm{i}\hat{W}_{\mathrm{p}}/2)|\psi^{\mathrm{p}}_\alpha(t)\rangle, \tag{51}$$

where now $\hat{W}_{\mathrm{p}}$ is a positive dissipative operator quadratic in the residual coupling $\hat{H}_F$ [27], given by

$$\hat{W}_{\mathrm{p}}(t)=(2/\hbar)\int_0^t \mathrm{d}t'\ \overline{\mathrm{tr}_{\mathrm{s}}[\hat{H}_F\hat{U}_0(t,t')\hat{H}_F\hat{\Gamma}^{\mathrm{s}}(t')]} \tag{52}$$

and the normalized p-amplitudes are $|\Psi^p_\alpha\rangle = |\psi^p_\alpha\rangle/\langle\psi^p_\alpha|\psi^p_\alpha\rangle$. This explicit form for the dissipative potential allows for its calculation or parameterization starting with an atomic model of the p-region. Additional details may be found in [27,28].

Instead of trying to describe the s-region in full detail, it is enough to follow its dynamics only to the extent needed to model the phenomena of interest in the p-region. This can be achieved using a description of the s-region in terms of time-dependent macroscopic variables $T(t)$ and $\bar{N}(t)$, the temperature, and number of electrons in the substrate, and of a reduced one-electron density operator $\hat{\gamma}^{(\mathrm{s})}(t)$, involving a subset of energy band states $\{\phi^{(\mathrm{s})}_\lambda\}$ of the s-region. Equations for the time-evolution of $T(t)$, $\bar{N}(t)$, and $\hat{\gamma}^{(\mathrm{s})}(t)$ can be derived from $\hat{\Gamma}^{(\mathrm{s})}(t)$. These quantities appear in the dissipative potential through (52), and therefore the p-region dynamics depends on their values.

Insofar the electronic relaxation in the metal is fast, dissipation can be assumed instantaneous, and the s-dissipative rate superoperator can be taken as the Lindblad form

$$\hat{\mathcal{L}}^{(D)}_{\mathrm{s}}\hat{\gamma}^{(\mathrm{s})}(t)=\sum_L\{\hat{C}^{(L)}_{\mathrm{s}}\hat{\gamma}^{(\mathrm{s})}(t)\hat{C}^{(L)\dagger}_{\mathrm{s}}-[\hat{C}^{(L)\dagger}_{\mathrm{s}}\hat{C}^{(L)}_{\mathrm{s}},\hat{\gamma}^{(\mathrm{s})}(t)]_+/2\}. \tag{53}$$

One way to implement this, already used in studies of photodesorption [42,45], is to make the choice $\hat{C}_{\mathrm{s}}^{(L)} = \sqrt{\kappa_{\lambda\to\lambda'}}|\phi_{\lambda'}^{(\mathrm{s})}\rangle\langle\phi_{\lambda}^{(\mathrm{s})}|$, where the transition rates $\kappa_{\lambda\to\lambda'}$, obtained from separate calculations or from experiment, can be used to construct the dissipative rate operator. This leads to an equation of motion for the reduced matrix $\boldsymbol{\gamma}^{(\mathrm{s})}(t)$ with elements $\gamma_{\lambda'\lambda}^{(\mathrm{s})}(t)$ in a basis of stationary s-states.

To summarize, the description of coupled p- and s-regions requires the solution of the following set of coupled differential equations.

$$
\begin{aligned}
\frac{\partial}{\partial t}\psi_{\alpha}^{p}(\boldsymbol{X},t) &= (\mathrm{i}\hbar)^{-1}[\hat{F}_{\mathrm{p}}(t) - \mathrm{i}\hat{W}_{\mathrm{p}}(t)/2]\psi_{\alpha}^{\mathrm{p}}(\boldsymbol{X},t) \ \ (\mathrm{A}),\\
\mathrm{d}T/\mathrm{d}t &= F[T(t),N(t)]\,, \qquad \mathrm{d}N/\mathrm{d}t = G[T(t),N(t)] \ \ (\mathrm{B}),\\
\mathrm{d}\boldsymbol{\gamma}^{(\mathrm{s})}/\mathrm{d}t &= (\mathrm{i}\hbar)^{-1}[\mathbf{F}_{\mathrm{s}}(t),\boldsymbol{\gamma}^{(\mathrm{s})}(t)] + \mathcal{L}_{\mathrm{s}}^{(D)}\boldsymbol{\gamma}^{(\mathrm{s})}(t) \ \ (\mathrm{C}),
\end{aligned} \tag{54}
$$

where $\boldsymbol{X}$ is the collection of atomic variables. Here the functions F and G can be obtained from treatments of near equilibrium processes and contain macroscopic parameters such as heat capacities, excitation rates, and relaxation rates [46]. The hamiltonian operator $\hat{F}_{\mathrm{p}}(t)$ and the matrix $\mathbf{F}_{\mathrm{s}}(t)$ of the effective hamiltonian in the s-region are shown to be time dependent, to allow for inclusion of couplings with an external light pulse of electric field $\mathcal{E}(t)$. The set of coupled equations in (A),(B),(C) must be solved coupled to each other.

To implement a numerical solution of these equations, it is further necessary to transform the partial differential equation of the p-density amplitudes into coupled ordinary differential equations in time. This can be done expanding the amplitudes in a basis of electronic states $\{|\Phi_{J}^{(\mathrm{el})}(\boldsymbol{X})\rangle\}$, for electronic states $J = g, e$, or more generally in a basis of vibronic states. In what follows the p-region variables have been assumed quantal in nature, with Z discretized on a grid, and (x,θ,ϕ) motions described with basis functions. Introducing a basis set of vibronic states

$$
|\Phi_{J\mathbf{v}}(Z,x,\theta,\phi)\rangle = |\phi_{J}^{(\mathrm{el})}(Z,x,\theta,\phi)\rangle\phi_{v_x}^{\mathrm{T}}(x)U_r(\theta)V_s(\phi), \tag{55}
$$

where the ket indicates an electronic state for fixed nuclear positions, and $\phi_{v_x}^{\mathrm{T}}$, U_r and V_s are basis functions suitable for the surface vibrational modes with quantum numbers $\mathbf{v} = (v_x, r, s)$, the p-amplitude $\psi_{g\mathbf{v}}$ is expanded as

$$
|\psi_{g\mathbf{v}}(Z,x,\theta,\phi,t)\rangle = \sum_{J,\mathbf{v}'} |\Phi_{J\mathbf{v}'}(Z,x,\theta,\phi)\rangle\psi_{J\mathbf{v}',g\mathbf{v}}^{(nu)}(Z,t) \tag{56}
$$

and the equation for the matrix $\boldsymbol{\psi}^{(nu)}(Z,t)$ of coefficient functions is

$$
\frac{\partial\psi^{(nu)}}{\partial t} = (\mathrm{i}\hbar)^{-1}[\hat{\mathbf{F}}_{\mathrm{p}} - \mathrm{i}\mathbf{W}_{p}/2 - \mathcal{E}_{\mathrm{p}}(t)\mathbf{D}_{\mathrm{p}}]\psi^{(nu)} \ \ (\mathrm{A}'). \tag{57}
$$

The equations in sets (A'), (B), and (C) are all coupled, but sets (B) and (C) can first be integrated over time to obtain the response of the s-region, and their results can be interpolated over time as needed to integrate the set (A'), where the effective field in the p-region must be obtained from that response.

The equations have been solved with a split-operator propagator [23] modified to include the dissipative potential term, and using a fast Fourier transform on a Z-grid of $N_G^{(\mathrm{p})}$ values. The effective electric field in the p-region, $\mathcal{E}_\mathrm{p}(t)$, has been obtained from the nonlinear response of the metal substrate as explained in [47]. The calculations were started with the system in the ground electronic state, and its vibrational motion along Z given by a wavefunction $\phi_{v_Z}(Z)$.

Desorption yields Y_α from initial vibrational-electronic state $\alpha = (g, v_Z, \mathbf{v})$, are obtained integrating the probabilities from a desorption distance Z_D to infinity, as

$$Y_\alpha(t) = \sum_{I\mathbf{v}'} \int_{Z_\mathrm{D}}^{\infty} \mathrm{d}Z \; |\psi^{(nu)}_{I\mathbf{v}',\alpha}(Z,t)|^2 , \tag{58}$$

which also provides the time evolution of the desorption yield as a pulse of light is applied.

Calculations have been done for 1-D, 2-D, and 3-D models, with variables Z, (Z, x), and (Z, x, θ), respectively. Comparison of results from the models with experimental data [48], are shown in Fig. 7. A single value of the yield was fit to experiment at a fluence of $3.5\,\mathrm{mJ\,cm^{-2}}$ [45, 47].

This comparison establishes that the treatment is realistic and that the 1-D model is useful for studies at low fluence. The 2-D models give similar results for the smaller fluence values, and are very close to the 1-D model. However as the fluence increases the model including the frustrated rotation

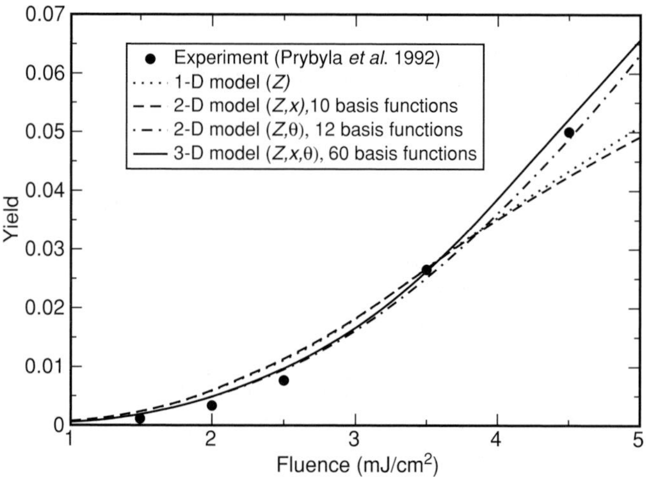

Fig. 7. Yield Y of CO desorbed from Cu(001) vs. the laser fluence for 1-D, 2-D, and 3-D models, compared to experiment [48]

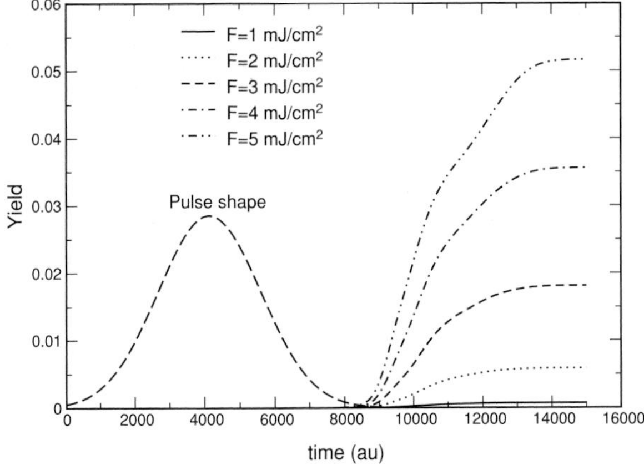

Fig. 8. Yield Y vs. time for several fluence values showing the shape of the exciting laser pulse and the delay in photodesorption calculated in agreement with experiment [48]

gives better agreement with experiment. The 3-D model is of course more realistic, and calculations at even higher fluence show that the 3-D model gives a flatter graph, due to increased de-excitation rates.

The models also display a delay between pulse arrival and photodesorption as observed in the experiments, calculated with the present model to be about 250 fs, and provide insight on the time evolution of desorption, as shown in Fig. 8. The delay is associated to the time it takes a wavepacket to build up its amplitude in the excited repulse potential leading to CO + Cu(001)*.

7.2 Adsorbate Vibrational Relaxation

An adsorbate at a surface may be vibrationally excited by collisions with species in a gas, or following relaxation to a ground electronic state after excitation by light. Here we treat the vibrational degrees of freedom of the adsorbate and substrate as quantal, with a hamiltonian $\hat{F}_{gg} = \hat{H}$ and a RDOp $\hat{\rho}_{gg} = \hat{\rho}$. We assume that the medium is at thermal equilibrium and that the coupling of adsorbate and substrate are small enough so that the memory superoperator can be calculated to second order in the coupling, in terms of the substrate correlation functions.

We start with the density operator $\hat{\Gamma}(t)$ for the whole system, composed of a species A interacting with the surface or reservoir R, taken here to be the p- and s-regions, respectively, and use the projection operator formalism mentioned at the end of Sect. 6.1. The RDOp $\hat{\rho}(t) = \mathrm{tr}_R[\hat{\Gamma}(t)]$, satisfies the equation

$$\frac{\mathrm{d}\hat{\rho}(t)}{\mathrm{d}t} = (\mathrm{i}\hbar)^{-1}[\hat{H}, \hat{\rho}] + \int_0^t \mathcal{K}(t,t')\hat{\rho}(t')\mathrm{d}t' \qquad (59)$$

in terms of a memory kernel superoperator $\mathcal{K}(t,t')$, of the form given in (44). This is a Volterra integro-differential equation that must be solved for the initial condition $\hat{\rho}(0) = \hat{\rho}_0$ corresponding to the preparation of the system before relaxation.

We consider the vibrational relaxation of a frustrated T-mode of the adsorbate (the primary region or A-subsystem) and treat it as a harmonic oscillator bilinearly coupled to the surface (the secondary region or R-subsystem), and model this as a reservoir of harmonic oscillators at a temperature T. The Hamiltonian for the total system is then $\hat{H} = \hat{H}_\mathrm{A} + \hat{H}_\mathrm{R} + \hat{H}_\mathrm{AR}$ with terms given in a second quantization notation by $\hat{H}_\mathrm{A} = \hbar\omega_0 \hat{a}^\dagger \hat{a}$, $\hat{H}_\mathrm{R} = \sum_j \hbar\omega_j \hat{b}_j^\dagger \hat{b}_j$ and $\hat{H}_\mathrm{AR} = \hat{q}\hat{B}$, where $\hat{q} = (\hat{a}^\dagger + \hat{a})/\sqrt{2}$, and with

$$\hat{B} = \hbar\sqrt{2}\sum_j \kappa_j (\hat{b}_j^\dagger + \hat{b}_j). \tag{60}$$

Here $\hat{a}$ and $\hat{a}^\dagger$ are the creation and annihilation operators for the frustrated T-vibrational mode of the adsorbate A with frequency ω_0, related to the vibrational displacement $\hat{q}$ and momentum $\hat{p}$, while $\hat{b}_j$ and $\hat{b}_j^\dagger$ are the creation and annihilation operators for the reservoir R excitations of frequencies ω_j. The κ_j are coupling strength coefficients. The operators $\hat{b}_j$ and $\hat{b}_j^\dagger$ have a spectral density per unit frequency $g(\omega) = \sum_j \delta(\omega - \omega_j)$ that depends on the nature of the reservoir R excitations. The bilinear coupling $\hat{H}_\mathrm{AR}$ leads to delayed dissipation when the range of the spectral density is close to the adsorbate vibrational frequency, and the dissipative memory kernel can be expressed in terms of the thermally averaged time-correlation function $C(t) = \langle\langle \hat{B}(t)\hat{B}(0)\rangle\rangle$. This includes the spectral function $J(\omega)$ given by $\omega^2 J(\omega) = 2\,g(\omega)|\kappa(\omega)|^2$, and it takes the form

$$C(t) = \int_0^\infty \left[\cos(\omega t)\coth\left(\frac{\hbar\omega}{2\,k_\mathrm{B}T}\right) - \mathrm{i}\sin(\omega t)\right]\omega^2 J(\omega)\mathrm{d}\omega. \tag{61}$$

The equation for $\hat{\rho}$ can be transformed into a matrix equation in the basis set $\{\phi_r\}$ of eigenstates of $\hat{H}_\mathrm{A}$, with eigenenergies $E_r = \hbar\omega_0(r + 1/2)$, where $r = 0, 1, \ldots$. Because $q_{sr} = \langle\phi_s|\hat{q}|\phi_r\rangle = 0$ for $r \neq s \pm 1$, there are no couplings in a two-state description between the diagonal elements of the density matrix corresponding to populations and the off-diagonal ones corresponding to quantum coherence, but couplings do appear with more than two states. Properties of the adsorbate varying over time can be obtained from the density matrix. In particular, the amount of energy left in the adsorbate motion after its initial excitation is obtained as $\Delta E_\mathrm{A} = E_\mathrm{A}(t) - E_\mathrm{A}(0)$, with $E_\mathrm{A}(t) = tr_\mathrm{A}[\hat{\rho}(t)\hat{H}_\mathrm{A}]$, which reduces in our model to $E_\mathrm{A}(t) = \hbar\omega_0 \sum_{r=0}^{1} \rho_{rr}(t)(r + 1/2)$ so that $\Delta E_\mathrm{A}(t) = -\hbar\omega_0\rho_{00}(t)/2$.

In our numerical method [49] we write the matrix version of (59) in a more compact form, as

$$\frac{\mathrm{d}\boldsymbol{\rho}(t)}{\mathrm{d}t} = \mathbf{f}[t, \boldsymbol{\rho}(t), \mathbf{z}(t)]\,, \quad \mathbf{z}(t) = \int_0^t \mathcal{K}[t, t', \boldsymbol{\rho}(t')]\mathrm{d}t'. \tag{62}$$

A generalized Runge–Kutta scheme then introduces time increments Δt and a sequence of $j = 1$ *to* m stages of iteration, with values $\mathbf{P}_{n,j} = \boldsymbol{\rho}(t_0 + n\Delta t)^{(j)}$ and $\mathbf{Z}_{n,i} = \mathbf{z}(t_0 + n\Delta t)^{(i)}$, in an algorithm which does not require the inverse of matrices and is applicable to many coupled states.

In what follows we concentrate on adsorbate relaxation due to coupling to phonons in the substrate. The couplings κ_j contain contributions both from direct coupling of vibrations and from their indirect coupling through short lived electron–hole excitations in the metal, and have been obtained from experiment [50]. The phonon frequencies ω_j may be considered to form a continuum with spectral density $g(\omega) = 18\pi N\omega^2/\omega_{\mathrm{D}}^3$, with $g(\omega) = 0$ for $\omega > \omega_{\mathrm{D}}$, and where N is the number of lattice atoms and ω_{D} is the Debye phonon cut-off frequency. We use a parameterization for $\kappa(\omega)$ in the neighborhood of ω_0 of the form $|\kappa(\omega)|^2 = [p + q(\omega - \omega_0)]/N$ where p and q are parameters which depend on the system, with values for CO/Cu(001) given in [50]. Figure 9 shows populations obtained from the diagonal elements of the RDM for the systems CO/Cu at 150 and at 300 K, starting with initial values $\rho_{11} = 1$, $\rho_{00} = 0$, and $\rho_{01} = 0$.

Results for instantaneous dissipation (given in [51]) have been obtained substituting $\hat{\rho}(t')$ with $\hat{\rho}(t)$ inside the integral of the Volterra equation, and they show that the Markovian approximation leads to the correct long time limit but is deficient at short times. Higher temperatures lead to decreased oscillation peaks and a faster relaxation to equilibrium, as expected. For CO/Cu, the population of the ground state $r = 0$ oscillates with a period around 2,000 au(T) at both temperatures. Comparing this with the decay time of the correlation function, one concludes that the correlation of reservoir vibrations does not decay rapidly enough to justify an approximation of instantaneous dissipation. From Fig. 9, the CO/Cu populations are found to relax within

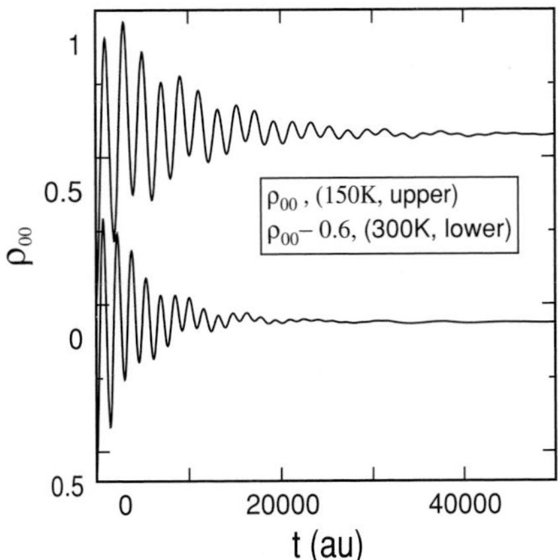

Fig. 9. The ground state population ρ_{00} vs. time for the CO/Cu(001) system at temperatures of 150 and 300 K from [51]

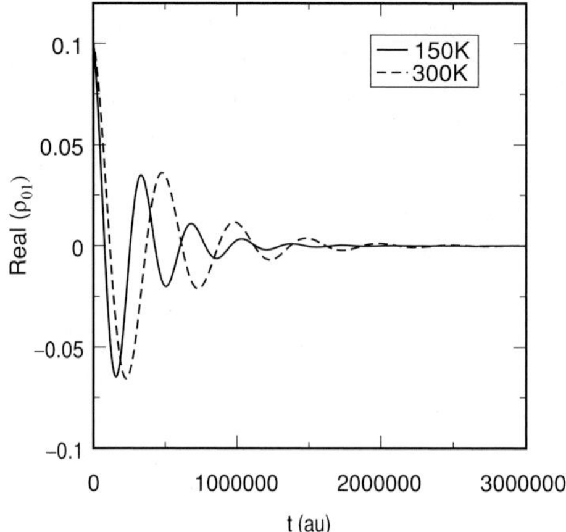

Fig. 10. Real part of the quantum coherence ρ_{01} vs. time for CO/Cu(001) at 150 and 300 K, for very long times, from [51]

about 4×10^4 au(T), or about 1.0 ps, at 150 K, with this time increasing at lower temperatures. This is in qualitative agreement with experimental results [52]. In the above case we set the initial quantum coherence ($\rho_{01} = \rho_{10}^*$) equal to zero, in which case it remains zero in our model. Figure 10 shows our results for the real part of ρ_{01} over a longer time range, using an initial value of $\rho_{01}(0) = 0.1 + 0.1\text{i}$.

The imaginary part of ρ_{01} shows a similar pattern for long times. Hence here again, the treatment of dissipation must incorporate memory effects.

8 Conclusion

A general formalism for quantum–classical systems, based on the density operator and the PWT, can be computationally implemented to deal with electronically excited systems. In our procedure, this has been done in a semiclassical limit that assumes short wavelengths in the phase space of classical-like variables, and introduces an effective potential for each initial condition in a set chosen from an initial quantum distribution in phase space. This has provided very good results for the photodissociation dynamics of NaI over several picoseconds, and also very good cross-sections for electronic excitation in Li + He collisions.

Dissipative dynamics arising in interactions with a medium can be described with a reduced density operator, and with dissipative potentials or rates related to atomic structure. We have briefly reviewed a derivation of dissipative potentials for self-consistently correlated primary and secondary regions of a complex molecular system, and its implementation for computational work. A procedure has also been described to calculate phenomena with delayed dissipation. These two developments have been applied to the femtosecond photodesorption of CO/Cu(001) and the vibrational relaxation of the same system after collisional excitation. The results of our models agree with experimental results and trends in both cases.

Acknowledments

The present work has been partly supported by the National Science Foundation of the USA.

References

1. R. C. Tolman, *The Principles of Statistical Mechanics* (Clarendon Press, Oxford, England, 1938)
2. J. von Neumann, *Mathematical Foundations of Quantum Mechanics* (Princeton University Press, Princeton, New Jersey, 1955)
3. U. Fano, Rev. Modern Phys. **29**, 74 (1957)
4. K. Blum, *Density Matrix Theory and Applications*, 2nd edn (Plenum Press, New York, 1981)
5. C. Cohen-Tannoudji, J. Dupont-Roc, and G. Grynberg, *Atom-Photon Interactions* (Wiley, New York, 1992)
6. S. Mukamel, *Principles of Nonlinear Optical Spectroscopy* (Oxford University Press, Oxford, England, 1995)
7. S. H. Lin, R. Alden, R. Islampour, H. Ma, and A. A. Villaeys, *Density Matrix Methods and Femtosecond Processes* (World Scientific, Singapore, 1991)
8. V. May and O. Kuhn, *Charge and Energy Transfer Dynamics in Molecular Systems*, 2nd edn (Wiley, New York, 2004)
9. R. Kubo, R. M. Toda, and N. Hashitsume, *Statistical Physics II*, 2nd edn (Springer, Berlin Heidelberg New York, 1991)
10. K. Lindenberg and B. West *The Non Equilibrium Statistical Mechanics of Open and Closed Systems* (VCH, New York, 1990)
11. D. A. Micha, J. Phys. Chem. **103**, 7562 (1999)
12. D. A. Micha, Adv. Quant. Chem. **41**, 139 (2002)
13. D. A. Micha and B. Thorndyke, Adv. Quant. Chem. **47**, 293 (2004)
14. M. Hillery, R. F. O'Connell, M. O. Scully, and E. P. Wigner, Phys. Rep. **106**, 121 (1984)
15. C. C. Martens, and J-Y. Fang, J. Chem. Phys. **106**, 4918 (1997)
16. R. Kapral, and G. Ciccotti, J. Chem. Phys. **110**, 8919 (1999)
17. D. A. Micha and B. Thorndyke, Int. J. Quant. Chem. **90**, 759 (2002)
18. A. Donoso, and C. C. Martens, J. Chem. Phys. **112**, 3980 (2000)
19. M. Santer, U. Manthe, and G. Stock, J. Chem. Phys. **114**, 2001 (2001)
20. A. W. Jasper, C. Zhu, S. Nangia, and D. G. Truhlar, Faraday Discuss. **127**, 1 (2004)
21. R. E. Wyatt *Quantum Dynamics with Trajectories* (Springer, Berlin, Heidelberg, New York, 2005)
22. V. Engel and H. Metiu, J. Chem. Phys. **90**, 6116 (1989)

23. B. Thorndyke and D. A. Micha, Chem. Phys. Lett. **403**, 280 (2005)
24. M. D. Feit, J. A. Fleck and A. Steiger, J. Comp. Phys. **47**, 412 (1982)
25. A. Reyes and D. A. Micha, J. Chem. Phys. **119**, 12308 (2003)
26. D. A. Micha, Int. J. Quant. Chem. **80**, 394 (2000)
27. D. Beksic and D. A. Micha, J. Chem. Phys. **103**, 3795 (1995)
28. Z. Yi, D. A. Micha, and J. Sund, J. Chem. Phys. **110**, 10562 (1999)
29. G. Lindblad, Commun. Math. Phys. **48**, 119 (1976)
30. V. Gorini, A. Kossakowski, and E. C. G. Sudarshan, J. Math. Phys. **17**, 821 (1976)
31. A. G. Redfield, Adv. Mag. Reson. **1**, 1 (1965)
32. W. T. Pollard, and R. A. Friesner, J. Chem. Phys. **100**, 5054 (1994)
33. R. W. Zwanzig: Statistical mechanics of irreversibility. In: *Lectures in Theoretical Physics* vol. III, ed by W. E. Brittin et al., (Wiley, New York, 1961), p 106
34. B. J. Berne, and D. Forster, Ann. Rev. Phys. Chem. **22**, 563 (1971)
35. J. T. Hynes and J. M. Deutch: Nonequilibrium problems - Projection operator techniques. In : *Physical Chemistry. An Advanced Treatise*, vol. 11B, ed by H. Eyring, W. Jost, and D. Henderson (Academic, New York, 1975), p 729
36. P. O. Lowdin, Int. J. Quant Chem., **QCS 16**, 485 (1982)
37. N. Makri, J. Phys. Chem. A **102**, 4414 (1998)
38. W. H. Miller, J. Phys. Chem. A **105**, 2942 (2001)
39. D. Kohen and D. J. Tannor, Adv. Chem. Phys. **111**, 219 (2000)
40. T. Mancal, and V. May, Eur. J. Phys. B **18**, 633 (2000)
41. I. Burghardt and K. B. Moller, J. Chem. Phys. **117**, 7409 (2002)
42. P. Saalfrank and R. Kosloff, J. Chem. Phys. **105**, 2441 (1996)
43. J. C. Tully, Annu. Rev. Phys. Chem. **51**, 153 (2000)
44. F. Hofmann and J. P. Toennies, Chem. Rev. **96**, 1307 (1996)
45. D. A. Micha, A. Santana, and A. Salam, J. Chem. Phys. **116**, 5173 (2002)
46. T. A. Germer, J. C. Stephenson, E. J. Heilweil, and R. R. Cavanagh, J. Chem. Phys. **101**, 1 (1994)
47. D. A. Micha and A. Santana, J. Phys. Chem. A **107** 7311 (2003)
48. J. A. Prybyla, H. W. K. Tom, and G. Aumiller, Phys. Rev. Lett. **68**, 503 (1992)
49. A. Leathers, and D. A. Micha, Chem. Phys. Lett. **415**, 46 (2005)
50. J. L. Vega, R. Guantes, S. Miret-Artes, and D. A. Micha, J. Chem. Phys. **121**, 8580 (2004)
51. A. Leathers and D. A. Micha, J. Phys. Chem. A **110**, 749 (2006)
52. J. Braun, A. P. Graham, F. Hofmann, and J. P. Toennies, J. Chem. Phys. **105**, 3258 (1996)

Quantum Dynamics of Ultrafast Molecular Processes in a Condensed Phase Environment

M. Thoss, I. Kondov, and H. Wang

Summary. The accurate description of quantum effects for reactions in a condensed phase environment continues to be a central issue in chemical dynamics. In this chapter, two recently proposed methods to simulate quantum dynamics in complex molecular systems are discussed – the multilayer version of the multiconfiguration time-dependent Hartree method and the self-consistent hybrid approach. The methods are applied to selected examples of ultrafast photoreactions in the condensed phase, including electron injection in the dye–semiconductor system coumarin 343 – TiO_2 and intervalence electron transfer in the mixed valence system $(NH_3)_5Ru^{III}NCRu^{II}(CN)_5^-$ in solution. Furthermore, we discuss the application of the methodology to simulate photoexcitation processes and time resolved optical spectra by including the coupling to the laser field explicitly in the calculation.

1 Introduction

Femtosecond laser spectroscopy has revealed that many photoinduced processes in complex molecular systems occur on a subpicosecond timescale [1–3]. Prominent examples include *cis–trans* photoisomerization reactions in proteins [4–7] and photoinduced charge transfer processes in solution or on surfaces [8–17]. The accurate description of quantum effects in such reactions continues to be a central issue in chemical reaction dynamics. Although at present there is no practical method that is capable of simulating quantum dynamics for a general, complex molecular system with arbitrary nuclear potentials, significant progress has been made recently in devising methods that allow accurate simulations of certain classes of quantum dynamical processes in large molecular systems or in the condensed phase.

Considering only methods that allow a numerically exact simulation of quantum dynamics in a condensed phase environment, two different strategies have been followed: First, path-integral methods based on the influence-functional technique [18], where the environment is formally integrated out. For example, numerical path-integral calculations based on this idea [19–27] have been used successfully to study the dynamics of the spin-boson model

[28, 29] — a two-level system interacting with a harmonic bath — which is a standard model for electron transfer (ET) in the condensed phase [30]. For cases where the influence functional is known analytically, e.g., for a harmonic or spin bath, this method allows the description of a problem with an (in principle) infinite number of degrees of freedom. For ultrafast molecular processes, however, one is typically only interested in the dynamics on a relatively short timescale (up to a few picoseconds). Since on this timescale only a limited number of degrees of freedom can be resolved, the infinite number of degrees of freedom of the bath (e.g., a solvent) can be represented by a finite number of degrees of freedom. In this case basis set methods for wave packet or density matrix propagation can be used to describe the corresponding dynamics, which represents the second, alternative, approach.

A particularly efficient method for simulating quantum dynamics in large systems is the multiconfiguration time-dependent Hartree (MCTDH) method [31–34]. The performance of this method has been demonstrated by numerous applications to gas-phase reactions of relatively large molecules in recent years [33–41]. Further applications have shown that this method can also be used to describe molecular systems in a dissipative environment with a moderate number of degrees of freedom (up to about ≈100) [42–46]. To extend its applicability to even larger and/or more complex systems, we have recently proposed two approaches: (1) The self-consistent hybrid method [47,48], where the accurate treatment of part of the overall system (the "core") is combined with an approximate description of the rest of the system (the "reservoir"). Due to the iterative optimization of the core-reservoir separation included in the self-consistent hybrid scheme, this method also allows (as the MCTDH method) an accurate (in principle numerically exact) treatment of the quantum dynamics. (2) A multilayer (ML) extension of the MCTDH method [49], which (as the original MCTDH method) is a rigorous quantum dynamical method.

The ML–MCTDH method and the self-consistent hybrid approach have so far been used to study a variety of ultrafast photoreactions in the condensed phase including various model studies of electron transfer (ET) reactions, photoinduced ET reactions in mixed-valence compounds in solution, heterogeneous ET reactions at dye–semiconductor interfaces, and photoisomerization reactions in a condensed phase environment [49–53]. In this article we review the basic ideas of the ML–MCTDH method and the self-consistent hybrid approach. To illustrate the performance of the methods, we discuss applications to two ultrafast photoreactions we have considered recently: (1) electron injection in the dye–semiconductor system coumarin 343 – TiO_2 and (2) intervalence electron transfer in the mixed valence system $(NH_3)_5Ru^{III}NCRu^{II}(CN)_5^-$. In addition, we will also discuss the application of the methodology to simulate photoexcitation processes and time resolved optical spectra.

2 Summary of Methodology

2.1 Hamiltonian and Observables of Interest

To study the dynamics of a molecular system in a condensed phase environment we consider the generic Hamiltonian

$$H = H_{\mathrm{s}} + H_{\mathrm{b}} + H_{\mathrm{sb}}, \tag{1}$$

where H_{s} and H_{b} denote the Hamiltonian of the system and environment (the "bath"), respectively, and H_{sb} their interaction. In the applications, we shall consider different dynamical observables which can be represented in form of correlation functions (throughout this paper we use atomic units in which $\hbar = 1$),

$$C_{AB}(t) = \mathrm{tr}\left\{\rho_{\mathrm{b}} A \mathrm{e}^{\mathrm{i}Ht} B \mathrm{e}^{-\mathrm{i}Ht}\right\}. \tag{2}$$

Here, A and B denote operators involving the "system" degrees of freedom that corresponds to some physical quantities (e.g., the reduced density matrix, dipole moment, etc.) and ρ_{b} is the initial density matrix for the "bath" degrees of freedom. To evaluate the trace we use a direct product basis $|n\rangle|j\rangle$, where the "bath" states $\{|n\rangle\}$ are the eigenstates of H_{b}, i.e.,

$$\rho_{\mathrm{b}} = \sum_n p_n |n\rangle\langle n|, \tag{3}$$

and the "system" states $\{|j\rangle\}$ are any convenient basis, in which operator A has the representation

$$A = \sum_j \sum_i a_{ij} |i\rangle\langle j|, \tag{4}$$

where $a_{ij} \equiv \langle i|A|j\rangle$. Using this basis to evaluate the trace leads to the following expression for $C_{AB}(t)$,

$$\begin{aligned} C_{AB}(t) &= \sum_n p_n \sum_j \sum_i a_{ij} \langle n|\langle j|\mathrm{e}^{\mathrm{i}Ht} B \mathrm{e}^{-\mathrm{i}Ht}|i\rangle|n\rangle \\ &= \sum_n p_n \sum_j \sum_i a_{ij} \langle \Psi_n^j(t)|B|\Psi_n^i(t)\rangle, \end{aligned} \tag{5}$$

where

$$|\Psi_n^i(t)\rangle = \mathrm{e}^{-\mathrm{i}Ht}|\Psi_n^i(0)\rangle = \mathrm{e}^{-\mathrm{i}Ht}|i\rangle|n\rangle. \tag{6}$$

Thus, the major computational task is to solve the time-dependent Schrödinger equations

$$i\frac{\partial}{\partial t}|\Psi_n^i(t)\rangle = H|\Psi_n^i(t)\rangle, \qquad n, i = 1, 2, ... \tag{7}$$

with initial conditions

$$|\Psi_n^i(0)\rangle = |n\rangle|i\rangle. \tag{8}$$

2.2 Multilayer Version of the Multiconfiguration Time-dependent Hartree Method

To solve the time-dependent Schrödinger equation we employ the multilayer (ML) version [49] of the multiconfiguration time-dependent Hartree method (MCTDH). To review this method, let us first briefly discuss the original (single-layer) MCTDH theory [31–34]. In this method, the overall wave function is expanded in terms of many time-dependent configurations

$$|\Psi(t)\rangle = \sum_J A_J(t)|\Phi_J(t)\rangle \equiv \sum_{j_1}\sum_{j_2}\cdots\sum_{j_M} A_{j_1 j_2 \dots j_M}(t) \prod_{k=1}^{M} |\phi_{j_k}^k(t)\rangle, \tag{9}$$

Here, $|\phi_{j_k}^k(t)\rangle$ is the "single-particle" (SP) function for the kth SP degree of freedom and M denotes the number of SP degrees of freedom. Each SP group usually contains several (Cartesian) degrees of freedom in our calculation, and for convenience the SP functions within the same SP degree freedom are chosen to be orthonormal.

Substituting the MCTDH ansatz, (9), into the Dirac-Frenkel variational principle [54] results in the following equations of motion [31]

$$i\dot{A}_J(t) = \langle\Phi_J(t)|\hat{H}|\Psi(t)\rangle = \sum_L \langle\Phi_J(t)|\hat{H}|\Phi_L(t)\rangle A_L(t), \tag{10}$$

$$i|\dot{\underline{\phi}}^k(t)\rangle = (1-\hat{P}^k)(\hat{\rho}^k)^{-1}\langle\hat{H}(t)\rangle^k|\underline{\phi}^k(t)\rangle, \tag{11}$$

where $|\underline{\phi}^k(t)\rangle = \{|\phi_1^k(t)\rangle, |\phi_2^k(t)\rangle, \dots\}^{\mathrm{T}}$ denotes the symbolic column vector of (the coefficients of) the SP functions for the kth SP degree of freedom, and $(\hat{\rho}^k)^{-1}$ denotes the pseudoinverse of the reduced density matrix. The mean-field operator $\langle\hat{H}(t)\rangle^k$ and the reduced density matrix $\hat{\rho}^k(t)$ are given by

$$\langle\hat{H}(t)\rangle_{nm}^k = \langle G_n^k(t)|\hat{H}|G_m^k(t)\rangle, \tag{12}$$

$$\rho_{nm}^k(t) = \langle G_n^k(t)|G_m^k(t)\rangle, \tag{13}$$

where the "single-hole" function, $|G_n^k(t)\rangle$, for the kth SP degree of freedom, is defined as [31–34]

$$|G_n^k(t)\rangle = \sum_{j_1}\cdots\sum_{j_{k-1}}\sum_{j_{k+1}}\cdots\sum_{j_M} A_{j_1\dots j_{k-1} n j_{k+1}\dots j_M}(t) \times|\phi_{j_1}^1(t)\rangle\dots|\phi_{j_{k-1}}^{k-1}(t)\rangle|\phi_{j_{k+1}}^{k+1}(t)\rangle|\phi_{j_M}^M(t)\rangle, \tag{14}$$

so that

$$|\Psi(t)\rangle = \sum_n |\phi_n^k(t)\rangle |G_n^k(t)\rangle. \tag{15}$$

The time-dependent projection operator $P^k(t)$ is defined in the subspace of SP functions as

$$P^k(t) = \sum_m |\phi_m^k(t)\rangle\langle\phi_m^k(t)|. \tag{16}$$

The main limitation of the MCTDH approach lies in its way of constructing the SP functions, which is based on a full configuration-interaction (FCI) expansion

$$|\phi_n^k(t)\rangle = \sum_I B_I^{k,n}(t)|u_I^k\rangle \equiv \sum_{i_1}\sum_{i_2}\cdots\sum_{i_{F(k)}} B_{i_1 i_2 \ldots i_{F(k)}}^{k,n}(t) \prod_{q=1}^{F(k)} |\varphi_{i_q}^{k,q}\rangle. \tag{17}$$

Here $F(k)$ is the number of Cartesian degrees of freedom within the kth SP group, and $|\varphi_{i_q}^{k,q}\rangle$ denotes the corresponding time-independent primitive basis functions for the qth Cartesian degree of freedom. The FCI-type expansion of the SP functions in (17) is usually limited to a few (∼10) degrees of freedom due to the exponential scaling of the number of basis functions versus the number of degrees of freedom in one SP group. Furthermore, the multiconfigurational expansion of the wave function in (9) is typically limited to ∼10 SP groups. As a result, a routine MCTDH calculation is limited to systems with a few tens of quantum degrees of freedom.

The recently proposed ML–MCTDH theory [49] circumvents this limitation by using a *dynamic* contraction of the basis functions that constitute the SP functions. Thereby, the FCI-type construction of the SP functions in (17) is replaced by a *time-dependent* multiconfigurational expansion

$$|\phi_n^k(t)\rangle = \sum_I B_I^{k,n}(t)|u_I^k(t)\rangle, \tag{18}$$

i.e., the basic MCTDH strategy is adopted to treat each SP function. For clarity we refer in the following to the SP defined in the original MCTDH approach as *level one* (L1) SP, which in turn contains several *level two* (L2) SPs

$$|u_I^k(t)\rangle = \prod_{q=1}^{Q(k)} |v_{i_q}^{k,q}(t)\rangle. \tag{19}$$

Similar to (9), the L1-SP function $|\phi_n^k(t)\rangle$ is thus expanded in the time-dependent L2-SP functions as

$$|\phi_n^k(t)\rangle = \sum_I B_I^{k,n}(t)\;|u_I^k(t)\rangle \equiv \sum_{i_1}\sum_{i_2}\cdots\sum_{i_Q(k)} B_{i_1 i_2 \ldots i_Q(k)}^{k,n}(t) \prod_{q=1}^{Q(k)} |v_{i_q}^{k,q}(t)\rangle. \tag{20}$$

Here, $Q(k)$ denotes the number of L2-SP degrees of freedom in the kth L1-SP and $|v_{i_q}^{k,q}(t)\rangle$ is the L2-SP function for the qth L2-SP degree of freedom. It is noted that both are in the context of the kth L1-SP group. The expansion of the overall wave function can thus be written in the form

$$|\Psi(t)\rangle = \sum_{j_1}\sum_{j_2}\dots\sum_{j_M} A_{j_1 j_2 \dots j_M}(t) \times \prod_{k=1}^{M}\left[\sum_{i_1}\sum_{i_2}\dots\sum_{i_{Q(k)}} B_{i_1 i_2 \dots i_{Q(k)}}^{k,j_k}(t) \prod_{q=1}^{Q(k)} |v_{i_q}^{k,q}(t)\rangle\right]. \tag{21}$$

The equations of motion within the ML–MCTDH approach can again be obtained from the Dirac–Frenkel variation principle [49]. For two layers, they are given by

$$i\Big|\dot{\Psi}(t)\Big\rangle_{\text{L1 coefficients}} = \hat{H}(t)\Big|\Psi(t)\Big\rangle, \tag{22}$$

$$i\Big|\dot{\underline{\phi}}^k(t)\Big\rangle_{\text{L2 coefficients}} = \left[1 - \hat{P}^k(t)\right]\left[\hat{\rho}^k(t)\right]^{-1}\Big\langle \hat{H}(t)\Big\rangle^k \Big|\underline{\phi}^k(t)\Big\rangle, \tag{23}$$

$$i\Big|\dot{\underline{v}}^{k,q}(t)\Big\rangle_{\text{L3 coefficients}} = \left[1 - \hat{P}_{\text{L2}}^{k,q}(t)\right]\left[\hat{\varrho}^{k,q}(t)\right]^{-1}\Big\langle \hat{\mathcal{H}}(t)\Big\rangle^{k,q} \Big|\underline{v}^{k,q}(t)\Big\rangle, \tag{24}$$

where the L2 mean-field operators and reduced densities are defined, similar to (22), in terms of the L2 single-hole functions $|g_{n,r}^{k,q}(t)\rangle$

$$\varrho_{rs}^{k,q}(t) = \sum_n \sum_m \rho_{nm}^k(t) \left\langle g_{n,r}^{k,q}(t) | g_{m,s}^{k,q}(t)\right\rangle, \tag{25}$$

$$\langle \hat{\mathcal{H}}(t)\rangle_{rs}^{k,q} = \sum_n \sum_m \left\langle g_{n,r}^{k,q}(t) \left| \langle \hat{H}(t)\rangle_{nm}^k \right| g_{m,s}^{k,q}(t)\right\rangle, \tag{26}$$

$$|\phi_n^k(t)\rangle = \sum_r |v_r^{k,q}(t)\rangle \, |g_{n,r}^{k,q}(t)\rangle. \tag{27}$$

The projection operator $\hat{P}_{\text{L2}}^{k,q}$ in L2-SP space is defined in a similar way as in (16) as

$$\hat{P}_{\text{L2}}^{k,q}(t) = \sum_l \left|v_l^{k,q}(t)\right\rangle\left\langle v_l^{k,q}(t)\right|. \tag{28}$$

The equations of motion for further layers are obvious extensions of (22)–(24).

The inclusion of several dynamically optimized layers in the ML–MCTDH method provides more flexibility in the variational functional, which significantly advances the capability of performing wave packet propagations in complex systems. This has been demonstrated by applications to several examples of photoreactions in the condensed phase including many degrees of freedom [49, 50, 53].

2.3 Self-Consistent Hybrid Method

The ML–MCTDH method, as well as the original MCTDH method, are rigorous (in principle numerically exact) quantum dynamical methods, i.e., if a sufficiently large number of SP functions are included, the solution of the equations of motion converges to the solution of the time-dependent Schrödinger equation. In many situations, however, there are parts of the overall system which do not require a rigorous quantum dynamical treatment. For example, slow solvent degrees of freedom can often also be accurately described using classical mechanics. A method that takes advantage of this fact without derogating the accuracy of the dynamical calculation is the self-consistent hybrid (SCH) method [47,48].

Basic Concept

The development of the SCH method was motivated by a variety of other dynamical hybrid approaches for simulating quantum dynamics in large systems, such as, for example, the classical Ehrenfest method [55–60] and the surface-hopping approach [61–66]. The major conceptual difference from these approaches is that in the SCH method an iterative convergence procedure is introduced in such a hybrid dynamical simulation. Thereby the overall system is first partitioned into a *core* and a *reservoir*, based on any convenient but otherwise rather arbitrary initial guess. A hybrid dynamical calculation is then carried out, with the core treated via a numerically exact quantum mechanical method and the reservoir treated via a more approximate method. Next, the size of the core is systematically increased, similar to increasing the number of basis functions in a basis-set calculation, and other variational parameters are adapted accordingly until convergence (usually to within 10% relative error) is reached for the overall quantum dynamics. The details of the convergence procedure have been discussed previously [47, 48]. The key concepts in the SCH method are thus the numerically exact treatment of the core and the systematic optimization of the core size, which makes the method variational in nature and ensures, at least in principle, convergence to the true quantum dynamical limit. In contrast to other commonly used hybrid methods, the SCH method entails no ambiguity in partitioning the overall system into the core and the reservoir parts – the true quantum dynamical result, by definition, is obtained when all the degrees of freedom are included in the core. In practice, however, convergence is achieved in many situations well before such a rigorous level of theory.

A variety of approaches can be adopted to treat the core/reservoir at a hybrid level. The essential requirement is that the quantum mechanical method used to treat the core should be both accurate (i.e., in principle numerically exact) and efficient. The approximate method to treat the reservoir should be easily implementable with reasonable accuracy. The former ensures that a moderately large number of core degrees of freedom (e.g., up to a few hundred degrees of freedom in model systems) can be treated in a numerically exact fashion, so that the converged result is approached in the full-core limit, whereas the latter ensures both numerical efficiency and the attainment of certain physical limits. Due to interactions between the core and the reservoir, the equations of motion for the two parts are coupled and solved simultaneously.

Currently, the most efficient, rigorous quantum dynamical method for treating the core part is the ML–MCTDH method outlined in Sect. 2.2. The flexible form of the variational functional in this method allows the quantum treatment of a much larger core subsystem than it is possible with other existing methods such as conventional wave packet propagation approaches and the original MCTDH method. Various approximate methods can be used to treat the reservoir, e.g., classical mechanics, semiclassical methods [51,67–69], or quantum perturbation theory. It is usually rather straightforward to select the most efficient one among these methods by examining the physical regimes of the reservoir. For example, if the reservoir has a rather low characteristic frequency, classical mechanics is often adequate to describe its dynamics for not too low temperatures. On the other hand, if the reservoir has a rather high characteristic frequency, one may use a perturbative quantum mechanical method to describe its impact on the core. It should be emphasized that the choice of these approximate methods, together with the core–reservoir partition, merely serves as a trial "initial guess." The central step of the method is to systematically include more degrees of freedom in the core for a rigorous treatment, i.e., a regular convergence test. Similar to situations in many other self-consistent variational calculations, the better the initial guess, the more easily the convergence is achieved. However, the converged result does not depend on the specific initial guess.

Practical Implementation

To discuss some details of implementation of the SCH method, we consider the correlation function (2) recast in the form

$$C_{AB}(t) = \mathrm{tr}\left[\hat{\rho}_{\mathrm{N}}\hat{A}\,\mathrm{e}^{\mathrm{i}\hat{H}t}\hat{B}\mathrm{e}^{-\mathrm{i}\hat{H}t}\right]. \tag{29}$$

Here, $\hat{H}$ is the Hamiltonian of the overall system, the density operator $\hat{\rho}_{\mathrm{N}}$ describes the initial state of the nuclear degrees of freedom, and $\hat{A}$ and $\hat{B}$ are observables of interest. In the SCH method the overall system is partitioned into a core and a reservoir. Accordingly, the total Hamiltonian is separated into a *core* and a *reservoir* part,

$$\hat{H} = H_{\mathrm{c}}(\hat{\mathbf{p}}_s, \hat{\mathbf{s}}) + H_{\mathrm{r}}(\hat{\mathbf{p}}, \hat{\mathbf{q}}) + H_{\mathrm{I}}(\hat{\mathbf{p}}_s, \hat{\mathbf{s}}; \hat{\mathbf{p}}, \hat{\mathbf{q}}), \tag{30}$$

where $H_{\mathrm{c}}(\hat{\mathbf{p}}_s, \hat{\mathbf{s}})$ and $H_{\mathrm{r}}(\hat{\mathbf{p}}, \hat{\mathbf{q}})$ represent the uncoupled Hamiltonian for the core and the reservoir, respectively, and $H_{\mathrm{I}}(\hat{\mathbf{s}}, \hat{\mathbf{p}}_s; \hat{\mathbf{p}}, \hat{\mathbf{q}})$ describes their interactions. The corresponding phase-space variables $(\mathbf{p}_s, \mathbf{s})$ and $(\mathbf{p}, \mathbf{q})$ belong to the core and the reservoir, respectively.

The core is treated rigorously by the ML–MCTDH method. As has been discussed earlier, various approximate methods can be used to treat the reservoir. The details of the implementation of the SCH method depend on the specific method used to treat the reservoir. For example, if the reservoir is treated by quantum perturbation theory, (29) is modified to a reduced trace over the core degrees of freedom [48]. Most applications of the SCH method have employed a classical treatment of the reservoir, similar to the classical Ehrenfest model [55–60], where the dynamics of the core and the reservoir are governed by different time-dependent Hamiltonians

$$\hat{H}_{\mathrm{c}}^{\mathrm{eff}}(t) = H_{\mathrm{c}}(\hat{\mathbf{p}}_s, \hat{\mathbf{s}}) + H_{\mathrm{I}}[\hat{\mathbf{p}}_s, \hat{\mathbf{s}}; \mathbf{p}_t, \mathbf{q}_t], \tag{31}$$

$$H_{\mathrm{r}}^{\mathrm{eff}}(t) = H_{\mathrm{r}}(\mathbf{p}_t, \mathbf{q}_t) + \langle \psi_{\mathrm{c}}(t) | H_{\mathrm{I}}(\hat{\mathbf{p}}_s, \hat{\mathbf{s}}; \mathbf{p}_t, \mathbf{q}_t) | \psi_{\mathrm{c}}(t) \rangle . \tag{32}$$

Here $|\psi_{\mathrm{c}}(t)\rangle$ represents the wavefunction for the core, and the Heisenberg operators $(\hat{\mathbf{p}}, \hat{\mathbf{q}})$ for the reservoir have been replaced by their corresponding (time-dependent) classical phase space variables $(\mathbf{p}_t, \mathbf{q}_t)$.

Within this mixed quantum–classical implementation of the SCH method, the quantum mechanical trace expression in (29) is modified as

$$C_{AB}(t) = \int \mathrm{d}\mathbf{p}_0 \int \mathrm{d}\mathbf{q}_0 \; \rho_N^{\mathrm{r}}(\mathbf{p}_0, \mathbf{q}_0) \; \mathrm{tr}\left\{ \hat{\rho}_N^{\mathrm{c}} \; \hat{A} \; \hat{B}(t) \right\}, \tag{33}$$

where $\hat{B}(t)$ denotes the Heisenberg operator which is obtained by time evolution from $\hat{B}$ using the time-dependent Hamiltonian $\hat{H}_{\mathrm{c}}^{\mathrm{eff}}(t)$. In (33) the trace is now only over the core degrees of freedom. The initial density matrix $\hat{\rho}_N$ is split into a core part, $\hat{\rho}_N^{\mathrm{c}}$, and a corresponding classical distribution ρ_N^{r} for the reservoir. In the applications discussed in Sect. 3, the initial phase space distribution $\rho_N^{\mathrm{r}}(\mathbf{p}_0, \mathbf{q}_0)$ is obtained based on a semiclassical prescription [70] by taking the Wigner transform [71] of the corresponding operator $\hat{\rho}_N^{\mathrm{r}}$

$$\rho_N^{\mathrm{r}}(\mathbf{p}_0, \mathbf{q}_0) = \frac{1}{(2\pi)^{N_{\mathrm{r}}}} \int \mathrm{d}\boldsymbol{\Delta}\mathbf{q} \; \mathrm{e}^{-\mathrm{i}\mathbf{p}_0 \cdot \boldsymbol{\Delta}\mathbf{q}} \left\langle \mathbf{q}_0 + \frac{\boldsymbol{\Delta}\mathbf{q}}{2} | \hat{\rho}_N^{\mathrm{r}} | \mathbf{q}_0 - \frac{\boldsymbol{\Delta}\mathbf{q}}{2} \right\rangle , \tag{34}$$

where N_{r} denotes the number of reservoir degrees of freedom. For cases where the Wigner transform is not available or difficult to evaluate, a purely classical Boltzmann distribution function can be used instead.

3 Applications

The ML–MCTDH method and the SCH approach have been applied to study a variety of ultrafast reactions in the condensed phase [49–53], including various model studies of electron transfer (ET) reactions, photoinduced ET reactions in mixed valence compounds in solution, heterogeneous ET reactions at dye–semiconductor interfaces, as well as photoisomerization reactions in a condensed phase environment. In this section we will consider two representative examples of ultrafast photoreactions in the condensed phase: (1) electron injection in the dye–semiconductor system coumarin 343 – TiO_2 and (2) intervalence electron transfer in the mixed valence system $(NH_3)_5Ru^{III}NCRu^{II}(CN)_5^-$. Furthermore, we will discuss the application of the methodology to simulate photoexcitation processes and time resolved optical spectra.

3.1 Electron Injection in the Dye–Semiconductor System Coumarin 343 – TiO_2

Photoinduced ET reactions at dye–semiconductor interfaces represent an interesting class of charge transfer processes. In particular, the process of electron injection from an electronically excited state of a dye molecule to a semiconductor substrate has been investigated in great detail experimentally in recent years [12–14, 72–84]. This process represents a key step for photonic-energy conversion in nanocrystalline solar cells [12, 75, 76, 85, 86]. Employing femtosecond spectroscopy techniques, it has been demonstrated that electron injection processes often take place on an ultrafast timescale. Electron injection as fast as 6 fs has been reported for alizarin adsorbed on TiO_2 nanoparticles [80]. For other sensitizing chromophores, e.g., coumarin 343 [14, 73, 75, 87, 88] or perylene [13, 89], injection times on the order of tens to hundreds of femtoseconds have been found. Studies of dye molecules with electron injection timescales on the order of a few tens to a few hundred femtoseconds also indicate that the coupling to the vibrational modes of the chromophore may have a significant impact on the injection dynamics [13, 89, 90]. In particular, the influence of coherent vibrational motion on the injection dynamics has been observed in studies of perylene adsorbed on TiO_2 nanoparticles [13, 89]. Other important effects that have been investigated experimentally are the influence of surface trap states [91,92] as well as bridging groups [93] on the kinetics of the electron injection process.

The theoretical modeling of ET at dye–semiconductor interfaces requires in principle a simulation of the electron injection dynamics. While for very fast injection processes (<10 fs), the dynamical influence of the nuclear degrees of freedom on the electron injection process is presumably of minor importance and one may consider the purely electronic injection dynamics [94], for ET reactions on the order of a few tens to a few hundred femtoseconds the coupling to the nuclear degrees of freedom has to be included in the dynamical simulation. Various theoretical approaches have been applied to study the electron injection dynamics at dye–semiconductor interfaces, including nonadiabatic molecular dynamics simulations based on the classical path or Ehrenfest model [95–100], Anderson–Newns type models of reduced dimensionality (taking into account typically a single reaction mode) [83,101–107], as well as Redfield theory [104]. In a recent model study, we have investigated in detail the influence of multidimensional coherent and dissipative vibrational motion on the electron injection dynamics [50] employing the SCH approach [47,48] in combination with the ML–MCTDH theory [49].

Here we will consider the electron injection dynamics of the dye–semiconductor system coumarin 343 – TiO_2. The ET dynamics in this system has been studied experimentally by a number of groups in the recent years [14,73,75,87,88]. Employing different techniques, injection times in the range 20–200 fs have been found. A recent theoretical study [108] based on electronic structure calculations for small complexes of titanium with coumarin 343 showed that the photoexcited state of this system is predominantly localized at the chromophore. Furthermore, this study also demonstrated that there is significant electronic–vibrational coupling which is distributed over a relatively large number of vibrational modes of coumarin. These results suggest that the coumarin 343 – TiO_2 system is particularly well suited to study the electron injection dynamics as well as the influence of the nuclear degrees of freedom on the ET process. Besides the fundamental interest in the electron injection mechanism, coumarin derivatives have also been investigated as alternative organic photosensitizers in nanocrystalline solar cells [109].

Model

To study the dynamics of ultrafast photoinduced electron injection from the electronically excited state of the chromophore coumarin 343 (in the following abbreviated as C343) into the conduction band of the semiconductor (TiO_2) substrate, we consider a generic model of heterogeneous ET based on an Anderson–Newns type Hamiltonian [50,110–113]. Within this model the Hamiltonian is represented in a basis of the following diabatic (charge localized) electronic states which are relevant for the photoreaction: the electronic ground state of the overall system $|\phi_\mathrm{g}\rangle$, the donor state of the ET process $|\phi_\mathrm{d}\rangle$ (which, in the limit of vanishing coupling between chromophore and semiconductor substrate, corresponds to the product of the first electronically excited state of C343 and an empty conduction band of the semiconductor), and the (quasi)continuum of acceptor states of the ET reaction $|\phi_k\rangle$ (corresponding in the zero coupling limit to the product of the cationic state of C343 and an electron with energy ϵ_k in the conduction band of the semiconductor substrate). Thus, the Hamiltonian reads

$$\begin{aligned} H = {} & |\phi_\mathrm{g}\rangle\epsilon_\mathrm{g}\langle\phi_\mathrm{g}| + |\phi_\mathrm{d}\rangle\epsilon_\mathrm{d}\langle\phi_\mathrm{d}| + \sum_k |\phi_k\rangle\epsilon_k\langle\phi_k| \\ & + \sum_k \left(|\phi_\mathrm{d}\rangle V_{\mathrm{d}k}\langle\phi_k| + |\phi_k\rangle V_{k\mathrm{d}}\langle\phi_\mathrm{d}|\right) + H_\mathrm{N}, \end{aligned} \tag{35}$$

where H_N denotes the part of the Hamiltonian which involves the nuclear degrees of freedom.

The electronic coupling matrix elements $V_{\mathrm{d}k}$ and the distribution of energies ϵ_k of the conduction band of the semiconductor can be specified by the energy-dependent decay width of the donor state

$$\Gamma(E) = 2\pi \sum_k |V_{\mathrm{d}k}|^2 \, \delta(E - \epsilon_k), \tag{36}$$

which describes the coupling-weighted density of states of the semiconductor substrate. In principle, the energy-dependent decay width $\Gamma(E)$ can be determined employing electronic structure theory calculations [114]. In the studies reported below, we have adopted a parameterization based on a tight-binding model which has been developed recently by Petersson et al. [115] to study the electron injection rate from an excited state of a dye molecule to a semiconductor substrate.

To study the influence of the nuclear degrees of freedom on the electron transfer reaction, we consider the vibrational degrees of freedom of C343 as well as the effect of a solvent surrounding the dye–semiconductor system

$$H_{\mathrm{N}} = H_m + H_{\mathrm{b}} + H_{\mathrm{sb}}. \tag{37}$$

The nuclear degrees of freedom of the chromophore are characterized using the normal modes of the electronic ground state of C343 as well as the gradients of the potential energy surfaces of the first optically excited state and the ground state of the cation of C343 resulting in

$$H_m = \frac{1}{2}\sum_l (P_l^2 + \Omega_l^2 Q_l^2) + \sum_l |\phi_{\mathrm{d}}\rangle \kappa_l^{\mathrm{d}} Q_l \langle\phi_{\mathrm{d}}| + \sum_k |\phi_k\rangle \sum_l \kappa_l^{\mathrm{a}} Q_l \langle\phi_k|. \tag{38}$$

The vibrational frequencies Ω_l as well as the gradient κ_l^{d}, κ_l^{a} have been determined by electronic structure calculations employing density functional theory [108].

To account for the influence of the surrounding solvent on the ET dynamics in our simulations, we employ a standard (outer sphere) linear response model [29, 116, 117] where the Hamiltonian of the dye–semiconductor system is coupled linearly to a bath of harmonic oscillators.

$$\begin{aligned} H_{\mathrm{b}} + H_{\mathrm{sb}} &= \frac{1}{2}\sum_j (p_j^2 + \omega_j^2 x_j^2) \\ &\quad + |\phi_{\mathrm{d}}\rangle \sum_j c_j^{\mathrm{d}} x_j \langle\phi_{\mathrm{d}}| + \sum_k |\phi_k\rangle \sum_j c_j^{\mathrm{a}} x_j \langle\phi_k|. \end{aligned} \tag{39}$$

The parameters of the solvent part of the Hamiltonian are characterized by the spectral densities

$$J^{\mathrm{d}}(\omega) = \frac{\pi}{2}\sum_j \frac{(c_j^{\mathrm{d}})^2}{\omega_j}\delta(\omega - \omega_j), \qquad J^{\mathrm{a}}(\omega) = \frac{\pi}{2}\sum_j \frac{(c_j^{\mathrm{a}})^2}{\omega_j}\delta(\omega - \omega_j), \tag{40}$$

in the donor and acceptor states, respectively. In the calculations considered below we have employed a model for relaxation in a polar solvent which includes a Gaussian part describing the fast (inertial) decay of the solvent polarization and a Debye part modeling the slower diffusive dynamics, i.e.

$$J^{\mathrm{d}}(\omega) = \sqrt{\pi}\frac{\lambda_G \omega}{\omega_G} \mathrm{e}^{-[\omega/(2\omega_G)]^2} + 2\lambda_D \frac{\omega \omega_D}{\omega^2 + \omega_D^2}, \tag{41}$$
$$J^{\mathrm{a}}(\omega) = \alpha^2 J^{\mathrm{d}}(\omega),$$

with $\omega_G = 144.54\ \mathrm{cm}^{-1}$, $\omega_D = 25\ \mathrm{cm}^{-1}$, $\lambda_D^{\mathrm{d}} = \lambda_G^{\mathrm{d}} = 700\ \mathrm{cm}^{-1}$, and $\alpha = -0.1$. The relaxation parameters ω_G, ω_D have been chosen in accordance with experimental results on solvation dynamics of C343 in water [118, 119] (for more details on the model see [120]). To illustrate the model parameters describing the nuclear degrees of freedom, Fig. 1 shows the reorganization energies of the intramolecular modes (with respect to the electron transfer transition $|\phi_{\mathrm{d}}\rangle \rightarrow |\phi_k\rangle$) as well as the spectral densities of the solvent. It is seen that the electronic–vibrational coupling is distributed over a relatively large number of intramolecular modes of C343. In the dynamical calculations presented later, 38 of the normal modes of C343 have been explicitly taken into account. These modes were selected according to their electronic–vibrational coupling strength.

Since in our dynamical simulations all degrees of freedom are treated explicitly, the continuum of electronic states (describing the conduction band of the semiconductor) as well as the continuous distribution of solvent modes are first discretized and represented by a finite number of states and modes, respectively. Thereby, the actual number of modes (states) necessary to represent the true continuum depends on the specific parameters considered and the timescale of interest, and serves as a convergence parameter. The

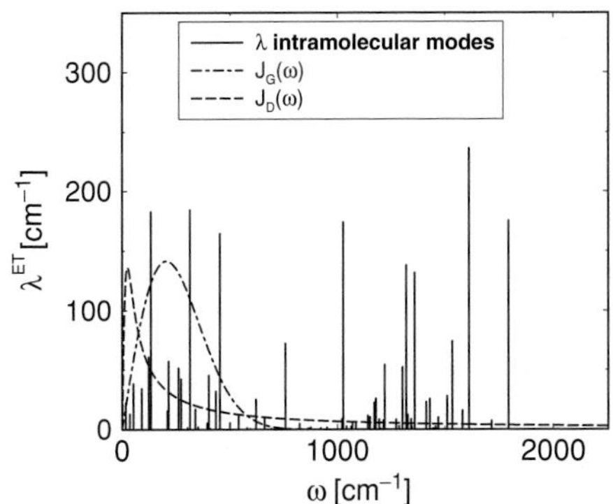

Fig. 1. Properties of the nuclear degrees of freedom of the ET model associated with the ET transition, i.e., the transition from the electronically excited state of neutral C343 to the electronic ground state of the cation of C343. Shown are the Gaussian (*dashed-dotted line*) and Debye (*dashed line*) part of the spectral density of the solvent environment as well as reorganization energies of the intramolecular modes of C343. The spectral densities have been scaled for better illustration

details of efficiently discretizing the continuum of electronic states and the continuous distribution of vibrational modes have been given previously in [47] and [50], respectively, and will not be repeated here. For the system considered here, the number of bath modes required varies between 30 and 50 and the number of electronic states between 200 and 400 (depending on the timescale of interest).

Results

The electron injection dynamics is described by the population of the donor state after photoexcitation,

$$P_{\rm d}(t) = {\rm tr}\{\rho_m \rho_b |\phi_{\rm d}\rangle\langle\phi_{\rm d}| {\rm e}^{{\rm i}Ht} |\phi_{\rm d}\rangle\langle\phi_{\rm d}| {\rm e}^{-{\rm i}Ht}\}, \tag{42}$$

where ρ_m and ρ_b denote the initial state of the nuclear degrees of freedom of the intramolecular modes of C343 and the solvent (described by the respective Boltzmann operators at $T = 300$ K), respectively. The dynamical simulations have been performed fully quantum mechanically employing the ML–MCTDH method. The result of the simulation shown in Fig. 2 exhibits an ultrafast injection of the electron from the electronically excited state of C343 chromophore into the semiconductor: More than 80% of the population of the initially prepared donor state decays within 20 fs into the conduction band of titanium oxide. This timescale is at the lower boundary of experimental results for the C343-TiO_2 system, where injection times in the range between 20 fs and 200 fs have been found using different techniques [14, 73, 75, 87, 88]. In addition to the dominating ultrafast injection, the simulation results also show a small component with slower injection dynamics and oscillatory structure superimposed on the decay. The comparison with a purely electronic

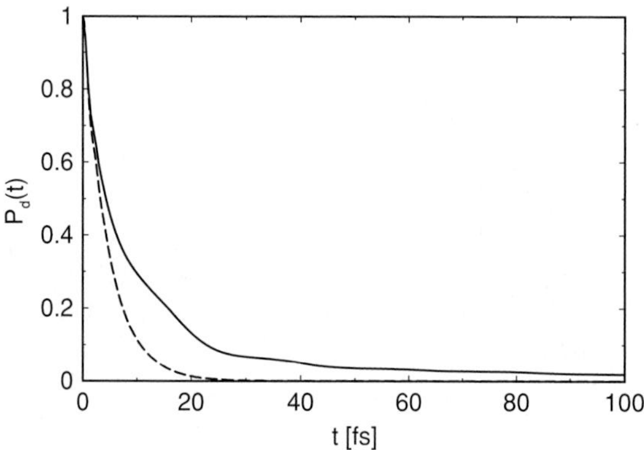

Fig. 2. Electron injection dynamics of the C343-TiO_2 system. Shown is the population of the donor state after photoexcitation with (*full line*) and without (*dashed line*) coupling to the nuclear degrees of freedom

calculation (dashed line in Fig. 2) (where the coupling to the nuclear degrees of freedom has been neglected) reveals that both the slowly decaying component and the oscillatory structures are caused by the coupling of the electron to the nuclear degrees of freedom.

We have analyzed in detail the mutual influence of electronic and nuclear degrees of freedom in the dynamics of the C343-TiO_2 system [120]. The analysis shows that the oscillatory structure is caused by vibrational motion of high-frequency modes of C343 which have oscillation periods similar to the fast ET time of $\approx$20 fs. While such a direct dynamical influence of vibrational coherence on the ET process is only possible for high-frequency (fast) modes, low-frequency (slow) modes exhibit another interesting effect due to electronic–vibrational coupling. As an example, Fig. 3 shows wave packet motion of the low-frequency vibration of the nitrogen group ($\Omega = 133$ cm^{-1}). This mode has negligible changes in equilibrium geometry with respect to photoexcitation but shows a rather large displacement with respect to the ET process, i.e., the transition from the photoexcited state to the cation. As a result, the photoexcitation prepares an (with respect to this mode) essentially stationary wave packet in the donor state. On the other hand, the wave packet in the acceptor states shows pronounced oscillatory motion. This motion is induced by the ultrafast ET process, the timescale of which ($\approx$20 fs) is more than one order of magnitude faster than the vibrational period of the mode ($t = 2\pi/\Omega \approx 251$ fs). Thus, similar to ultrafast photoexcitation, the ultrafast ET process prepares a coherent wave packet on the potential energy surface of the acceptor states which then (due to the relatively large reorganization energy) starts to oscillate. This process of ET-induced vibrational motion is beyond the traditional Marcus theory of ET [30] which assumes that vibrational relaxation is faster than ET so that ET proceeds from a vibrationally relaxed initial state. Here, ET is so fast that it can induce coherent vibrational motion.

Overall, the analysis of the electronic–vibrational dynamics in the C343-TiO_2 system demonstrates that within the same molecule – depending on the

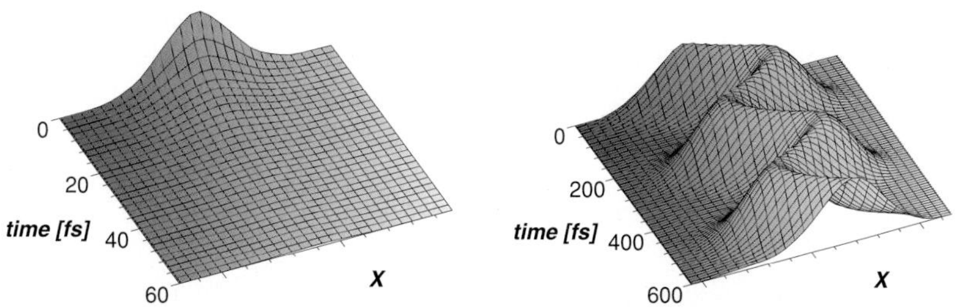

Fig. 3. Wave packet dynamics of one of the vibrational modes of C343 after photoexcitation. Shown is the reduced density of a low-frequency vibration of the nitrogen group (Ω = 133 cm^{-1}) in the donor state (*left*) and averaged over the acceptor states (*right*), respectively

timescale of nuclear motion – electronic–vibrational coupling can result in ET driven by coherent vibrational motion as well as vibrational motion induced by ET. Thus, the heterogeneous ET reaction in the C343-TiO_2 system is another example of a photoreaction which cannot be described by traditional rate theories because the fundamental assumption of a timescale separation between ET dynamics and nuclear motion is not fulfilled.

3.2 Intervalence Electron Transfer Reaction in $(NH_3)_5Ru^{III}NCRu^{II}(CN)_5^-$

The reaction considered in the previous Section represents an example for an ultrafast ET reaction on a timescale of ≈20 fs. In this section, we consider an example for a somewhat slower (yet still very fast) ET reaction. To this end, we consider an intervalence ET reaction in the mixed-valence compound $(NH_3)_5Ru^{III}NCRu^{II}(CN)_5^-$ (for simplicity we denote this compound as RuRu). The generic metal–metal charge transfer (MMCT) process in this system can be schematically represented as

$$(NH_3)_5Ru^{III}NCRu^{II}(CN)_5^- \underset{\text{ET}}{\overset{h\nu}{\rightleftarrows}} (NH_3)_5Ru^{II}NCRu^{III}(CN)_5^- . \tag{43}$$

Upon photoexcitation into the MMCT band an electron is transferred from one metal center to another. This process is followed by an ultrafast internal conversion, resulting in the back transfer of the electron on a subpicosecond timescale [121–123] . The study of this particular ET reaction is of interest for both experimental and theoretical reasons: First, in femtosecond pump-probe studies of this ET reaction in solution coherent oscillations have been observed on timescales longer than the ET time, suggesting that the coherence of the vibrational motion is maintained during the ET process [121–123]. Second, the observation of multiple timescales in optical signals indicates that the ET process cannot be characterized by a simple exponential decay. This is often attributed to the existence of multiple timescales in the solvation dynamics, as well as the influence of the strongly bound ligand modes. Therefore, golden-rule type approaches may not be applicable [124, 125]. Finally, modelings of these type of ET reactions are relatively straightforward since they correspond to direct optical ET and, therefore, most of the parameters required can be obtained from the analysis of absorption and resonance Raman spectra.

Model

The details of the model used to study the ET reaction in RuRu have been described in [124, 125]. Briefly, the study is based on a Hamiltonian of the form

$$H = |\phi_1\rangle E_1 \langle\phi_1| + |\phi_2\rangle E_2 \langle\phi_2| + |\phi_1\rangle V \langle\phi_2| + |\phi_2\rangle V \langle\phi_1| + H_{\mathrm{N}}, \tag{44}$$

where $|\phi_1\rangle$ and $|\phi_2\rangle$ denote the electronic ground state and the charge-transfer state, which results from the photoinduced ET between the metal centers, respectively, and V is the donor–acceptor coupling matrix element. Similar as earlier, the nuclear degrees of freedom comprise inner-sphere intramolecular modes of the RuRu complex as well as an outer-sphere solvent, described by the nuclear Hamiltonian

$$H_{\mathrm{N}} = H_m + H_{\mathrm{b}} + H_{\mathrm{sb}} \tag{45}$$

with

$$H_m = \frac{1}{2}\sum_l \left[P_l^2 + \Omega_l^2\left(X_l - |\phi_2\rangle\frac{c_l}{\Omega_l^2}\langle\phi_2|\right)^2\right]. \tag{46}$$

and

$$H_{\mathrm{b}} + H_{\mathrm{sb}} = \frac{1}{2}\sum_j \left[p_j^2 + \omega_j^2\left(x_j - |\phi_2\rangle\frac{d_j}{\omega_j^2}\langle\phi_2|\right)^2\right]. \tag{47}$$

The modeling of the former is based on the analysis of experimental lineshapes [123,126], where nine Raman-active modes are taken into account. The solvent is modeled by the bimodal spectral density, (41), with $\lambda_G = 2,240$ cm^{-1}, $\omega_G = 100$ cm^{-1}, $\lambda_D = 960$ cm^{-1}, and $\omega_D = 10$ cm^{-1}. All other parameters have been given in [124, 125].

Results

The dynamics of the back transfer of the electron after photoexcitation is directly reflected by the population of the charge-transfer state,

$$P(t) = \mathrm{tr}\{\rho_m\rho_b|\phi_2\rangle\langle\phi_2|\mathrm{e}^{\mathrm{i}Ht}|\phi_2\rangle\langle\phi_2|\mathrm{e}^{-\mathrm{i}Ht}\}, \tag{48}$$

where ρ_m and ρ_b denote the initial state of the nuclear degrees of freedom of the intramolecular modes and the solvent (described by the respective Boltzmann operators), respectively. Figure 4 shows $P(t)$ for a temperature of $T = 300$ K. The results have been obtained employing the SCH method, whereby all intramolecular modes and 10% (high-frequency) bath modes were included in the core (for details, see [124,125]). Overall, the population exhibits a bimodal decay: a fast component on a timescale of ≈120 fs which accounts for approximately 70% of the ET, and a slower component on the timescale of 1–2 picoseconds. This bimodal decay is in qualitative agreement with the experimental results of Barbara et al., where similar timescales were found in pump-probe signals [121, 123]. Several oscillatory structures superimposed on the population decay are observed, the timescale of which can be roughly divided in three groups: the frequency of the very fast oscillation, which can only be seen in the first 10 fs in the inset of Fig. 4, corresponds to the Rabi

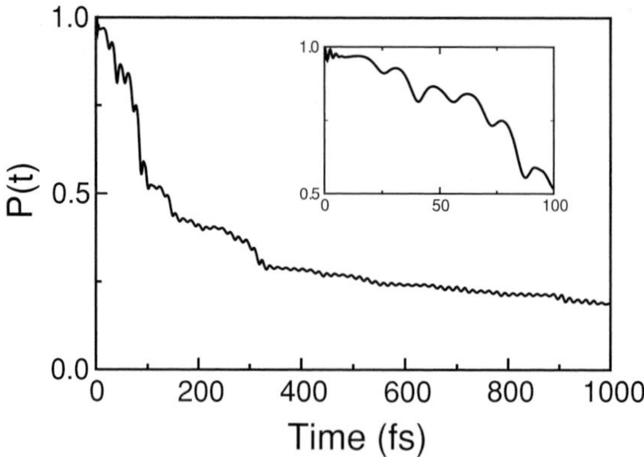

Fig. 4. Time-dependent population of the charge-transfer state (at $T = 300$ K) for the RuRu system. The inset shows $P(t)$ for the first 100 fs. Adapted from [125]

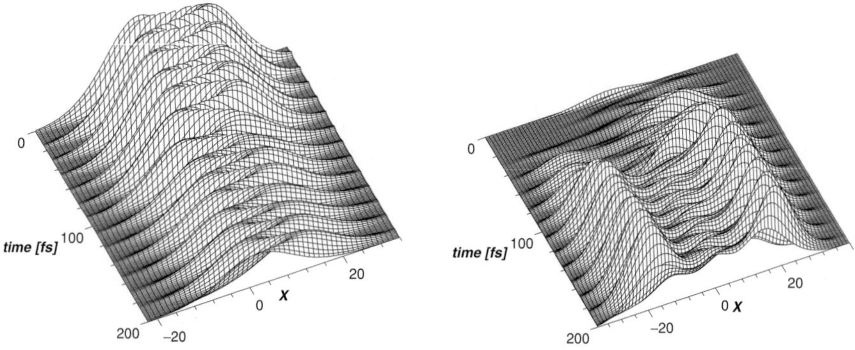

Fig. 5. Vibrational dynamics of the RuRu system. Shown is the time-dependent density of a CN-stretch vibration ($\Omega = 2,118$ cm^{-1}) in the charge transfer (*left*) and the ground (*right*) state, respectively. Adapted from [51].

frequency of the bare electronic two state system and is therefore a remnant of electronic coherence. The oscillations on a timescale of about 16 fs reflect the vibrational motion of the two high-frequency intramolecular modes included in the model (both high-frequency modes have been assigned to CN-stretch vibrations [127]). Finally, several step-like structures on a longer timescale can be seen. The electronic dynamics thus indicates that the ET reaction exhibits significant vibrational coherence effects.

This finding is substantiated by the vibrational dynamics. In [51] we have analyzed in detail the vibrational motion accompanying the back ET reaction in RuRu. As a representative example, Fig. 5 shows the wave packet dynamics of a high-frequency vibration (which has been assigned to the CN-stretch mode of the bridge ligand) in the ground and charge-transfer state. The analysis reveals correlated electronic–vibrational dynamics: whenever the wave packet

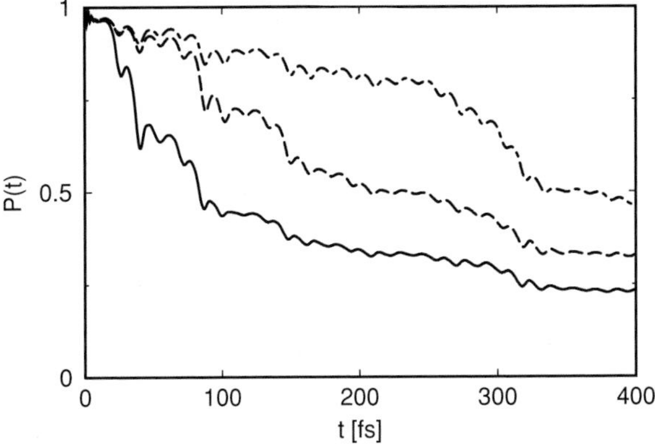

Fig. 6. Dependence of the ET dynamics in the RuRu system on the relaxation parameters of the solvent: $\omega_G = 200\ \mathrm{cm}^{-1}, \omega_D = 20\ \mathrm{cm}^{-1}$ (*full* line); $\omega_G = 50\ \mathrm{cm}^{-1}, \omega_D = 5\ \mathrm{cm}^{-1}$ (*dashed line*); $\omega_G = 20\ \mathrm{cm}^{-1}$, $\omega_D = 2\ \mathrm{cm}^{-1}$ (*dashed-dotted line*). Adapted from [124]

passes the crossing between the potential energy surfaces of the ground and excited electronic states, part of the wave packet is transfered to the ground state. This mechanism results in a modulation of the electronic population dynamics and thus indicates vibrationally coherent ET.

Another interesting mechanism in ultrafast ET reactions are dynamic solvent effects, i.e., the influence of the relaxation timescale of the solvent on the ET reaction. Experimentally it has been shown that the ET dynamics in the RuRu system depends on the solvent, e.g., the average ET time for the RuRu compound in water, a fast relaxing solvent, was found to be $\tau_{\mathrm{ET}} = 100$ fs, which is faster compared to $\tau_{\mathrm{ET}} = 220$ fs in the slower relaxing solvent ethylene glycol [121]. To investigate if this change in ET time is due to the different relaxation timescale of the respective solvent, we have performed simulations for models where only the relaxation parameter of the solvent was changed. The results in Fig. 6 show indeed that the ET becomes slower with longer solvent relaxation times, thus demonstrating the dynamical solvent effect.

3.3 Simulation of Photoexcitation Processes and Time-resolved Optical Spectra

In the two examples for ultrafast photoreactions discussed earlier the photoexcitation process was modeled by starting the calculation in the photoexcited state (corresponding to an infinitely short laser pulse). Both the ML–MCTDH method and the SCH approach also allow a rather straightforward direct simulation of the photoexcitation process by including the coupling to the laser field in the calculation. Thus, the overall (material + laser field) Hamiltonian reads

$$H_{\text{tot}}(t) = H - \mu E(t), \tag{49}$$

where μ denotes the dipole operator and $E(t)$ the electric field of the laser pulse. Including several laser pulses, the methodology can also be used to simulate time-dependent nonlinear spectra [52], such as for example pump-probe or photon-echo spectra.

In this section, we present results of ML–MCTDH simulations for photoexcitation processes and pump-probe spectra for a model that represents qualitatively a class of intervalence ET reactions such as the RuRu system studied earlier. The model Hamiltonian is given by (44–47). The model comprises four intramolecular modes and a solvent environment described by the spectral density in (41) with parameters $\lambda_G = 2,250\ \text{cm}^{-1}$, $\lambda_D = 1,250\ \text{cm}^{-1}$, $\omega_G = 500\ \text{cm}^{-1}$, $\omega_D = 500\ \text{cm}^{-1}$. All other parameters of the model are given in Table 1.

The observable of interest is the time-dependent population of the photoexcited charge-transfer state,

$$P(t) = \frac{1}{\text{tr}[\text{e}^{-\beta H}]} \text{tr}\left[\text{e}^{-\beta H} U^\dagger(t) |\phi_2\rangle\langle\phi_2| U(t)\right], \tag{50}$$

which is analogous to (48) but includes explicitly the laser field in the simulation,

$$\text{i}\frac{\partial}{\partial t} U(t) = [H - \mu E(t)]\, U(t). \tag{51}$$

Another important physical quantity is the overall polarization induced by the laser field

$$I(t) = \frac{1}{\text{tr}[\text{e}^{-\beta H}]} \text{tr}\left[\text{e}^{-\beta H}\, U^\dagger(t)\, \mu\, U(t)\right]. \tag{52}$$

Fourier decomposition of $I(t)$ along different directions gives different types of time-resolved nonlinear spectra [52, 128, 129]. In (50) and (52), the material system is initially at thermal equilibrium, represented by the Boltzmann

Table 1. Parameters of the model Hamiltonian (44) considered in Sect. 3.3, including vibrational frequencies Ω_l and reorganization energies $\lambda_l = c_l^2/(2\Omega_l^2)$ of the intramolecular modes as well as electronic energies and coupling strength. All quantities are given in cm^{-1}

l	Ω_l	λ_l
1	2,100	750
2	650	750
3	400	750
4	150	750
$E_2 - E_1 = 6{,}500$		$V = 500$

operator $e^{-\beta H}$, until the laser field induces the electronic excitation and the subsequent ET process. Noting the similarity between the Boltzmann operator and the time-evolution operator e^{-iHt}, it is apparent that the ML–MCTDH method is also applicable to evaluate $e^{-\beta H}$ by replacing the real time t with the imaginary time $\tau = -i\beta$ [130].

Figure 7 depicts results of simulations of $P(t)$ corresponding to a weak (upper panel) and strong (lower panel) laser field, respectively. Similar to the dynamics exhibited by the RuRu system, the results show that the population dynamics of the charge transfer state after photoexcitation is nonexponential and has stepwise structures. This indicates the influence of strongly-coupled vibrational modes on the ultrafast ET process. It is also interesting to note that the qualitative behavior of $P(t)$ is very similar for weak and strong laser field. The stronger laser field essentially results in a larger population transfer to the excited charge-transfer state and additional field-induced (Rabi) oscillations of the population at short times.

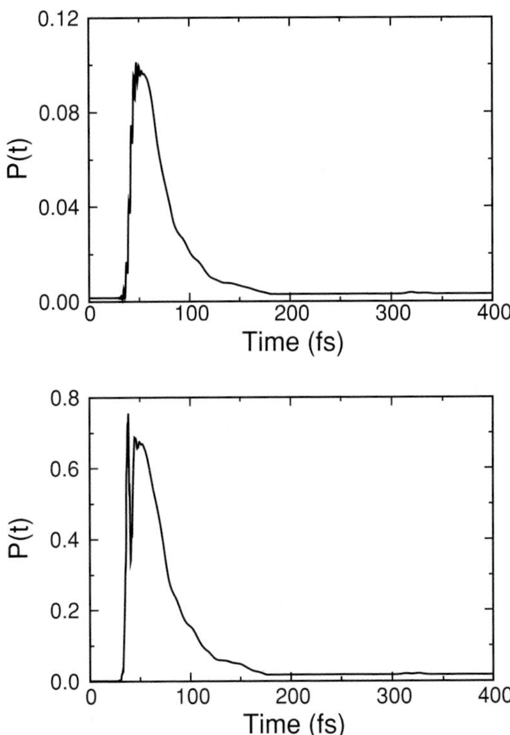

Fig. 7. Time-dependent population of the photoexcited charge-transfer state at $T = 25$ K. The parameters of the material system have been given in the text, and the parameters for the laser pulse are: pulse carrier frequency $\omega = 13,000$ cm^{-1}, center of the pulse $t_d = 40$ fs, pulse duration $\tau = 10$ fs. The strength of the laser pulse is weak in the upper panel, and strong in the lower panel

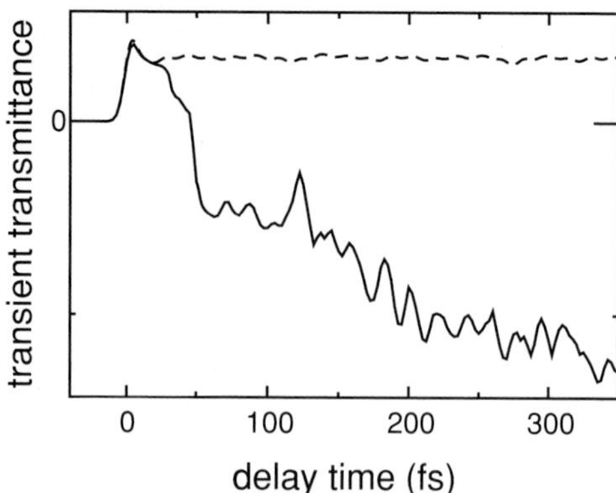

Fig. 8. Pump-probe transient transmittance spectrum corresponding to the ET system considered in the upper panel of Fig. 7. The *dashed line* depicts results for the same ET system except that the ET coupling matrix element V has been set to zero. The carrier frequency and the duration of the probe pulse are the same as those of the pump pulse in Fig. 7, upper panel

The pump-probe (transient transmittance) spectrum for the model ET system is displayed in Fig. 8 [52]. To discuss the manifestation of ET dynamics in this time-resolved spectrum, Fig. 8 also depicts the pump-probe spectrum for the same set of parameters except that the ET coupling matrix element V has been set to zero (dashed line). A more detailed analysis reveals (data not shown) that the pump-probe spectrum for the system without ET is characterized by an ultrafast decaying (stimulated emission) component at short delay times and a long time stimulated Raman contribution. The results for the system including the ET process indicate an additional absorptive process, which results in an overall negative transient transmittance signal at longer times. This additional absorptive contribution to the signal corresponds to absorption from vibrationally excited states in the electronic ground state which are formed in the ET process, and is thus a manifestation of the ET dynamics.

4 Concluding Remarks

In this article, we have studied two examples of ultrafast photoreactions in a condensed phase environment: electron injection in the dye–semiconductor system coumarin 343-TiO_2 and intervalence electron transfer in the mixed valence system $(NH_3)_5Ru^{III}NCRu^{II}(CN)_5^-$. To describe the dynamics in these reactions, we have employed the multilayer formulation of the MCTDH method and the self-consistent hybrid approach. Both methods allow accurate simulations of quantum dynamics in complex molecular systems.

The results demonstrate the pronounced influence of electronic–nuclear coupling, which – depending on the relative timescale – may result in electron transfer driven by coherent vibrational motion or vibrational motion induced by ultrafast electron transfer. The simulation for the mixed valence system, furthermore, reveals the influence of the relaxation timescale of the surrounding solvent on the electron transfer dynamics (dynamic solvent effects).

In addition, we have also discussed the application of the methodology to simulate photoexcitation processes and time resolved optical spectra by including the coupling to the laser field explicitly in the calculation. As has been shown for a model of intervalence electron transfer reactions, the simulation of time resolved optical spectra allows to investigate the manifestation of electronic and nuclear dynamics in experimentally accessible spectra. Since the dynamical method is based on a nonperturbative treatment of the material–laser field coupling, it can also be applied to investigate the possibility of coherent control in complex molecular systems in the condensed phase.

Acknowledgments

The generous allocation of computing time by the National Energy Research Scientific Computing Center (NERSC) and the Leibniz Rechenzentrum, Munich, is gratefully acknowledged. This work has been supported by the National Science Foundation (NSF) CAREER award CHE-0348956 (HW), the Deutsche Forschungsgemeinschaft (DFG), and a collaborative research grant of NSF and DAAD (German Academic Exchange Service).

References

1. *Femtosecond Chemistry*, edited by J. Manz and L. Wöste (VCH, New York, 1995)
2. A. H. Zewail, J. Phys. Chem. A **104**, 5660 (2000)
3. A. Douhal and J. Santamaria, *Femtochemistry and Femtobiology: Ultrafast Dynamics in Molecular Science* (World Scientific, Singapore, 2002)
4. J. Michl and V. Bonačić-Koutecký, *Electronic Aspects of Organic Photochemistry* (Wiley, New York, 1990)
5. G. Orlandi, F. Zerbetto, and M. Zgierski, Chem. Rev. **91**, 867 (1991)
6. L. A. Peteanu, R. W. Schoenlein, R. A. Mathies, and C. V. Shank, Proc. Natl. Acad. Sci. USA **90**, 11762 (1993)
7. F. Bernardi, M. Olivucci, and M. A. Robb, Chem. Soc. Rev. **25**, 321 (1996)
8. P. F. Barbara, G. C. Walker, and T. P. Smith, Science **256**, 975 (1992)
9. B. Wolfseder, L. Seidner, G. Stock, W. Domcke, M. Seel, S. Engleitner, and W.Zinth, Chem. Phys. **233**, 323 (1998)
10. I. V. Rubtsov and K. Yoshihara, J. Phys. Chem. A **103**, 10202 (1999)
11. J. Jortner and M. Bixon, *Electron Transfer: From Isolated Molecules to Biomolecules, Dynamics and Spectroscopy Adv. Chem. Phys. Vols. 106-107* (Wiley, New York, 1999)
12. A. Hagfeldt and M. Grätzel, Chem. Rev. **95**, 49 (1995)

13. C. Zimmermann, F. Willig, S. Ramakrishna, B. Burfeindt, and B. Pettinger, J. Phys. Chem. B **105**, 9245 (2001)
14. R. Huber, J. E. Moser, M. Grätzel, and J. Wachtveitl, Chem. Phys. Lett. **285**, 39 (2002)
15. T. Fiebig, M. Chachisvilis, M. Manger, A. Zewail, A. Douhal, I. Garcia-Ochoa, and D. de La Hoz Ayso, J. Phys. Chem. A **103**, 7419 (1999)
16. S. Lochbrunner, A. J. Wurzer, and E. Riedle, J. Chem. Phys. **112**, 10699 (2000)
17. N. Ernsting, S. Kovalenko, T. Senyushkina, J. Saam, and V. Farztdinov, J. Phys. Chem. A **105**, 3443 (2001)
18. R. P. Feynman and A. R. Vernon, Ann. Phys. **24**, 118 (1963)
19. R. D. Coalson, J. Chem. Phys. **86**, 6823 (1986)
20. C. H. Mak and D. Chandler, Phys. Rev. A **44**, 2352 (1991)
21. R. Egger and U. Weiss, Z. Phys. B **89**, 97 (1992)
22. R. Egger and C. H. Mak, Phys. Rev. B **50**, 15210 (1994)
23. N. Makri and D. E. Makarov, J. Chem. Phys. **102**, 4600 (1995)
24. M. Winterstetter and W. Domcke, Chem. Phys. Lett. **236**, 445 (1995)
25. J. Stockburger and C. H. Mak, Phys. Rev. Lett. **80**, 2657 (1998)
26. L. Mühlbacher and R. Egger, J. Chem. Phys. **118**, 179 (2003)
27. M. Thorwart, E. Paladino, and M. Grifoni, Chem. Phys. **296**, 333 (2004)
28. A. J. Leggett, S. Chakravarty, A. T. Dorsey, M. P. A. Fisher, A. Garg, and W. Zwerger, Rev. Mod. Phys. **59**, 1 (1987)
29. U. Weiss, *Quantum Dissipative Systems*, 2nd ed. (World Scientific, Singapore, 1999)
30. R. A. Marcus and N. Sutin, Biochim. Biophys. Acta **811**, 265 (1985)
31. H.-D. Meyer, U. Manthe, and L. S. Cederbaum, Chem. Phys. Lett. **165**, 73 (1990)
32. U. Manthe, H.-D. Meyer, and L. S. Cederbaum, J. Chem. Phys. **97**, 3199 (1992)
33. M. H. Beck, A. Jäckle, G. A. Worth, and H.-D. Meyer, Phys. Rep. **324**, 1 (2000)
34. H.-D. Meyer and G. A. Worth, Theor. Chem. Acc. **109**, 251 (2003)
35. A. Raab, G. Worth, H.-D. Meyer, and L. Cederbaum, J. Chem. Phys. **110**, 936 (1999)
36. F. Huarte-Larrañaga and U. Manthe, J. Chem. Phys. **113**, 5115 (2000)
37. S. Mahapatra, G. A. Worth, H. D. Meyer, L. S. Cederbaum, and H. Köppel, J. Phys. Chem. A **105**, 5567 (2001)
38. C. Cattarius, G. A. Worth, H.-D. Meyer, and L. S. Cederbaum, J. Chem. Phys. **115**, 2088 (2001)
39. H. Nauendorf, G. Worth, H.-D. Meyer, and O. Kühn, J. Phys. Chem. **106**, 719 (2002)
40. M. D. Coutinho-Neto, A. Viel, and U. Manthe, J. Chem. Phys. **121**, 9207 (2004)
41. T. Wu, H.-J. Werner, and U. Manthe, Science **306**, 2227 (2004)
42. G. A. Worth, H.-D. Meyer, and L. S. Cederbaum, J. Chem. Phys. **109**, 3518 (1998)
43. H. Wang, J. Chem. Phys. **113**, 9948 (2000)
44. D. Egorova, M. Thoss, W. Domcke, and H. Wang, J. Chem. Phys. **119**, 2761 (2003)
45. M. Nest and H.-D. Meyer, J. Chem. Phys. **119**, 24 (2003)
46. I. Burghardt, M. Nest, and G. Worth, J. Chem. Phys. **119**, 5364 (2003)

47. H. Wang, M. Thoss, and W. H. Miller, J. Chem. Phys. **115**, 2979 (2001)
48. H. Wang and M. Thoss, Israel J. Chem. **42**, 167 (2002)
49. H. Wang and M. Thoss, J. Chem. Phys. **119**, 1289 (2003)
50. M. Thoss, I. Kondov, and H. Wang, Chem. Phys. **304**, 169 (2004)
51. M. Thoss, W. Domcke, and H. Wang, Chem. Phys. **296**, 217 (2004)
52. H. Wang and M. Thoss, Chem. Phys. Lett. **389**, 43 (2004)
53. M. Thoss and H. Wang, Chem. Phys. (2005), **131**, 243
54. J. Frenkel, *Wave Mechanics* (Clarendon, Oxford, 1934)
55. N. F. Mott, Proc. Cambridge Phil. Soc. **27**, 553 (1931)
56. J. B. Delos, W. R. Thorson, and S. K. Knudson, Phys. Rev. A **6**, 709 (1972)
57. G. D. Billing, Chem. Phys. Lett. **30**, 391 (1975)
58. R. B. Gerber, V. Buch, and M. A. Ratner, J. Chem. Phys. **77**, 3022 (1982)
59. D. A. Micha, J. Chem. Phys. **78**, 7138 (1983)
60. R. Graham and M. Höhnerbach, Z. Phys. B **57**, 233 (1984)
61. W. H. Miller and T. F. George, J. Chem. Phys. **56**, 5637 (1972)
62. J. C. Tully, J. Chem. Phys. **93**, 1061 (1990)
63. M. F. Herman, J. Chem. Phys. **76**, 2949 (1982)
64. F. J. Webster, P. J. Rossky, and R. A. Friesner, Comput. Phys. Commun. **63**, 494 (1991)
65. S. Chapman, Adv. Chem. Phys. **82**, 423 (1992)
66. D. F. Coker and L. Xiao, J. Chem. Phys. **102**, 496 (1995)
67. M. A. Sepúlveda and F. Grossmann, Adv. Chem. Phys. **96**, 191 (1996)
68. W. H. Miller, J. Phys. Chem. A **105**, 2942 (2001)
69. K. Kay, Annu. Rev. Phys. Chem. **56**, 255 (2005)
70. H. Wang, X. Song, D. Chandler, and W. Miller, J. Chem. Phys. **110**, 4828 (1999)
71. E. Wigner, Phys. Rev. **40**, 749 (1932)
72. J. E. Moser and M. Grätzel, Chem. Phys. **176**, 493 (1993)
73. J. M. Rehm, G. L. McLendon, Y. Nagasawa, K. Yoshihara, J. Moser, and M. Grätzel, J. Phys. Chem. **100**, 9577 (1996)
74. I. Martini, J. Odak, G. V. Hartland, and P. V. Kamat, J. Chem. Phys. **107**, 8064 (1997)
75. J. Wachtveitl, R. Huber, S. Spörlein, J. Moser, and M. Grätzel, Int. J. Photoenergy **1**, 131 (1999)
76. J. Asbury, E. Hao, Y. Wang, H. N. Ghosh, and T. Lian, J. Phys. Chem. B **105**, 4545 (2001)
77. G. Ramakrishna and H. N. Gosh, J. Phys. Chem. A **106**, 2545 (2002)
78. J. Schnadt et al., Nature **418**, 620 (2002)
79. K. A. Walters, D. A. Gaal, and J. T. Hupp, J. Phys. Chem. B **106**, 5139 (2002)
80. R. Huber, J. E. Moser, M. Grätzel, and J. Wachtveitl, J. Phys. Chem. B **106**, 6494 (2002)
81. J. Kallioinen, G. Benkö, V. Sundström, J. Korrpi-Tommola, and A. Yartsev, J. Phys. Chem. B **106**, 4396 (2002)
82. K. Takeshita, Y. Sasaki, M. Kobashi, Y. Tanaka, S. Maeda, A. Yamakata, T. Ishibashi, and H. Onishi, J. Phys. Chem. B **107**, 4156 (2003)
83. L. Wang, R. Ernstorfer, F. Willig, and V. May, **109**, 9589 (2005)
84. N. Anderson and T. Lian, Annu. Rev. Phys. Chem. **56**, 491 (2005)
85. A. Hagfeldt and M. Grätzel, Acc. Chem. Res. **33**, 269 (2000)
86. M. Grätzel, Nature **414**, 338 (2001)

87. K. Murakoshi, S. Yanagida, M. Capel, and E. Castner, in *Interfacial Electron Transfer Dynamics of Photosensitized Zinc Oxide Nanoclusters*, edited by M. Moskovits (ACS, Washington DC, 1997)
88. H. N. Gosh, J. B. Asbury, Y. Wang, and T. Lian, J. Phys. Chem. B **102**, 10208 (1998)
89. F. Willig, C. Zimmermann, S. Ramakrishna, and W. Storck, Electrochim. Acta **45**, 4565 (2000)
90. G. Benkö, J. Kallioinen, J. Korppi-Tommola, A. Yartsev, and V. Sundström, J. Am. Chem. Soc. **124**, 489 (2002)
91. R. Huber, S. Spörlein, J. Moser, M. Grätzel, and J. Wachtveitl, J. Phys. Chem. B **104**, 8995 (2002)
92. E. Hao, N. Anderson, J. Asbury, and T. Lian, J. Phys. Chem. B **106**, 10191 (2002)
93. J. Asbury, E. Hao, Y. Wang, and T. Lian, J. Phys. Chem. B **104**, 11957 (2000)
94. L. Rego and V. Batista, J. Am. Chem. Soc. **125**, 7989 (2003)
95. W. Stier and O. Prezhdo, Israel J. Chem. **42**, 213 (2002)
96. W. Stier and O. Prezhdo, J. Phys. Chem. B **106**, 8047 (2002)
97. W. Stier and O. Prezhdo, Adv. Mat. **16**, 240 (2004)
98. W. Duncan, W. Stier, and O. Prezhdo, J. Am. Chem. Soc. **127**, 7941 (2005)
99. W. Duncan, W. Stier, and O. Prezhdo, J. Phys. Chem. B **109**, 365 (2005)
100. L. Rego and V. Batista, J. Chem. Phys. **122**, 154709 (2005)
101. S. Ramakrishna, F. Willig, and V. May, Phys. Rev. B **62**, 16330 (2000)
102. S. Ramakrishna and F. Willig, J. Phys. Chem. B **104**, 68 (2000)
103. S. Ramakrishna, F. Willig, and V. May, J. Chem. Phys. **115**, 2743 (2001)
104. M. Schreiber, I. Kondov, and U. Kleinekathöfer, J. Lumin. **94**, 471 (2001)
105. S. Ramakrishna, F. Willig, and V. May, Chem. Phys. Lett. **351**, 242 (2002)
106. S. Ramakrishna, F. Willig, and V. May, J. Phys. Chem. B **107**, 607 (2003)
107. L. Wang and V. May, J. Chem. Phys. **121**, 8039 (2004)
108. I. Kondov, M. Thoss, and H. Wang, Int. J. Quant. Chem. **106**, 1291 (2006)
109. K. Hara, K. Sayama, Y. Ohga, A. Shinpo, S. Suga, and H. Arakawa, Chem. Comm. **6**, 569 (2001)
110. D. Newns, Phys. Rev. **178**, 1123 (1969)
111. J. Muscat and D. Newns, in *Progress in Surface Science*, edited by S. Davison (Pergamon, Oxford, 1978), Vol. 9
112. K. Sebastian, J. Chem. Phys. **90**, 5056 (1989)
113. Y. Boroda and G. A. Voth, J. Chem. Phys. **104**, 6168 (1996)
114. I. Kondov, H. Wang, and M. Thoss, in preparation
115. A. Petersson, M. Ratner, and H. Karlsson, J. Phys. Chem. B **104**, 8498 (2000)
116. N. Makri, J. Phys. Chem. B **103**, 2823 (1999)
117. Y. Georgievskii, C.-P. Hsu, and R. A. Marcus, J. Chem. Phys. **110**, 5307 (1999)
118. R. Jimenez, G. R. Fleming, P. V. Kumar, and M. Maroncelli, Nature **369**, 471 (1994)
119. R. M. Stratt and M. Maroncelli, J. Phys. Chem. **100**, 12981 (1996)
120. I. Kondov, M. Thoss, and H. Wang, J. Phys. Chem. A **110**, 1364 (2006)
121. P. Kambhampati, D. H. Son, T. W. Kee, and P. F. Barbara, J. Phys. Chem. A **104**, 10637 (2000)
122. D. H. Son, P. Kambhampati, T. W. Kee, and P. F. Barbara, J. Phys. Chem. A **106**, 4561 (2002)
123. K. Tominaga, D. A. V. Kliner, A. E. Johnson, N. E. Levinger, and P. F. Barbara, J. Chem. Phys. **98**, 1228 (1993)

124. M. Thoss and H. Wang, Chem. Phys. Lett. **358**, 298 (2002)
125. H. Wang and M. Thoss, J. Phys. Chem. A **107**, 2126 (2003)
126. G. C. Walker, P. F. Barbara, S. K. Doorn, Y. Dong, and J. T. Hupp, J. Phys. Chem. **95**, 5712 (1991)
127. S. K. Doorn and J. T. Hupp, J. Am. Chem. Soc. **111**, 1142 (1989)
128. S. Mukamel, *Principles of Nonlinear Optical Spectroscopy* (University Press, Oxford, 1995)
129. W. Domcke and G. Stock, Adv. Chem. Phys. **100**, 1 (1997)
130. H. Wang and M. Thoss, J. Chem. Phys. **124**, 034114 (2006)

Part II

New Methods for Quantum Molecular Dynamics in Large Systems

II.1 Semiclassical Methods

Decoherence in Combined Quantum Mechanical and Classical Mechanical Methods for Dynamics as Illustrated for Non-Born–Oppenheimer Trajectories

Donald G. Truhlar

Summary. This chapter discusses the role of decoherence in mixed quantum–classical approaches to electronically nonadiabatic chemical dynamics. The correlation of electronic and nuclear motion, which is not included in the semiclassical Ehrenfest or time-dependent Hartree method, induces decoherence in the reduced electronic density matrix, and the chapter shows how this can be modeled by adding algorithmic demixing to the Liouville-von Neumann equation. The resulting mixed quantum–classical equations of motion involve stochastically controlled, smooth, and continuous surface switching coupled to coherent propagation through each region of strong interaction of the electronic states. The chapter also reviews test results that show good agreement with fully quantum mechanical results for a diverse set of atom–diatom test cases.

1 Introduction

The coupling of quantum mechanics to classical mechanics is a recurring theme in the treatment of complex systems because a full quantum mechanical treatment is usually possible only for simple systems. The coupling may occur in the generation of potential energy surfaces, as in combined quantum mechanical and molecular mechanical methods [1–3] or it may occur in the dynamics step, as when quantum mechanical nuclear-motion effects are combined with transition state theory or molecular dynamics simulations [2, 4–7]. Conventional molecular dynamics simulations themselves, even when the nuclear motion is only treated classically, involve using quantum mechanics, explicitly or implicitly, to derive the Born–Oppenheimer [8] potential energy surface and then treating nuclear motion classically [7, 9–12]. This kind of joining of the two mechanics raises fewer theoretical questions than the first two. However, if we allow for Born–Oppenheimer breakdown, that is, electronic nonadiabaticity, then a number of conceptual issues arise [7, 13–31]. Very similar issues arise in Born–Oppenheimer processes if some nuclear degrees of freedom are treated quantum mechanically and others classically [32–41]. The present article is concerned with this problem and especially with elucidating the important role of decoherence in shaping a physically correct form for the equations of motion of the quantal and classical subsystems. Furthermore we seek a practical algorithm that allows us to simulate systems in which decoherence plays an important role. Although one could illustrate the theory by any problem in which some degrees of freedom are treated as quantal but are not simply adiabatic and other degrees of freedom are treated as classical, we use the problem of electronic nonadiabaticity as our illustrative example, with all nuclear degrees of freedom classical. Furthermore, we start with a Hartree approximation (also called the Ehrenfest approximation [14, 21], the time-dependent self-consistent-field approximation [20, 37, 40], or the self-consistent eikonal approximation [15]), which assumes only a mean-field (uncorrelated) coupling of the electronic and nuclear degrees of freedom, and we show how adding correlation effects leads to decoherence. Since the nuclear degrees of freedom are coupled to the electronic ones, we will see that they require some quantum mechanical elements for their description.

2 Theory

The quantum mechanical time-dependent Hartree approximation for coupled electronic and nuclear motion is

$$\Psi = \phi^{\text{elec}}\left(\mathbf{r}, t\right) \psi^{\text{nuc}}\left(\mathbf{R}, t\right). \tag{1}$$

The factors in (2) satisfy an electronic mean-field Schrödinger equation

$$\mathrm{i}\hbar \frac{\partial}{\partial t} \phi^{\text{elec}} = \langle \psi^{\text{nuc}} \left| H \right| \psi^{\text{nuc}} \rangle_{\mathbf{R}} \, \phi^{\text{elec}}(\mathbf{R}, t) \tag{2}$$

and a nuclear mean-field Schrödinger equation

$$\mathrm{i}\hbar \frac{\partial}{\partial t} \psi^{\text{nuc}} = \left\langle \psi^{\text{elec}} \left| H \right| \psi^{\text{elec}} \right\rangle_{\mathbf{r}} \psi^{\text{nuc}}(\mathbf{R}, t), \tag{3}$$

where $\mathrm{i} = \sqrt{-1}$, $\hbar$ is Planck's constant divided by 2π, $\mathbf{r}$ and $\mathbf{R}$ denote the electronic and nuclear coordinates, respectively, and t is time.

Now we approximate ψ^{nuc} by an ensemble of trajectories, which yields a semiclassical time-dependent Hartree approximation [37]. The nuclear mean-field wave packet is replaced by an ensemble of classical trajectories propagating under the influence of the self-consistent potential

$$U^{\text{SCP}} = \left\langle \psi^{\text{elec}} \left| H \right| \psi^{\text{elec}} \right\rangle_{\mathbf{r}}. \tag{4}$$

The electronic mean-field Schrödinger equation becomes

$$\mathrm{i}\hbar \frac{\partial}{\partial t} \phi^{\text{elec}} = \langle H \rangle_{\text{nuclear ensemble}} \, \phi^{\text{elec}}(\mathbf{r}, t). \tag{5}$$

This treatment neglects important correlations between electronic and nuclear motion. A better starting point than (1) is a multiconfigurational wave packet [20, 42–45]. A wave packet in a multielectronic-state molecular system may be written

$$\Psi = \sum_{\text{states } \alpha} c_\alpha(t)\, \phi_\alpha^{\text{elec}}\left(\mathbf{r}, \mathbf{R}(t)\right) \psi_\alpha^{\text{nuc}}\left(\mathbf{R}, t\right), \tag{6}$$

where $\phi_\alpha^{\text{elec}}$ is a normalized component of the electronic wave function, ψ_α^{nuc} is a normalized component of the nuclear-motion wave packet, and c_α is a time-dependent coefficient.

To improve upon the mean-field approximation in the semiclassical treatment, we add correlation by making the independent-trajectory approximation [15, 40]. (A quantum wave packet analog is the "independent first generation" approximation [45].) This replaces (5) by

$$\mathrm{i}\hbar \frac{\partial}{\partial t} \phi^{\text{elec}} = H\left(\mathbf{R}(t)\right) \phi^{\text{elec}}\left(\mathbf{r}, \mathbf{R}, (t)\right) \tag{7}$$

for each trajectory. The combination of the independent-trajectory approximation and the semiclassical time-dependent Hartree approximation is called the semiclassical Ehrenfest approximation [23, 26–30].

Next we choose an electronic basis

$$\phi^{\text{elec}} = \sum_\alpha c_\alpha(t) \phi_\alpha^{\text{el}}\left(r, \mathbf{R}(t)\right), \tag{8}$$

where c_α is a coefficient, and ϕ_α^{el} is an antisymmetrized many-electron configuration state function in either the adiabatic [8] or a diabatic [26] representation. Furthermore we make the semiclassical replacement

$$\frac{\partial \phi_\alpha^{\text{el}}}{\partial t} = \frac{\mathrm{d}\mathbf{R}}{\mathrm{d}t} \frac{\partial \phi_\alpha^{\text{el}}}{\partial \mathbf{R}}. \tag{9}$$

Substituting (8) and (9) into (7) yields the following time-dependent Schrödinger equation for the coefficients along the trajectory [20]

$$\mathrm{i}\hbar \frac{\partial c_\alpha}{\partial t} = \sum_\beta c_\beta(t) \left[-\mathrm{i}\hbar \dot{\mathbf{R}} \cdot \mathbf{d}_{\alpha\beta} + U_{\alpha\beta}\left(\mathbf{R}(t)\right)\right], \tag{10}$$

where

$$U_{\alpha\beta} \equiv \left\langle \phi_\alpha^{\text{el}} \left| H_{\text{elec}} \right| \phi_\beta^{\text{el}} \right\rangle_{\mathbf{r}} \tag{11}$$

and

$$\mathbf{d}_{\alpha\beta} \equiv \left\langle \phi_\alpha^{\text{el}} \left| \nabla_{\mathbf{R}} \right| \phi_\beta^{\text{el}} \right\rangle_{\mathbf{r}}. \tag{12}$$

In (12), H_{elec} is the so called electronic Hamiltonian, which also includes nuclear repulsion. It is defined by

$$H_{\text{elec}} = H - T_{\mathbf{R}}, \tag{13}$$

where $T_{\mathbf{R}}$ is the nuclear kinetic energy. Since $\mathbf{d}_{\alpha\beta}$ is anti-Hermitian, its diagonal elements vanish identically. Note that if an adiabatic representation is used in (8), $\mathbf{U}$ is diagonal [13b], whereas if a diabatic representation is used in (8), $\mathbf{d}_{\alpha\beta}$ is assumed to be negligible and is neglected. The diagonal elements of $\mathbf{U}$ are called potential energy surfaces, and U_{ii} is often denoted as V_i. In well established but somewhat inconsistent conventions, the off-diagonal elements of $\mathbf{U}$ are called the diabatic couplings, and $\mathbf{d}_{\alpha\beta}$ is called the nonadiabatic coupling. It is convenient to reformulate (10) in terms of the reduced electronic density matrix, which is defined by its matrix elements as follows:

$$\rho_{\alpha\beta} \equiv c_\alpha c_\beta^* . \tag{14}$$

Substituting (14) into (10) yields a unitary Liouville-von Neumann equation [46], which in our case can be written as:

$$i\hbar \frac{\partial \rho_{\alpha\beta}}{\partial t} = -\sum_\gamma \left(\left[-i\hbar \dot{\mathbf{R}} \cdot \mathbf{d}_{\gamma\beta} + U_{\gamma\beta} \right] \rho_{\alpha\gamma} - \{\text{permute indices}\} \right) . \tag{15}$$

Equation (15) is also called a unitary quantum Liouville equation. It is the quantum mechanical analog of Liouville's theorem in classical mechanics, and it is equivalent to the time-dependent Schrödinger equation [47, 48].

In the semiclassical Ehrenfest method one solves the coupled quantum mechanical equation (15) [or the equivalent equation (10)] for the electrons and simultaneously the classical equations of motion with the effective potential of equation (4) for the nuclear motion. Because we made the independent trajectory approximation, we repeat this calculation for an ensemble of initial conditions in the classical phase space (which may be sampled classically [11, 12] or quasiclassically [9, 10, 12a]), we average over initial conditions, and we sum over final states. The semiclassical Ehrenfest method shares with the exact solution of the Schrödinger equation that the results are independent of the representation (adiabatic, diabatic, or intermediate) used for the quantum subsystem. In fact this is true for each individual trajectory, not just for the ensemble average. But there is a serious defect in this method, namely that the system ends in an unphysical final state. Consider, for example, a collision or a photodissociation event where the final state is a diatomic molecule AB and an atom C. Suppose that the total energy is 2.5 eV above the classical potential energy of ground-state products, and that the products have one excited electronic state with an electronic excitation energy of 2.0 eV. The accurate quantum mechanical distribution of nuclear-motion energies will be bimodal: systems in the ground-electronic state will have 2.5 eV of nuclear energy, and systems in the excited electronic state will have 0.5 eV of nuclear energy. One might, under certain circumstances, even find a 50:50 distribution of these states. However, because a semiclassical Ehrenfest trajectory propagates on an average potential energy surface (4), it might end with an average nuclear energy of 1.25 eV or 1.5 eV (or in fact any energy in the range 0–2.5 eV), rather than being restricted to one of the two quantally allowed values.

Why is the semiclassical Ehrenfest method wrong? Because $\rho_{\alpha\beta}$ fails to tend to $\delta_{\alpha\beta}$ (a kronecker delta) as $t \to \infty$. And why does that failure occur? Because (10) and (15), being equivalent to the time-dependent Schrödinger equation, are wrong for a subsystem.

There is only one system governed by the Schrödinger equation, namely the entire universe. All other systems are subsystems and satisfy a quantum master equation, in particular, a nonunitary Liouville-von Neumann equation with dissipation and dephasing. In our example of non-Born–Oppenheimer trajectories, the nuclei serve as a "bath" or "environment" for the electronic subsystem [30,49–53]. To understand the effect of this bath, consider the wave packet of (6). In our example, there are two electronic states, corresponding to $\alpha = 1$ and $\alpha = 2$. The component of the nuclear wave packet corresponding to the lower-energy electronic state ($\alpha = 1$) moves faster, as does the trajectory subensemble corresponding to this subpacket. Therefore the two terms in (6) get out of phase, and they become subpackets in different regions of space; for these reasons their overlap tends to zero. As a consequence, $\rho_{\alpha\beta} \to 0$.

When a semiclassical Ehrenfest trajectory finishes a non-Born–Oppenheimer event, $\rho_{\alpha\beta}$ for $\alpha \neq \beta$ is not zero and $\rho_{\alpha\alpha}$ is neither zero nor unity. That is, the trajectory does not decohere to a pure state. Physically, dephasing would cause the off-diagonal elements to decay

$$\begin{pmatrix} \rho_{\alpha\beta} & \rho_{\alpha\beta} \\ \rho_{\beta\alpha} & \rho_{\beta\beta} \end{pmatrix} \to \begin{pmatrix} \rho'_{\alpha\alpha} & 0 \\ 0 & \rho'_{\beta\beta} \end{pmatrix} . \tag{16}$$

Algorithmically, we want our statistical ensemble of trajectories to "demix" to an ensemble of trajectories with quantized electronic states, schematically

$$\begin{pmatrix} \rho'_{\alpha\alpha} & 0 \\ 0 & \rho'_{\beta\beta} \end{pmatrix} \to \rho'_{\alpha\alpha} \begin{pmatrix} 1 & 0 \\ 0 & 0 \end{pmatrix} + \rho'_{\beta\beta} \begin{pmatrix} 0 & 0 \\ 0 & 1 \end{pmatrix} . \tag{17}$$

To achieve this we add algorithmic decay to the unitary Liouville-von Neumann equation such that each trajectory, at any given time, decoheres toward a given state, called the "decoherent state," in such a way that the distribution of states (averaged over an ensemble of trajectories) is self-consistent with the density matrix. The resulting nonunitary Liouville-von Neumann equation, also called a quantum master equation, has the form

$$\frac{\mathrm{d}\rho_{\alpha\beta}}{\mathrm{d}t} = \left[\frac{\mathrm{d}\rho_{\alpha\beta}}{\mathrm{d}t}\right]_{\text{unitary}} + \left[\frac{\mathrm{d}\rho_{\alpha\beta}}{\mathrm{d}t}\right]_{\text{decoherent}} , \tag{18}$$

where the first term on the right-hand side is from (15) and generates dynamics equivalent to the Schrödinger Equation, and the second term is an algorithmic control term added to simulate the effect of decoherence. Both terms conserve total energy and total angular momentum of the combined quantal and classical subsystems. However energy is transferred between the two subsystems; when energy is transferred from the quantal subsystem to the classical one, this may be considered to be a form of dissipation.

The density matrix of (18) is a reduced density matrix, that is, a density matrix of a subsystem traced over its environment. In the present case, it is the electronic density matrix obtained by tracing over the nuclear degrees of freedom. This matrix, being Hermitian, can be diagonalized in any basis. In which basis does it become diagonal and stay diagonal? That basis is called the pointer basis [54], and the selection of this basis by the decoherent process is called environment-induced superselection or einselection [55]. The pointer basis is determined by the interaction of the subsystem with its environment; this interaction is sometimes called the measuring process. For example if the system is a spin-$\frac{1}{2}$ particle ($S = \frac{1}{2}$), and its interaction is to encounter a detector properly designed to measure S_z, the pointer basis will be the eigenvectors of the operator $\hat{S}_z$. If, however, the interaction with the environment is to encounter a detector properly designed to measure S_x, the pointer basis will be the eigenvector of $\hat{S}_x$. More generally, if the subsystem behaves adiabatically (such as when the frequencies of the environment are much lower than those of the subsystem), the pointer basis will be the adiabatic energy states of the subsystem [56], which is fully in accord with intuition. In the limit where the self-Hamiltonian is negligible compared to the subsystem-environment interaction, the eigenvectors of the interaction becomes the pointer state [57]. In the general case the pointer basis is unknown. The analog of the pointer basis in our algorithm is the basis used to express the decoherent states; we may call this the algorithmic pointer basis. Since the physical pointer basis is not easy to predict and may change with time as the system explores different regions of nuclear configuration space (i.e., as the electronic subsystem explores different aspects of its nuclear environment), our goal is to find an algorithm whose accuracy does not depend strongly on the choice of algorithmic pointer basis. In practice this means we seek an algorithm that yields good results in both the adiabatic and diabatic representations. Not only must we choose an algorithmic pointer basis, we must also choose the decoherent state, which will be labeled K. Thus $\alpha = K$ for the state toward which the system is decohering at a particular time along a particular trajectory.

To derive a form for the second term of (18) we make the reasonable assumption that Re c_α and Im c_α (in (10)) decay by a pure first-order process at the same rate in the algorithmic pointer basis [23]; this conserves the electronic phase angle, that is, it conserves arctan (Im c_α/Re c_α). Then we obtain, for example for the $\alpha\beta = \alpha K$ element [27]

$$\left(\frac{\mathrm{d}\rho_{\alpha K}}{\mathrm{d}t}\right)_{\text{decoherent}} = \frac{1}{2}\left[\frac{1}{\rho_{KK}}\left(\sum_{\gamma \neq K}\frac{\rho_{\gamma\gamma}}{\tau_{K\gamma}}\right) - \frac{1}{\tau_{\alpha K}}\right]\rho_{\alpha K}. \tag{19}$$

The more common assumption is that the master equation is linear, which yields:

$$\left(\frac{\mathrm{d}\rho_{\alpha\beta}}{\mathrm{d}t}\right)_{\text{decoherent}} = -\frac{1}{\tau_{\alpha\beta}}\rho_{\alpha\beta}, \quad \alpha \neq \beta. \tag{20}$$

To which state does the system decohere? We determine this stochastically by Tully's [17] fewest switches algorithm, which was originally proposed for use in surface hopping calculations. Trajectory surface hopping calculations [13, 16–18] stochastically switch the state in which the system propagates (i.e., the potential energy surface governing nuclear motion) to keep the ensemble of nuclear trajectories as consistent as possible with the quantal evolution of the quantal subsystem governed by the unitary Liouville-von Neumann equation. In contrast, our algorithm [29, 30] stochastically switches the decoherent state to keep the nuclear ensemble consistent with the unitary Liouville-von Neumann equation over each passage through a strong interaction of the electronic states, which is called *coherent switching*. At the same time the nuclei propagate on a potential energy surface consistent with the nonunitary Liouville-von Neumann equation incorporating *decay of mixing*. The algorithm is therefore called coherent switches with decay of mixing (CSDM). Because the boundaries of the coherent switching regions introduce time nonlocality, the algorithm is non-Markovian.

In summary, the CSDM algorithm introduces decoherence into the electronic reduced density matrix such that in the strong interaction region the potential energy surface governing nuclear motion has the desirable (representation-independent) properties of the semiclassical Ehrenfest potential, whereas in the asymptotic or weakly coupled regions the effective potential reduces to that of the decoherent state in the pointer basis. But the decoherent state switches stochastically in a coherent way for each complete passage through a strong interaction region. Thus we evolve two density matrices, one (evolved with decay of mixing) controls the effective potential energy surface for nuclear motion, and the other (evolved coherently through strong-interaction regions) controls stochastic switching of the decoherent state.

The decoherence process is first-order with rate constant τ^{-1}. For example, for a diagonal element $\rho_{\alpha\alpha}$ of the density matrix, with $\alpha \neq K$, we have

$$\frac{\mathrm{d}\rho_{\alpha\alpha}}{\mathrm{d}t} = \left(\frac{\mathrm{d}\rho_{\alpha\alpha}}{\mathrm{d}t}\right)_{\text{unitary}} - \frac{\rho_{\alpha\alpha}}{\tau_{\alpha K}}, \tag{21}$$

where the first term on the right is associated with coherent Ehrenfest propagation and the second term causes demixing. There are similar equations for other density matrix elements, except that they are nonlinear for off-diagonal elements $\rho_{\alpha\beta}$. We call τ the decoherence time or the demixing time.

Since our algorithmic demixing is analogous to but not identical to physical decoherence, it is reasonable that our demixing time should be similar to but not identical to the physical decoherence time. We therefore base our choice of the demixing time on three principles:

1. The semiclassical limit of a wave function is the sum of WKB-like trajectories associated with minimum wave packets, and decoherence of the superposition is faster than decoherence of the individual packets [57]. Nuclear wave packets move at different speeds on different surfaces, causing

dephasing and decay of overlap, and this leads to decay of off-diagonal elements of the density matrix [53].
2. The pointer basis is the one in which decoherence is fastest [57].
3. Decoherence slows down when the momentum component in the nonadiabatic coupling direction is small [29].

Using the first two principles we derived [53] an approximate expression for the physical decoherence rate constant for electronically nonadiabatic chemistry:

$$\frac{1}{\tau} = \frac{1}{\tau^{\Delta F}} + \sqrt{\left(\frac{1}{\tau^{\Delta p}}\right)^2 + \left(\frac{1}{\tau^{\Delta F}}\right)^2}, \tag{22}$$

where $\tau^{\Delta p}$ is a complicated expression associated with the wave packet having different momenta p_α and p_β on two surfaces V_α and V_β, and $\tau^{\Delta F}$ is a complicated expression associated with the wave packet experiencing different forces on the two surfaces. For parallel surfaces in one dimension,

$$\tau = \tau^{\Delta p} = \frac{\hbar}{|V_\alpha - V_\beta|}\sqrt{\frac{4\pi^2 |p_\alpha - p_\beta|}{\bar{p}}}, \tag{23}$$

where $\bar{p}$ is the average momentum. The first factor ("prefactor") on the right-hand side of (23) is the fastest time scale in the system.

For our purposes, the "correct" rate of algorithmic demixing is whatever makes the ensemble average with the *independent-trajectory approximation* best simulate the rate of change of populations and final-state distributions. We found that the following works well

$$\tau = \frac{\hbar}{|V_\alpha - V_\beta|}\left(1 + \frac{E_0}{(\mathbf{p}\cdot\hat{s})^2 / 2\mu}\right), \tag{24}$$

where $\mathbf{p}$ is nuclear momentum, $\hat{s}$ is the direction of the nonadiabatic coupling, μ is the nuclear reduced mass ($\mathbf{p}$ and μ both correspond to isoinertial coordinates scaled to a single reduced mass), and E_0 is a parameter that we set equal to 0.1 hartree. The final factor in (24) is motivated by principle no. 3 above and by the fact [55] that the fastest time scale in the system provides a lower bound on the physical decoherence time. Although our experience indicates that the performance of (24) can be improved by making the prefactor larger, we find that with the current form the results are reasonably insensitive to E_0 and that (24) works well for a diverse set of non-Born–Oppenheimer processes [29, 30].

Two further issues need to be considered. First is the direction of decoherent energy release and decoherent energy uptake (these energy exchanges are required because the potential energy surface is self-consistent with the decohering density matrix). We formulated the decoherence term such that the direction of the nuclear momentum in which energy is exchanged as the system decoheres (as the pointer state is einselected) is the direction of the nonadiabatic coupling vector when nonadiabatic coupling is large and is in the direction of the vibrational momentum when the nonadiabatic coupling is small. The latter is motivated by the existence of a "small" but nonremovable component of the nonadiabatic coupling associated with any motion of the nuclei [26].

The final issue to be considered is the criterion for a strong coupling region over which the density matrix that controls stochastic switching evolves coherently. For calculations in the adiabatic representation we take the boundaries of strong-coupling regions as the minima of the magnitude of the nonadiabatic coupling. For calculations in the diabatic representation, we take these boundaries as the minima of the diabatic level spacing (gap); using the maximum gap turned out to be slightly less accurate on average. At boundaries between strong-coupling regions, the switch-controlling coherent density matrix is synched to the relaxing one that controls the effective potential. This key element of the method differs from all previous trajectory surface hopping and decoherence algorithms; as a result the amount of decoherence introduced at strong-interaction-region boundaries depends on the length of the strong coupling region and the relaxation rates controlled by the decoherence times.

We emphasize that the DM potential energy surface switches gradually and smoothly between the various electronic surfaces; no hops are invoked, and therefore no frustrated hops arise. In the DM formalism, we preserve Ehrenfest-like motion in strong interaction regions or when the decay times are long. In the limit of short decay times, the DM formalism is similar to surface hopping in having instantaneous decay of the reduced density matrix, but surface hopping has no synching.

3 Tests

We validated the CSDM method for a variety of test cases for which we computed [58–60] accurate quantum mechanical transition probabilities by methods developed earlier [61] for converged quantum mechanical scattering theory. All of the test systems have the form

$$\mathrm{A}^* + \mathrm{BC} \rightarrow \begin{cases} \mathrm{AB} + \mathrm{C} \\ \mathrm{A} + \mathrm{BC} \end{cases}, \tag{25}$$

where * denotes electronic excitation; A, B, and C are atoms; and the collision occurs in full three-dimensional space with total angular momentum zero. The masses of the atoms in atomic mass units are denoted m_A, m_B, and m_C. The full set of tests are presented in [29, 30], and here I give only a survey of the results of those studies.

First we consider a case with $m_\mathrm{A} = 10$, $m_\mathrm{B} = 1.00783$, $m_\mathrm{C} = 6$, and a potential energy surface resembling that for $\mathrm{Br}^* + \mathrm{H}_2$. This is a case of weakly coupled surfaces that do not cross in either representation; the gap between the adiabatic surfaces is about 0.36 eV throughout the whole important region, and the diabatic coupling is a constant, 0.20 eV [59]. Table 1 shows results for a case with a total energy of 1.10 eV where the initial vibrational and rotational quantum numbers of BC are respectively $v = 0, j = 6$. In the table, P_R denotes the probability of reaction (top product in (25)), and P_Q is the probability of nonreactive quenching (electronic-to-vibrational energy transfer; bottom product in (25)). Table 1 shows the actual calculated probabilities, and Table 2 shows the dependence on representation.

Tables 1 and 2 show that trajectory surface hopping has a very strong dependence on representation. In simple cases like this weakly coupled atom–diatom collision, it is not too difficult to recognize which representation provides a better description (in this case it is the adiabatic one). However, for systems with complex potential energy surfaces, it is not always possible to know which representation is more appropriate [22]. There may be systems with some initial conditions for which the adiabatic representation is more accurate and other initial conditions for which the diabatic representation is

Table 1. Test results for a weakly coupled case

method	representation	P_Q	P_R
Trajectory surface hopping methods			
Parlant-Gislason (PG)[a]	adiabatic	0.01	0.002
	diabatic	0.55	0.359
Tully's fewest switches[b]	adiabatic	0.18	0.025
	diabatic	0.40	0.161
Fewest switches with time uncertainty[c]	adiabatic	0.18	0.015
	diabatic	0.33	0.044
Self-Consistent-potential methods			
Semiclassical Ehrenfest (SE)	either	0.003	0.000
CSDM	adiabatic	0.15	0.021
	diabatic	0.18	0.012
Accurate			
Quantum scattering	either	0.14	0.26

[a]Method of [16]
[b]Original TFS+ method of [17] with frustrated hops ignored
[c]FSTU gradV method of [24] and [25]

Table 2. Representation dependence for the weakly coupled case of Table 1

type of methods	method	P (diabatic)/P (adiabatic) or P (adiabatic/P (diabatic) Quenching	Reaction
TSH	PG	55	180
	TFS+	2.1	6
	FSTU gradV	1.8	3
SCP	SE	1.0	–[a]
	CSDM	1.2	1.7

[a]Cannot compute because no reaction was observed due to qualitatively incorrect Ehrenfest potentials in the reactive exit valley

more appropriate. Furthermore, and even more serious, is that for systems with complex coupled potential energy surfaces, there may be regions of configuration space where the diabatic representation is more suitable and other regions or product valleys where the adiabatic representation is more suitable. Thus it may not be possible to find a good zero-order description that remains valid for a whole trajectory; this was one of the original motivations for trying to incorporate the representation independence of the semiclassical Ehrenfest method into our scheme. Tables 1 and 2 do show that the results obtained by the semiclassical Ehrenfest method are independent of representation; unfortunately though the results are too inaccurate to be useful. The CSDM method reduces the representation dependence to factors of 1.2 and 1.7 for the two probabilities, and the results are reasonably accurate in both representations, especially when we consider that the weak coupling case is especially difficult for semiclassical methods.

We carried out similar comparisons for additional test cases. In particular, we considered three kinds of systems, all of the form of (25) [28–30, 58–60]. We considered three cases of the weak coupling type already discussed, nine test cases with energetically accessible avoided crossings (where the diabatic potentials cross, but the adiabatic ones do not), and five test cases with energetically accessible conical intersections. The weak coupling cases include two strengths of diabatic coupling, one of which is studied with two different initial conditions. The avoided crossing cases consist of three different couplings (varying in strength and extent of delocalization), each studied with three different initial rotational states. The conical intersection cases correspond to five different coupling functions. The results [28–30] are shown in Table 3, which presents mean unsigned percentage errors in the probabilities of quenching and reaction, in the total nonadiabatic transition probability ($P_{\mathrm{N}} \equiv P_{\mathrm{Q}} + P_{\mathrm{R}}$), and in the final internal energy distributions of the diatomic fragments. The means are computed by logarithmic averaging [62] so as to give equal weight to overestimates and underestimates, and the results are averaged over the diabatic and adiabatic calculations. The CSDM method leads to uniformly good results for all three kinds of systems. In fact the errors are comparable to the

Table 3. Mean unsigned percentage errors of semiclassical methods for non-Born–Oppenheimer trajectories tested against accurate quantal results for 17 test cases averaged over diabatic and adiabatic representations

kind of method	method	kind of system: weak coupling	avoided crossing	conical intersection	averaged over kinds of systems
TSH	PG	298	107	52	152
	TFS+	195	58	44	99
	FSTU gradV	74	40	44	53
SCP	SE	–[a]	65	55	–
	CSDM	24	21	31	25

[a]Cannot compute mean error because there is no reaction, and hence there are no reactive products for which to compute mean internal energies

accuracy attainable [21] by trajectory methods for single-surface problems of this nature.

4 Concluding Remarks

We have shown that decoherence is essential for modeling the quantum mechanical electronic subsystem in the simulation of electronically nonadiabatic chemical dynamics. We have developed an improved self-consistent-potential method called Coherent Switches with Decay of Mixing (CSDM) by writing the time derivative of each density matrix element as the sum of a coherent Ehrenfest-like term and a demixing term. The demixing terms control the decay of the system from a mixed state to a stochastically selected pure state called the decoherent state. The form of the equations was determined by requiring:

- Conservation of total energy and angular momentum;
- Conservation of electronic phase angle;
- The decoherent state switches to maintain self-consistency, but it is otherwise chosen as coherently as possible for each complete passage through a strong interaction region (corresponding to non-Markovian decoherence);
- The direction in which energy is exchanged between the classical vibrational degrees of freedom and the quantal electronic degrees of freedom as the system decoheres is chosen physically based on the nature of the nonadiabatic coupling.

The CSDM algorithm provides a semiclassical version of the multiconfigurational self-consistent-field method that puts mixed quantum/classical dynamics for non-Born–Oppenheimer systems on a comparable footing with BO dynamics. In particular the accuracy is comparable to that attainable when trajectory methods are applied to single-surface problems. Furthermore the classical subsystem experiences no discontinuities in momenta, coordinates, or potentials, there is relatively little dependence on representation, and the cost of the calculation is similar to that for single-surface trajectories.

A key advantage of the semiclassical SCDM algorithm is that it is more practical than a fully quantal multiconfigurational quantum master equation [63–67] for applications to complex systems.

Acknowledgments

The work reviewed here was carried out in collaboration with Michael Hack, Ahren Jasper, Shikha Nangia, and Chaoyuan Zhu, and their contributions to all stages of the project are gratefully acknowledged. This work was supported in part by the National Science Foundation through grant no. CHE03-49122.

References

1. Gao J, Thompson MA (eds) (1998) Combined quantum mechanical and molecular mechanical methods. American Chemical Society, Washington
2. Gao J, Truhlar DG (2002) Quantum mechanical methods for enzyme kinetics. Annu Rev Phys Chem 53: 467–505
3. Lin H, Truhlar DG QM/MM: What have we learned, where are we, and where do we go from here? Theor Chem Acc, in press
4. Truhlar DG, Garrett BC (1980) Variational transition state theory. Acc Chem Res 13: 440–448
5. Truhlar DG, Isaacson AD, Garrett BC (1985) Generalized transition state theory. In: Baer M (ed) Theory of chemical reaction dynamics. CRC Press, Boca Raton, Vol. 4, pp. 65–137
6. Guo Y, Thompson DL (1998) A multidimensional semiclassical approach for treating tunneling within classical trajectory simulations. In: Thompson DL (ed) Modern methods for multidimensional dynamics calculations in chemistry. World Scientific, Singapore, pp. 713–737
7. Marx D, Hutter J (2000) *Ab initio* molecular dynamics: theory and implementation. In: Grotendorst J (ed) Modern methods and algorithms of quantum chemistry. John von Neuman Institute for Computing, Jülich, pp. 329–477
8. Born M, Haung K (1954) The dynamical theory of crystal lattices. Oxford University Press, London
9. Truhlar DG, Muckermann JT (1979) Reactive scattering cross sections: Quasiclassical and semiclassical methods. In: Bernstein RB (ed) Atom-molecule collision theory. Plenum, New York pp. 505–566
10. Raff LM, Thompson DL (1985) The classical trajectory approach to reactive scattering. In: Baer M (ed) Theory of chemical reaction dynamics. CRC Press, Boca Raton, Vol. 3, pp. 1–121
11. Rapoport DC (1995) The art of molecular dynamics simulation. Cambridge University Press, Cambridge

12. (a) Schatz GC, Horst M, Takayanagi T (1998) Computational methods for polyatomic bimoleclar reactions. In: Thompson DL (ed) Modern methods for multidimensional dynamics calculations in chemistry. World Scientific, Singapore, pp. 1–33. (b) Benjamin I, Molecular dynamics methods for studying liquid interfacial phenomena. In: Thompson DL (ed) Modern methods for multidimensional dynamics calculations in chemistry. World Scientific, Singapore, pp. 101–142. (c) Bolton K, Hase WL, Peslherbe GH, Direct dynamics simulations of reactive systems. In: Thompson DL (ed) Modern methods for multidimensional dynamics calculations in chemistry. World Scientific, Singapore, pp. 143–189. (d) Stanton RV, Miller JL Kollman DA, Macromolecular dynamics, In: Thompson DL (ed) Modern methods for multidimensional dynamics calculations in chemistry. World Scientific, Singapore, pp. 355–383, (e) Brady JW, Molecular dynamics simulations of carbohydrate solvation. In: Thompson DL (ed) Modern methods for multidimensional dynamics calculations in chemistry. World Scientific, Singapore, pp. 384–400. (f) Rice BM, Molecular simulations of detonation. In: Thompson DL (ed) Modern methods for multidimensional dynamics calculations in chemistry. World Scientific, Singapore, pp. 472–528
13. (a) Tully JC, Preston RK (1971) Trajectory surface hopping approach to nonadiabatic molecular collisions: the reaction of H^+ with D_2. J Chem Phys 55: 562–572. (b) Truhlar DG, Duff JW, Blais NC, Tully JC, Garrett BC (1982) The Quenching of $Na(3^2P)$ by H_2: Interactions and dynamics. J Chem Phys 77: 764–766. (c) Blais, NC, Truhlar DG (1983) Trajectory-surface-hopping study of $Na(3p\ ^2P) + H_2 \rightarrow Na(3s\ ^2S) + H_2(v', j', \theta)$. J Chem Phys 79: 1334–1342
14. Meyer HD, Miller WH (1979) A classical analog for electronic degrees of freedom in nonadiabatic collision processes. J Chem Phys 70: 3214–3223
15. Micha DA (1983) A self-consistent eikonal treatment of electronic transitions in molecular collisions. J Chem Phys 78: 7138–7145
16. Parlant G, Gislason EA (1989) An exact trajectory surface hopping procedure: comparison with exact quantal calculations. J Chem Phys 91: 4416–4418
17. Tully JC (1990) Molecular dynamics with electronic transitions. J Chem Phys 93: 1061–1071
18. Chapman S (1992) The classical trajectory–surface hopping approach to charge-transfer processes. Adv Chem Phys 82: 423–483
19. Muller U, Stock G (1997) Surface-hopping modeling of photoinduced relaxation dynamics on coupled potential-energy surfaces. J Chem Phys 107: 6230–6245
20. Tully JC (1998) Nonadiabatic dynamics. In: Thompson DL (ed) Modern methods for multidimensional dynamics calculations in chemistry, World Scientific, Singapore, pp. 34–72
21. Topaler MS, Allison TC, Schwenke DW, Truhlar DG (1998) What is the best semiclassical method for photochemical dynamics in systems with conical intersectioins? J Chem Phys 109: 3321–3345, (1999) 110: 687–688(E), (2000) 113: 3928(E)
22. Hack MD, Truhlar DG (2000) Semiclassical trajectories at an exhibition. J Phys Chem A 104: 7917–7926
23. Hack MD, Truhlar DG (2001) A natural decay of mixing algorithm for non-Born Oppenheimer trajectories. J Chem Phys 114: 9305–9314
24. Jasper AW, Stechmann SN, Truhlar DG (2002) Fewest switches with time uncertainty: A modified trajectory surface hopping algorithm with better accuracy for classically forbidden electronic transitions. J Chem Phys 116: 5424–5431, 117: 1024(E)

25. Jasper AW , Truhlar DG (2003) Improved treatment of momentum at classically forbidden electronic transitions in trajectory surface hopping calculations. Chem Phys Lett 369: 60–67
26. Jasper AW, Kendrick BK, Mead CA, Truhlar DG (2004) Non-Born–Oppenheimer chemistry: Potential surfaces, couplings, and dynamics. Adv Ser Phys Chem 14: 329–391
27. Zhu C, Jasper AW, Truhlar DG (2004) Non-Born–Oppenheimer trajectories with self-consistent decay of mixing. J Chem Phys 120: 5543–5557
28. Jasper AW, Zhu C, Nangia S, Truhlar DG (2004) Introductory lecture: Nonadiabatic effects in chemical dynamics. Discuss Faraday Soc 127: 1–22
29. Zhu C, Nangia S, Jasper AW, Truhlar DG (2004) Self-consistent decay of mixing and trajectory surface hopping with coherent complete passages. J Chem Phys 121: 7658–7670
30. Zhu C, Jasper AW, Truhlar DG (2005) Non-Born–Oppenheimer Liouville-von Neumann dynamics. Evolution of a subsystem controlled by linear and populations-driven decay of mixing with decoherent and coherent switching. J Chem Theory Comput 1: 527–540
31. Stock G, Thoss M (2005) Classical description of nonadiabatic quantum dynamics. Adv Chem Phys 131: 243–375
32. Rapp D, Sharp TE (1963) Vibrational energy transfer in molecular collisions involving large transition probabilities. J Chem Phys 38: 2641–2648
33. Sharp TE, Rapp D (1965) Evaluation of approximations used in the calculation of excitation by collision. I. Vibrational excitation of molecules. J Chem Phys 43: 1233–1244
34. Muckermann JT, Rusinek I, Roberts RE, Alexander M (1976) Probabilities for classically forbidden transitions using classical and classical path methods. J Chem Phys 65: 2416–2428
35. Billing Sorensen G (1974) Semiclassical three-dimensional inelastic scattering theory. J Chem Phys 61: 3340–3343
36. Gentry WR (1979) Vibrational excitation II: classical and semiclassical methods. In Bernstein RB (ed) Atom-molecule collision theory. Plenum, New York, pp. 391–425
37. Gerber RB, Buch V, Ratner MA (1982) Time-dependent self-consistent field approximation for intramolecular energy transfer. I. Formulation and application to dissociation of van der Waals molecules. J Chem Phys 77: 3022–3030
38. Halcomb LL, Diestler DJ (1985) Dynamics of vibrational predissociation of van der Waals molecules. In: Numrich RW (ed) Supercomputer applications. Plenum, New York, pp. 273–288
39. Billing GD (1987) Rate constants for vibrational transitions in diatom-diatom collisions. Comput Phys Commun 44: 121–136
40. Garcia-Vela A, Gerber RB, Imre DG (1992) Mixed quantum wave packet/classical trajectory treatment of the photodissociation process ArHCl $\rightarrow$ Ar+H+Cl. J Chem Phys 97:7242–7250
41. Han H, Brumer P (2005) Decoherence effects in reactive scattering. J Chem Phys 122: 144316/1–8
42. Sawada S, Metiu H (1986) A multiple trajectory theory for curve crossing problems obtained by using a Gaussian wave packet representation of the nuclear motion. J Chem Phys 84: 227–238
43. Meyer H-D, Manthe U, Cederbaum LS (1990) The multi-configurational time-dependent Hartree approximation. Chem Phys Lett 165: 73–78

44. Kotler Z, Neria E, Nitzan A (1991) Multi-configurational time-dependent self-consistent-field approximations in the numerical solution of quantum dynamical problems. Comput Phys Commun 63: 234–258
45. Hack MD, Wensmann AM, Truhlar DG, Ben-Nun M, Martinez TJ (2001) Comparison of full multiple spawning, Trajectory surface hopping, and converged quantum mechanics for electronically nonadiabatic dynamics. J Chem Phys 115: 1172–1186
46. Neumann JV (1955) Mathematical foundations of quantum mechanics, Princeton University Press, Princeton [originally published in German in 1932: Mathematische Grundlagen der Quantenmechanik, Springer, Berlin Heidelberg New York]
47. Sakurai JJ (1985) Modern quantum mechanics. Addison-Wesley, Redwood City, CA p. 181
48. Gottfried K, Yan TM (2003) Quantum mechanics: Fundamentals. 2nd edn, Springer, Berlin Heidelberg New York, p. 64
49. Schwartz BJ, Bittner ER, Prezhdo OV, Rossky PJ (1996) Quantum decoherence and the isotope effect in condensed phase nonadiabatic molecular dynamics simulations. J Chem Phys 104: 5942–5955
50. Bittner ER, Rossky PJ (1995) Quantum decoherence in mixed quantum-classical systems: nonadiabatic processes. J Chem Phys 103: 8130–8143
51. Bittner ER, Rossky PJ (1997) Decoherent histories and nonadiabatic quantum molecular dynamics simulations. J Chem Phys 107: 8611–8618
52. Preszhdo OV, Rossky PJ (1997) Evaluation of quantum transition rates from quantum-classical molecular dynamics simulations. J Chem Phys 107: 5863–5878
53. Jasper AW, Truhlar DG (2005) Electronic decoherence time for non-Born–Oppenheimer trajectories. J Chem Phys 123: 64103/1–4
54. Zurek WH (1981) Pointer basis of quantum apparatus: Into what mixture does the wave packet collapse? Phys Rev D 24: 1516–1525
55. Zurek WH (2003) Decoherence, einselection, and the quantum origins of the classical. Rev Mod Phys 75: 715–775
56. Paz JP, Zurek H (1999) Quantum limit of decoherence: environment induced superselection of energy eigenstates. Phys Rev Lett 82: 5181–5185
57. Paz JP, Habib S, Zurek WH (1993) Reduction of the wave packet: preferred observable and decoherence time scale. Phys Rev D 47: 488–501
58. Volobuev YL, Hack MD, Topaler MS, Truhlar DG (2000) Continuous surface switching: An improved time-dependent self-consistent-field method for nonadiabatic dynamics. J Chem Phys 112: 9716–9726
59. Jasper AW, Hack MD, Truhlar DG (2001) The treatment of classically forbidden electronic transitions in semiclassical trajectory surface hopping calculations. J Chem Phys 115: 1804–1816
60. Jasper AW, Truhlar DG (2005) Conical intersections and semiclassical trajectories: Comparison to accurate quantum dynamics and analyses of the trajectories. J Chem Phys 122: 44101/1–16
61. (a) Staszewska G, Truhlar, DG (1987) Convergence of L^2 methods for scattering problems. J Chem Phys 86: 2793–2804. (b) Schwenke DW, Haug K, Truhlar DG, Sun Y, Zhang JZH, Kouri DJ (1987) Variational basis-set calculations of accurate quantum mechanical reaction probabilities. J Phys Chem 91: 6080–6082. (c) Sun Y, Kouri DJ, Truhlar DG, Schwenke DW (1990) Dynamical basis sets for

algebraic variational calculations in quantum mechanical scattering theory. Phys Rev A 41: 4857–4862. (d) Schwenke DW, Mielke SL, Truhlar DG. (1991) Variational reactive scattering calculations: computational optimization strategies. Theoret Chim Acta 79: 241–269. (e) Tawa GJ, Mielke SL, Truhlar DG, Schwenke DW (1994) Algebraic variational and propagation formalisms for quantal dynamics calculations of electronic-to-vibrational, rotational energy transfer and application to the quenching of the 3p state of sodium by hydrogen molecules. J Chem Phys 100: 5751–5777. (f) Tawa GJ, Mielke SL, Truhlar DG, Schwenke DW (1994) Linear algebraic formulation of reactive scattering with general basis functions. Adv Mol Vib Coll Dyn 2B: 45–116

62. Allison TC, Truhlar DG (1998) Testing the accuracy of practical semiclassical methods: Variational transition state theory with optimized multidimensional tunneling. In: Thompson DL (ed) Modern methods for multidimensional dynamics calculations in chemistry. World Scientific, Singapore, pp. 618–712
63. Stock G (1995) Nonperturbative generalized master equation for the spin-boson problem. Phys Rev E 51: 3038–3044
64. Pesce L, Saalfrank P (1998) The coupled channel density matrix method for open quantum systems: Formulation and application to the vibrational relaxation of molecules scattering from nonrigid surfaces. J Chem Phys 108: 3045–3056
65. Pesce L, Gerdts T, Manthe U, Saalfrank P (1998) Variational wave packet method for dissipative photodesorption problems. Chem Phys Lett 288: 383–390
66. Burghardt I (2001) Reduced dynamics with initial correlations: multiconfigurational approach. J Chem Phys 114: 89–101
67. Burghardt I, Nest M, Worth GA (2003) Multiconfigurational system-bath dynamics using Gaussian wave packets: energy relaxation and decoherence induced by a finite-dimensional bath. J Chem Phys 119: 5364–5378

Time-Dependent, Direct, Nonadiabatic, Molecular Reaction Dynamics

Y. Öhrn and E. Deumens

Summary. Electron Nuclear Dynamics (END) is a time-dependent, nonadiabatic, direct, theory of molecular processes. It has a hierarchical structure which permits the theory to be applied at levels of increasing sophistication and accuracy. Each rank in this hierarchy is defined by the choice of wave function for participating electrons and atomic nuclei. The dynamical variables of the theory are the wave function parameters, which carry the time-dependence. The time-dependent variational principle is employed to derive the equations of motion.

We describe the simplest level of approximation, which we call minimal END, and also give a brief account of how this level of theory can be extended. Various applications are discussed with results from calculations of integral and differential cross section for binary molecular encounters.

1 Introduction

Concepts and pictures that shape much of the modeling and simulations of molecular events have their origin in the separation of electronic and nuclear dynamics. The time-independent Schrödinger equation for the total molecular system

$$H\Psi = E\Psi \tag{1}$$

is supplemented with the electronic Schrödinger equation

$$H_{\mathrm{el}}\Phi_k(x;X) = U_k(X)\Phi_k(x;X), \tag{2}$$

solutions of which are sought for fixed nuclear positions X. When electronic structure calculations are carried out for a large enough set of nuclear geometries so-called Born–Oppenheimer potential energy surfaces (PESs) $U_k(X)$ are obtained. The corresponding stationary electronic states Φ_k then are employed as a basis for the total molecular state, i.e.

$$\Psi(x,X) = \sum_k \Phi_k(x;X)\chi_k(X). \tag{3}$$

When this expansion is substituted into (1), this expression multiplied by a particular electronic state vector Φ_l^*, and integrated over all electronic degrees of freedom the following expression obtains:

$$\left[T_n + U_l(X) + \int \Phi_l^*(x;X) T_n \Phi_l(x;X) \mathrm{d}x - E\right] \chi_l(X) = -\sum_{k \neq l} \int \Phi_l^*(x;X) T_n \Phi_k(x;X) \mathrm{d}x \chi_k(X), \tag{4}$$

where T_n is the kinetic energy operator and χ_k the wave functions for the nuclei. This equation is traditionally expressed for a fixed total angular momentum and in a set of coordinates internal to the molecular system, which often leads to quite involved expressions for T_n. The integral on the left is the so-called adiabatic correction to the Born–Oppenheimer PES U_l and the right-hand side has the nonadiabatic coupling terms.

Although there are today a number of different approaches of how to extract useful information from these equations, such as the use of density functional methods for the electronic degrees of freedom, the direct use of reduced density matrices, etc. the basic pictures and the rendering of dynamics stem from the use of stationary electronic states and the associated PESs. Also many time-dependent approaches to molecular processes interpret results in terms of this picture of electronic potentials as the source of the forces that drive the nuclear dynamics.

An alternative view of molecular dynamical processes is offered by focusing on the evolving total molecular state vector ψ in a time-dependent formulation using general bases, i.e. not necessarily those provided by the stationary electronic states. Electron nuclear dynamics (END) theory [1, 2] is such an approach.

The starting point is the action

$$A = \int_{t_1}^{t_2} L(\psi, \psi^*) \mathrm{d}t \tag{5}$$

in terms of the quantum mechanical Lagrangian ($\hbar = 1$)

$$L = \left\langle \psi \middle| H - \mathrm{i}\frac{\partial}{\partial t} \middle| \psi \right\rangle / \langle \psi | \psi \rangle . \tag{6}$$

The time-dependence is carried by a number of wave function parameters $q(t)$, such as average nuclear positions and momenta, and molecular orbital coefficients, etc. The principle of least action or the time-dependent variational principle $\delta A = 0$ yields the Euler–Lagrange equations

$$\frac{\mathrm{d}}{\mathrm{d}t} \frac{\partial L}{\partial \dot{q}} = \frac{\partial L}{\partial q} . \tag{7}$$

Should the wave function be so general that its variations can reach all parts of Hilbert space, then the Euler–Lagrange equations would become the time-dependent Schrödinger equation. However, for all problems of chemical interest the, necessarily, approximate wave function form for the molecular system will yield a set of coupled first-order differential equations in the time parameter t, which in a variational sense optimally approximates the time-dependent Schrödinger equation. The wave function parameters $q(t)$ that carry the time-dependence play the role of dynamical variables and it becomes important to choose a form of evolving state vector with parameters that are continuous and differentiable. Generalized coherent states are useful in this context [2, 3].

2 Minimal END

END theory can be viewed as a hierarchical approach to molecular processes. The various possible choices of families of molecular wave functions representing the participating electrons and atomic nuclei can be arranged in an array of increasing complexity ranging from a single determinantal description of the electrons and classical nuclei to a multi-configurational quantum representation of both electrons and nuclei [4]. The simplest level of END theory is implemented in a program package [5] that includes efficient molecular integral routines and well tested propagation algorithms to solve the system of coupled END equations.

This minimal END employs a wave function

$$|\psi(t)\rangle = |R(t), P(t)\rangle |z(t), R(t), P(t)\rangle, \tag{8}$$

where

$$\langle X|R(t), P(t)\rangle = \prod_k \exp[-\frac{1}{2}\left(\frac{\mathbf{X}_k - \mathbf{R}_k}{b}\right)^2 + \mathrm{i}\mathbf{P}_k \cdot (\mathbf{X}_k - \mathbf{R}_k)] \tag{9}$$

and

$$\langle x|z(t), R(t), P(t)\rangle = \det \chi_i(\mathbf{x}_j) \tag{10}$$

with the spin orbitals

$$\chi_i = u_i + \sum_{j=N+1}^{K} u_j z_{ji}(t) \tag{11}$$

expanded in terms of atomic spin orbitals

$$\{u_i\}_1^K, \tag{12}$$

which in turn are expanded in a basis of traveling Gaussians

$$(x - R_x)^l (y - R_y)^m (z - R_z)^n \exp\left[-a(\mathbf{x} - \mathbf{R})^2 - \frac{\mathrm{i}}{\hbar M}\mathbf{P} \cdot (\mathbf{x} - \mathbf{R})\right] \tag{13}$$

centered on the average nuclear positions $\mathbf{R}$ and moving with velocity $\mathbf{P}/M$.

In the narrow nuclear wave packet limit, $a \to 0$, the Lagrangian may be expressed as

$$L = \sum_{j,l} \left\{ \left[P_{jl} + \frac{\mathrm{i}}{2} \left(\frac{\partial \ln S}{\partial R_{jl}} - \frac{\partial \ln S}{\partial R'_{jl}} \right) \right] \dot{R}_{jl} + \frac{\mathrm{i}}{2} \left(\frac{\partial \ln S}{\partial P_{jl}} - \frac{\partial \ln S}{\partial P'_{jl}} \right) \dot{P}_{jl} \right\}$$
$$+ \frac{\mathrm{i}}{2} \sum_{p,h} \left(\frac{\partial \ln S}{\partial z_{ph}} \dot{z}_{ph} - \frac{\partial \ln S}{\partial z^*_{ph}} \dot{z}^*_{ph} \right) - E \qquad (14)$$

with $S = \langle z, R', P' | z, R, P \rangle$ and

$$E = \sum_{jl} P_{jl}^2 / 2M_l + \langle z, R', P' | H_{\mathrm{el}} | z, R, P \rangle / \langle z, R', P' | z, R, P \rangle . \qquad (15)$$

Here H_{el} is the electronic Hamiltonian including the nuclear–nuclear repulsion terms, P_{jl} is a Cartesian component of the momentum and M_l the mass of nucleus l. One should note that the bra depends on z^* while the ket depends on z and that the primed R and P equal their unprimed counterparts and the prime simply denotes that they belong to the bra.

The Euler–Lagrange equations

$$\frac{\mathrm{d}}{\mathrm{d}t} \frac{\partial L}{\partial \dot{q}} = \frac{\partial L}{\partial q} \qquad (16)$$

can now be formed for the dynamical variables

$$q = R_{jl}, P_{jl}, z_{ph}, z^*_{ph} \qquad (17)$$

and collected into a matrix equation

$$\begin{bmatrix} \mathrm{i}\mathbf{C} & \mathbf{0} & \mathrm{i}\mathbf{C}_R & \mathrm{i}\mathbf{C}_P \\ \mathbf{0} & -\mathrm{i}\mathbf{C}^* & -\mathrm{i}\mathbf{C}_R^* & -\mathrm{i}\mathbf{C}_P^* \\ \mathrm{i}\mathbf{C}_R^\dagger & -\mathrm{i}\mathbf{C}_R^{\mathrm{T}} & \mathbf{C}_{RR} & -\mathbf{I} + \mathbf{C}_{RP} \\ \mathrm{i}\mathbf{C}_P^\dagger & -\mathrm{i}\mathbf{C}_P^{\mathrm{T}} & \mathbf{I} + \mathbf{C}_{PR} & \mathbf{C}_{PP} \end{bmatrix} \begin{bmatrix} \dot{\mathbf{z}} \\ \dot{\mathbf{z}}^* \\ \dot{\mathbf{R}} \\ \dot{\mathbf{P}} \end{bmatrix} = \begin{bmatrix} \partial E / \partial \mathbf{z}^* \\ \partial E / \partial \mathbf{z} \\ \partial E / \partial \mathbf{R} \\ \partial E / \partial \mathbf{P} \end{bmatrix} , \qquad (18)$$

where the dynamical metric contains the elements

$$(C_{XY})_{ik;jl} = -2 \mathrm{Im} \left. \frac{\partial^2 \ln S}{\partial X_{ik} \partial Y_{jl}} \right|_{R'=R, P'=P} , \qquad (19)$$

$$(C_{X_{ik}})_{ph} = (C_X)_{ph;ik} = \left. \frac{\partial^2 \ln S}{\partial z^*_{ik} \partial X_{ik}} \right|_{R'=R, P'=P} , \qquad (20)$$

which are the nonadiabatic coupling terms, and

$$C_{ph;qg} = \frac{\partial^2 \ln S}{\partial z^*_{ph} \partial z_{qg}} \Big|_{R'=R, P'=P} . \qquad (21)$$

In this minimal END approximation the electronic basis functions are centered on the average nuclear positions, which are dynamical variables. In the limit of classical nuclei these are conventional basis functions used in molecular electronic structure theory, and they follow the dynamically changing nuclear positions. As can be seen from the equations of motion discussed above the evolution of the nuclear positions and momenta is governed by Newton-like equations with Hellman–Feynman forces, while the electronic dynamical variables are complex molecular orbital coefficients which follow equations that look like those of the time-dependent-Hartree–Fock (TDHF) approximation [6]. The coupling terms in the dynamical metric are the well-known nonadiabatic terms due to the fact that the basis moves with the dynamically changing nuclear positions.

The time evolution of molecular processes in the END formalism employs a Cartesian laboratory frame of coordinates. This means that in addition to the internal dynamics overall translation and rotation of the molecular system are treated. The six extra degrees of freedom add work, but become a smaller part of the total effort as the complexity of the system grows. The advantage is that the kinetic energy terms are simple. This means that the effect of small kinetic energy terms, such as mass polarization, often neglected using internal coordinates, is included. Furthermore, the complications of having to choose different internal coordinates for product channels exhibiting different fragmentations are not present. One can treat all product channels on an equal footing in the same laboratory frame. Since the fundamental invariance laws with respect to overall translation and rotation are satisfied within END [2] it is straightforward to extract the internal dynamics at any time during the evolution.

3 Electrons as Reactants and Products

Molecular collision processes with free electrons either as reactants or as products have for historical reasons [7] been treated theoretically somewhat separate from those problems in atomic and molecular collision dynamics where only bound electrons appear. This seems to be due in part to the fact that the interaction of electron projectiles with the bound state electrons of molecular targets involves the complications of permutation symmetry of identical particles and partially due to that the treatment of continuum states in molecular quantum mechanics is far less developed than that of bound states. It is interesting to note that even modern texts [8] on electron–molecule scattering approaches the general electron–molecule collision via potential scattering, treats exchange collisions as a special problem, and deals separately with elastic and various inelastic processes.

In order to treat molecular reactive processes involving ionization, electron scattering, and recombination END can be augmented to deal with combined electron–molecule dynamics. This is done employing a coherent state based approach to describe freely moving electron wave packets. This is a natural extension of the common approach to center basis functions on nuclear centers. It also relates to the approach using floating Gaussians [9] in electronic structure calculations. The basic idea is to add to the parameters describing the electronic wave function the position and momentum of centers carrying basis functions and to define a consistent dynamics for these degrees of freedom.

Considering a single free center with basis functions

$$(x-\rho_x)^l(y-\rho_y)^m(z-\rho_z)^n \exp\left[-c(\mathbf{x}-\rho)^2 - \frac{\mathrm{i}}{\hbar}\pi\cdot(\mathbf{x}-\rho)\right], \tag{22}$$

where $\mathbf{x} = (x, y, z)$ is an electron coordinate, $\rho = (\rho_x, \rho_y, \rho_z)$ is the center coordinate and π the center momentum. The Euler–Lagrange equations are then constructed with the dynamical variables $q = R, \rho, P, \pi, z, z^*$ and we can write

$$\begin{bmatrix} \mathrm{i}\mathbf{C} & \mathbf{0} & \mathrm{i}\mathbf{C}_R & \mathrm{i}\mathbf{C}_\rho & \mathrm{i}\mathbf{C}_P & \mathrm{i}\mathbf{C}_\pi \\ \mathbf{0} & -\mathrm{i}\mathbf{C}^* & -\mathrm{i}\mathbf{C}_R^* & -\mathrm{i}\mathbf{C}_\rho^* & -\mathrm{i}\mathbf{C}_P^* & -\mathrm{i}\mathbf{C}_\pi^* \\ \mathrm{i}\mathbf{C}_R^\dagger & -\mathrm{i}\mathbf{C}_R^\mathrm{T} & \mathbf{C}_{RR} & \mathbf{C}_{R\rho} & -\mathbf{I}+\mathbf{C}_{RP} & \mathbf{C}_{R\pi} \\ \mathrm{i}\mathbf{C}_\rho^\dagger & -\mathrm{i}\mathbf{C}_\rho^\mathrm{T} & \mathbf{C}_{\rho R} & \mathbf{C}_{\rho\rho} & \mathbf{C}_{\rho P} & \mathbf{C}_{\rho\pi} \\ \mathrm{i}\mathbf{C}_P^\dagger & -\mathrm{i}\mathbf{C}_P^\mathrm{T} & \mathbf{I}+\mathbf{C}_{PR} & \mathbf{C}_{P\rho} & \mathbf{C}_{PP} & \mathbf{C}_{P\pi} \\ \mathrm{i}\mathbf{C}_\pi^\dagger & -\mathrm{i}\mathbf{C}_\pi^\mathrm{T} & \mathbf{C}_{\pi R} & \mathbf{C}_{\pi\rho} & \mathbf{C}_{\pi P} & \mathbf{C}_{\pi\pi} \end{bmatrix} \begin{bmatrix} \dot{\mathbf{z}} \\ \dot{\mathbf{z}}^* \\ \dot{\mathbf{R}} \\ \dot{\rho} \\ \dot{\mathbf{P}} \\ \dot{\pi} \end{bmatrix} = \begin{bmatrix} \partial E/\partial \mathbf{z}^* \\ \partial E/\partial \mathbf{z} \\ \partial E/\partial \mathbf{R} \\ \partial E/\partial \rho \\ \partial E/\partial \mathbf{P} \\ \partial E/\partial \pi \end{bmatrix} \tag{23}$$

with definitions of the elements analogous to those in (19), (20), and (21).

4 Cross-Sections

END trajectories for a molecular process are obtained by integrating the equations (18) and (23) from suitable initial conditions for the reactants to a time where the products are well separated or no further change occurs in the system. In the case of a binary molecular reactive collision, minimal END, which uses classical nuclei, requires that for each trajectory the reactants are given some initial relative orientation. One of the reactant moieties is considered the target and placed stationary at the origin of the laboratory cartesian coordinate system while the other collision partner, considered the projectile, is placed sufficiently distant so the interaction with the target is negligible. A Thouless determinant in a suitable basis is constructed for, say, the ground electronic state of the entire system. The projectile is given an impact parameter b and a momentum commensurate with the chosen collision energy E.

Each set of initial conditions leads to a particular set of product fragments and states. The final evolved state $|\psi\rangle$ may be projected against a number of possible final stationary electronic states $|f\rangle$ expressed in the same basis as that of the initial state to yield a transition probability $P_{fo}(b, E, \varphi) = |\langle f|\psi\rangle|^2$, which is a function of the collision energy E, the relative initial orientations, and the scattering angles (θ, φ) or impact parameter and angle (b, φ).

The classical differential cross-section for a particular product channel with probability P_{fo} is

$$\frac{\mathrm{d}\sigma_{fo}(E,\theta,\varphi)}{\mathrm{d}\Omega} = \sum_j P_{fo}(b_j, E, \varphi) \frac{b_j}{\sin\theta |\mathrm{d}\Theta/\mathrm{d}b_j|}, \tag{24}$$

where the sum runs over all impact parameters b_j leading to the same scattering direction (θ, φ) for the fragment going to the detector. In this expression $\Theta(b)$ is the deflection function, which, for the first branch of the scattering region, satisfies $|\Theta| = \theta$.

For randomly oriented reactants, as is the case in gas phase reactions, trajectories for a sufficient number of initial relative orientations are used to produce an angular grid to calculate orientaionally averaged cross sections [10, 11]

$$\frac{\mathrm{d}\sigma_{\mathrm{f}}(E,\theta,\varphi)}{\mathrm{d}\Omega} = \left\langle \frac{\mathrm{d}\sigma_{\mathrm{f}o}}{\mathrm{d}\Omega} \right\rangle_o \tag{25}$$

The well-known deficiencies of the classical cross-section in (24) that occur for small angle scattering and at so-called rainbow angles, where $\frac{\mathrm{d}\theta}{\mathrm{d}b_j} = 0$, as well as the lack of interference effects between the various trajectories in the sum, can be removed with semiclassical corrections such as the uniform Airy [12, 13] or the Schiff approximations [14].

The Schiff approximation [11, 14] for small angle scattering yields a scattering amplitude (for oriented reactants)

$$f_o(\boldsymbol{k}_{\mathrm{f}}, \boldsymbol{k}_0) = \mathrm{i}k_0 \int_0^\infty J_0(qb)(1 - \mathrm{e}^{\mathrm{i}\delta_o(b,\varphi)}) b \, \mathrm{d}b \tag{26}$$

with $q = |\boldsymbol{k}_f - \boldsymbol{k}_0|$ the magnitude of the momentum transfer, $\delta_o(b, \varphi)$ the semiclassical phase shift, and J_0 the Bessel function of order 0.

The semiclassical phase shift is connected to the deflection function or scattering angle, which is obtained directly from each END trajectory. The expression [7]

$$\Theta(b, \varphi) = \frac{1}{k_0} \frac{\mathrm{d}\delta_o(b,\varphi)}{\mathrm{d}b} \tag{27}$$

can then be integrated to yield the phase shift. Also q, which depends on the angle between incoming projectile wave vector $\boldsymbol{k}_0$ and the final wave vector $\boldsymbol{k}_{\mathrm{f}}$ pointing toward the detector, is obtained from the END trajectories.

The form

$$\frac{\mathrm{d}\sigma_{fo}}{\mathrm{d}\Omega} = P_{\mathrm{f}o}(E,\theta,\varphi) \frac{k_f}{k_0} |f_o(\boldsymbol{k}_{\mathrm{f}}, \boldsymbol{k}_0)|^2 \tag{28}$$

has been used with some success in predicting absolute direct and charge transfer differential cross sections for a number of ion–atom and ion–molecule collisions at energies ranging from tens of eV to 100 keV [11, 15–24].

However, this form of differential cross-section still does not include interference effects and in order to remedy this an augmented Schiff approximation has been introduced such that

$$f_o(\boldsymbol{k}_{\mathrm{f}}, \boldsymbol{k}_0) = \mathrm{i}k_0 \int_0^\infty \langle f|\psi\rangle_o J_0(qb)(1 - \mathrm{e}^{\mathrm{i}\delta_o(b,\varphi)})b\,\mathrm{d}b, \tag{29}$$

i.e. the complex amplitude $\langle f|\psi\rangle_o$ is inserted obtained by projecting the evolved state vector against a particular final state f for each orientation of the reactants. In spite of the difficulty in tracking the phase from one trajectory to the next this approach has had some success for simple ion-atom systems [11] and even yielded accurate state-state cross-sections.

Integral cross-sections are less sensitive to the fine details of the collision dynamics and the classical cross sections have produced accurate predictions down to energies of about 0.5 eV [25] for $H_2^+ + H_2$ reactions, and also allowed to shed some light on reaction mechanisms for molecular beam studies of $D_2 + NH_3^+$ [26–28]. Recent calculations covering a wide range of collision energies for the test case $H^+ + H$ [29] show excellent agreement with experiments.

It can be shown [13, 30, 31] that by employing coherent states a posteriori information about rovibrational resolution of cross-sections can be obtained from dynamics with classical nuclei.

5 Examples of Applications of Minimal END

Absolute integral and differential cross-section for direct as well as charge transfer processes in simple ion–atom and ion–molecule collisions have been calculated with minimal END with good agreement with the best experiments. We show in Fig. 1 one of the early simple test cases of direct cross-sections for

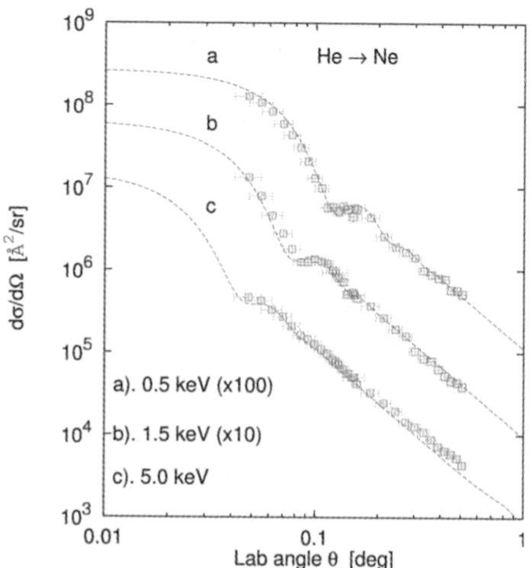

Fig. 1. Direct differential cross-sections for He/Ne collisions [15] compared to the absolute differential cross sections measured by Gao et al. [32]

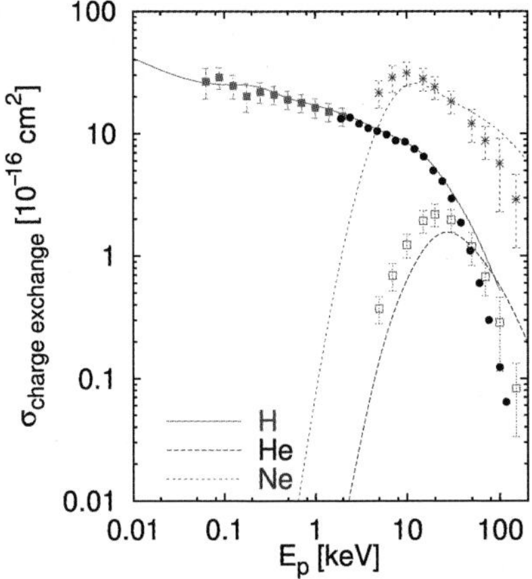

Fig. 2. Total cross-sections for collisions of protons with atomic hydrogen, helium, and neon [16] compared with measured values by ∘ [33], • [34], and (⊔,∗) [35]

collisions between helium and neon atoms at three different energies in the keV range.

Also charge transfer cross sections are accurately predicted with minimal END for collisions over a range of energies in the eV to keV range. In Fig. 2 we show total cross-sections for proton collisions with atomic hydrogen, helium, and neon.

Some advantages of a method such as END include that it proceeds without precalculated potential energy surfaces, it accounts for nonadiabatic coupling terms, and all allowed processes are treated at the same level of approximation. This is particularly useful for hyperthermal reactions involving polyatomic systems where often several electronic states may be involved. We illustrate this with the calculated fragmentation cross-sections of ethane colliding with energetic protons [36]. In Fig. 3 we show the dominant fragmentation channel, and in Fig. 4 the fragmentation cross sections of the remaining allowed channels in this energy range are depicted.

Experimental data on fragmentation cross sections of polyatomic molecules are scarce. There are some results for smaller systems. For instance, the fragmentation of methane by 30 eV protons was studied experimentally by Toennies group [37] and END was applied to this system [10] with interesting agreements with results and some alternative suggestions to those of experimental conclusions.

These results suggest that product branching ratios for collisions can be affected and perhaps controlled by choice of projectiles and their energies.

The rendering of results from a time-dependent treatment of reaction dynamics is a problem that leads to movies and time-lapse pictures of dynamical events. Obviously a great variety of properties and dynamical quantities can be displayed in this manner. We have found it quite useful to use such techniques

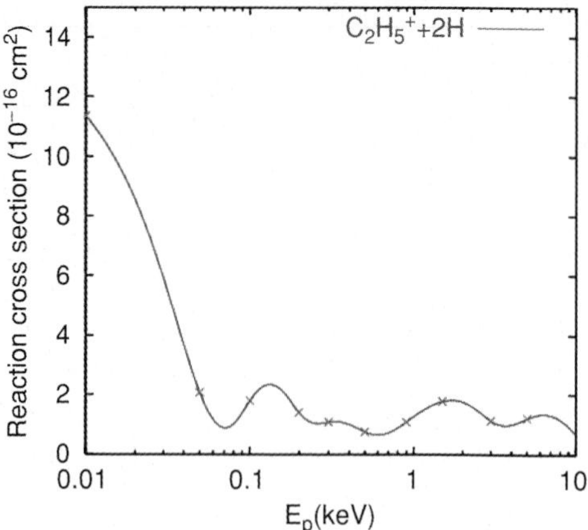

Fig. 3. Calculated reaction cross-section [36] for the dominant fragmentation channel of energetic protons colliding with ethane molecules in the given energy range

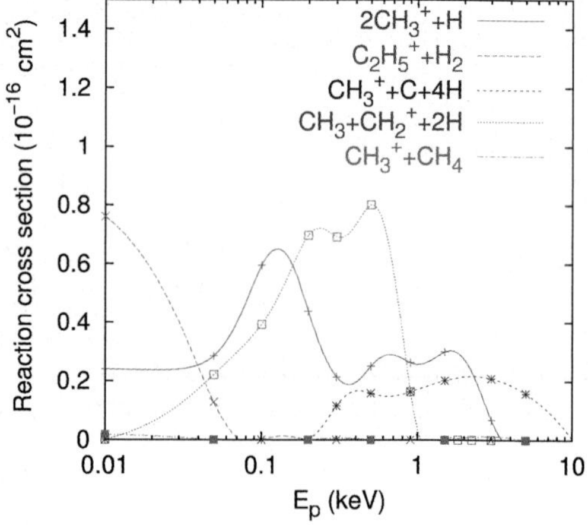

Fig. 4. Calculated reaction cross-sections [36] for all but the dominant allowed fragmentation channels in the considered energy range

also for discovering errors. It is often a lot easier to discover flaws in the dynamics from a pictorial representation of the massive amounts of data than it is to search in tables of numbers.

As an example of rendering of a reaction with dynamically active electrons we display in Fig. 5 six panels of a typical trajectory for the $H + D_2$ reaction leading to HD products. The electrons are shown as a sphere around each nucleus with the size proportional to the time-dependent population on that atom. This result comes from a study of the time-delay effects in the formation

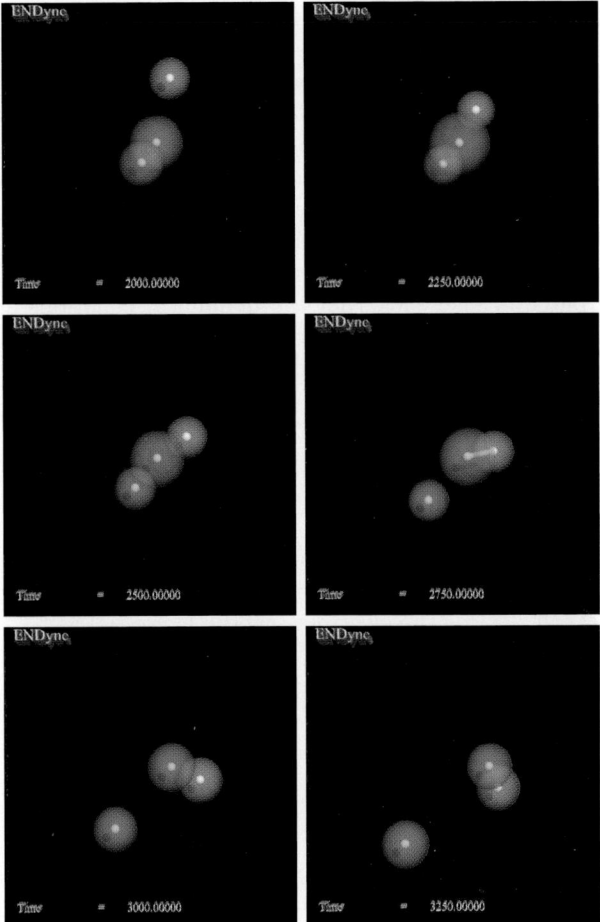

Fig. 5. Six snap shots of $H + D_2 \rightarrow D + HD$ with the dynamical electrons represented by a sphere around each nucleus with the size proportional to the electronic population on each atom. The H approaches from above in the first frame, and polarizes the $D - D$ bond in the second frame. The third to sixth frames show the products departing rovibrationally excited

of HD products when H projectiles collide with D_2 targets at 1.64 eV in the center of mass. Time-delays of about 25 femtoseconds between the backward and forward scattered product molecules were put forth as a signature of resonance in this reaction [38, 39]. END calculations [24] also bore this out.

It is of course not quite accurate to display the electron cloud around H in a different color to that around D, since in END all the electrons of the reacting system are treated as indistinguishable. However, it is helpful to do so in aiding the eye in following the time evolution.

6 Extensions of Minimal END

Even if many molecular processes, in particular at higher energies, can be accurately described with classical nuclei, it is sometimes, particularly at lower energies, and for light nuclei, such as protons, necessary to account for quantum effects on the instantaneous nuclear dynamics. This can be done in the END framework [4].

Minimal END is derived as the narrow wave packet limit of frozen nuclear gaussian wave packets. From the dynamical equations one can discern, as was first pointed out by Heller [40], that the dynamical equations are the same whether the zero width limit is achieved or not. Only a slight renormalization of the matrix elements is required, but the qualitative dynamics is the same.

For quantum nuclei END considers a basis centered at the parametric positions R and with nuclear momentum factors just as is the case for the traveling gaussians for electrons with momentum P in (13). Such basis functions are denoted $|s, R, P\rangle$ for a set of s-type orbitals, $|jk, R, P\rangle$ $(j = x, y, z)$ for p-type orbitals and k labeling the particular nucleus, etc. A molecular wave function, similar to a Born–Huang series corresponding to a single Thouless determinant, $|z, R, P\rangle$, for the electrons, can be expressed as

$$v|s, R, P\rangle|z, R, P\rangle + \sum_{j,k} v_{jk}|jk, R, P\rangle \frac{\partial}{\partial R_{jk}}|z, R, P\rangle$$

$$+ \sum_{j,k,i,l} v_{jkil}|jkil, R, P\rangle \frac{\partial^2}{\partial R_{jk}\partial R_{il}}|z, R, P\rangle + \cdots . \tag{30}$$

The END equations for this type of molecular wave function has been studied [4], and is being included in the ENDyne code.

The single Thouless determinant description of the electronic part of the molecular wave function also clearly has its shortcomings. Although sufficiently accurate results for cross sections and transition probabilities have been obtained for reactive collisions at higher energies, it is clear that at lower energies when a single PES is dominating the dynamics a better electronic description is needed to provide accurate barriers and gradients. This will be achieved by a so-called vector Hartree–Fock wave function. This construction is a complete active space multi-configurational (CAS-MC) wave function [41] parameterized as a vector coherent state [42–44].

Such a wave function is a linear combination of Thouless determinants, which means that not only are the spin orbitals non-orthogonal but so are the determinants. It is anticipated that rather short expansions of this kind will allow correct spin symmetry and permit sufficient accuracy of the electron dynamics such that low energy processes can be studied. This option is being included in the ENDyne program package.

Acknowledgments

This work was completed with support of NSF grant 00057476.

References

1. Y. Öhrn *et al.*, in *Time-Dependent Quantum Molecular Dynamics*, edited by J. Broeckhove and L. Lathouwers (Plenum, New York, 1992), pp. 279–292
2. E. Deumens, A. Diz, R. Longo, and Y. Öhrn, Rev. Mod. Phys **66**, 917 (1994)
3. J. R. Klauder and B.-S. Skagerstam, *Coherent States, Applications in Physics and Mathematical Physics* (World Scientific, Singapore, 1985)
4. E. Deumens and Y. Öhrn, J. Phys. Chem. **105**, 2660 (2001)
5. E. Deumens *et al.*, *ENDyne version 5 Software for Electron Nuclear Dynamics*, Quantum Theory Project, University of Florida, Gainesville FL 32611-8435, http://www.qtp.ufl.edu/endyne.html, 2002
6. E. Deumens and Y. Öhrn, Int. J. Quant. Chem.: Quant. Chem. Symp. **23**, 31 (1989)
7. N. F. Mott and H. S. W. Massey, *The Theory of Atomic Collisions* (Oxford University, Oxford, 1965)
8. E. W. McDaniel, *Atomic Collisions; Electron and Photon Projectiles* (Wiley, New York, 1989)
9. A. A. Frost, J. Chem. Phys. **47**, 3707 (1967)
10. D. Jacquemin, J. A. Morales, E. Deumens, and Y. Öhrn, J. Chem. Phys. **107**, 6146 (1997)
11. R. Cabrera-Trujillo, J. R. Sabin, E. Deumens, and Y. Öhrn, Adv. Quant. Chem. **47**, 253 (2004)
12. K. W. Ford and J. A. Wheeler, Ann. Phys. **7**, 259 (1959)
13. J. A. Morales, A. C. Diz, E. Deumens, and Y. Öhrn, J. Chem. Phys. **103**, 9968 (1995)
14. L. I. Schiff, Phys. Rev. **103**, 443 (1956)
15. R. Cabrera-Trujillo, J. R. Sabin, Y. Öhrn, and E. Deumens, Phys. Rev. A **61**, 032719 (2000)
16. R. Cabrera-Trujillo, J. R. Sabin, Y. Öhrn, and E. Deumens, Phys. Rev. Lett. **84**, 5300 (2000)
17. R. Cabrera-Trujillo, Y. Öhrn, E. Deumens, and J. R. Sabin, Phys. Rev. A **62**, 052714 (2000)
18. R. Cabrera-Trujillo, E. Deumens, Y. Öhrn, and J. R. Sabin, Nucl. Instrum. Methods B **168**, 484 (2000)
19. R. Cabrera-Trujillo, J. R. Sabin, E. Deumens, and Y. Öhrn, in *Application of Accelerators in Research and Industry, The Sixteen International Conference*, edited by J. L. Duggan and I. L. Morgan (American Institute of Physics, College Park, MD, 2001), p. 3
20. R. Cabrera-Trujillo, Y. Öhrn, J. R. Sabin, and E. Deumens, Phys. Rev. A **65**, 024901 (2002)
21. R. Cabrera-Trujillo, Y. Öhrn, E. Deumens, and J. R. Sabin, J. Chem. Phys. **116**, 2783 (2002)
22. R. Cabrera-Trujillo *et al.*, Phys. Rev. A **66**, 042712 (2002)

23. R. Cabrera-Trujillo, J. R. Sabin, Y. Öhrn, and E. Deumens, J. Elec. Spectr. **129**, 303 (2003)
24. R. Cabrera-Trujillo, Y. Öhrn, E. Deumens, and J. R. Sabin, J. Phys. Chem. A **108**, 8935 (2004)
25. Y. Öhrn, J. Oreiro, and E. Deumens, Int. J. Quant. Chem. **58**, 583 (1996)
26. R. J. S. Morisson, W. E. Conaway, and R. N. Zare, Chem. Phys. Lett. **113**, 435 (1985)
27. J. C. Poutsma *et al.*, Chem. Phys. Lett. **305**, 343 (1999)
28. M. Coutinho-Neto, E. Deumens, and Y. Öhrn, J. Chem. Phys. **116**, 2794 (2002)
29. B. Killian, R. Cabrera-Trujillo, E. Deumens, and Y. Öhrn, J. Phys. B: **37**, 1 (2004)
30. J. A. Morales, E. Deumens, and Y.Öhrn, J. Math. Phys. **40**, 766 (1999)
31. A. Blass, E. Deumens, and Y. Öhrn, J. Chem. Phys. **115**, 8366 (2001)
32. R. S. Gao *et al.*, Phys. Rev. A **36**, 3077 (1987)
33. M. W. Gealy and B. V. Zyl, Phys. Rev. A **36**, 3091 (1987)
34. G. W. McClure, Phys. Rev. **148**, 47 (1966)
35. M. E. Rudd *et al.*, Phys. Rev. A **28**, 3244 (1983)
36. R. Cabrera-Trujillo, J. R. Sabin, E. Deumens, and Y. Öhrn, Phys. Rev. A **71**, 044702 (2005)
37. Y.-N. Chiu *et al.*, J. Chem. Phys. **88**, 6814 (1988)
38. F. Fernández-Alonso *et al.*, Angew. Chem. Int. Ed. **39**, 2748 (2000)
39. S. Althorpe *et al.*, Nature **416**, 67 (2002)
40. E. J. Heller, J. Chem. Phys. **62**, 1544 (1975)
41. E. Deumens, Y. Öhrn, and B. Weiner, J. Math. Phys. **32**, 1166 (1991)
42. D. J. Rowe, G. Rosensteel, and R. Gilmore, J. Math. Phys. **26**, 2787 (1985)
43. K. Hecht, *The Vector Coherent State Method and its Applications to Problems of Higher Symmetries* (Springer, New York, 1987)
44. D. J. Rowe, R. LeBlanc, and K. T. Hecht, J. Math. Phys. **29**, 287 (1988)

The Semiclassical Initial Value Series Representation of the Quantum Propagator

Eli Pollak

Summary. One of the central open challenges of the 21st century is the computation of real time quantum dynamics for systems with "many" degrees of freedom. A promising approach for obtaining approximate real time quantum dynamics is through the use of the semiclassical initial value approximation for the exact quantum propagator. The main drawback of this class of approximations was its ad hoc nature, it was in many senses an uncontrolled approximation scheme. This drawback has been recently remedied by showing that the semiclassical initial value representation (SCIVR) propagator is just a leading order term in a formally exact series representation of the true quantum propagator. In this review we present the SCIVR series representation, its successes and future challenges in applications to "large" systems. In addition, a new interaction representation initial value series representation for the exact quantum propagator is formulated.

1 Introduction

Classical dynamics is a well established numerical method for obtaining real time information on the evolution of systems with many thousands of degrees of freedom. Although, even today there are limitations to classical molecular dynamics, most notably the extent of time for which one can get converged results, these limitations are trivial when compared to the difficulty of carrying out real time quantum mechanics computations in systems with many degrees of freedom. The central stumbling block is that the real time quantum propagator is complex. Its path integral representation at time t is a sum over all paths leading from the configuration space point $\mathbf{x}$ to $\mathbf{x}'$ weighted by $\exp(\mathrm{i}S/\hbar)$, where S is the action along the path. It is this oscillatory weight function which causes any direct Monte Carlo estimate of the propagator or its matrix elements to be exponentially expensive and so unrealistic except for systems with only a very few degrees of freedom.

From its inception, a central problem in quantum mechanics, was to relate it to classical dynamics. The formal theory, applicable to systems of arbitrary size was developed during the second half of the 20th century. The central object here is the so called van Vleck semiclassical propagator [1], which represents the stationary phase limit of the quantum propagator. In this stationary phase limit, instead of a sum over all paths, one has at most a countable set of classical paths that lead from $\mathbf{x}$ to $\mathbf{x}'$ at time t:

$$K_{VV}(\mathbf{x},\mathbf{x}';t) = \sum_{\text{all traj's}} \left((2\pi \mathrm{i}\hbar)^N \frac{\partial \mathbf{x}'_t}{\partial \mathbf{x}} \right)^{-1/2} \exp\left(\frac{\mathrm{i}S(\mathbf{x},\mathbf{x}';t)}{\hbar} \right), \tag{1}$$

where $S(\mathbf{x},\mathbf{x}';t)$ denotes the action along the trajectory including the "Maslov index" and N is the number of degrees of freedom. This form was very successful in explaining quantum superposition [2], obtaining the semiclassical quantization rules for quasiperiodic classical systems [3] and classically chaotic systems [4]. As a practical tool though it is not very useful. There are two main deficiencies. One is the necessity of solving a double ended boundary condition for the appropriate trajectories. The second problem is that the propagator diverges whenever the denominator in the prefactor goes through zero. This divergence may be "fixed" by using uniform semiclassical approximations [5], but the methodology becomes rather tedious and is not readily amenable to solution for large systems.

To overcome some of these problems, Miller suggested [6] to use an initial value form of the van Vleck propagator. By changing variables from the initial and final point in configuration space to the initial point in phase space, Miller wrote down for the first time a semiclassical initial value representation of the propagator:

$$K_{\mathrm{M}} = \int \mathrm{d}\mathbf{p}\mathrm{d}\mathbf{q}\, |\mathbf{q}_t\rangle\langle\mathbf{q}| \left(\frac{1}{(2\pi \mathrm{i}\hbar)^N} \frac{\partial \mathbf{q}_t}{\partial \mathbf{p}} \right)^{1/2} \exp\left(\frac{\mathrm{i}S(\mathbf{q},\mathbf{p};t)}{\hbar} \right). \tag{2}$$

This form addresses the two major difficulties appearing in the van Vleck form. One does not need to solve a double ended boundary condition problem, rather one needs to propagate from any point in phase space a classical trajectory until time t. Furthermore, the derivative with respect to the initial conditions is now in the numerator, so even if it vanishes, it does not introduce any infinities as in the Van Vleck form. From a numerical point of view though, one remains with a difficult sign problem. Even though the sum over all paths is reduced to a sum over all classical paths, the weight is highly oscillatory and so difficult to converge using a Monte Carlo scheme.

Some years later, Heller suggested [7] a "frozen Gaussian" approximation to the propagator, which also has the "nice" property of being an initial value representation, moreover it has a built in Gaussian weighting function in the form of coherent states

$$K_{\mathrm{H}}(t) = \int \frac{\mathrm{d}\mathbf{p}\mathrm{d}\mathbf{q}}{(2\pi\hbar)^N} \exp\left(\frac{\mathrm{i}\tilde{S}(\mathbf{p},\mathbf{q};t)}{\hbar} \right) |g(\mathbf{p},\mathbf{q};t)\rangle\langle g(\mathbf{p},\mathbf{q};0)|, \tag{3}$$

where the $|g\rangle$'s are coherent states, whose coordinate representation is Gaussian

$$\langle \mathbf{x}|g(\mathbf{p},\mathbf{q};0)\rangle = \left(\frac{\det(\Gamma)}{\pi^N}\right)^{1/4} \exp\left(-\frac{1}{2}(\mathbf{q}-\mathbf{x})^T\Gamma(\mathbf{q}-\mathbf{x}) + \frac{\mathrm{i}}{\hbar}\mathbf{p}(\mathbf{x}-\mathbf{q})\right), \tag{4}$$

where for a system with N degrees of freedom, Γ is a width parameter matrix (with positive eigenvalues) with dimension $N \times N$. If the width matrix is chosen as constant in time, one talks about a frozen Gaussian approximation, if it is time dependent, one uses the terminology thawed Gaussian approximation. The action appearing in the exponent of (3) is no longer "simple" but involves the action along a coherent state weighted average of the classical potential. The frozen and thawed Gaussian forms have the advantage that now the coherent states provide a natural Gaussian weighting function, so that the multidimensional phase space integral can at least in principle be converged using Monte Carlo methods. The Gaussian weighting causes a natural decay which leads to convergence. However, this representation has a major problem, it is not unitary, as time evolves, the initial unitarity property is rapidly lost and the results become seemingly meaningless. Heller's form raises another difficulty. There is almost arbitrary freedom in the choice of the width parameter matrix. So, one has an approximation which is uncontrolled. For different width parameters one gets different results and there is no objective criterion by which to choose the width parameter.

The unitarity problem was solved to a large extent by Herman and Kluk. In their now famous paper [8], they showed that what was missing in Heller's frozen Gaussian was a prefactor which depends on elements of the monodromy matrix. When including the prefactor, the "frozen Gaussian" propagator, now called the Herman–Kluk SCIVR propagator was approximately unitary for rather long times. How about the arbitrariness of the width parameter matrix? Kluk et al. [9] showed that there was a large region in parameter space, for which the results were not too sensitive to changes in the width parameters, so that from a practical point of view, this problem was not too acute.

The early numerical and analytical studies of the Herman–Kluk SCIVR propagator opened the way for a large body of numerical studies of ever increasing complexity (for recent reviews see [10–14]). There were a number of important milestones in the applications of the theory. One was the invention of the forward–backward simplification of the SCIVR expression for thermal correlation functions [15–18]. This led to a significant decrease in the oscillatory nature of the phases in the SCIVR expression. Numerical convergence was then achieved with a significantly smaller sample of Monte Carlo points. A second trick which is used extensively is Filinov filtering of the SCIVR expression for correlation functions [20–22]. This too significantly reduces the phase oscillations, however, it must be applied with care, since, especially in the presence of classically chaotic systems it can also degrade the signal, leading to erroneous results [23].

The derivation of the HK SCIVR has also been a subject of intense debate [12, 24, 25]. However, lately, especially through the studies of Martin-Fierro and Gomez-Llorente [26] and Kay [14] it is evident that the HK SCIVR is a valid semiclassical expression.

Although the Herman Kluk SCIVR expression for the propagator and its variants may be considered as important progress in our ability to compute quantum phenomena in complex systems, there remain some severe drawbacks. There are in principle, an infinite number of SCIVR expressions for the propagator, all of which are exact for harmonic systems and all of which reduce to the van Vleck semiclassical limit in a stationary phase sense. This ambiguity should be resolved. Even for a given SCIVR expression, there remains ambiguity in the choice of the width parameters appearing in the coherent states. Finally, when the number of degrees of freedom increases, or when the time interval becomes "long" the phase oscillations in the SCIVR expressions become unmanageable. It is not an accident that to date the SCIVR based expressions have not been applied to systems with more than a few dozen degrees of freedom. In Sect. 2, we will review how the SCIVR series representation [27, 28] has helped in overcoming some of these deficiencies.

2 The SCIVR Series Method

2.1 The Formalism

The exact quantum propagator $\hat{K}(t)$ obeys the equation of motion

$$\mathrm{i}\hbar\frac{\partial}{\partial t}\hat{K}(t)=\hat{H}\hat{K}(t), \quad \hat{K}(0)=\hat{I}. \tag{5}$$

The SCIVR propagator $\hat{K}_0(t)$ is (except for harmonic systems) not identical to the exact propagator so that $\left(\mathrm{i}\hbar\frac{\partial}{\partial t}-\hat{H}\right)\hat{K}_0(t)\neq 0$. We can therefore define the residue to be a "correction operator" [29] so that:

$$\mathrm{i}\hbar\frac{\partial}{\partial t}\hat{K}_0(t)=\hat{H}\hat{K}_0(t)+\hat{C}(t), \qquad \hat{K}_0(0)=\hat{I}, \tag{6}$$

where we have imposed the condition that at least at the initial time the SCIVR propagator is the identity operator. The correction operator is determined uniquely by the choice made for the SCIVR propagator. Given $\hat{K}_0(t)$ one knows exactly what the correction operator is. Computation of its matrix elements is typically not more or less difficult than computation of the SCIVR propagator itself. Explicit expressions for the correction operator associated with the HK propagator may be found in [30]. More general formulae for the correction operator, for families of SCIVR propagators may be found in [31].

Equation (6) may be formally solved [28] as:

$$\hat{K}_0(t-t_i)=\hat{K}(t-t_i)-\frac{\mathrm{i}}{\hbar}\int_{t_i}^{t}\mathrm{d}t'\hat{K}(t-t')\hat{C}(t'). \tag{7}$$

The numerical experience with the various SCIVR propagators implies that typically they are not far off the mark. This implies that we may consider the correction operator to be "small" in some sense and so justifies a perturbative expansion of the formal solution given in (7). Representing the exact propagator in terms of a series, in which the jth element is of the order of $\hat{C}(t)^j$

$$\hat{K}(t - t_i) = \sum_{j=0}^{\infty} \hat{K}_j(t - t_i); \tag{8}$$

inserting this expansion into the formal solution (7) gives the recursion relation

$$\hat{K}_{j+1}(t - t_i) = \frac{\mathrm{i}}{\hbar} \int_{t_i}^{t} \mathrm{d}t' \hat{K}_j(t - t')\hat{C}(t'). \tag{9}$$

This recursion relation together with the known form of the correction operator provides a series representation of the exact propagator. Experience has shown that this series typically converges rapidly [28, 30, 32–35]. This rapid convergence is a reflection of the fact that the zeroth order SCIVR term does incorporate in it much of the quantum phenomena. However, an analytic theory of the convergence properties remains unknown at present.

The SCIVR series representation of the propagator is exact. It immediately provides a pathway for removal of some of the difficulties associated with the SCIVR approximation, mentioned in the Introduction. The series representation turns the ambiguity in the definition of the SCIVR propagator to a strength as discussed Sect. 2.2. The SCIVR series representation is actually just a generalization of time dependent perturbation theory. In time dependent perturbation theory one divides the Hamiltonian into two parts $\hat{H} = \hat{H}_0 + \hat{H}_1$ and chooses as the zeroth order propagator the supposedly known propagator for $\hat{H}_0$: $\hat{K}_0(t) = \exp(-\frac{\mathrm{i}}{\hbar}\hat{H}_0 t)$. Since $\mathrm{i}\hbar \frac{\partial \hat{K}_0(t)}{\partial t} = \hat{H}_0 \hat{K}_0(t) = (\hat{H}_0 + \hat{H}_1)\hat{K}_0(t) - \hat{H}_1 \hat{K}_0(t)$ one finds that the "correction operator" is just $-\hat{H}_1 \hat{K}_0(t)$ and the series representation of the exact propagator is just the standard interaction picture representation. For example, the first-order correction term is $\hat{K}_1(t) = \frac{\mathrm{i}}{\hbar} \int_0^t \mathrm{d}t' \exp(-\frac{\mathrm{i}}{\hbar}\hat{H}_0(t - t'))\hat{H}_1 \exp(-\frac{\mathrm{i}}{\hbar}\hat{H}_0 t')$.

2.2 Properties of the Correction Operator

Viewing the SCIVR series representation as a generalization of perturbation theory implies that one should minimize the correction operator in some sense. For example, when considering an overlap function of the form $c(t) = \left\langle \Psi \middle| \hat{K}(t) \middle| \Psi \right\rangle = \sum_{j=0}^{\infty} c_j(t)$ with $c_j(t)$ being of the order $\hat{C}^j$ in the correction operator, then in the spirit of perturbation theory the width parameters of the coherent states appearing in the SCIVR propagator should be chosen so that the first-order term in the series $c_1(t) = \frac{\mathrm{i}}{\hbar} \left\langle \Psi \middle| \int_0^t \mathrm{d}t' \hat{K}_0(t - t')\hat{C}(t') \middle| \Psi \right\rangle$ is minimal. We have shown in a number of examples that indeed this leads to improved convergence of the SCIVR series [28,32,34]. However, the computation of this first-order term is more expensive than the computation of the zeroth order term since it involves a product of two operators. It turned out therefore to be of practical convenience to minimize instead the expectation value of $\left\langle \Psi \middle| \hat{C}(t) \middle| \Psi \right\rangle$ [30]. Here, one needs to compute only a single operator and the numerical effort is the same as the computation of $c_0(t)$, instead of a product of two operators as in the full first-order expression. This cheaper minimization led to very similar parameters, as compared to the full minimization of the first-order term.

The SCIVR series representation also opens the way for understanding the advantage of the HK type SCIVR as compared to the Miller type of SCIVR propagator. If one takes the time derivative of the prefactor in the Miller type SCIVR propagator one sees that

$$\frac{\partial}{\partial t}\left(\frac{\partial \mathbf{q}_t}{\partial \mathbf{p}}\right)^{1/2} = \frac{1}{2}\left(\frac{\partial \mathbf{q}_t}{\partial \mathbf{p}}\right)^{-1/2} \frac{\partial \mathbf{p}_t}{\partial \mathbf{p}}.$$

In other words, in the associated correction operator, the derivative with respect to initial conditions in the prefactor moves from the numerator to the denominator and so whenever it vanishes it leads to a divergence in the correction operator. However, if one uses the Herman Kluk SCIVR propagator, the prefactor itself is typically a complex factor which does not vanish at any time and so this problem does not occur. More specifically the SCIVR propagator can be written without loss of generality as:

$$\hat{K}_0(t) = \int_{-\infty}^{\infty} \prod_{j=1}^{N} \left(\frac{\mathrm{d}p_j \mathrm{d}q_j}{2\pi\hbar}\right) R(\mathbf{p},\mathbf{q},t) \mathrm{e}^{\frac{\mathrm{i}}{\hbar} S(\mathbf{p},\mathbf{q},t)} |g(\mathbf{p},\mathbf{q},t)\rangle\langle g(\mathbf{p},\mathbf{q},0)|. \quad (10)$$

The prefactor $R(\mathbf{p},\mathbf{q},t)$ for the Herman Kluk SCIVR is

$$R_{\mathrm{HK}}(\mathbf{p},\mathbf{q},t) = \left(\det\left[\frac{1}{2}\left(\frac{\partial \mathbf{q}_t}{\partial \mathbf{q}} + \mathbf{\Gamma}^{-1}\frac{\partial \mathbf{p}_t}{\partial \mathbf{p}}\mathbf{\Gamma} - \mathrm{i}\hbar\frac{\partial \mathbf{q}_t}{\partial \mathbf{p}}\mathbf{\Gamma} + \frac{\mathrm{i}}{\hbar}\mathbf{\Gamma}^{-1}\frac{\partial \mathbf{p}_t}{\partial \mathbf{q}}\right)\right]\right)^{\frac{1}{2}} \quad (11)$$

with $\mathbf{\Gamma}$ being the constant width parameter matrix as it appears in the coherent states, see (4). For example, for a one-dimensional harmonic oscillator with frequency ω, the HK prefactor is just $\left[\cos(\omega t) - \mathrm{i}\sin(\omega t)\frac{1}{2}\left(\frac{\omega}{\hbar\mathbf{\Gamma}} + \frac{\hbar\mathbf{\Gamma}}{\omega}\right)\right]^{1/2}$ and this prefactor never vanishes while for the Miller van Vleck form one has that $\frac{\partial \mathbf{p}_t}{\partial \mathbf{q}} = \sin(\omega t)/\omega$ and this evidently vanishes whenever $\omega t = n\pi$. It should be stressed that the Miller SCIVR propagator is exact for the harmonic oscillator, so that after integration over the phase space, the correction operator vanishes identically, however, this phase space integration would be difficult to converge when using a Monte Carlo algorithm. The same is not true for the Herman Kluk form.

To obtain the general form of the correction operator associated with the SCIVR propagator as given in (11) it is necessary to specify the time evolution of the coordinates and momenta. For any specified time evolution one would derive an associated correction operator. It is therefore of interest to present a general formalism for possible dynamics and the associated family of correction operators. For this purpose we introduced [31] a normalized averaging function $f(\mathbf{x}-\mathbf{q})$ which is even with respect to the argument. Given a quantum Hamiltonian with form $\hat{H} = \frac{\hat{\mathbf{p}}_q^2}{2} + V(\hat{\mathbf{q}})$ one considers a classical dynamics which is governed by the classical Hamiltonian

$$\widetilde{H}_{\mathrm{cl}} = \frac{\mathbf{p}_q^2}{2} + \widetilde{V}(\mathbf{q}) \tag{12}$$

and $\widetilde{V}(\mathbf{q})$ is the averaged potential

$$\widetilde{V}(\mathbf{q}) = \int_{-\infty}^{\infty} \mathrm{d}\mathbf{x} f(\mathbf{q}-\mathbf{x}) V(\mathbf{x}). \tag{13}$$

The classical (mass weighted) coordinates and momenta $\mathbf{q}(t)$ and $\mathbf{p}(t)$ obey Hamilton's equations of motion *on the averaged potential*

$$\dot{q}_j(t) = \frac{\partial \widetilde{H}_{\mathrm{cl}}}{\partial p_j} = p_j(t), \tag{14}$$

$$\dot{p}_j(t) = -\frac{\partial \widetilde{H}_{\mathrm{cl}}}{\partial q_j} = -\frac{\partial \widetilde{V}[\mathbf{q}(t)]}{\partial q_j}, \tag{15}$$

where the dot denotes time differentiation.

The correction operator may now be written down as [32]:

$$\hat{C}(t) = \int_{-\infty}^{\infty} \prod_{j=1}^{N} \left(\frac{\mathrm{d}p_j \mathrm{d}q_j}{2\pi\hbar} \right) R(\mathbf{p},\mathbf{q},t) \mathrm{e}^{\frac{\mathrm{i}}{\hbar} S(\mathbf{p},\mathbf{q},t)} \Delta V(\hat{\mathbf{q}},t) |g(\mathbf{p},\mathbf{q},t)\rangle\langle g(\mathbf{p},\mathbf{q},0)| \tag{16}$$

with the "potential difference" operator having the form:

$$\begin{aligned} \Delta V(\hat{\mathbf{q}},t) = {} & \nabla\widetilde{V}\left[\mathbf{q}(t)\right] \cdot (\hat{\mathbf{q}}-\mathbf{q}(t)) - V(\hat{\mathbf{q}}) - \frac{\partial S}{\partial t} + \frac{\mathbf{p}^{\mathrm{T}} \cdot \mathbf{p}}{2} \\ & + \mathrm{i}\hbar \frac{\dot{R}}{R} + \frac{\mathrm{i}\hbar}{4} \frac{\partial \log[\det \Gamma_{\mathrm{r}}(t)]}{\partial t} - \frac{\hbar^2}{2} \mathrm{Tr}\left[\Gamma(t)\right] \\ & + \frac{\hbar^2}{2} (\hat{\mathbf{q}}-\mathbf{q}(t))^{\mathrm{T}} \Gamma(t)\Gamma(t) (\hat{\mathbf{q}}-\mathbf{q}(t)) \\ & - \frac{\mathrm{i}\hbar}{2} (\hat{\mathbf{q}}-\mathbf{q}(t))^{\mathrm{T}} \left[\frac{\partial}{\partial t}\Gamma(t)\right] (\hat{\mathbf{q}}-\mathbf{q}(t)) . \end{aligned} \tag{17}$$

Here we have assumed that the width matrix may be time dependent and complex. The real and imaginary parts of the matrix are denoted by the subscripts "r" and "i" ($\Gamma_{\mathrm{r}}(t), \Gamma_{\mathrm{i}}(t)$) and for the sake of brevity we have omitted the fact that the matrices may be dependent on the initial time phase space variables. At time $t = 0$, the imaginary part will vanish and the real part has only positive eigenvalues.

This form of the correction operator can be used to define families of useful SCIVR propagators. At this point, neither the width matrices nor the prefactor have been defined. A "good" SCIVR propagator is one for which the correction operator is "small." Ideally, one would want to make the potential difference operator vanish, however, this is impossible in general. Instead, one may demand for example, that the potential difference operator vanish on the average [32]. Limiting oneself to a real constant width parameter matrix, choosing the "averaging function" $f(\mathbf{x} - \mathbf{q}) = \delta(\mathbf{x} - \mathbf{q})$ and averaging the potential difference operator over the coherent state such that $\langle g(\mathbf{p}, \mathbf{q}; t) \left| \Delta V(\hat{\mathbf{q}}, t) \right| g(\mathbf{p}, \mathbf{q}; t) \rangle = 0$ leads immediately to the Heller frozen Gaussian form of the SCIVR propagator as given in (3). In [31] we have shown how one can use this methodology to generalize the thawed Gaussian propagator of Baranger et al. [12].

2.3 Renormalization

The frozen Gaussian SCIVR has a major numerical advantage over the Herman Kluk SCIVR since one does not need to compute the monodromy matrix elements. However, there is a price to pay. While the HK SCIVR is approximately unitary for rather long times, the prefactor free SCIVR loses unitarity rather rapidly [36]. Here too though, the SCIVR series approach provides a route for overcoming the loss of unitarity [32]. Consider for example the overlap function $c_0(t) = \left\langle \Psi \left| \hat{K}_0(t) \right| \Psi \right\rangle$. Associated with it is a normalization function $N(t) = \left\langle \Psi \left| \hat{K}_0^\dagger(t) \hat{K}_0(t) \right| \Psi \right\rangle$. One may now define a renormalized SCIVR propagator $\hat{K}_{N0}(t) = \hat{K}_0(t)/\sqrt{N(t)}$ so that by definition $\left\langle \Psi \left| \hat{K}_{N0}^\dagger(t) \hat{K}_{N0}(t) \right| \Psi \right\rangle = 1$. The correction operator associated with the renormalized propagator is then modified, one readily finds that

$$\hat{C}_N(t) = \hat{C}(t)/\sqrt{N(t)} - \frac{1}{2} \hat{K}_{N0}(t) \frac{\mathrm{d} \ln N(t)}{\mathrm{d}t} .$$

The SCIVR series representation of the renormalized propagator follows through using the renormalized correction operator $\hat{C}_N(t)$. Similar strategies may be employed when computing correlation functions in dissipative systems. The computation of the normalization function is more expensive than the computation of the overlap function since it involves a product of two propagators. However, it is still less expensive than the computation of the second term in the SCIVR series and usually needs to be carried out for internal consistency checks. It is also noteworthy that the suggested renormalization is not general. For different wavepackets $|\Psi\rangle$ one would get different normalization functions. The renormalization to be used will thus depend on the specific problem to be solved.

2.4 IVR in the Interaction Representation

The SCIVR series methodology can be also used to develop IVR approximations within the interaction representation of quantum mechanics. The Hamiltonian is divided into two parts as before $\hat{H} = \hat{H}_0 + \hat{H}_1$ and $\hat{K}_0(t) = \exp(-\frac{\mathrm{i}}{\hbar}\hat{H}_0 t)$ is assumed to be known exactly. In the interaction picture, the propagator is represented as a product of two propagators $\hat{K}(t) = \hat{K}_0(t)\hat{K}_I(t)$, where $\hat{K}_I(t) = \exp_+\left(-\frac{\mathrm{i}}{\hbar}\int_0^t \mathrm{d}t'\hat{H}_1(t')\right)$ is the time ordered exponential with $\hat{H}_1(t) = \hat{K}_0^\dagger(t)\hat{H}_1\hat{K}_0(t)$. An Interaction Representation IVR (IRIVR) would be an IVR representation for the interaction propagator $\hat{K}_I(t)$. A pioneering work on the SCIVR interaction representation was presented by Shao and Makri [37]. They replaced the classical dynamics appearing in the Herman–Kluk SCIVR propagator with forward–backward dynamics, that is each trajectory is propagated classically forward in time under the action of the full Hamiltonian and then propagated classically backward to time zero under the action of the zeroth order Hamiltonian. The resulting SCIVR approximation for $\hat{K}_I(t)$ has the desired properties that $\hat{K}_{I0}(0) = I$ and that it reduces to the identity operator I if $\hat{H}_1 = 0$. Especially the first property implies that here too one can apply the correction operator formalism, the correction operator would now be defined as $\hat{C}_I(t) = \mathrm{i}\hbar\frac{\partial \hat{K}_{I0}(t)}{\partial t} - \hat{H}_1(t)\hat{K}_{I0}(t)$ and (7)–(9) follow through readily.

A different suggestion for an IRIVR would be:

$$\hat{K}_{I0}(t) = \int_{-\infty}^{\infty}\prod_{j=1}^{N}\left(\frac{\mathrm{d}p_j\mathrm{d}q_j}{2\pi\hbar}\right)\exp\left(-\frac{\mathrm{i}}{\hbar}\int_0^t \mathrm{d}t'\langle g(\mathbf{p},\mathbf{q})|\hat{H}_1(t')|g(\mathbf{p},\mathbf{q})\rangle\right) \times |g(\mathbf{p},\mathbf{q})\rangle\langle g(\mathbf{p},\mathbf{q})|. \tag{18}$$

Note that here the time dependence does not come from classical mechanics, but rather from time propagation with the known zeroth order propagator $\hat{K}_0(t)$. The IRIVR suggested in (18) also reduces to the identity operator both at the initial time as well as when the perturbation part of the Hamiltonian $\hat{H}_1$ vanishes.

The equation of motion for the interaction representation propagator is

$$\mathrm{i}\hbar\frac{\partial \hat{K}_I(t)}{\partial t} = \hat{H}_1(t)\hat{K}_I(t). \tag{19}$$

The correction operator in this interaction representation is defined as:

$$\begin{aligned}\hat{C}_I(t) &= \mathrm{i}\hbar\frac{\partial \hat{K}_{I0}(t)}{\partial t} - \hat{H}_1(t)\hat{K}_{I0}(t)\\ &= \int_{-\infty}^{\infty}\prod_{j=1}^{N}\left(\frac{\mathrm{d}p_j\mathrm{d}q_j}{2\pi\hbar}\right)\left(\langle g|\hat{H}_1(t)|g\rangle - \hat{H}_1(t)\right)\mathrm{e}^{-\frac{\mathrm{i}}{\hbar}\int_0^t \mathrm{d}t'\langle g|\hat{H}_1(t')|g\rangle}|g\rangle\langle g|,\end{aligned} \tag{20}$$

where for the sake of brevity we wrote everywhere g instead of $g(\mathbf{p}, \mathbf{q})$. Evidently, this correction operator has the property that at every point in phase space, when averaged over the coherent state, it vanishes. This is similar to the condition which leads to the prefactor free SCIVR as described earlier. Given the correction operator, one may repeat the same algebra as before and obtain a series representation of the interaction propagator in terms of the correction operator and $\hat{K}_0(t)$. The usefulness of representations of this sort remains an open topic for future investigation, however, even at this point, one sees from here the potential and power of the series method for obtaining practical ways of computing real time quantum dynamics using Monte Carlo methods.

3 Applications

The SCIVR series method has been applied thus far to a number of cases. The first application was to the overlap function $c(t) = \left\langle \Psi \left| \hat{K}(t) \right| \Psi \right\rangle$ for a Gaussian wavefunction evolving on a quartic potential Hamiltonian [28]. Optimization of the width parameter, as described in the previous section led to a converged correlation function using only the first two terms in the SCIVR series, for times up to three periods of motion in the well of the quartic potential [30]. As in any time dependent perturbation theory, it is necessary to go to higher order in the perturbation series as the time interval increases.

The next application was to the double slit scattering model studied by Miller and coworkers [32]. This two dimensional problem was solved using two different variants of the SCIVR series method. Optimization of the width parameter and use of the Herman–Kluk propagator led to accurate results for the interference pattern of the scattered wavepacket using only the first two terms in the SCIVR series. Use of the Heller frozen Gaussian method with renormalization, led to convergence with the first three terms in the SCIVR series. The numerical effort involved was similar in both cases, here the great advantage of the prefactor free Heller SCIVR did not play a major role, since the number of degrees of freedom in this model scattering problem is only two.

One of the interesting challenges to the SCIVR method was the computation of deep tunneling probabilities. A partially successful attempt at describing tunneling with the aid of classical trajectories is based on Wigner transformation of the initial wave-packet followed by classical propagation of the phase space variables. The wavepacket has a tail whose energy is larger than the barrier height and classical trajectories propagated at these high energies will lead one over the barrier. The probability for transmission is exponentially small and so one has a qualitative description of tunneling via classical trajectories [38]. This classical path description of tunneling is exact for a parabolic barrier. However, as noted by Maitra and Heller [39], it fails for "deep" tunneling. When the barrier goes to a constant value at plus or minus infinity, the classical path contribution becomes too small. Tunneling is dominated by nonclassical paths connecting two manifolds of classical trajectories lying close to the separatrix between transmitted and reflected trajectories. Kay [40] showed that one may improve upon the classical path description with the aid of an SCIVR propagator provided that the coherent states have a complex time dependent width. Grossmann [41] estimated the tunneling probability by splitting the propagator into a product of two equal time propagators and then estimating each half time propagator using an SCIVR propagator. Burant and Batista continued in this vein using time slicing of the propagator [42].

We have shown that deep tunneling is accounted for when using the SCIVR series method [28]. The nth term in the series involves a product of $n+1$ propagators, each related to the other by an overlap matrix element of coherent states. Thus, the nth term involves $n+1$ trajectory segments, whose final and initial points are related through a Gaussian overlap function. Such a path was termed a coherent classical path. We found that "deep" tunneling is well described by such coherent classical paths, and that only a finite small number of segments (typically less than 5) are needed to accurately describe the tunneling process. This analysis worked well both for thermal and energy dependent scattering.

As already discussed, one of the central problems in application of the SCIVR method to "large" systems is that the cost of propagating all the elements in the monodromy matrix is rather high in cpu time. One way of overcoming this is to use the prefactor free methods. A different way is to use a hybrid method, that is construct the prefactor but using only the system variables while treating the bath variables with the prefactor free formalism [34]. This hybrid method was applied to the relaxation of an anharmonic oscillator bilinearly coupled to a bath of five harmonic oscillators. Comparison with numerically exact results using basis set methods showed that it again sufficed to use only the first two terms in the SCIVR series to obtain convergence.

4 Discussion

The SCIVR series method has led to significant conceptual progress. The various ambiguities associated with SCIVR propagators was resolved through the understanding that the SCIVR propagator is just a zeroth order term in a perturbation series. The "best" SCIVR propagator is the one that leads to the quickest convergence. For different problems this can mean different choices. The analysis of the correction operator has led to new families of SCIVR operators, has provided new physical insight into Heller's frozen Gaussian propagator and has shown the way to construct general forms of prefactor free and hybrid propagators with favorable numerical convergence properties.

There are though obstacles that need to be addressed. In chaotic systems, the Herman–Kluk prefactor diverges exponentially in time making a Monte Carlo evaluation of even the HK SCIVR propagator impossible [43]. This may be remedied by using prefactor free hybrid or thawed Gaussian propagators. In all these cases either the prefactor is a constant, or the effect of chaos is limited or the monodromy matrix elements appear in the denominator instead of the numerator so that the contribution of the highly chaotic trajectories is exponentially small at long times. There is perhaps another strategy and that is to replace the classical Wigner dynamics by the so called Q representation of the quantum Hamiltonian wherein the quantum Hamiltonian is averaged over coherent states. The Herman–Kluk form of the SCIVR propagator in this Q representation has been derived by Martin-Fierro and Gomez-Llorente [26]. In the Q representation, the Gaussian averaging of the potential softens it and so reduces the hard chaos. To date though, none of these options have been studied seriously, so that chaos in the SCIVR approach remains an open question.

Perhaps, though the most serious challenge facing the SCIVR series method is the old phase problem. As the number of degrees of freedom increases, so do the variations in the phase in the integrand of the SCIVR propagator and it becomes increasingly difficult to converge the results using Monte Carlo methods. It is not an accident that to date one does not find converged SCIVR results for systems with more than a few dozen degrees of freedom. Even then, results are obtained for relatively short times. This highlights the need for extracting information from short time quantum propagation. One way of doing this is with the filter diagonalization method [44, 45]. We have recently shown how one may use a time dependent Rayleigh-Ritz variational theorem to extract tunneling probabilities and wave functions in a symmetric double well potential [35]. Further studies of the efficacy of these short time methods will be needed to turn the SCIVR series method into a viable tool for computing eigenvalues.

The initial value series representation in the interaction representation presented in this paper is an additional twist on the use of IVR methods to obtain the exact propagator with the aid of Monte Carlo methods. Here too the series method assures that at least in principle, if the series converges it will lead to the exact propagator. The interaction representation has the advantage that it "pulls out" of the IVR portion a large part of the phase, thus hopefully leading to easier convergence when using Monte Carlo methods. The utility of the interaction representation IVR series remains a topic for future studies.

In summary, the SCIVR series method has been shown thus far to be viable for a variety of problems, but the proof of the pudding is its applicability to systems with at least dozens of degrees of freedom.

Acknowledgment

This work has been supported by grants of the US Israel Binational Science Foundation, the Petroleum Research Foundation, the Israel Science Foundation and the German Israel Foundation for Basic Research.

References

1. J.H. Van Vleck, Proc. Natl. Acad. Sci. USA **14**, 178 (1928)
2. W.H. Miller, Adv. Chem. Phys. **25**, 69 (1974)
3. M.V. Berry and M. Tabor, Proc. R. Soc. London A **356**, 375 (1977)
4. M.C. Gutzwiller, J. Math. Phys. **12**, 343 (1971)
5. M.S. Child, *Molecular Collision Theory*, (Academic, London, 1974)
6. W.H. Miller, J. Chem. Phys. **53**, 3578 (1970)
7. E. Heller, J. Chem. Phys. **75**, 2923 (1981)
8. M.F. Herman and E. Kluk, Chem. Phys. **91**, 27 (1984)
9. E. Kluk, M.F. Herman and H.L. Davis, J. Chem. Phys. **84**, 326 (1986)
10. F. Grossmann, Comments At. Mol. Phys. **34**, 141 (1999)
11. D.J. Tannor and S. Garaschuk, Annu. Rev. Phys. Chem. **51**, 553 (2000)
12. M. Baranger, M.A.M. de Aguiar, F. Keck, H.J. Korsch and B. Schellhaas, J. Phys. A: Math. Gen. **34**, 7227 (2001)
13. W.H. Miller, J. Phys. Chem. A **105**, 2942 (2001)
14. K.G. Kay, "Semiclassical initial value treatments of atoms and molecules", Annu. Rev. Phys. Chem. **56**, 255 (2005)
15. N. Makri and K. Thompson, Chem. Phys. Lett. **291**, 101 (1998)
16. J. Shao and N. Makri, J. Phys. Chem. A, **103**, 7753 (1999)
17. X. Sun and W.H. Miller, J. Chem. Phys. **103**, 7753 (1999)
18. M. Thoss, H. Wang and W.H. Miller, J. Chem. Phys. **114**, 9220 (2001)
19. V.S. Filinov, Nucl. Phys. B **271**, 717 (1986)
20. B.W. Spath and W.H. Miller, J. Chem. Phys. **104**, 95 (1996)
21. A.R. Walton and D.E. Manolopoulos, Mol. Phys. **87**, 961 (1996)
22. E.A. Coronado, V.S. Batista and W.H. Miller, J. Chem. Phys. **112** 5566(2000)
23. M. Spanner, V.S. Batista and P. Brumer, J. Chem. Phys. **122**, 084111 (2005)
24. F. Grossmann and M.F. Herman, J. Phys. A: Math. Gen. **35**, 9489 (2002)
25. M. Baranger, M.A.M. de Aguiar, F. Keck, H.J. Korsch and B. Schellhaas, J. Phys. A: Math. Gen. **34**, 9493 (2002)
26. E. Martin Fierro and J.M. Gomez Llorente, Chem. Phys., vol. 322, p. 13 (2006)
27. E. Pollak and J. Shao, J. Phys. Chem. A **107**, 7112 (2003)
28. S. Zhang and E. Pollak, Phys. Rev. Lett. **91**, 190201 (2003)
29. J. Ankerhold, M. Saltzer and E. Pollak, J. Chem. Phys. **116**, 5925 (2002)
30. S. Zhang and E. Pollak, J. Chem. Phys. **119**, 11058 (2003)
31. E. Pollak and S. Miret-Artes, J. Phys. A. **37**, 9669 (2004)
32. S. Zhang and E. Pollak, J. Chem. Phys. **121**, 3384 (2004)
33. D. H. Zhang and E. Pollak, Phys. Rev. Lett. **93**, 140401 (2004)
34. S. Zhang and E. Pollak, J. Chem. Theor. Comp. **1**, 345 (2005)
35. M. Saltzer and E. Pollak, J. Chem. Theor. Comp. **1**, 439 (2005)
36. C. Harabati, J.M. Rost and F. Grossmann, J. Chem. Phys. **120**, 26 (2004)
37. J. Shao and N. Makri, J. Chem. Phys. **113**, 3681 (2000)
38. S. Keshavamurthy and W.H. Miller, Chem. Phys. Lett. **218**, 189 (1994)
39. N.T. Maitra and E.J. Heller, Phys. Rev. Lett. **78**, 3035 (1997)
40. K.G. Kay, J. Chem. Phys. **107**, 2313 (1997)
41. F. Grossmann, Phys. Rev. Lett. **85**, 903 (2000)
42. J.C. Burant and V.S. Batista, J. Chem. Phys. **116**, 2748 (2002)
43. N.T. Maitra, J. Chem. Phys. **112**, 531 (2000)
44. M.R. Wall and D. Neuhauser, J. Chem. Phys. **102**, 8011 (1995)
45. V.A. Mandelshtam, Prof. Nucl. Magn. Res. Spect. **38**, 159 (2001)

II.2 Mixed Quantum-Classical Statistical Mechanics Methods

Quantum Statistical Dynamics with Trajectories

G. Ciccotti, D.F. Coker, and Raymond Kapral

Summary. In this chapter we review the key issues in the construction of a mixed quantum–classical statistical mechanics. Two approaches are outlined: First, the construction of a formally consistent quantum–classical scheme which entails modified dynamical equations, and a modified equilibrium distribution that is the stationary state of the quantum–classical dynamics. Second, an approach which starts from the exact quantum correlation functions and introduces approximations for both the dynamics and (possibly separately) the equilibrium distribution. The first scheme is internally consistent, but inconsistencies arise in the properties of the quantum–classical correlation functions including the fact that time translation invariance and the Kubo identity are only valid to $\mathcal{O}(\hbar)$. The second scheme does not address the consistency issues explicitly, but has to provide suitable criteria for approximations for both the dynamics and the equilibrium distribution. Two approaches to the practical implementation of this second scheme are presented (1) a mixed quantum–classical propagation, closely related to the first scheme and (2) a linearized path integral approach.

1 Introduction

Consider quantum systems which can be partitioned into two subsystems, one of which behaves almost classically while the other requires a full quantum description. It is reasonable to assume that the overall quantum behavior of the total system will be simplified due to the presence of this almost classical component. This fact has motivated the development of mixed quantum–classical methods, an idea which has attracted considerable interest for a number of years. In spite of the simplicity of this idea, the formulation of a mixed quantum–classical dynamics is not a simple problem, and many conceptual difficulties arise making this a very active area of research.

A great deal of effort in this area has been devoted to the development of approximate quantum–classical dynamical schemes while much less effort has been invested in exploring statistical mechanical issues. In the physical sciences one is interested in the calculation of time-dependent expectation values and (equilibrium) correlation functions. In order to compute these quantities one needs not only mechanics but also statistical mechanics.

One can imagine investigating statistical mechanical issues from two perspectives (1) construct a fictitious world in which one formulates statistical mechanics based on an underlying quantum–classical dynamics or (2) begin with the full quantum statistical mechanical description of the real world and make approximations to the quantum dynamics that lead to a representation in terms of trajectories. There are advantages and disadvantages to both schemes. As we shall see it is difficult to construct a consistent mixed quantum–classical formulation, however, if scheme (1) could be carried out one would have a consistent statistical mechanical formulation in the fictitious quantum–classical world. The essential issue then would be to determine the extent to which this fictitious world is a faithful model of the corresponding real one defined above. In scheme (2) the starting point is the correct statistical mechanical description of the quantum world but approximations are used to reduce the dynamics to trajectories. These approximations introduce inconsistencies in the formulation. In particular the full quantum equilibrium distribution is not stationary under the approximate quantum evolution.

In this chapter we will address some of these issues and illustrate the ideas by considering specific examples of methods which construct approximate trajectory descriptions of quantum evolution. In Sect. 2 we describe the formulation of a statistical mechanics based on quantum–classical equations of motion and point out some difficulties that arise in carrying out this program. In Sect. 3 we consider formulations based on approximations to the dynamics in the full quantum statistical mechanical expressions for time correlation functions. The ideas are illustrated by considering two examples that approximate the dynamics in terms of trajectories. Finally we conclude with some observations and perspectives for future research.

2 Quantum–Classical Worlds

We begin by formulating the quantum laws underlying dynamics and statistical mechanics. Let $\hat{B}$ be an observable of the system, then the Heisenberg equation describing its motion is

$$\frac{\mathrm{d}\hat{B}(t)}{\mathrm{d}t} = \frac{\mathrm{i}}{\hbar}[\hat{H}, \hat{B}(t)] \,. \tag{1}$$

The Liouville–von Neuman equation of motion for the density matrix $\hat{\rho}$ is given instead by

$$\frac{\partial \hat{\rho}(t)}{\partial t} = -\frac{\mathrm{i}}{\hbar}[\hat{H}, \hat{\rho}(t)] \,, \tag{2}$$

and the equation for the average value of an observable can be written in either of the following two equivalent forms obtained using cyclic permutation of the trace:

$$\overline{B(t)} = \mathrm{Tr}\hat{B}\hat{\rho}(t) = \mathrm{Tr}\hat{B}(t)\hat{\rho}(0) \,. \tag{3}$$

As described in the Sect. 1 we partition our system into two subsystems: the first subsystem contains n degrees of freedom representing particles with mass m and coordinate operators $\hat{q}$; the second subsystem comprises N degrees of freedom describing particles of mass M and coordinate operators $\hat{Q}$. The hamiltonian operator may be written as

$$\hat{H} = \frac{\hat{P}^2}{2M} + \frac{\hat{p}^2}{2m} + \hat{V}(\hat{q}, \hat{Q}) \equiv \frac{\hat{P}^2}{2M} + \hat{h}(\hat{Q}) , \tag{4}$$

where $\hat{p}$ and $\hat{P}$ are momentum operators, $\hat{V}(\hat{q}, \hat{Q})$ is the total potential energy, and $\hat{h}$ is the hamiltonian of the first subsystem in the field of the second subsystem with fixed coordinates. We employ a condensed notation such that $\hat{q} = (\hat{q}_1, \hat{q}_2, \ldots, \hat{q}_n)$ and $\hat{Q} = (\hat{Q}_1, \hat{Q}_2, \ldots, \hat{Q}_N)$, with an analogous notation for $\hat{p}$ and $\hat{P}$.

Let us now consider the partial Wigner transformation [1] of the density matrix with respect to the subset of Q coordinates [2],

$$\hat{\rho}_{\mathrm{W}}(R, P) = (2\pi\hbar)^{-N} \int \mathrm{d}Z \mathrm{e}^{\mathrm{i}P\cdot Z/\hbar} \langle R - \frac{Z}{2} | \hat{\rho} | R + \frac{Z}{2} \rangle . \tag{5}$$

In this representation the quantum Liouville equation is

$$\begin{aligned} \frac{\partial \hat{\rho}_{\mathrm{W}}(R, P, t)}{\partial t} &= -\frac{\mathrm{i}}{\hbar} \left((\hat{H}\hat{\rho})_{\mathrm{W}} - (\hat{\rho}\hat{H})_{\mathrm{W}} \right) \\ &= -\frac{\mathrm{i}}{\hbar} \left(\hat{H}_{\mathrm{W}} \mathrm{e}^{\hbar\Lambda/2\mathrm{i}} \hat{\rho}_{\mathrm{W}}(t) - \hat{\rho}_{\mathrm{W}}(t) \mathrm{e}^{\hbar\Lambda/2\mathrm{i}} \hat{H}_{\mathrm{W}} \right) , \end{aligned} \tag{6}$$

where the partially Wigner transformed hamiltonian is

$$\hat{H}_{\mathrm{W}}(R, P) = \frac{P^2}{2M} + \frac{\hat{p}^2}{2m} + \hat{V}_{\mathrm{W}}(\hat{q}, R) , \tag{7}$$

and Λ is the negative of the Poisson bracket operator, $\Lambda = \overleftarrow{\nabla}_P \cdot \overrightarrow{\nabla}_R - \overleftarrow{\nabla}_R \cdot \overrightarrow{\nabla}_P$, where the direction of an arrow indicates the direction in which the operator acts. To obtain this equation we used the definition of the partial Wigner transform of an observable

$$\hat{A}_{\mathrm{W}}(R, P) = \int \mathrm{d}Z \mathrm{e}^{-\mathrm{i}P\cdot Z/\hbar} \langle R + \frac{Z}{2} | \hat{A} | R - \frac{Z}{2} \rangle , \tag{8}$$

and the fact that the partial Wigner transform of a product of operators is [3]

$$(\hat{A}\hat{B})_{\mathrm{W}}(R, P) = \hat{A}_{\mathrm{W}}(R, P) \mathrm{e}^{\hbar\Lambda/2\mathrm{i}} \hat{B}_{\mathrm{W}}(R, P) . \tag{9}$$

Suppose now that the subsystem comprising the particles with masses M is taken to represent an environment or bath and assume that $M \gg m$. In this limit it can be shown that $\mathrm{e}^{\hbar\Lambda/2\mathrm{i}}$ can be replaced by $(1 + \hbar\Lambda/2\mathrm{i})$

and the full von Neuman equation reduces to the quantum–classical Liouville equation [2, 4–10]

$$\begin{aligned}\frac{\partial \hat{\rho}_{\mathrm{W}}(R,P,t)}{\partial t} &= -\frac{\mathrm{i}}{\hbar}[\hat{H}_{\mathrm{W}},\hat{\rho}_{\mathrm{W}}(t)] + \frac{1}{2}\left(\left\{\hat{H}_{\mathrm{W}},\hat{\rho}_{\mathrm{W}}(t)\right\} - \left\{\hat{\rho}_{\mathrm{W}}(t),\hat{H}_{\mathrm{W}}\right\}\right)\\ &\equiv -\mathrm{i}\hat{\mathcal{L}}\hat{\rho}_{\mathrm{W}}(t) \equiv -(\hat{H}_{\mathrm{W}},\hat{\rho}_{\mathrm{W}}(t))\ . \end{aligned} \tag{10}$$

Here [,] is a commutator, while { , } indicates a Poisson parenthesis on the R and P variables. The second line of this equation defines the quantum–classical Liouville operator $\hat{\mathcal{L}}$ and the quantum–classical bracket. The quantum–classical equation of motion for a dynamical variable $\hat{B}_{\mathrm{W}}$ can be written in a similar form as

$$\frac{\mathrm{d}\hat{B}_{\mathrm{W}}(t)}{\mathrm{d}t} = \mathrm{i}\hat{\mathcal{L}}\hat{B}_{\mathrm{W}}(t) \equiv (\hat{H}_{\mathrm{W}},\hat{B}_{\mathrm{W}}(t))\ . \tag{11}$$

Equation (10) is the quantum–classical Liouville equation describing the coupled evolution of our two subsystems. It can be shown that as a result of the coupling a purely newtonian description of bath dynamics is no longer possible [11]. However, it is possible to express the solution of this equation in terms of an ensemble of surface hopping trajectories [2].

Even though this evolution is well defined, quantum–classical dynamics does not possess a Lie algebraic structure like quantum or classical mechanics since the Jacobi identity is violated by the quantum–classical bracket [12, 13]

$$(\hat{A}_{\mathrm{W}},(\hat{B}_{\mathrm{W}},\hat{C}_{\mathrm{W}})) + (\hat{C}_{\mathrm{W}},(\hat{A}_{\mathrm{W}},\hat{B}_{\mathrm{W}})) + (\hat{B}_{\mathrm{W}},(\hat{C}_{\mathrm{W}},\hat{A}_{\mathrm{W}})) \neq 0\ . \tag{12}$$

This leads to pathologies in the general formulation of quantum–classical dynamics and statistical mechanics [12, 13].

A fundamental ingredient of statistical mechanics is the equilibrium density which is the stationary solution of the Liouville equation. The well known form of the quantum canonical equilibrium density is $\hat{\rho}_e = Z_Q^{-1}\exp(-\beta\hat{H})$ which, expressed in terms of the partial Wigner transform, can be written as

$$\hat{\rho}_{\mathrm{We}}(R,P) = (2\pi\hbar)^{-N}\int \mathrm{d}Z e^{\mathrm{i}P\cdot Z/\hbar}\langle R - \frac{Z}{2}|\hat{\rho}_{\mathrm{e}}|R + \frac{Z}{2}\rangle\ . \tag{13}$$

This quantity is not stationary under quantum–classical dynamics. So the equilibrium density of the quantum–classical approach has to be determined by solving the equation

$$\mathrm{i}\hat{\mathcal{L}}\hat{\rho}_{\mathrm{We}} = 0. \tag{14}$$

An explicit solution for this equation has not been found although a recursive one, obtained by developing the density matrix in a power series in $\hbar$ or the mass ratio, $\mu = m/M$, in the partial Wigner representation, can be written down. While it is difficult to find the full solution to any order in $\hbar$, it is not difficult to find the solution analytically to $\mathcal{O}(\hbar)$. To this order the result agrees with that of the partial Wigner transform of the exact canonical quantum equilibrium density. This expression for the equilibrium density matrix to $\mathcal{O}(\hbar)$ can be useful for testing the validity of approximate calculations of time correlation functions.

To complete the presentation of this approach we now define the quantum–classical forms of equilibrium time correlation functions and their associated transport coefficients. The issue we address is the construction of a nonequilibrium statistical mechanics in a world obeying quantum–classical dynamics. To carry out this program we begin by constructing a linear response theory for quantum–classical dynamics [12]. The formalism parallels that for quantum (or classical) systems. We suppose the quantum–classical system with hamiltonian $\hat{H}_{\mathrm{W}}$ is subjected to a time dependent external force that couples to the observable $\hat{A}_{\mathrm{W}}$, so that the total hamiltonian is

$$\hat{\mathbf{H}}_{\mathrm{W}}(t) = \hat{H}_{\mathrm{W}} - \hat{A}_{\mathrm{W}} F(t) \; . \tag{15}$$

The evolution equation for the density matrix takes the form

$$\frac{\partial \hat{\rho}_{\mathrm{W}}(t)}{\partial t} = -(\mathrm{i}\hat{\mathcal{L}} - \mathrm{i}\hat{\mathcal{L}}_A F(t))\hat{\rho}_{\mathrm{W}}(t) \; , \tag{16}$$

where $\mathrm{i}\hat{\mathcal{L}}_A$ has a form analogous to $\mathrm{i}\hat{\mathcal{L}}$ with $\hat{A}_{\mathrm{W}}$ replacing $\hat{H}_{\mathrm{W}}$, $\mathrm{i}\hat{\mathcal{L}}_A = (\hat{A}_{\mathrm{W}},)$. The formal solution of this equation is found by integrating from t_0 to t,

$$\hat{\rho}_{\mathrm{W}}(t) = \mathrm{e}^{-\mathrm{i}\hat{\mathcal{L}}(t-t_0)}\hat{\rho}_{\mathrm{W}}(t_0) + \int_{t_0}^{t} \mathrm{d}t' \; \mathrm{e}^{-\mathrm{i}\hat{\mathcal{L}}(t-t')}\mathrm{i}\hat{\mathcal{L}}_A \hat{\rho}_{\mathrm{W}}(t') F(t') \; . \tag{17}$$

We now choose, as usual, $\hat{\rho}_{\mathrm{W}}(t_0)$ to be the equilibrium density matrix, $\hat{\rho}_{\mathrm{We}}$. As discussed above $\hat{\rho}_{\mathrm{We}}$ is the invariant solution of the quantum–classical dynamics, $\mathrm{i}\hat{\mathcal{L}}\hat{\rho}_{\mathrm{We}} = 0$. In this case the first term on the right-hand side of (17) reduces to $\hat{\rho}_{\mathrm{We}}$ and is independent of t_0. We may assume that the system with hamiltonian $\hat{H}_{\mathrm{W}}$ is in thermal equilibrium at $t_0 = -\infty$, and with this boundary condition, to first order in the external force, (17) is

$$\hat{\rho}_{\mathrm{W}}(t) = \hat{\rho}_{\mathrm{We}} + \int_{-\infty}^{t} \mathrm{d}t' \; \mathrm{e}^{-\mathrm{i}\hat{\mathcal{L}}(t-t')}\mathrm{i}\hat{\mathcal{L}}_A \hat{\rho}_{\mathrm{We}} F(t') \; . \tag{18}$$

Then, computing $\overline{B_{\mathrm{W}}(t)} = \mathrm{Tr}' \int \mathrm{d}R \; \mathrm{d}P \, \hat{B}_{\mathrm{W}} \hat{\rho}_{\mathrm{W}}(t)$, where Tr' is the partial trace over the quantum degrees of freedom, to obtain the response function, we find

$$\begin{aligned} \overline{B_{\mathrm{W}}(t)} &= \int_{-\infty}^{t} \mathrm{d}t' \; \mathrm{Tr}' \int \mathrm{d}R \, \mathrm{d}P \; \hat{B}_{\mathrm{W}} \mathrm{e}^{-\mathrm{i}\hat{\mathcal{L}}(t-t')}\mathrm{i}\hat{\mathcal{L}}_A \hat{\rho}_{\mathrm{We}} F(t') \\ &= \int_{-\infty}^{t} \mathrm{d}t' \; \langle (\hat{B}_{\mathrm{W}}(t-t'), \hat{A}_{\mathrm{W}}) \rangle F(t') \equiv \int_{-\infty}^{t} \mathrm{d}t' \; \phi_{BA}^{\mathrm{QC}}(t-t') F(t') \; . \end{aligned} \tag{19}$$

Thus, the quantum–classical form of the response function is

$$\phi_{BA}^{\mathrm{QC}}(t) = \langle(\hat{B}_{\mathrm{W}}(t), \hat{A}_{\mathrm{W}})\rangle_{\mathrm{QC}} = \mathrm{Tr}' \int \mathrm{d}R\, \mathrm{d}P\, \hat{B}_{\mathrm{W}}(t)(\hat{A}_{\mathrm{W}}, \hat{\rho}_{\mathrm{We}}) , \tag{20}$$

where in writing the second equality in (20), we have used cyclic permutations under the trace and integrations by parts. The derivation of linear response theory in the quantum–classical world is completely analogous to that in quantum mechanics up to (20). However, the simplifications that are easily derived in the full quantum, or classical worlds are not available at the moment for the quantum–classical world. In particular, the calculation of the response function in the quantum–classical approach should be performed using (20) and cannot be started using well-known standard time correlation function expressions (notice that the Kubo transformed form can be shown to differ from the expression given in (20) by terms of $\mathcal{O}(\hbar^2)$ [12]).

At this point we have all the ingredients for the computation of transport properties and expectation values of dynamical variables in a quantum–classical world. The equilibrium time correlation function in (20) entails evolution of $\hat{B}_{\mathrm{W}}(t)$ under quantum–classical dynamics, evaluation of the quantum–classical bracket of $\hat{A}_{\mathrm{W}}$ and $\hat{\rho}_{\mathrm{We}}$, and an integration over the classical phase space coordinates and trace over the quantum states.

While this statistical mechanical formulation is complete, it is worth remarking that some aspects of the quantum mechanical calculation do not carry over to the quantum–classical world. These concern time translation invariance and alternate forms for the time correlation function expressions for transport coefficients. The first issue we examine is time translation invariance of the equilibrium time correlation functions [11]. A quantum mechanical response function can be written in the two equivalent forms

$$\phi_{BA}(t) = \langle \frac{\mathrm{i}}{\hbar}[\hat{B}(t), \hat{A}]\rangle = \langle \frac{\mathrm{i}}{\hbar}[\hat{B}(t+\tau), \hat{A}(\tau)]\rangle , \tag{21}$$

as is easily seen using stationarity of the canonical equilibrium density matrix and cyclic permutations under the trace. This property is not exactly satisfied by the correlation function in quantum–classical response function (20). To see this we may write (20) more explicitly as

$$\begin{aligned} \phi_{BA}^{\mathrm{QC}}(t) &= \langle(\hat{B}_{\mathrm{W}}(t), \hat{A}_{\mathrm{W}})\rangle_{\mathrm{QC}} \qquad (22) \\ &= \frac{\mathrm{i}}{\hbar}\Big(\langle \hat{B}_{\mathrm{W}}(t)\,(1+\hbar\Lambda/2\mathrm{i})\,\hat{A}_{\mathrm{W}}\rangle_{\mathrm{QC}} - \langle \hat{A}_{\mathrm{W}}\,(1+\hbar\Lambda/2\mathrm{i})\,\hat{B}_{\mathrm{W}}(t)\rangle_{\mathrm{QC}}\Big), \end{aligned}$$

Using cyclic permutations under the trace, integration by parts and the fact that $\hat{\rho}_{\mathrm{We}}$ is invariant under quantum–classical dynamics, one may show that

$$\langle \hat{B}_{\mathrm{W}}(t)\,(1+\hbar\Lambda/2\mathrm{i})\,\hat{A}_{\mathrm{W}}\rangle_{\mathrm{QC}} = \langle \mathrm{e}^{\mathrm{i}\mathcal{L}\tau}(\hat{B}_{\mathrm{W}}(t)\,(1+\hbar\Lambda/2\mathrm{i})\,\hat{A}_{\mathrm{W}})\rangle_{\mathrm{QC}} \,. \tag{23}$$

However, the evolution of a composite operator in quantum–classical dynamics cannot be written exactly in terms of the quantum–classical evolution of its constituent operators, but only to terms $\mathcal{O}(\hbar)$. To see this, consider the action of the quantum–classical Liouville operator on the composite operator $\hat{C}_{\rm W} = \hat{B}_{\rm W}(1 + \hbar\Lambda/2{\rm i})\hat{A}_{\rm W}$. A straightforward calculation shows that

$$
{\rm i}\hat{\mathcal{L}}\hat{C}_{\rm W} = ({\rm i}\hat{\mathcal{L}}\hat{B}_{\rm W})\left(1 + \frac{\hbar\Lambda}{2{\rm i}}\right)\hat{A}_{\rm W} + \hat{B}_{\rm W}\left(1 + \frac{\hbar\Lambda}{2{\rm i}}\right)({\rm i}\hat{\mathcal{L}}\hat{A}_{\rm W}) + \mathcal{O}(\hbar)\,. \quad (24)
$$

It follows that

$$
\begin{aligned}
\hat{C}_{\rm W}(\tau) = {\rm e}^{{\rm i}\hat{\mathcal{L}}\tau}\hat{C}_{\rm W} &= \left({\rm e}^{{\rm i}\hat{\mathcal{L}}\tau}\hat{B}_{\rm W}\right)\left(1 + \frac{\hbar\Lambda}{2{\rm i}}\right)\left({\rm e}^{{\rm i}\hat{\mathcal{L}}\tau}\hat{A}_{\rm W}\right) + \mathcal{O}(\hbar) \\
&= \hat{B}_{\rm W}(\tau)\left(1 + \frac{\hbar\Lambda}{2{\rm i}}\right)\hat{A}^{\dagger}_{\rm W}(\tau) + \mathcal{O}(\hbar)\,. \quad (25)
\end{aligned}
$$

Therefore, the quantum–classical correlation function satisfies standard time translation invariance only to $\mathcal{O}(\hbar)$,

$$
\phi^{\rm QC}_{BA}(t) = \langle(\hat{B}_{\rm W}(t), \hat{A}_{\rm W})\rangle_{\rm QC} = \langle(\hat{B}_{\rm W}(t+\tau), \hat{A}_{\rm W}(\tau))\rangle_{\rm QC} + \mathcal{O}(\hbar)\,, \quad (26)
$$

although its most strict form, (23), is surely satisfied.

Next, we consider alternate forms for correlations that are commonly used in computations. The quantum mechanical response functions can be written in an equivalent form using the Kubo identity. The quantum–classical version of the Kubo identity holds only to $\mathcal{O}(\hbar)$ [12],

$$
(\hat{A}_{\rm W}, \hat{\rho}_{\rm We}) = \int_0^{\beta} {\rm d}\lambda\; \hat{\rho}_{\rm We}(1 + \frac{\hbar\Lambda}{2{\rm i}})\dot{\hat{A}}_{\rm W}(-{\rm i}\hbar\lambda) + \mathcal{O}(\hbar)\,. \quad (27)
$$

Since the quantum–classical form of the Kubo identity is valid only to $\mathcal{O}(\hbar)$, the various autocorrelation function expressions for transport coefficients, to which we are accustomed, are equivalent only to $\mathcal{O}(\hbar)$. The results of comparisons of computations of both forms of the correlation functions can provide information about the reduction to the quantum–classical limit.

A discussion of the scheme used to simulate quantum–classical dynamics is postponed to next section and we simply remark here that statistical mechanical quantities may be computed within the quantum–classical framework. A limitation is a lack of knowledge of the exact form of the equilibrium density matrix, so we cannot compute exact quantum-classical time correlation functions. Note however that relaxation from given initial density matrices can be computed without any approximation other than that on the dynamics. So, for example, since quantum–classical dynamics is exact for the spin-boson model, it is possible to compute the exact relaxation, given an initial nonequilibrium distribution. Simulations of this model have confirmed the utility of surface-hopping algorithms for its study.

3 Approximations to the Real World

A very different route is to begin with any rigorous expression for the quantum mechanical response, e.g., in terms of quantum time correlation functions – since we know that they are all fundamentally equivalent – and make approximations to either or both the dynamics and equilibrium density. This approach implicitly avoids questions of consistency but they exist. These inconsistencies make these treatments invalid. However, if by a stroke of luck or design the inconsistencies are numerically small, these methods can often be very useful. With these approaches we can in principle independently approximate the equilibrium structure or the propagator so that we have more freedom than with the mixed quantum–classical statistical mechanical approach of the previous section.

The quantum time correlation function of two operators of the system is defined as

$$\begin{aligned} C_{AB}(t;\beta) &\equiv \langle \hat{A}\hat{B}(t)\rangle = \mathrm{Tr}\ \hat{A}\hat{B}(t)\hat{\rho}_{\mathrm{e}} \\ &= \frac{1}{Z_Q}\mathrm{Tr}\ \hat{A}\mathrm{e}^{\frac{\mathrm{i}}{\hbar}t\hat{H}}\hat{B}\mathrm{e}^{-\frac{\mathrm{i}}{\hbar}t\hat{H}}\mathrm{e}^{-\beta\hat{H}}\ . \end{aligned} \tag{28}$$

There are many different ways described in the literature to construct approximations to this correlation function [14–19]. Here we will illustrate how such approximations are implemented using two example approaches that we have explored:

(1) Mixed Wigner representation approach

The first approach [20,21] we consider uses the ingredients of the quantum–classical Liouville dynamics discussed in Sect. 2. We begin by introducing the coordinate representation of the operators so that the correlation function becomes

$$C_{AB}(t;\beta) = \mathrm{Tr}'\int \mathrm{d}Q_1\,\mathrm{d}Q_2\ \langle Q_1|\hat{B}(t)|Q_2\rangle\langle Q_2|\hat{\rho}_{\mathrm{e}}\hat{A}|Q_1\rangle\ . \tag{29}$$

Making use of the change of variables, $Q_1 = R - Z/2$ and $Q_2 = R + Z/2$, this equation may be written in the equivalent form

$$\begin{aligned} C_{AB}(t;\beta) = \mathrm{Tr}'\int \mathrm{d}R\,\mathrm{d}Z\ &\langle R-\frac{Z}{2}|\hat{B}(t)|R+\frac{Z}{2}\rangle \\ &\times\langle R+\frac{Z}{2}|\hat{\rho}_{\mathrm{e}}\hat{A}|R-\frac{Z}{2}\rangle\ . \end{aligned} \tag{30}$$

The next step in the calculation is to replace the coordinate space matrix elements of the operators with their representation in terms of Wigner transformed quantities. The partial Wigner transform of an operator, $\hat{O}$, is defined in (8) while the inverse transform is

$$\langle R+\frac{Z}{2}|\hat{O}|R-\frac{Z}{2}\rangle = \frac{1}{(2\pi\hbar)^N}\int \mathrm{d}P\mathrm{e}^{\frac{\mathrm{i}}{\hbar}P\cdot Z}\hat{O}_{\mathrm{W}}(R,P)\ . \tag{31}$$

For simplicity we write $X \equiv (R, P)$. It is convenient to consider a representation of such operators in basis of eigenfunctions, here we consider an adiabatic basis to make connection with surface-hopping dynamics. The partial Wigner transformed hamiltonian can be written as $\hat{H}_{\rm W} = P^2/2M + \hat{h}_{\rm W}(R)$. The last equality defines the hamiltonian $\hat{h}_{\rm W}(R)$ for the light mass subsystem in the presence of fixed particles of the heavy mass subsystem. The adiabatic basis is determined from the solutions of the eigenvalue problem, $\hat{h}_{\rm W}(R)|\alpha; R\rangle = E_\alpha(R)|\alpha; R\rangle$. The adiabatic representation of $\hat{O}_{\rm W}(X)$ is

$$\hat{O}_{\rm W}(X) = \sum_{\alpha\alpha'} |\alpha; R\rangle O_{\rm W}^{\alpha\alpha'}(X)\langle\alpha'; R| \, , \tag{32}$$

where $O_{\rm W}^{\alpha\alpha'}(X) = \langle\alpha; R|\hat{O}_{\rm W}(X)|\alpha'; R\rangle$.

By inserting (32) into (31) we can express the coordinate representation of the operator $\hat{O}$ as

$$\langle R - \frac{Z}{2}|\hat{O}|R + \frac{Z}{2}\rangle = \frac{1}{(2\pi\hbar)^N} \sum_{\alpha\alpha'} \int {\rm d}P \, {\rm e}^{\frac{\rm i}{\hbar}P\cdot Z} |\alpha; R\rangle (O)_{\rm W}^{\alpha\alpha'}(X)\langle\alpha'; R| \, . \tag{33}$$

Using this result in (30), we obtain

$$C_{AB}(t;\beta) = \sum_{\alpha,\alpha'} \int {\rm d}X \, (\hat{B}(t))_{\rm W}^{\alpha\alpha'}(X)(\hat{\rho}_{\rm e}\hat{A})_{\rm W}^{\alpha'\alpha}(X) \, . \tag{34}$$

This equation is still formaly exact but now we approximate the dynamics using the quantum–classical evolution given in (11).

In the adiabatic basis the quantum–classical Liouville operator defined in (10) takes the form

$${\rm i}\mathcal{L}_{\alpha'\alpha,\beta'\beta}(X) = \Big({\rm i}\omega_{\alpha'\alpha}(R) + {\rm i}L_{\alpha'\alpha}(X)\Big)\delta_{\alpha'\beta'}\delta_{\alpha\beta} - J_{\alpha'\alpha,\beta'\beta}(X) \, , \tag{35}$$

where $\omega_{\alpha\alpha'}(R) = (E_\alpha(R) - E_{\alpha'}(R))/\hbar$ and

$${\rm i}L_{\alpha'\alpha}(X) = \frac{P}{M}\cdot\frac{\partial}{\partial R} + \frac{1}{2}\left(F_{\rm W}^{\alpha'}(R) + F_{\rm W}^{\alpha}(R)\right)\cdot\frac{\partial}{\partial P} \, , \tag{36}$$

is the classical Liouville operator involving the mean of the Hellmann–Feynman forces where $F_{\rm W}^{\alpha} = -\langle\alpha; R|\frac{\partial\hat{V}_{\rm W}(\hat{q},R)}{\partial R}|\alpha; R\rangle = -\langle\alpha; R|\frac{\partial\hat{H}_{\rm W}(R)}{\partial R}|\alpha; R\rangle$. Quantum transitions and bath momentum changes are described by

$$\begin{aligned} J_{\alpha'\alpha,\beta'\beta}(X) = & -\frac{P}{M}\cdot d_{\alpha'\beta'}\left(1 + \frac{1}{2}S_{\alpha'\beta'}(R)\cdot\frac{\partial}{\partial P}\right)\delta_{\alpha\beta} \\ & -\frac{P}{M}\cdot d_{\alpha\beta}\left(1 + \frac{1}{2}S_{\alpha\beta}(R)\cdot\frac{\partial}{\partial P}\right)\delta_{\alpha'\beta'} \, , \end{aligned} \tag{37}$$

where $S_{\alpha\beta} = (E_\alpha - E_\beta)d_{\alpha\beta}(\frac{P}{M}\cdot d_{\alpha\beta})^{-1}$ and $d_{\alpha\beta} = \langle\alpha; R|\nabla_R|\beta; R\rangle$ is the nonadiabatic coupling matrix element.

In this approximation the correlation function is then given by

$$C_{AB}(t;\beta) = \sum_{\alpha,\alpha'} \int \mathrm{d}X\; B_{\mathrm{W}}^{\alpha\alpha'}(X,t)(\hat{\rho}_{\mathrm{e}}\hat{A})_{\mathrm{W}}^{\alpha'\alpha}(X), \tag{38}$$

where

$$B_{\mathrm{W}}^{\alpha\alpha'}(X,t) = (\mathrm{e}^{\mathrm{i}\mathcal{L}t}\hat{B}_{\mathrm{W}}(X))^{\alpha\alpha'} . \tag{39}$$

Various ways of simulating nonadiabatic transitions in quantum–classical dynamics have been devised, as well as schemes for computing the evolution operator [2, 9, 20–26]. These schemes typically employ an ensemble of surface hopping trajectories with classical trajectory segments [21, 25, 26]. Approximations must also be introduced to evaluate the equilibrium density matrix. In making these approximations the consistency problem is not necessarily the most serious. In the desire to achieve consistency one could use the quantum–classical equilibrium density matrix, however, there would remain two problems. (1) This quantity is not known in closed form therefore expressions based on approximations for it would leave the consistency unattained. (2) As mentioned, (38) is not an admissible form for the quantum–classical response as given in (20) (although it can be related to $\mathcal{O}(\hbar)$) and therefore would not result in a consistent, interesting, quantum–classical object. Chemical rate coefficients written in terms of Kubo transformed correlation functions have been computed using this strategy [21, 27]. In the case of spin boson type models for reaction rates the fact that one knows the exact Wigner transformed equilibrium bath density can be exploited along with a quadratic approximation near the barrier top to obtain an estimate of the reaction rate that includes quantum equilibrium effects [21]. In more complex systems, like models for proton transfer in the condensed phase, one can exploit the high temperature limit to obtain suitable approximation to the equilibrium density [28]. See Chapter 13 in this volume for a discussion of this approach applied to reaction rate problems.

Although algorithms have been developed that have allowed one to simulate chemical reaction rates and short time relaxation processes, further algorithmic development is needed to simulate quantum–classical dynamics for long times.

(2) Linearized path integral approach

An alternative to the calculation of quantum time correlation functions is offered by the so-called linearized path integral approach [17, 18, 29–33]. In developing this approach it is simplest to work with a basis defined as the tensor product, $|Q\alpha\rangle$, of bath position states $|Q\rangle$ and a quantum subsystem basis $|\alpha\rangle$ which, for convenience, we choose to be independent of bath configuration (see Chap. 14 in this volume for a discussion of how this formalism changes when the adiabatic basis is used). The quantum time correlation function in this representation becomes

$$\langle \hat{A}\hat{B}(t)\rangle = \sum_{\alpha\beta,\alpha'\beta'} \int \mathrm{d}Q_0\, \mathrm{d}Q_N\, \mathrm{d}Q'_0\, \mathrm{d}Q'_N \langle Q_0\alpha|\hat{\rho}_{\mathrm{e}}\hat{A}|Q'_0\alpha'\rangle \tag{40}$$

$$\times \langle Q'_0\alpha'|\mathrm{e}^{\frac{\mathrm{i}}{\hbar}\hat{H}t}|Q'_N\beta'\rangle\langle Q'_N\beta'|\hat{B}|Q_N\beta\rangle\langle Q_N\beta|\mathrm{e}^{-\frac{\mathrm{i}}{\hbar}\hat{H}t}|Q_0\alpha\rangle .$$

Here the hamiltonian is the usual one defined in (4), while the quantum subsystem hamiltonian has matrix elements $h_{\alpha\beta}(\hat{Q}) = \langle\alpha|\hat{h}(\hat{Q})|\beta\rangle$.

A convenient representation to account for the effects of the quantum subsystem transitions on the bath degrees of freedom is offered by the mapping hamiltonian formalism [34–39]. The core of this idea is to replace the quantum subsystem with a system of fictitious harmonic oscillators which can take only a restricted set of excitations representing the states of the basis. Therefore the states of the real system are mapped onto states of the fictitious harmonic oscillator system according to

$$|\alpha\rangle \to |m_\alpha\rangle = |0_1, \ldots, 1_\alpha, ..0_n\rangle . \tag{41}$$

This prescription maps the Hilbert space spanned by the original n quantum subsystem states into one coinciding with a subspace of n-oscillators of unit mass with at most one quantum of excitation in a single specific oscillator. Under these conditions the hamiltonian of the fictitious system is obtained by requiring that its matrix elements are equal to those of the corresponding physical states $\langle m_\alpha|\hat{h}_m(\hat{Q})|m_\beta\rangle = \langle\alpha|\hat{h}(\hat{Q})|\beta\rangle$. So that

$$\hat{h}_m(\hat{Q}) = \frac{1}{2}\sum_{\lambda} h_{\lambda,\lambda}(\hat{Q})(\hat{q}_\lambda^2 + \hat{p}_\lambda^2 - \hbar) + \frac{1}{2}\sum_{\lambda,\lambda'} h_{\lambda,\lambda'}(\hat{Q})(\hat{q}_{\lambda'}\hat{q}_\lambda + \hat{p}_{\lambda'}\hat{p}_\lambda) \tag{42}$$

where $\hat{q}_\lambda$ and $\hat{p}_\lambda$ are the λth mapping oscillator's position and momentum operators reconstructed from the creation and annhilation operators of the occupation number representation. Then the total hamiltonian of the system becomes $\hat{H}_m = \hat{P}^2/2M + \hat{h}_m(\hat{Q})$ and the propagator matrix elements of the real system are given by the mapping propagator matrix elements

$$\langle Q_N\beta|\mathrm{e}^{-\frac{\mathrm{i}}{\hbar}\hat{H}t}|Q_0\alpha\rangle = \langle Q_N m_\beta|\mathrm{e}^{-\frac{\mathrm{i}}{\hbar}\hat{H}_m t}|Q_0 m_\alpha\rangle . \tag{43}$$

To proceed, we now apply standard discrete path integral techniques to express the right-hand side of (43) as a functional integral over bath subsystem paths of an integrand containing the quantum subsystem transition amplitude evaluated along each path. This result parallels that of Pechukas [40] thus

$$\langle Q_N m_\beta|\mathrm{e}^{-\frac{\mathrm{i}}{\hbar}\hat{H}_m t}|Q_0 m_\alpha\rangle = \int \prod_{k=1}^{N-1} \mathrm{d}Q_k \frac{\mathrm{d}P_k}{2\pi\hbar}\frac{\mathrm{d}P_N}{2\pi\hbar}\mathrm{e}^{\frac{\mathrm{i}}{\hbar}S} \tag{44}$$

$$\times \langle m_\beta|\mathrm{e}^{-\frac{\mathrm{i}}{\hbar}\epsilon\hat{h}_m(Q_N)} \ldots \mathrm{e}^{-\frac{\mathrm{i}}{\hbar}\epsilon\hat{h}_m(Q_1)}|m_\alpha\rangle ,$$

where

$$S = \epsilon \sum_{k=1}^{N} \left[P_k \frac{(Q_k - Q_{k-1})}{\epsilon} - \frac{P_k^2}{2M} \right] \tag{45}$$

and $\epsilon = t/N$ is the time slice.

The transition amplitude $\langle m_\beta | \mathrm{e}^{-\frac{\mathrm{i}}{\hbar}\epsilon \hat{h}_m(Q_N)} \cdots \mathrm{e}^{-\frac{\mathrm{i}}{\hbar}\epsilon \hat{h}_m(Q_1)} | m_\alpha \rangle$ contains a discrete time ordered propagator that evolves the initial mapping subsystem state according to the time dependent mapping hamiltonian where the time-dependence arises because of the changing configuration of the bath along the path $(Q_1, \ldots, Q_N)$. For any given specification of the bath subsystem path, the quadratic nature of the mapping hamiltonian in the mapping subsystem variables in (42) allows us to obtain an exact expression for the mapping transition amplitude. A particularly convenient expression for the transition amplitude can be obtained using semiclassical methods which are exact for quadratic hamiltonians with time-dependent coefficients (see [41] for details of the manipulations). The result is

$$\begin{aligned} &\langle m_\beta | \mathrm{e}^{-\frac{\mathrm{i}}{\hbar}\epsilon \hat{h}_m(Q_N)} \cdots \mathrm{e}^{-\frac{\mathrm{i}}{\hbar}\epsilon \hat{h}_m(Q_1)} | m_\alpha \rangle \\ &\quad \sim \int \mathrm{d}q_0 \, \mathrm{d}p_0 \, (q_{\beta t} + \mathrm{i} p_{\beta t})(q_{\alpha 0} - \mathrm{i} p_{\alpha 0}) \times \exp\left\{ -\frac{1}{2} \sum\nolimits_\lambda (q_{\lambda 0}^2 + p_{\lambda 0}^2) \right\} \end{aligned} \tag{46}$$

Here $(q_t, p_t) = (q_{1t}, \ldots, q_{nt}, p_{1t}, \ldots, p_{nt})$ is the mapping phase space point that evolves *classically* from the initial sampled point (q_0, p_0) to time t according to the given realization of the discrete time-dependent hamiltonian $(\hat{h}_m(Q_N), \ldots, \hat{h}_m(Q_1))$.

This expression for the transition amplitude can now be conveniently re-written by introducing a polar representation of the complex polynomials appearing in the above result, thus

$$\begin{aligned} \langle m_\beta | \mathrm{e}^{-\frac{\mathrm{i}}{\hbar}\epsilon \hat{h}_m(Q_N)} \cdots \mathrm{e}^{-\frac{\mathrm{i}}{\hbar}\epsilon \hat{h}_m(Q_1)} | m_\alpha \rangle &= \int \mathrm{d}q_0 \, \mathrm{d}p_0 \, r_{t,\beta}(\{Q_k\}) \mathrm{e}^{-\mathrm{i}\Theta_{t\beta}(\{Q_k\})} \\ &\quad \times r_{0\alpha} \mathrm{e}^{\mathrm{i}\Theta_{0,\alpha}} G_0 . \end{aligned} \tag{47}$$

Here $G_0 = \exp\left\{ -\frac{1}{2} \sum_\lambda (q_{0,\lambda}^2 + p_{0,\lambda}^2) \right\}$, $r_{t,\beta}(\{Q_k\}) = \sqrt{q_{t,\beta}^2(\{Q_k\}) + p_{t,\beta}^2(\{Q_k\})}$, and

$$\begin{aligned} \Theta_{t,\beta}(\{Q_k\}) &= \tan^{-1}\left(\frac{p_{0,\beta}}{q_{0,\beta}} \right) + \int_0^t \mathrm{d}\tau \, h_{\beta,\beta}(Q_\tau) \\ &\quad + \int_0^t \mathrm{d}\tau \sum_{\lambda \neq \beta} \left[h_{\beta,\lambda}(Q_\tau) \frac{(p_{\tau\beta} p_{\tau\lambda} + q_{\tau\beta} q_{\tau\lambda})}{(p_{\tau\beta}^2 + q_{\tau\beta}^2)} \right] \\ &= \tan^{-1}\left(\frac{p_{0,\beta}}{q_{0,\beta}} \right) + \int_0^t \theta_\beta(Q_\tau) \mathrm{d}\tau . \end{aligned} \tag{48}$$

Equation (48) defines the function $\theta_\beta(Q)$.

Substituting (47) and its analogue for the backward propagator (primed quantities) into the expression for the correlation function we finally obtain

$$\langle \hat{A}\hat{B}(t)\rangle = \sum_{\alpha\beta,\alpha'\beta'} \int \mathrm{d}Q_0\,\mathrm{d}Q_0' \int \prod_{k=1}^{N} \mathrm{d}Q_k \frac{\mathrm{d}P_k}{2\pi\hbar} \int \prod_{k=1}^{N} \mathrm{d}Q_k' \frac{\mathrm{d}P_k'}{2\pi\hbar} \int \mathrm{d}q_0\,\mathrm{d}p_0\,\mathrm{d}q_0'\,\mathrm{d}p_0'$$
$$\times \mathrm{e}^{\frac{\mathrm{i}}{\hbar}(S-S')} r_{0\alpha'}' \mathrm{e}^{-\mathrm{i}\Theta_{0,\alpha'}'} G_0' r_{0\alpha} \mathrm{e}^{\mathrm{i}\Theta_{0,\alpha}} G_0 \langle Q_0\alpha|\hat{\rho}_\mathrm{e}\hat{A}|Q_0'\alpha'\rangle$$
$$\times \langle Q_N'\beta'|B|Q_N\beta\rangle r_{t,\beta}(\{Q_k\}) \mathrm{e}^{-\mathrm{i}\Theta_{t\beta}(\{Q_k\})} r_{t,\beta'}'(\{Q_k'\}) \mathrm{e}^{\mathrm{i}\Theta_{t\beta'}'(\{Q_k'\})} . \tag{49}$$

Here we employ a shorthand notation labeling the mapping oscillator states with their state index, e.g., $m_\alpha \equiv \alpha$.

All manipulations performed so far are exact, and the nuclear evolution is still described at the full quantum level. To proceed to a computable expression [17,41], we now change bath subsystem variables to mean, $\bar{R}_k = (Q_k + Q_k')/2$, and difference, $Z_k = Q_k - Q_k'$, coordinates (with similar transformation for the bath momenta, $\bar{P}_k = (P_k + P_k')/2$ and $Y_k = P_k - P_k'$ say) and Taylor series expand the phase in (49). Truncating this expansion to linear order in the difference variables we obtain the following approximate expression for the correlation function

$$\langle \hat{A}\hat{B}(t)\rangle = \sum_{\alpha\beta,\alpha'\beta'} \int \mathrm{d}q_0\,\mathrm{d}p_0\,\mathrm{d}q_0'\,\mathrm{d}p_0'\, r_{0\alpha'}' \mathrm{e}^{-\mathrm{i}\Theta_{0,\alpha'}'} G_0' r_{0\alpha} \mathrm{e}^{\mathrm{i}\Theta_{0,\alpha}} G_0$$
$$\times \int \mathrm{d}\bar{R}_0\,\mathrm{d}Z_0 \int \prod_{k=1}^{N} \mathrm{d}\bar{R}_k \frac{\mathrm{d}\bar{P}_k}{2\pi\hbar} \int \prod_{k=1}^{N} \mathrm{d}Z_k \frac{\mathrm{d}Y_k}{2\pi\hbar}$$
$$\times \langle \bar{R}_0 + \frac{Z_0}{2}\alpha|\hat{\rho}_\mathrm{e}\hat{A}|\bar{R}_0 - \frac{Z_0}{2}\alpha'\rangle \mathrm{e}^{-\mathrm{i}\bar{P}_1 Z_0}$$
$$\times \langle \bar{R}_N - \frac{Z_N}{2}\beta'|B|\bar{R}_N + \frac{Z_N}{2}\beta\rangle \mathrm{e}^{\mathrm{i}\bar{P}_N Z_N}$$
$$\times \mathrm{e}^{-\mathrm{i}\epsilon\{[\nabla\theta_\beta(\bar{R}_N)+\nabla\theta_{\beta'}'(\bar{R}_N)]/2\}Z_N}$$
$$\times r_{t,\beta}(\{\bar{R}_k\}) r_{t,\beta'}'(\{\bar{R}_k\}) \mathrm{e}^{-\mathrm{i}\epsilon\sum_{k=1}^{N}[\theta_\beta(\bar{R}_k)-\theta_{\beta'}'(\bar{R}_k)]}$$
$$\times \mathrm{e}^{-\mathrm{i}\epsilon\sum_{k=1}^{N-1}\{(\bar{P}_{k+1}-\bar{P}_k)/\epsilon+[\nabla\theta_\beta(\bar{R}_k)+\nabla\theta_{\beta'}'(\bar{R}_k)]/2\}Z_k}$$
$$\times \mathrm{e}^{-\mathrm{i}\epsilon\sum_{k=1}^{N}\{\bar{P}_k/M-(\bar{R}_k-\bar{R}_{k-1})/\epsilon\}Y_k} \tag{50}$$

The integrals over the end-point difference coordinates Z_0 and Z_N in this linearized approximate form can be performed defining the Wigner transformed operators

$$(\hat{\rho}_\mathrm{e}\hat{A})_\mathrm{W}^{\alpha,\alpha'}(\bar{R}_0,\bar{P}_1) = \int \mathrm{d}Z_0 \langle \bar{R}_0 + \frac{Z_0}{2}\alpha|\hat{\rho}_\mathrm{e}\hat{A}|\bar{R}_0 - \frac{Z_0}{2}\alpha'\rangle \mathrm{e}^{-\mathrm{i}\bar{P}_1 Z_0} \tag{51}$$

and in the limit of $\epsilon \to 0$

$$(\hat{B})_{\mathrm{W}}^{\beta',\beta}(\bar{R}_N, \bar{P}_N) = \int \mathrm{d}Z_N \langle \bar{R}_N + \frac{Z_N}{2}\beta'|\hat{B}|\bar{R}_N - \frac{Z_N}{2}\beta\rangle \mathrm{e}^{-\mathrm{i}\bar{P}_N Z_N}. \quad (52)$$

All integrals over the difference coordinates, Z_k, and difference momenta, Y_k for $0 < k < N$ can also be performed as they are integral representations of delta functions, so the linearized approximation for the time correlation function can finally be expressed as

$$\begin{aligned}\langle \hat{A}\hat{B}(t)\rangle = &\sum_{\alpha\beta,\alpha'\beta'} \int \mathrm{d}\bar{R}_0\, \mathrm{d}q_0\, \mathrm{d}p_0\, \mathrm{d}q_0'\, \mathrm{d}p_0'\, r'_{0\alpha'} \mathrm{e}^{-\mathrm{i}\Theta'_{0,\alpha'}} G_0' r_{0\alpha} \mathrm{e}^{\mathrm{i}\Theta_{0,\alpha}} G_0 \\ &\times \int \prod_{k=1}^{N} \mathrm{d}\bar{R}_k \frac{\mathrm{d}\bar{P}_k}{2\pi\hbar} (\hat{\rho}_{\mathrm{e}}\hat{A})_{\mathrm{W}}^{\alpha,\alpha'}(\bar{R}_0, \bar{P}_1)(\hat{B})_{\mathrm{W}}^{\beta',\beta}(\bar{R}_N, \bar{P}_N) \\ &\times r_{t,\beta}(\{\bar{R}_k\}) r'_{t,\beta'}(\{\bar{R}_k\}) \mathrm{e}^{-\mathrm{i}\epsilon \sum_{k=1}^{N}(\theta_\beta(\bar{R}_k) - \theta'_{\beta'}(\bar{R}_k))} \\ &\times \prod_{k=1}^{N-1} \delta\left(\frac{\bar{P}_{k+1} - \bar{P}_k}{\epsilon} - F_k^{\beta,\beta'}\right) \prod_{k=1}^{N} \delta\left(\frac{\bar{P}_k}{M} - \frac{\bar{R}_k - \bar{R}_{k-1}}{\epsilon}\right), \end{aligned} \quad (53)$$

where

$$\begin{aligned} F_k^{\beta,\beta'} &= -\frac{1}{2}\{\nabla_{\bar{R}_k} h_{\beta,\beta}(\bar{R}_k) + \nabla_{\bar{R}_k} h_{\beta',\beta'}(\bar{R}_k)\} \\ &- \frac{1}{2}\sum_{\lambda \neq \beta} \nabla_{\bar{R}_k} h_{\beta,\lambda}(\bar{R}_k) \left\{ \frac{(p_{\beta k}p_{\lambda k} + q_{\beta k}q_{\lambda k})}{(p_{\beta k}^2 + q_{\beta k}^2)} \right\} \\ &- \frac{1}{2}\sum_{\lambda \neq \beta'} \nabla_{\bar{R}_k} h_{\beta',\lambda}(\bar{R}_k) \left\{ \frac{(p'_{\beta' k}p'_{\lambda k} + q'_{\beta' k}q'_{\lambda k})}{p'^2_{\beta' k} + q'^2_{\beta' k})} \right\}. \end{aligned} \quad (54)$$

The product of δ-functions in (53) amounts to a time-stepping prescription in which the mean path evolves classically. As the motion of the mapping variables is already classical, the calculation of the time correlation function has been reduced to a two step procedure (1) sampling a set of initial conditions for the bath variables from a probability distribution related to the partial Wigner transform of the thermal density times the operator $\hat{A}$, i.e., the factor $(\hat{\rho}_{\mathrm{e}}\hat{A})_{\mathrm{W}}^{\alpha,\alpha'}(\bar{R}_0, \bar{P}_1)$ in (53), and a Gaussian distribution, $G_0'G_0$ as defined under (47), for the mapping subsystem variables; (2) integration of a set of coupled classical equations of motion for the mapping and bath variables. The first of these tasks can be accomplished only approximately using recently developed local harmonic approximate methods for sampling the Wigner density for complex systems [17,18]. The second task of evolving the classical dynamics is straightforward. However, we note that depending on the specific term of the correlation function which is being evaluated, the forces in (54) are determined by different time-dependent linear combinations of pairs of diagonal, and off-diagonal elements of the quantum subsystem hamiltonian. The diagonal terms are identified by the *final* states in the propagators appearing in the original expression for the correlation function, while the off-diagonal terms are responsible for the feedback between bath motion and changes in the quantum subsystem state occupations. The latter are affected by the bath propagation through the parametric dependence of the classical counterpart of (42), but the coupling mechanism is not deducible from a single hamiltonian. In spite of this unusual characteristic, all propagations required in this approximate evaluation of the correlation function are classical and local in time and maintain the usual properties of classical, or quantum, mechanics e.g., time reversibility).

To highlight the basic similarities between the two approximate approaches we have outlined here for computing time correlation functions in mixed quantum–classical systems, (53) can be put into the form of (38) by making the following identification

$$\begin{aligned} B_{\mathrm{W}}^{\alpha\alpha'}(X,t) &= \sum_{\beta\beta'} \int \mathrm{d}q_0\, \mathrm{d}p_0\, \mathrm{d}q_0'\, \mathrm{d}p_0'\, r_{0\alpha'}' \mathrm{e}^{-\mathrm{i}\Theta_{0,\alpha'}'} G_0' r_{0\alpha} \mathrm{e}^{\mathrm{i}\Theta_{0,\alpha}} G_0 \\ &\times \int \mathrm{d}\bar{R}_1 \prod_{k=2}^{N} \mathrm{d}\bar{R}_k \frac{\mathrm{d}\bar{P}_k}{2\pi\hbar} (\hat{B})_{\mathrm{W}}^{\beta',\beta}(X_t(X)) \\ &\times r_{t,\beta}(\{\bar{R}_k\}) r_{t,\beta'}'(\{\bar{R}_k\}) \mathrm{e}^{-\mathrm{i}\epsilon \sum_{k=1}^{N} (\theta_\beta(\bar{R}_k) - \theta_{\beta'}'(\bar{R}_k))} \\ &\times \prod_{k=1}^{N-1} \delta\left(\frac{\bar{P}_{k+1} - \bar{P}_k}{\epsilon} - F_k^{\beta,\beta'} \right) \prod_{k=1}^{N} \delta\left(\frac{\bar{P}_k}{M} - \frac{\bar{R}_k - \bar{R}_{k-1}}{\epsilon} \right), \end{aligned} \tag{55}$$

where now the initial phase space point is $X = (\bar{R}_0, \bar{P}_1)$, and the terminal point, $X_t = (\bar{R}_N, \bar{P}_N)$, is an implicit function of X determined by sequentially evaluating the δ-function integrals and classically time stepping the propagation of the bath. Comparing this result with the expressions at the end of the previous section it is clear that the basic features of these two approaches are similar but that the underlying dynamics is very different. These differences stem in part from the different representations employed but also result from different approximations made in the derivations. It is beyond the scope of this chapter to present a detailed comparison of these two approaches. As mentioned earlier and outlined below, they both yield good results for model condensed phase systems so exploring the connections between these different ideas may prove fruitful in developing algorithms for implementing mixed quantum–classical methods for computing time correlation functions.

The central approximation of the linearized path integral approach to nonadiabatic dynamics outlined here is that the Taylor expansion of the phase of the integrand in the path integral expression for the correlation function can be truncated at low order. One could imagine computing higher order corrections with significant additional computational effort beyond the linearized approach. This lowest order approximation, however, has proved particularly reliable in various model test calculations [41]. With the spin-boson model, for example, calculations of the time dependent expectation value of the spin population difference, starting from a nonequilibrium initial condition in which the coupling between an excited spin and an independent, thermal equilibrium harmonic oscillator bath is turned on at $t = 0$, gave results that were in excellent agreement with exact calculations [42] over a wide range of friction and temperature. Small deviations between exact results and those of the linearized approximate approach are observed at low temperature and high friction. Under these conditions the assumption that the only important contributions to the correlation function (or time-dependent expectation value) come from pairs of forward and backward paths that remain "close" to one another (keeping only terms to linear order in the path difference) is violated since at low temperatures the initial bath density has larger off-diagonal elements so forward and backward bath paths which differ significantly can begin to make contributions and these are ignored in the linearized scheme. The linearized path integral approach, however, is found to converge very quickly with trajectory ensemble size for these nonadiabatic problems, requiring fewer than 1,000 trajectories to converge these spin-boson calculations. This feature makes these methods promising for realistic model condensed applications in future studies.

In general the appeal of these methods is that they require Monte Carlo sampling and trajectories, features that scale favorably with the dimensions of the system, especially when compared with basis set methods. Unfortunately a quantitative assessment of this favorable conjecture is far from evident.

4 Conclusion

We have seen that it is possible to develop a consistent approach to equilibrium and nonequilibrium quantum–classical statistical mechanics. However, due to the different algebraic structures of the exact quantum bracket in the Wigner representation and its quantum–classical counterpart described here, the formulation contains one unpleasant feature: the nonassociative property of the product. This feature leads to a violation of the Jacobi identity so that in contrast to both quantum and classical mechanics, the quantum–classical approach does not have a Lie algebraic structure. This in turn leads to the fact that the Onsager reciprocal relation and the Kubo identity are valid only to order $\mathcal{O}(\hbar)$. It is conceivable that this approach can be improved by developing a quantum–classical bracket that satisfies the Lie algebraic structure. This is a challenge worthy of future research. Even if such a program could be carried out one would be left with the task of testing the fidelity of this quantum–classical world as a model of the real quantum world in the limit discussed in this chapter.

In the other approach considered here we saw that one could start with the full quantum statistical mechanical structure of the time correlation function and develop approximations to both the quantum evolution and equilibrium density. This type of approach readily leads to promising results as demonstrated in applications to models. The major drawback of such an approach is that the consistency between the quantum equilibrium structure and the approximate dynamics is lost, although one has gained the possibility to consider independently approximations to the evolution and the equilibrium structure. Examples of the utility of being able to make independent approximations to the evolution and equilibrium structure in which the consistency problem does not seem to matter much include applications reported in various references [21, 27, 28, 33, 41]. These local successes, however, do not justify a general statement and we do not yet know what physical conditions need to be satisfied to guarantee that the inconsistency problem will not be crucial. In the context of the approaches described in Sect. 3 one can also attempt to consistently approximate the equilibrium structure and dynamics although it is unclear at the present time how such consistency could be achieved.

In contrasting the two approaches we should not lose sight of the fact that the ultimate aim is to compare theoretical predictions with rigorous results for the real problem. We have seen that approximations enter both schemes in various ways. As we noted earlier we must ascertain the validity of quantum–classical worlds as models of the real world. In fact, since a consistent quantum–classical world has not yet been constructed we have the residual task of testing the validity of predictions of this model. This would be instrumental not only in realizing its limitations and give ways to improve the approach but also in establishing a preliminary test of the correspondence between the quantum–classical and the real worlds. Simulations on model systems indicate that violations of the Lie algebraic structure and its consequences may be minor for many applications [27], and thus scheme (1) may have practical utility. In the approach that begins with exact quantum equilibrium time correlation function, the freedom to approximate both the equilibrium density and the dynamics, separately or together, provides one with additional possibilities. Some of these approximations could indeed be unfaithful to the real world and highly inconsistent, while others may provide results much closer to the real quantum world. Of course there is nothing unique about the approaches we have discussed here, and other fresh ideas can come from many alternative formulations of quantum mechanics and/or ways to go to the semiclassical limit of quantum mechanics and, moreover, there is no indication that these alternative approaches will be less successful [43, 44]. So the way ahead is open and, at this point, it is unclear which alternative will prevail. Thus, it is worth pursuing these programs of research in all directions.

Acknowledgments

We are pleased to thank the referee for useful comments that clarified the text. The research of RK was supported in part by a grant from the Natural Sciences and Engineering Research Council of Canada. DFC acknowledges support for this research from the US National Science Foundation under grant number CHE-0316856, as well as the hospitality of the Chemical Physics Theory Group (CPTG) at the University of Toronto. GC wishes to thank RK and DFC for providing research travel funds and also acknowledges the hospitality of the CPTG.

References

1. E. Wigner: Phys. Rev. **40**, 749 (1932)
2. R. Kapral and G. Ciccotti: J. Chem. Phys. **110**, 8919 (1999)
3. K. Imre, E. Özizmir, M. Rosenbaum and P. F. Zwiefel: J. Math. Phys. **5**, 1097 (1967); M. Hillery, R. F. O'Connell, M. O. Scully and E. P. Wigner: Phys. Repts. **106**, 121 (1984)
4. I. V. Aleksandrov: Z. Naturforsch. **36a**, 902 (1981)
5. V. I. Gerasimenko: Theor. Math. Phys. **50**, 77 (1982); D. Ya. Petrina, V. I. Gerasimenko and V. Z. Enolskii: Sov. Phys. Dokl. **35**, 925 (1990)
6. W. Boucher and J. Traschen: Phys. Rev. D **37**, 3522 (1988)
7. W. Y. Zhang and R. Balescu: J. Plasma Phys. **40**, 199 (1988); R. Balescu and W. Y. Zhang: J. Plasma Phys. **40**, 215 (1988)
8. O. V. Prezhdo and V. V. Kisil: Phys. Rev. A **56**, 162 (1997)
9. C. C. Martens and J.-Y. Fang: J. Chem. Phys. **106**, 4918 (1996); A. Donoso and C. C. Martens: J. Phys. Chem. **102**, 4291 (1998); D. Kohen and C. C. Martens: J. Chem. Phys. **111**, 4343 (1999); **112**, 7345 (2000)
10. I. Horenko, C. Salzmann, B. Schmidt and C. Schütte: J. Chem. Phys. **117**, 11075 (2002)
11. R. Kapral and G. Ciccotti: *A Statistical Mechanical Theory of Quantum Dynamics in Classical Environments*, in *Bridging Time Scales: Molecular Simulations for the Next Decade*, 2001, eds. P. Nielaba, M. Mareschal and G. Ciccotti, (Springer, Berlin Heidelberg New York, 2003), p. 445
12. S. Nielsen, R. Kapral and G. Ciccotti: J. Chem. Phys. **115**, 5805 (2001)
13. J. Caro and L. L. Salcedo: Phys. Rev. A **60**, 842 (1999)
14. E. Rabani and D. Reichman: J. Chem. Phys. **120**, 1458 (2004)
15. S. A. Egorov, E. Rabani and B. J. Berne: J. Phys. Chem. B **103**, 10978 (1999)
16. A. A. Golosov, D. R. Reichman and E. Rabani: J. Chem. Phys. **118**, 457 (2003)
17. J. A. Poulsen, G. Nyman, and P. J. Rossky: J. Chem. Phys. **119**, 12179 (2003); J. Phys. Chem A **108**, 8743 (2004); Proc. Natl Acad. Sci. **102**, 6709 (2005)
18. Q. Shi and E. Geva: J. Chem. Phys. **118**, 8173 (2003); J. Chem. Phys. **120**, 10647 (2004); Q. Shi and E. Geva: J. Phys. Chem. A **107**, 9059 (2003); J. Phys. Chem. A **107**, 9070 (2003)
19. Q. Shi and E. Geva: J. Phys. Chem. A **108**, 6109 (2004); J. Chem. Phys. **121**, 3393 (2004)
20. A. Sergi and R. Kapral: J. Chem. Phys. **121**, 7565 (2004)
21. H. Kim and R. Kapral: J. Chem. Phys. **122**, 214105 (2005); J. Chem. Phys. **123**, 194108 (2005)
22. D. Mac Kernan, G. Ciccotti and R. Kapral: J. Chem. Phys. **106** (2002)
23. C. Wan and J. Schofield: J. Chem. Phys. **113**, 7047 (2000)
24. M. Santer, U. Manthe, G. Stock: J. Chem. Phys. **114**, 2001 (2001)

25. D. Mac Kernan, G. Ciccotti and R. Kapral: J. Phys. Condens. Matter **14**, 9069 (2002)
26. A. Sergi, D. Mac Kernan, G. Ciccotti and R. Kapral: Theor. Chem. Acc. **110**, 49 (2003)
27. A. Sergi and R. Kapral: J. Chem. Phys. **118**, 8566 (2003)
28. G. Hanna and R. Kapral: J. Chem. Phys. **122**, 244505 (2005)
29. R. Hernandez and G. Voth: Chem. Phys. Letts. **223**, 243 (1998)
30. W. H. Miller: J. Phys. Chem. A **105**, 2942 (2001); H. Wang, X. Sun and W. H. Miller: J. Chem. Phys. **108**, 9726 (1998); X. Sun, H. Wang and W. H. Miller: J. Chem. Phys. **109**, 7064 (1998); M. Thoss, H. Wang and W. H. Miller: J. Chem. Phys. **114**, 47 (2001)
31. S. Zhang and E. Pollak: J. Chem. Phys. **118**, 4357 (2003)
32. M. F. Herman and D. F. Coker: J. Chem. Phys. **111**, 1801 (1999)
33. S. Causo, G. Ciccotti, D. Montemayor, S. Bonella, and D. F. Coker: J. Phys. Chem. B **109**, 6855 (2005)
34. W. H. Miller and C. W. McCurdy: J. Chem. Phys. **69**, 5163 (1978)
35. C. W. McCurdy, H. D. Meyer and W. H. Miller: J. Chem. Phys. **70**, 3177 (1979)
36. X. Sun and W. H. Miller: J. Chem. Phys. **106**, 6346 (1997)
37. G. Stock and M. Thoss: Phys. Rev. Lett. **78**, 578 (1997)
38. G. Stock and M. Thoss: Phys. Rev. A **59**, 64 (1999)
39. S. Bonella and D. F. Coker: J. Chem. Phys. **118**, 4370 (2003); J. Chem. Phys. **114**, 7778 (2001); Chem. Phys. **268**, 323–334 (2001)
40. P. Pechukas: Phys. Rev. **181**, 166, 174 (1969)
41. S. Bonella and D. F. Coker: J. Chem. Phys. **122**, 194102 (2005); Proc. Natl Acad. Sci. **102**, 6715 (2005); Comp. Phys. Commun. **169**, 267 (2005)
42. N. Makri and K. Thompson: Chem. Phys. Lett. **291**, 101 (1998); K. Thompson and N. Makri: J. Chem. Phys. **110**, 1343 (1999); N. Makri: J. Phys. Chem. B **103**, 2823 (1999)
43. I. Burghardt: J. Chem. Phys. **122**, 094103 (2005)
44. O. V.Prezhdo: Theor. Chem. Acc. **114**, (2005)

Quantum–Classical Reaction Rate Theory

G. Hanna, H. Kim, and R. Kapral

Summary. A correlation function formalism for the calculation of rate constants in mixed quantum–classical systems is presented. The full quantum equilibrium density is retained in the rate expressions and quantum–classical Liouville dynamics is used to propagate the species variables in time. Results for a model two-level system coupled to a nonlinear oscillator that is coupled to a harmonic bath and for a proton transfer reaction in a polar liquid solvent are presented. The rate coefficients for these systems are computed using surface-hopping dynamics based on the solution of the quantum–classical Liouville equation.

1 Introduction

A knowledge of the rates of condensed phase chemical reactions is necessary for an understanding of many problems in chemistry and biology. If one is interested in the reactive dynamics of a light particle immersed in an environment of heavy molecules, a quantum rate theory is required to correctly describe this dynamics. Consider a proton transfer occurring in a solvent or large molecule. Due to its light mass, the proton's thermal de Broglie wavelength is comparable in length to the distance over which it travels. As a result, the proton must be treated quantum mechanically and the importance of such quantum effects is well documented. Experimental evidence suggests that hydrogen tunneling is important in enzyme catalysis under physiological conditions [1]. The magnitude of such quantum effects can be gauged by comparing the measured or calculated deuterium kinetic isotope effect for these reactions with that predicted by classical transition state theory. In addition, quantum effects in the environment surrounding the proton may be significant. Quantum phenomena exist in the solvent dynamics associated with the transfer of excess protons in liquid water and can explain the anomalously high mobility of these protons [2, 3].

Although it is not difficult to write a correlation function expression for the time-dependent rate coefficient of a reacting quantum system [4], a full quantum dynamical simulation of a condensed phase system containing a large number of degrees of freedom is not computationally feasible. Calculations of rate constants for reactive processes occurring in many-body environments, which incorporate quantum effects, have been performed using a variety of computational techniques. The techniques used include influence-functional [5, 6] and real-time path integral methods [7, 8], methods based on the stochastic Schrödinger equation [9, 10], centroid dynamics [11], golden rule and Fokker–Planck formulations [12], mode coupling theories [13, 14], techniques based on the initial value representation [15–22], mapping Hamiltonian methods [23,24], nonadiabatic statistical methods [25], surface-hopping schemes [26–30], multi–configuration time-dependent Hartree methods [31, 32], and methods based on the quantum-classical Liouville equation [33–39].

In this chapter, we consider systems for which a description in terms of quantum-classical dynamics is appropriate [37], i.e., systems in which a subset of the degrees of freedom are treated quantum mechanically while the dynamics of the remainder of the degrees of freedom can be adequately described by classical mechanics. We first derive expressions for the quantum mechanical rate coefficient of a general reaction $A \rightleftharpoons B$ and then obtain their quantum-classical analogs. Next, we consider the choice of a reaction coordinate and the specification of species variables used to monitor the progress of a quantum reaction and discuss the rate expressions which arise from such a choice. We apply this quantum–classical rate theory to a two-level quantum system coupled to a classical nonlinear oscillator which is in turn coupled to a classical harmonic bath, and to the more realistic situation of a proton transfer reaction occurring in a polar solvent.

2 Rate Theory

Quantum–classical expressions for rate coefficients have been derived [40,41], and computed for model systems [40–42] and proton transfer reactions [43]. An alternate approach to the calculation of quantum transport properties was described recently [44,45]. The starting point of this approach is the full quantum mechanical expression for a transport property; however, the evolution of dynamical variables is carried out in the quantum–classical limit. This scheme has the advantage that the full quantum mechanical equilibrium structure of the system, described by a spectral density function, is retained; only the quantum mechanical time evolution is replaced by quantum–classical time evolution. The calculation of the quantum equilibrium structure, although a difficult problem, is far more tractable than that of the quantum time evolution of a many-body system. Exact expressions for the reaction rate coefficient have been derived in this more general context [45]. In many cases, one may take advantage of convenient features of the system to make approximations which simplify the computation of these expressions. For each system, the most applicable reaction coordinate must be identified, along with the dynamical variables which characterize the microscopic species involved in the reaction.

In this section we shall derive a series of quantum mechanical expressions for the rate coefficient of a general interconversion reaction $A \rightleftharpoons B$ starting from the flux–flux quantum correlation function. By taking the quantum–classical limit of these expressions, we obtain formulas that can be computed using quantum–classical surface-hopping dynamics.

2.1 Quantum Mechanical Rate Expressions

For a quantum mechanical system in thermal equilibrium undergoing a transformation $A \rightleftharpoons B$, a rate constant k_{AB} may be calculated from the time integral of a flux–flux correlation function [46],

$$k_{AB} = \frac{1}{n_A^{\mathrm{eq}}} \int_0^\infty \mathrm{d}t \langle \hat{j}_A ; \hat{j}_B(t) \rangle = \frac{1}{\beta n_A^{\mathrm{eq}}} \int_0^\infty \mathrm{d}t \langle \frac{i}{\hbar} [\hat{j}_B(t), \hat{A}] \rangle \, , \tag{1}$$

where $\hat{A} = \hat{N}_A$ is the A species operator, n_A^{eq} is the equilibrium density of species A, $\hat{j}_A = \dot{\hat{A}} = (i/\hbar)[\hat{H}, \hat{A}]$ is the flux of $\hat{A}$ with Hamiltonian $\hat{H}$, with an analogous expression for $\hat{j}_B$, $[\cdot,\cdot]$ is the commutator and the angular brackets $\langle \hat{A}; \hat{B} \rangle = \frac{1}{\beta} \int_0^\beta \mathrm{d}\lambda \langle \mathrm{e}^{\lambda \hat{H}} \hat{A} \mathrm{e}^{-\lambda \hat{H}} \hat{B} \rangle$ denote a Kubo transformed correlation function, with $\beta = (k_B T)^{-1}$. The equilibrium quantum canonical average is $\langle \cdots \rangle = Z_Q^{-1} \mathrm{Tr} \cdots \mathrm{e}^{-\beta \hat{H}}$, where Z_Q is the partition function. The time evolution of the reactive flux is given by projected dynamics. In simulations it is often convenient to consider the time-dependent rate coefficient defined as the finite time integral of the flux–flux correlation function,

$$\begin{aligned} k_{AB}(t) &= \frac{1}{n_A^{\mathrm{eq}}} \int_0^t \mathrm{d}t' \langle \hat{j}_A ; \hat{j}_B(t') \rangle = \frac{1}{n_A^{\mathrm{eq}}} \langle \dot{\hat{A}} \, ; \hat{B}(t) \rangle \\ &= \frac{1}{\beta n_A^{\mathrm{eq}}} \left\langle \frac{i}{\hbar} [\hat{B}(t), \hat{A}] \right\rangle \, , \end{aligned} \tag{2}$$

where we have replaced projected dynamics by ordinary dynamics and assumed $[\hat{B}, \hat{A}] = 0$.

Writing the second equality in (2) in detail and inserting arbitrary time variables t_1 and t_2, we can write the rate coefficient $k_{AB}(t)$ as,

$$k_{AB}(t) = \frac{1}{n_A^{\mathrm{eq}} \beta Z_Q} \int_0^\beta \mathrm{d}\lambda \mathrm{Tr} \Big(\dot{\hat{A}}(t_1 - \mathrm{i}\hbar\lambda) \mathrm{e}^{\frac{\mathrm{i}}{\hbar} \hat{H} t'} \hat{B}(t_2) \mathrm{e}^{-\frac{\mathrm{i}}{\hbar} \hat{H} t'} \mathrm{e}^{-\beta \hat{H}} \Big) \, , \tag{3}$$

where $t' \equiv t + t_1 - t_2$. To insert the times t_1 and t_2, we used the fact that the time evolution of an operator $\hat{O}$ is given by $\hat{O}(t) = \mathrm{e}^{\frac{\mathrm{i}}{\hbar} \hat{H} t} \hat{O} \mathrm{e}^{-\frac{\mathrm{i}}{\hbar} \hat{H} t}$.

We partition the entire quantum system into a subsystem $\mathcal{S}$ plus environment $\mathcal{E}$ so that the Hamiltonian is the sum of the kinetic energy operators of the subsystem and environment and the potential energy of the entire system, $\hat{H} = \hat{P}^2/2M + \hat{p}^2/2m + \hat{V}(\hat{q},\hat{Q})$, where lower and upper case symbols refer to the subsystem and environment, respectively. In the next subsection we shall show how the rate coefficients for a system partitioned in this way can be evaluated in the quantum–classical limit. For the present, however, it is convenient to first make a Wigner transform over all degrees of freedom, subsystem plus environment, and later single out the subsystem and environmental degrees of freedom for different treatments. Introducing a coordinate representation $\{\mathcal{Q}\} = \{q\}\{Q\}$ of the operators in (3) (calligraphic symbols denote variables for the entire system), making the change of variables $\mathcal{Q}_1 = \mathcal{R}_1 - \mathcal{Z}_1/2$, $\mathcal{Q}_2 = \mathcal{R}_1 + \mathcal{Z}_1/2$, etc., and then expressing the matrix elements of the operators in terms of their Wigner transforms, we obtain

$$\begin{aligned} k_{AB}(t) = & \frac{1}{\beta n_A^{\mathrm{eq}}} \int_0^\beta \mathrm{d}\lambda \int \mathrm{d}\mathcal{X}_1 \mathrm{d}\mathcal{X}_2 (\dot{A})_{\mathrm{W}}(\mathcal{X}_1, t_1) B_{\mathrm{W}}(\mathcal{X}_2, t_2) \\ & \times \frac{1}{(2\pi\hbar)^{2\nu} Z_Q} \int \mathrm{d}\mathcal{Z}_1 \mathrm{d}\mathcal{Z}_2 \mathrm{e}^{-\frac{\mathrm{i}}{\hbar}(\mathcal{P}_1\cdot\mathcal{Z}_1 + \mathcal{P}_2\cdot\mathcal{Z}_2)} \\ & \times \left\langle \mathcal{R}_1 + \frac{\mathcal{Z}_1}{2} \right| \mathrm{e}^{\frac{\mathrm{i}}{\hbar}\hat{H}(t'+\mathrm{i}\hbar\lambda)} \left| \mathcal{R}_2 - \frac{\mathcal{Z}_2}{2} \right\rangle \\ & \times \left\langle \mathcal{R}_2 + \frac{\mathcal{Z}_2}{2} \right| \mathrm{e}^{-\beta\hat{H} - \frac{\mathrm{i}}{\hbar}\hat{H}(t'+\mathrm{i}\hbar\lambda)} \left| \mathcal{R}_1 - \frac{\mathcal{Z}_1}{2} \right\rangle , \end{aligned} \tag{4}$$

where $Z_Q = (2\pi\hbar)^{-\nu} \int \mathrm{d}\mathcal{X} (\mathrm{e}^{-\beta\hat{H}})_{\mathrm{W}}(\mathcal{X})$. In writing this equation we used the fact that the matrix element of an operator $\hat{O}(t)$ can be expressed in terms of its Wigner transform $O_{\mathrm{W}}(\mathcal{X}, t)$ as

$$\left\langle \mathcal{R} - \frac{\mathcal{Z}}{2} \right| \hat{O}(t) \left| \mathcal{R} + \frac{\mathcal{Z}}{2} \right\rangle = \frac{1}{(2\pi\hbar)^\nu} \int \mathrm{d}\mathcal{P} \mathrm{e}^{-\frac{\mathrm{i}}{\hbar}\mathcal{P}\cdot\mathcal{Z}} O_{\mathrm{W}}(\mathcal{X}, t), \tag{5}$$

where ν is the coordinate space dimension and

$$O_{\mathrm{W}}(\mathcal{X}, t) = \int \mathrm{d}\mathcal{Z} \mathrm{e}^{\frac{\mathrm{i}}{\hbar}\mathcal{P}\cdot\mathcal{Z}} \left\langle \mathcal{R} - \frac{\mathcal{Z}}{2} \right| \hat{O}(t) \left| \mathcal{R} + \frac{\mathcal{Z}}{2} \right\rangle , \tag{6}$$

defines the Wigner transform. We use the notation $\mathcal{R} = (r, R)$, $\mathcal{P} = (p, P)$ and $\mathcal{X} = (r, R,\ p, P)$, where again the lower case symbols refer to the subsystem and the upper case symbols refer to the environment.

We define the spectral density by

$$\begin{aligned} W(\mathcal{X}_1, \mathcal{X}_2, t) = & \frac{1}{(2\pi\hbar)^{2\nu} Z_Q} \int \mathrm{d}\mathcal{Z}_1 \mathrm{d}\mathcal{Z}_2 \mathrm{e}^{-\frac{\mathrm{i}}{\hbar}(\mathcal{P}_1\cdot\mathcal{Z}_1 + \mathcal{P}_2\cdot\mathcal{Z}_2)} \\ & \times \left\langle \mathcal{R}_1 + \frac{\mathcal{Z}_1}{2} \right| \mathrm{e}^{\frac{\mathrm{i}}{\hbar}\hat{H}t} \left| \mathcal{R}_2 - \frac{\mathcal{Z}_2}{2} \right\rangle \\ & \times \left\langle \mathcal{R}_2 + \frac{\mathcal{Z}_2}{2} \right| \mathrm{e}^{-\beta\hat{H} - \frac{\mathrm{i}}{\hbar}\hat{H}t} \left| \mathcal{R}_1 - \frac{\mathcal{Z}_1}{2} \right\rangle . \end{aligned} \tag{7}$$

If we let

$$\overline{W}(\mathcal{X}_1,\mathcal{X}_2,t)=\frac{1}{\beta}\int_0^{\beta}\mathrm{d}\lambda W(\mathcal{X}_1,\mathcal{X}_2,t+\mathrm{i}\hbar\lambda)$$
$$=\frac{2}{\beta}\int_0^{\frac{\beta}{2}}\mathrm{d}\lambda \mathrm{Re}W(\mathcal{X}_1,\mathcal{X}_2,t+\mathrm{i}\hbar\lambda), \tag{8}$$

we can write the rate coefficient as

$$k_{AB}(t)=\frac{1}{n_A^{\mathrm{eq}}}\int \mathrm{d}\mathcal{X}_1\mathrm{d}\mathcal{X}_2(\dot{A})_{\mathrm{W}}(\mathcal{X}_1,t_1)B_{\mathrm{W}}(\mathcal{X}_2,t_2)\overline{W}(\mathcal{X}_1,\mathcal{X}_2,t+t_1-t_2)\,. \tag{9}$$

We may choose the times t_1 and t_2 to yield various forms for the correlation function. Since the time evolution of the operator is usually more convenient than that of the spectral density, we set $t_1=0$ and $t_2=t$ to give

$$k_{AB}(t)=\frac{1}{n_A^{\mathrm{eq}}}\int \mathrm{d}\mathcal{X}_1\mathrm{d}\mathcal{X}_2(\mathrm{i}L_{\mathrm{W}}(\mathcal{X}_1)A_{\mathrm{W}}(\mathcal{X}_1))B_{\mathrm{W}}(\mathcal{X}_2,t)\overline{W}(\mathcal{X}_1,\mathcal{X}_2,0). \tag{10}$$

The quantum Liouville operator in Wigner-transformed form is $\mathrm{i}L_{\mathrm{W}}=\frac{2}{\hbar}H_{\mathrm{W}}(\mathcal{X})\sin(\hbar\Lambda/2)$, where Λ is the negative of the Poisson bracket operator. We can rewrite (10) as

$$k_{AB}(t)=\frac{1}{n_A^{\mathrm{eq}}}\int \mathrm{d}\mathcal{X}B_{\mathrm{W}}(\mathcal{X},t)\overline{W}_{A'}(\mathcal{X},0), \tag{11}$$

where[1]

$$\overline{W}_{A'}(\mathcal{X},t)=\int \mathrm{d}\mathcal{X}'(\mathrm{i}L_{\mathrm{W}}(\mathcal{X}')A_{\mathrm{W}}(\mathcal{X}'))\overline{W}(\mathcal{X}',\mathcal{X},t). \tag{12}$$

From the last equality in (2), we can obtain an alternative form of the rate coefficient involving the commutator of $\hat{A}$ and $\hat{B}(t)$. Performing a set of manipulations similar to those used above, we may show that $k_{AB}(t)$ is also given by

$$k_{AB}(t)=\frac{i}{\hbar\beta n_A^{\mathrm{eq}}}\int \mathrm{d}\mathcal{X}_1\mathrm{d}\mathcal{X}_2A_{\mathrm{W}}(\mathcal{X}_1)B_{\mathrm{W}}(\mathcal{X}_2,t)$$
$$\times[W(\mathcal{X}_1,\mathcal{X}_2,\mathrm{i}\hbar\beta)-W(\mathcal{X}_1,\mathcal{X}_2,0)]$$
$$=\frac{2}{\hbar\beta n_A^{\mathrm{eq}}}\int \mathrm{d}\mathcal{X}_1\mathrm{d}\mathcal{X}_2A_{\mathrm{W}}(\mathcal{X}_1)B_{\mathrm{W}}(\mathcal{X}_2,t)\mathrm{Im}W(\mathcal{X}_1,\mathcal{X}_2,0), \tag{13}$$

[1] Here, $\overline{W}_{A'}$ corresponds exactly to $\overline{W}_A$ defined in [47].

where Im stands for the imaginary part. Using the definition

$$W_A(\mathcal{X},t) = \int \mathrm{d}\mathcal{X}' A_{\mathrm{W}}(\mathcal{X}') W(\mathcal{X}',\mathcal{X},t), \tag{14}$$

we can rewrite $k_{AB}(t)$ as

$$k_{AB}(t) = \frac{2}{\hbar\beta n_A^{\mathrm{eq}}} \int \mathrm{d}\mathcal{X} B_{\mathrm{W}}(\mathcal{X},t) \mathrm{Im} W_A(\mathcal{X},0). \tag{15}$$

So far, both (11) and (15) for the time-dependent rate coefficient are exact.

We find that the following symmetry relations hold for W:

$$W(\mathcal{X}_1,\mathcal{X}_2,t)^* = W(\mathcal{X}_2,\mathcal{X}_1,-t)\ , \tag{16}$$

$$W(\mathcal{X}_1,\mathcal{X}_2,t+\mathrm{i}\hbar\lambda)^* = W(\mathcal{X}_1,\mathcal{X}_2,t+\mathrm{i}\hbar(\beta-\lambda))\ . \tag{17}$$

Note that $W(\mathcal{X}',\mathcal{X},t+\mathrm{i}\hbar\lambda)$ is real only for $\lambda=\beta/2$; namely,

$$W\left(\mathcal{X}_1,\mathcal{X}_2,t+\frac{\mathrm{i}\hbar\beta}{2}\right)^* = W\left(\mathcal{X}_1,\mathcal{X}_2,t+\frac{\mathrm{i}\hbar\beta}{2}\right). \tag{18}$$

This corresponds to the first order term when $\overline{W}$ is expanded in terms of β,

$$\overline{W}(\mathcal{X}_1,\mathcal{X}_2,0) = W\left(\mathcal{X}_1,\mathcal{X}_2,\frac{\mathrm{i}\hbar\beta}{2}\right) + \mathcal{O}(\beta^2). \tag{19}$$

In the high temperature limit, the higher order terms in β become negligible. Note that the symmetry relations above also hold for $W_{A'}(\mathcal{X},t)$.

In the long time limit, the time-dependent rate coefficient, $k_{AB}(t)$, decays to zero. However, if there is a large difference between the time scales of the chemical reaction and the transient microscopic dynamics, the rate coefficient first decays to a plateau from which the rate constant can be extracted. If absorbing boundaries are introduced to prevent escape of the trajectory from the metastable states once they are reached from the barrier top, the rate coefficient will no longer decay to zero and will assume a constant value at long times. This can be achieved more rigorously by formulating the rate expressions using projection operator techniques [46].

2.2 Quantum–Classical Rate Expressions

In this section we show how to take the quantum–classical limit of the general expressions for the rate coefficient, which treat the system plus environment fully quantum mechanically. By taking the quantum–classical limit [37] of these expressions we can obtain rate coefficient expressions that are amenable to solution using surface-hopping methods. The computation of the initial value of W is still a challenging problem but far less formidable than the solution of the time-dependent Schrödinger equation for the entire quantum system.

To make a connection with the surface-hopping representation of the solution of the quantum–classical Liouville equation [37], we first observe that $A_{\mathrm{W}}(\mathcal{X})$ can be written as

$$A_{\mathrm{W}}(\mathcal{X}) = \int \mathrm{d}z\, \mathrm{e}^{\frac{\mathrm{i}}{\hbar}p\cdot z} < r - \frac{z}{2}|\hat{A}_{\mathrm{W}}(X)|r + \frac{z}{2} > , \tag{20}$$

where $\hat{A}_{\mathrm{W}}(X)$ is the *partial* Wigner transform of $\hat{A}$, defined as in (6), but with the transform taken only over the environmental degrees of freedom. The partial Wigner transform of the Hamiltonian is $\hat{H}_{\mathrm{W}} = P^2/2M + \hat{p}^2/2m + \hat{V}_{\mathrm{W}}(\hat{q}, R) \equiv P^2/2M + \hat{h}_{\mathrm{W}}(R)$, where $\hat{h}_{\mathrm{W}}(R)$ is the Hamiltonian of the subsystem in the fixed field of the environment. The adiabatic eigenstates are the solutions of the eigenvalue problem, $\hat{h}_{\mathrm{W}}(R)|\alpha; R >= E_\alpha(R)|\alpha; R >$. We may now express $A_{\mathrm{W}}(\mathcal{X})$ in the adiabatic basis to obtain,

$$A_{\mathrm{W}}(\mathcal{X}) = \sum_{\alpha\alpha'} \int \mathrm{d}z\, \mathrm{e}^{\frac{\mathrm{i}}{\hbar}p\cdot z} < r - \frac{z}{2}|\alpha; R > A_{\mathrm{W}}^{\alpha\alpha'}(X) < \alpha'; R|r + \frac{z}{2} > , \tag{21}$$

where $A_{\mathrm{W}}^{\alpha\alpha'}(X) =< \alpha; R|\hat{A}_{\mathrm{W}}(X)|\alpha'; R >$.

Inserting this expression and its analog for $B_{\mathrm{W}}(\mathcal{X}_2)$ into (10), we obtain

$$k_{AB}(t) = \frac{1}{n_A^{\mathrm{eq}}} \sum_{\alpha\alpha'} \int \mathrm{d}X B_{\mathrm{W}}^{\alpha\alpha'}(X, t) W_{A'}^{\alpha'\alpha}\left(X, \frac{\mathrm{i}\hbar\beta}{2}\right) , \tag{22}$$

using the approximation prescribed by (19). The matrix elements of $W_{A'}$ in the adiabatic basis are given by

$$W_{A'}^{\alpha'\alpha}\left(X, \frac{i\hbar\beta}{2}\right) = \sum_{\alpha_1\alpha_1'} \int \mathrm{d}X' \left(\mathrm{i}\mathcal{L}(X')A_{\mathrm{W}}(X')\right)^{\alpha_1\alpha_1'} W^{\alpha_1'\alpha_1\alpha'\alpha}\left(X', X, \frac{\mathrm{i}\hbar\beta}{2}\right) , \tag{23}$$

where

$$W^{\alpha_1'\alpha_1\alpha'\alpha}\left(X', X, \frac{\mathrm{i}\hbar\beta}{2}\right) = \frac{1}{(2\pi\hbar)^{2\nu} Z_Q} \int \mathrm{d}Z \mathrm{d}Z' \mathrm{e}^{-\frac{\mathrm{i}}{\hbar}(P\cdot Z + P'\cdot Z')}$$
$$\times < \alpha'; R| < R + \frac{Z}{2}|\mathrm{e}^{-\frac{\beta}{2}\hat{H}}|R' - \frac{Z'}{2} > |\alpha_1; R' >$$
$$\times < \alpha_1'; R'| < R' + \frac{Z'}{2}|\mathrm{e}^{-\frac{\beta}{2}\hat{H}}|R - \frac{Z}{2} > |\alpha; R > . \tag{24}$$

From (15), the alternative form of the rate coefficient can be obtained

$$k_{AB}(t) = \frac{2}{\hbar\beta n_A^{\mathrm{eq}}}\sum_{\alpha\alpha'}\int \mathrm{d}\mathcal{X}\mathrm{Im}[B_{\mathrm{W}}^{\alpha\alpha'}(X,t)W_A^{\alpha'\alpha}(X,0)], \qquad (25)$$

where

$$W_A^{\alpha'\alpha}(X,0) = \sum_{\alpha_1\alpha_1'}\int \mathrm{d}X' A_{\mathrm{W}}^{\alpha_1\alpha_1'}(X')W^{\alpha_1'\alpha_1\alpha'\alpha}(X',X,0). \qquad (26)$$

In the quantum–classical limit, $B_{\mathrm{W}}^{\alpha'\alpha}(X,t)$ satisfies the quantum–classical Heisenberg equation:

$$\frac{\mathrm{d}}{\mathrm{d}t}B_{\mathrm{W}}^{\alpha'\alpha}(X,t) = \sum_{\beta\beta'}\mathrm{i}\mathcal{L}_{\alpha'\alpha,\beta'\beta}(X)B_{\mathrm{W}}^{\beta'\beta}(X,t). \qquad (27)$$

The quantum–classical Liouville operator, $\mathrm{i}\mathcal{L}$, in the adiabatic basis is given by $\mathrm{i}\mathcal{L}_{\alpha\alpha',\beta\beta'}(X) = [\mathrm{i}\omega_{\alpha\alpha'}(R) + \mathrm{i}L_{\alpha\alpha'}(X)]\delta_{\alpha\beta}\delta_{\alpha'\beta'} - J_{\alpha\alpha',\beta\beta'}(X)$ [37], where the classical evolution operator is defined by

$$\mathrm{i}L_{\alpha\alpha'} = \frac{P}{M}\frac{\partial}{\partial R} + \frac{1}{2}\left[F_{\mathrm{W}}^{\alpha}(R) + F_{\mathrm{W}}^{\alpha'}(R)\right]\frac{\partial}{\partial P}, \qquad (28)$$

with

$$J_{\alpha\alpha',\beta\beta'}(X) = -\frac{P}{M}\mathrm{d}_{\alpha\beta}\left[1 + \frac{1}{2}S_{\alpha\beta}(R)\frac{\partial}{\partial P}\right]\delta_{\alpha'\beta'}$$

$$-\frac{P}{M}\mathrm{d}_{\alpha'\beta'}^{*}\left[1 + \frac{1}{2}S_{\alpha'\beta'}^{*}(R)\frac{\partial}{\partial P}\right]\delta_{\alpha\beta}. \qquad (29)$$

Here the frequency is $\omega_{\alpha\alpha'}(R) = [E_\alpha(R) - E_{\alpha'}(R)]/\hbar$, the Hellmann–Feynman force is $F_{\mathrm{W}}^{\alpha} = -\left\langle \alpha; R \left| \partial\hat{V}_{\mathrm{W}}(\hat{q},R)/\partial\hat{R} \right| \alpha; R\right\rangle$, the nonadiabatic coupling matrix element is $\mathrm{d}_{\alpha\beta} = \langle \alpha; R|\nabla_R|\beta; R\rangle$, and $S_{\alpha\beta} = (E_\alpha - E_\beta)\mathrm{d}_{\alpha\beta}[(P/M)\cdot d_{\alpha\beta}]^{-1}$.

It should be noted that $\overline{W}_{A'}^{\alpha'\alpha}(X,t)$ and $W_A^{\alpha'\alpha}(X,t)$ satisfy the following symmetry relations:

$$\overline{W}_{A'}^{\alpha'\alpha}(X,t)^* = \overline{W}_{A'}^{\alpha\alpha'}(X,t), \qquad (30)$$

$$W_A^{\alpha'\alpha}(X,t+\mathrm{i}\hbar\lambda)^* = W_A^{\alpha\alpha'}(X,t+\mathrm{i}\hbar(\beta-\lambda)). \qquad (31)$$

It follows that

$$\{\overline{W}_{A'}^{\alpha'\alpha}(X,t) + \overline{W}_{A'}^{\alpha\alpha'}(X,t)\}^* = \overline{W}_{A'}^{\alpha'\alpha}(X,t) + \overline{W}_{A'}^{\alpha\alpha'}(X,t), \qquad (32)$$

and

$$\{W_A^{\alpha'\alpha}(X,t+\mathrm{i}\hbar\lambda)+W_A^{\alpha\alpha'}(X,t+\mathrm{i}\hbar\lambda)\}^* \\ = W_A^{\alpha'\alpha}(X,t+\mathrm{i}\hbar(\beta-\lambda))+W_A^{\alpha\alpha'}(X,t+\mathrm{i}\hbar(\beta-\lambda)). \tag{33}$$

Using these properties, we may write $k_{AB}(t)$ from (22) and (25) as

$$k_{AB}(t)=\frac{1}{n_A^{\mathrm{eq}}}\sum_{\alpha}\sum_{\alpha'\geq\alpha}(2-\delta_{\alpha'\alpha})\int \mathrm{d}X\ \mathrm{Re}\left[B_{\mathrm{W}}^{\alpha\alpha'}(X,t)W_{A'}^{\alpha'\alpha}\left(X,\frac{\mathrm{i}\hbar\beta}{2}\right)\right], \tag{34}$$

or

$$k_{AB}(t)=\frac{2}{\hbar\beta n_A^{\mathrm{eq}}}\sum_{\alpha}\int \mathrm{d}X\big(B_{\mathrm{W}}^{\alpha\alpha}(X,t)\mathrm{Im}W_A^{\alpha\alpha}(X,0) \\ +\sum_{\alpha'>\alpha}\mathrm{Im}[B_{\mathrm{W}}^{\alpha\alpha'}(X,t)\{W_A^{\alpha'\alpha}(X,0)-W_A^{\alpha\alpha'}(X,0)^*\}]\big). \tag{35}$$

These rate coefficient expressions involve quantum–classical evolution of the matrix element $B_{\mathrm{W}}^{\alpha\alpha'}(X,t)$ but retain the full quantum equilibrium structure of the system. We now derive specific forms of the rate coefficient based on different choices of dynamical variables $B_{\mathrm{W}}(X,t)$.

3 Species Variables

We now have to choose specific forms of the dynamical variables A_{W} and B_{W} which characterize the chemical species in the reacting mixture, but first we need some insight into how to choose them. This will be the topic of the next subsection.

3.1 Reaction Coordinate and Free Energy

To illustrate how one chooses a particular species variable, we consider a two-level quantum subsystem coupled to an environment with many degrees of freedom. This is an interesting case since many features of condensed phase proton and electron transfer processes can often be captured by such two-level models. In many situations, due to the nature of the coupling between the quantum and classical degrees of freedom, one may choose a reaction coordinate, $\xi(R)$, which depends solely (either directly or parametrically) on the classical coordinates. In such a case, reactive events in the quantum subsystem are reflected by changes in a function of the classical coordinates. The reaction coordinate must be appropriate in the sense that it will be able to detect the formation of the various chemical species in the reacting mixture, if monitored along the course of a reaction. The guide to the specification of the relevant species variables for our two-level model is provided by the structure of the ground and first excited state free energy profiles along $\xi(R)$.

The free energy along the reaction coordinate can be obtained analytically for simple two-level systems [40] or, more generally, generated from long constant temperature trajectories on the different adiabatic surfaces. The free energy corresponding to adiabatic surface α is given by

$$\beta W_\alpha(\xi') = -\ln \frac{P_\alpha(\xi')}{P_\mathrm{u}}, \tag{36}$$

where P_u is the uniform probability density of ξ, and

$$P_\alpha(\xi') = \frac{\int \mathrm{d}R\,\mathrm{d}P\,\delta(\xi(R)-\xi')\mathrm{e}^{-\beta H_\alpha}}{\sum_\alpha \int \mathrm{d}R\,\mathrm{d}P\,\mathrm{e}^{-\beta H_\alpha}}, \tag{37}$$

is the probability density for finding the numerical value ξ' of $\xi(R)$ when the system is in adiabatic state α with Hamiltonian $H_\alpha = \sum_i P_i^2/(2M_i) + E_\alpha(R)$, where the sum runs over all classical particles i, and P_i and M_i are the momentum and mass of the ith particle, respectively. We may then write the free energy as

$$\beta W_\alpha(\xi') = -\ln \frac{\langle\delta(\xi(R)-\xi')\rangle_\alpha}{P_\mathrm{u}} - \ln \frac{p_\alpha}{p_1}, \tag{38}$$

where $\langle\delta(\xi(R)-\xi')\rangle_\alpha$ is defined by,

$$\langle\delta(\xi(R)-\xi')\rangle_\alpha = \frac{\int \mathrm{d}R\,\mathrm{d}P\,\delta(\xi(R)-\xi')\mathrm{e}^{-\beta H_\alpha}}{\int \mathrm{d}R\,\mathrm{d}P\,\mathrm{e}^{-\beta H_\alpha}}, \tag{39}$$

and can be estimated by binning $\xi(R)$ along a long trajectory on adiabatic surface α. The probability that the system is in state α is $p_\alpha = \int \mathrm{d}\xi'\, P_\alpha(\xi')$, and therefore

$$\frac{p_\alpha}{p_1} = \frac{\int \mathrm{d}R\,\mathrm{d}P \mathrm{e}^{-\beta(E_\alpha - E_1)}\mathrm{e}^{-\beta H_1}}{\int \mathrm{d}R\,\mathrm{d}P \mathrm{e}^{-\beta H_1}}. \tag{40}$$

This factor is related to the relative probability that the system is in state α (regardless of the value of ξ), and can be determined from a long adiabatic trajectory on the ground state surface.

Figure 1 schematically shows two sets of free energy profiles for a two-level system; they correspond to systems in which there is weak (left panel) and strong (right panel) coupling between the quantum subsystem and the reaction coordinate, respectively. In both, the ground state surface has two minima corresponding to two stable species separated by a high barrier at $\xi(R) = \xi^\ddagger$. In the left panel, the excited state surface is nearly parallel to the ground state surface, whereas in the right panel it has a single minimum. Since transitions between the two stable species will occur on a long time scale (due to the high barrier and excitations to higher states), we may identify the

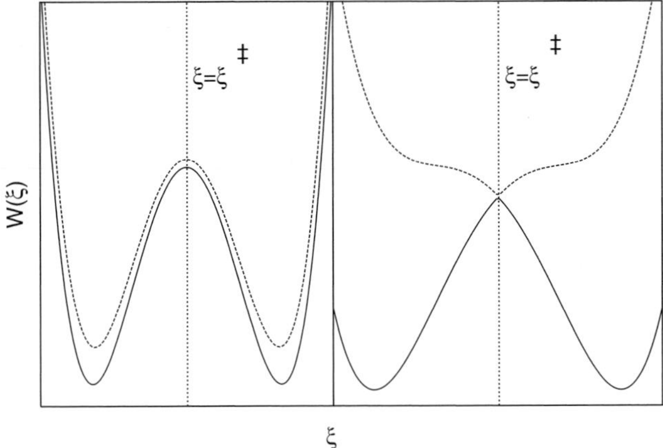

Fig. 1. A schematic illustration of two contrasting sets of free energy (W) profiles along a reaction coordinate ξ. The *left* and *right* panels, respectively, depict situations of weak and strong coupling between the quantum subsystem and reaction coordinate. The *dotted* lines at $\xi = \xi^{\ddagger}$ indicate the position of the barrier top

values of $\xi(R)$ greater than and less than $\xi^{\ddagger}$ with species A and B, respectively. Hence, we may use the Heaviside functions $\theta(\xi(R) - \xi^{\ddagger})$ and $\theta(\xi^{\ddagger} - \xi(R))$ as variables which correspond to species A and B, respectively.

Let us consider a system in which only one classical coordinate, R_0, is directly coupled to the quantum subsystem. In this case, the progress of the quantum reaction can be simply monitored by the reaction coordinate $\xi(R) = R_0$. For the remainder of this section, all the derivations are carried out using this reaction coordinate because the mathematical manipulations are less cumbersome using this reaction coordinate.

3.2 Reactive Flux Operator

The A and B species operators may be defined as $\hat{A}_{\mathrm{W}} = \theta(-R_0)$ and $\hat{B}_{\mathrm{W}} = \theta(R_0)$, where θ is the Heaviside function and the dividing surface is located at $\xi^{\ddagger} = 0$. For this choice of species variable, $W_{A'}^{\alpha'\alpha}(X, \frac{i\hbar\beta}{2})$ defined in (23), can be simplified by taking advantage of the fact that integrations over all X' coordinates can be performed to obtain,

$$W_{A'}^{\alpha'\alpha}\left(X, \frac{i\hbar\beta}{2}\right) = \frac{1}{(2\pi\hbar)^{\nu} Z_Q} \frac{i\hbar}{M_0} \int dZ\, dZ_0' (\partial\delta(Z_0')/\partial Z_0') e^{-\frac{i}{\hbar}P\cdot Z}$$
$$\times < \alpha'; R_0| \left\langle R + \frac{Z}{2} \right| e^{-\frac{\beta}{2}\hat{H}} \left| -\frac{Z_0'}{2} \right\rangle$$
$$\times \left\langle \frac{Z_0'}{2} \right| e^{-\frac{\beta}{2}\hat{H}} \left| R - \frac{Z}{2} \right\rangle |\alpha; R_0 > . \quad (41)$$

In this equation the adiabatic eigenstates depend parametrically only on R_0 since the subsystem $\mathcal{S}$ couples directly only to the coordinate R_0.

In order to compute the rate, we need to carry out quantum–classical evolution of $B_{\mathrm{W}}^{\alpha\alpha'}(X,t)$, as dictated by (27), and sample from an initial quantum distribution with weights determined by $W_{A'}^{\alpha'\alpha}(X,\frac{i\hbar\beta}{2})$. The imaginary time propagators in $W_{A'}^{\alpha'\alpha}(X,\frac{i\hbar\beta}{2})$ can, in principle, be computed using quantum path integral methods [48] or approximations such as linearization methods [23,24,49,50]. Below we show how one may construct approximate analytical expressions for this quantity, which will be used to obtain the numerical results in the next section.

Parabolic potential in barrier region

In activated rate processes a knowledge of the dynamics of a system in the vicinity of its potential energy barrier is crucial for the calculation of the rate constant. In many situations the potential is locally parabolic in the barrier region and such harmonic barrier approximations have been employed frequently in the study of quantum and classical reaction rates [17,51–55]. Here we show how the local harmonic character of the barrier along the reaction coordinate R_0 can be exploited to construct an approximate form for $W_A^{\alpha'\alpha}\left(X,\frac{i\hbar\beta}{2}\right)$, which is useful for the situation depicted in the left panel of Fig. 1.

To proceed with the analytical calculation, we first partition the Hamiltonian into $\hat{H} = \hat{H}_{sn} + \hat{H}_{b(n)}$, where $\hat{H}_{sn} = \hat{H}_s + \hat{H}_n + \hat{V}_{sn}$ is the Hamiltonian of the subsystem plus a subset of degrees of freedom $\mathcal{N}$ plus the coupling between them, and $\hat{H}_{b(n)}$ is the Hamiltonian of the bath $\mathcal{B}$ in the field of $\mathcal{N}$. For our model the subset $\mathcal{N}$ is just that associated with the R_0 coordinate. Then, we assume that the imaginary time propagator may be written as $\exp(-\beta\hat{H}/2) \approx \exp(-\beta\hat{H}_{sn}/2)\exp(-\beta\hat{H}_{b(n)}/2)$, so that (41) for $W_{A'}^{\alpha'\alpha}(X,\frac{i\hbar\beta}{2})$ is given by

$$\begin{aligned} W_{A'}^{\alpha'\alpha}\left(X,\frac{i\hbar\beta}{2}\right) &= \frac{1}{Z_Q}\frac{i}{2\pi M_0}\int \mathrm{d}Z_0\mathrm{d}Z_0'\delta'(Z_0')\mathrm{e}^{-\frac{i}{\hbar}P_0\cdot Z_0} \\ &\times <\alpha';R_0|\left\langle R_0+\frac{Z_0}{2}\right|\mathrm{e}^{-\frac{\beta}{2}\hat{H}_{sn}}\left|-\frac{Z_0'}{2}\right\rangle \\ &\times \left\langle\frac{Z_0'}{2}\right|\mathrm{e}^{-\frac{\beta}{2}\hat{H}_{sn}}\left|R_0-\frac{Z_0}{2}\right\rangle|\alpha;R_0>\rho_b(P_b,R_b;R_0), \end{aligned} \tag{42}$$

where $\rho_b(P_b,R_b;R_0)$ is proportional to the Wigner transform of the canonical equilibrium density matrix for the bath in the field of the R_0 coordinates,

$$\begin{aligned} \rho_b(P_b,R_b;R_0) &= \frac{1}{(2\pi\hbar)^{\nu-1}}\int \mathrm{d}Z_b\mathrm{e}^{-\frac{i}{\hbar}P_b\cdot Z_b} \\ &\times \left\langle R_b+\frac{Z_b}{2}\right|\mathrm{e}^{-\beta\hat{H}_{b(n)}}\left|R_b-\frac{Z_b}{2}\right\rangle. \end{aligned} \tag{43}$$

Next, we single out the barrier region around $R_0 = 0$ for special consideration. Separating the Hamiltonian $\hat{H}_{sn}$ into a harmonic term $\hat{H}_{h0} = P_0^2/2M_0 - \frac{1}{2}M_0\omega^{\ddagger 2}R_0^2$ (where $\omega^{\ddagger}$ is the frequency at the barrier top) and remainder terms $\hat{h}_{sn}$, we can write $\hat{H}_{sn} = \hat{H}_{h0} + \hat{h}_{sn}$. The eigenstates of $\hat{h}_{sn}$ are $|\alpha; R_0>$ as above but the eigenvalues, denoted by $\varepsilon_\alpha(R_0)$, are related to the $E_\alpha(R_0)$ introduced earlier by $\varepsilon_\alpha(R_0) = E_\alpha(R_0) + \frac{1}{2}M_0\omega^{\ddagger 2}R_0^2$. Taking $\exp(-\beta\hat{H}_{sn}/2) \approx \exp(-\beta\hat{H}_{h0}/2)\exp(-\beta\hat{h}_{sn}/2)$, the matrix elements in (42) can then be written as

$$
\begin{aligned}
&< \alpha'; R_0| < R_0 + \frac{Z_0}{2}|\mathrm{e}^{-\frac{\beta}{2}\hat{H}_{sn}}| - \frac{Z_0'}{2} >< \frac{Z_0'}{2}|\mathrm{e}^{-\frac{\beta}{2}\hat{H}_{sn}}|R_0 - \frac{Z_0}{2} > |\alpha; R_0 > \\
&=< R_0 + \frac{Z_0}{2}|\mathrm{e}^{-\frac{\beta}{2}\hat{H}_{h0}}| - \frac{Z_0'}{2} >< \frac{Z_0'}{2}|\mathrm{e}^{-\frac{\beta}{2}\hat{H}_{h0}}|R_0 - \frac{Z_0}{2} > \\
&\quad \times < \alpha'; R_0|\mathrm{e}^{-\frac{\beta}{2}\hat{h}_{sn}(R_0+\frac{Z_0}{2})}\mathrm{e}^{-\frac{\beta}{2}\hat{h}_{sn}(R_0-\frac{Z_0}{2})}|\alpha; R_0 > . \qquad (44)
\end{aligned}
$$

Using the representation of $\hat{h}_{sn}$ in the adiabatic basis, $\mathrm{e}^{-\frac{\beta}{2}\hat{h}_{sn}(R_0)} = \sum_\alpha |\alpha; R_0 > \mathrm{e}^{-\frac{\beta}{2}\varepsilon_\alpha(R_0)} < \alpha; R_0|$, expressing the matrix element in a Taylor series in Z_0 and retaining up to first order terms in Z_0, we find

$$
\begin{aligned}
&< \alpha'; R_0|\mathrm{e}^{-\frac{\beta}{2}h_{sn}(R_0+\frac{Z_0}{2})}\mathrm{e}^{-\frac{\beta}{2}h_{sn}(R_0-\frac{Z_0}{2})}|\alpha; R_0 > \\
&= \mathrm{e}^{-\beta\varepsilon_\alpha(R_0)}\left[\delta_{\alpha\alpha'} + \frac{Z_0}{2}O_{\alpha'\alpha}(R_0)\mathrm{d}_{\alpha'\alpha}(R_0) + \mathcal{O}(Z_0^2)\right], \qquad (45)
\end{aligned}
$$

where $O_{\alpha'\alpha}(R_0) = \left(1 - \mathrm{e}^{-\frac{\beta}{2}\varepsilon_{\alpha'\alpha}(R_0)}\right)^2$ and $\varepsilon_{\alpha'\alpha} = \varepsilon_{\alpha'} - \varepsilon_\alpha$.

Finally, using the expression for the matrix elements of the harmonic oscillator imaginary time propagator,

$$
\begin{aligned}
< R_0|\mathrm{e}^{-\frac{\beta}{2}\hat{H}_{h0}}|R_0' >= &\sqrt{\frac{2aM_0u}{\pi\sin u}}\exp\Big[-aM_0u \\
&\left\{-(R_0+R_0')^2\tan\frac{u}{2} + (R_0-R_0')^2\cot\frac{u}{2}\right\}\Big], \qquad (46)
\end{aligned}
$$

where $u = \beta\hbar\omega^{\ddagger}/2$ and $a = (2\beta\hbar^2)^{-1}$, and carrying out the integrations over Z_0 and Z_0', we have

$$
\begin{aligned}
W_{A'}^{\alpha'\alpha}\left(X, \frac{\mathrm{i}\hbar\beta}{2}\right) = &\frac{1}{2\pi\hbar Z_Q}\frac{1}{\cos^2 u}\sqrt{\frac{2M_0u'}{\beta\hbar^2\pi}}\mathrm{e}^{-\frac{2M_0u'}{\beta\hbar^2}R_0^2} \\
&\times\frac{P_0}{M_0}\mathrm{e}^{-\frac{\beta P_0^2}{2M_0u'}}F_{\alpha'\alpha}(R_0)\rho_b(P_b, R_b; R_0), \qquad (47)
\end{aligned}
$$

where $u' \equiv u\cot u$ and

$$
F_{\alpha'\alpha}(R_0) = \mathrm{e}^{-\beta\varepsilon_\alpha(R_0)}\left(\delta_{\alpha'\alpha} + \frac{1}{2}\left(1 - \frac{\beta P_0^2}{M_0u'}\right)\frac{\mathrm{i}\hbar}{P_0}\mathrm{d}_{\alpha'\alpha}O_{\alpha'\alpha}\right). \qquad (48)
$$

The off-diagonal contribution to $W_{A'}$ is imaginary and therefore, from (34), only the imaginary part of $B_{\mathrm{W}}^{\alpha\alpha'}(X,t)$ contributes to the rate.

Partitioning of the propagator

When the ground and excited states have different structures in the barrier region (as in the right panel of Fig. 1), the parabolic approximation used above is no longer valid and another approximation must be made. In this connection, instead of singling out the harmonic part of $\hat{H}_{sn}$ in the barrier region, one may partition $\hat{H}_{sn}$ into kinetic plus potential terms as $\hat{H}_{sn} = \hat{P}_0^2/2M_0 + \hat{h}_0$. Then approximating the propagator in (42) as $\mathrm{e}^{\beta \hat{H}_{sn}/2} \approx \mathrm{e}^{\beta \hat{P}_0^2/4M_0} \mathrm{e}^{\beta \hat{h}_0/2}$, and carrying out a series of calculations similar to those outlined above, we obtain

$$W_{A'}^{\alpha'\alpha}\left(X, \frac{\mathrm{i}\hbar\beta}{2}\right) = \frac{1}{2\pi\hbar Z_Q}\sqrt{\frac{2M_0}{\beta\hbar^2\pi}}\mathrm{e}^{-\frac{2M_0}{\beta\hbar^2}R_0^2} \times \frac{P_0}{M_0}\mathrm{e}^{-\frac{\beta P_0^2}{2M_0}}\mathcal{F}_{\alpha'\alpha}(R_0)\rho_b(P_b, R_b; R_0), \tag{49}$$

where $\mathcal{F}_{\alpha'\alpha}$ has a definition similar to that of $F_{\alpha'\alpha}$ but with $\varepsilon_\alpha(R_0)$ replaced by $E_\alpha(R_0)$,

$$\mathcal{F}_{\alpha'\alpha}(R_0) = \mathrm{e}^{-\beta E_\alpha(R_0)}\left(\delta_{\alpha'\alpha} + \frac{1}{2}\left(1 - \frac{\beta P_0^2}{M_0}\right)\frac{\mathrm{i}\hbar}{P_0}\mathrm{d}_{\alpha'\alpha}\mathcal{O}_{\alpha'\alpha}\right). \tag{50}$$

Likewise, $\mathcal{O}_{\alpha'\alpha}$ has a definition analogous to that of $O_{\alpha'\alpha}$ with $\varepsilon_\alpha(R_0)$ replaced by $E_\alpha(R_0)$.

The advantages of the two methods based on (47) and (49) are worth noting. Equation (49) does not assume a particular form for the potential in the barrier region, while in (47), a harmonic form is assumed. However, (47) retains the quantum effects resulting from the coupling between the potential and kinetic terms unlike (49).

Classical treatment of environmental coordinates

Making the high temperature approximation $\lim_{\beta\to 0}\sqrt{\frac{a}{\beta\pi}}\mathrm{e}^{-\frac{a}{\beta}R_0^2} = \delta(R_0)$ and using the classical analog of (43), $\rho_b^{\mathrm{cl}}(P_b, R_b; R_0) = \frac{\mathrm{e}^{-\beta H_{b(n)}}}{(2\pi\hbar)^{\nu-1}}$, (49) reduces to

$$\begin{aligned} W_{A'}^{\alpha'\alpha}\left(X, \frac{\mathrm{i}\hbar\beta}{2}\right) &= \frac{1}{2\pi\hbar Z_Q}\delta(R_0)\frac{P_0}{M_0}\mathrm{e}^{-\frac{\beta P_0^2}{2M_0}}\mathcal{F}_{\alpha'\alpha}(R_0)\rho_b^{\mathrm{cl}}(P_b, R_b; R_0) \\ &= \frac{1}{(2\pi\hbar)^\nu Z_Q}\delta(R_0)\frac{P_0}{M_0}\mathrm{e}^{-\beta H_\alpha(X)} \\ &\quad \times \left(\delta_{\alpha'\alpha} + \frac{1}{2}\left(1 - \frac{\beta P_0^2}{M_0}\right)\frac{\mathrm{i}\hbar}{P_0}d_{\alpha'\alpha}\mathcal{O}_{\alpha'\alpha}\right), \end{aligned} \tag{51}$$

where $H_\alpha(X) = H_{b(n)} + E_\alpha(R_0)$. This result may be substituted into (34) to obtain an expression for the rate coefficient:

$$k_{AB}(t) = k_{AB}^{\mathrm{d}}(t) + k_{AB}^{o}(t), \tag{52}$$

where the diagonal contribution is

$$k_{AB}^{\mathrm{d}}(t) = \frac{-1}{n_A^{\mathrm{eq}}(2\pi\hbar)^{\nu} Z_Q} \sum_{\alpha} \int \mathrm{d}X B_{\mathrm{W}}^{\alpha\alpha}(X,t)\delta(R_0)\frac{P_0}{M_0}\mathrm{e}^{-\beta H_\alpha(X)}, \tag{53}$$

and the off-diagonal contribution is

$$k_{AB}^{o}(t) = \frac{1}{n_A^{\mathrm{eq}}(2\pi\hbar)^{\nu} Z_Q} \sum_{\alpha'>\alpha} \int \mathrm{d}X \mathrm{Im}\{B_{\mathrm{W}}^{\alpha\alpha'}(X,t)\}\delta(R_0)\frac{P_0}{M_0}\mathrm{e}^{-\beta H_\alpha(X)} \times \left(1 - \frac{\beta P_0^2}{M_0}\right)\frac{\hbar}{P_0} d_{\alpha'\alpha}\mathcal{O}_{\alpha'\alpha}. \tag{54}$$

The diagonal contribution agrees with the result obtained earlier using quantum–classical linear response theory [40], while the off-diagonal contribution does not due to the inherent differences in the approximations made. For a general reaction coordinate, $\xi(R)$, the high temperature approximation leads to

$$k_{AB}^{\mathrm{d}}(t) = \frac{-1}{n_A^{\mathrm{eq}}(2\pi\hbar)^{\nu} Z_Q} \sum_{\alpha} \int \mathrm{d}X \, \frac{P}{M} \cdot \nabla_R \xi(R) B_{\mathrm{W}}^{\alpha\alpha}(X,t) \times \delta(\xi(R) - \xi^{\ddagger})\mathrm{e}^{-\beta H_\alpha(X)}, \tag{55}$$

and

$$k_{AB}^{o}(t) = \frac{1}{n_A^{\mathrm{eq}}(2\pi\hbar)^{\nu} Z_Q} \sum_{\alpha'>\alpha} \int \mathrm{d}X \mathrm{Im}\{B_{\mathrm{W}}^{\alpha\alpha'}(X,t)\}\delta(\xi(R) - \xi^{\ddagger})\mathrm{e}^{-\beta H_\alpha(X)} \times \left(\sum_j \frac{\nabla_{R_j}\xi(R)}{M_j}\left[d_{\alpha'\alpha}^{j} - \beta P_j\left(\frac{P}{M}\cdot d_{\alpha'\alpha}\right)\right]\right)\hbar\mathcal{O}_{\alpha'\alpha}. \tag{56}$$

4 Applications

In order to compute the rate constants of processes such as proton and electron transport in condensed phases, one must account for the effects of the environmental degrees of freedom. The theory presented in the previous sections provides a convenient framework in which a rate study of such systems can be performed.

In this section, we show the results of a rate coefficient calculation for a proton transfer reaction occurring in a linear hydrogen-bonded complex dissolved in a polar solvent. The proton is treated quantum mechanically and the remainder of the degrees of freedom is treated classically. Valuable insight into such rate processes can also be obtained, more efficiently, by studying simple transfer reaction models which simulate the effect of a condensed phase environment on a reaction coordinate. In this connection, we first show rate results for a two-level quantum system coupled to a classical nonlinear oscillator that is in turn coupled to classical harmonic bath. For these two applications, an appropriate choice of reaction coordinate and species variables is made and quantum–classical Liouville dynamics is used to evolve the species variables.

4.1 Two-level Model for Transfer Reactions

Spin-boson-type models, where a two-level quantum system is bilinearly coupled to a bath of independent harmonic oscillators, have often been used to compute nonadiabatic reaction rates [8, 16, 23, 48, 56, 57]. For such spin-boson systems quantum–classical dynamics is exact and our simulation algorithms that employ quantum–classical trajectories have been shown [58] to reproduce the exact quantum results [56]. The rate constant for such spin-boson systems, when computed using quantum–classical dynamics and sampling from quantum initial states, corresponds to that obtained in a full quantum treatment [45, 47]. Here we consider a more complex model involving coupling between the two-level system, a nonlinear oscillator and a bath of harmonic oscillators as a more realistic model for quantum particle transfer in the condensed phase. No exact results are available for this model.

Model

The model system we consider has the Hamiltonian operator, expressed in the diabatic basis $\{|L\rangle, |R\rangle\}$ [40]

$$\mathbf{H} = \begin{pmatrix} V_n(R_0) + \hbar\gamma_0 R_0 & -\hbar\Omega \\ -\hbar\Omega & V_n(R_0) - \hbar\gamma_0 R_0 \end{pmatrix} + \left(\frac{P_0^2}{2M_0} + \sum_{j=1}^{N} \frac{P_j^2}{2M_j} + \sum_{j=1}^{N} \frac{M_j}{2}\omega_j^2 \left(R_j - \frac{c_j}{M_j\omega_j^2} R_0 \right)^2 \right) \mathbf{I} \,. \tag{57}$$

In this model, a two-level system is coupled to a classical nonlinear oscillator with mass M_0 and phase space coordinates (R_0, P_0). This coupling is given by $\hbar\gamma_0 R_0 = \hbar\gamma(R_0)$. The nonlinear oscillator, which has a quartic potential energy function $V_n(R_0) = aR_0^4/4 - M_0\omega^{\ddagger 2}R_0^2/2$, is then bilinearly coupled to a bath of N independent harmonic oscillators. From the first matrix in (57), we see that the diabatic energies are given by $E_{1,2}^d(R_0) = V_n(R_0) \pm \hbar\gamma_0 R_0$ and the coupling between the diabatic states is $-\hbar\Omega$. The bath harmonic oscillators labelled $j = 1, ..., N$ have masses M_j and frequencies ω_j. The bilinear coupling is characterized by an Ohmic spectral density [56, 57], $J(\omega) = \pi \sum_{j=1}^{N} (c_j^2/(2M_j\omega_j^2)\delta(\omega - \omega_j)$, where $c_j = (\xi\hbar\omega_0 M_j)^{1/2}\omega_j$, $\omega_j = -\omega_c \ln(1 - j\omega_0/\omega_c)$, and $\omega_0 = \frac{\omega_c}{N}\left(1 - \mathrm{e}^{-\omega_{max}/\omega_c}\right)$, with ω_c a cut-off frequency.

The adiabatic states obtained from the diagonalization of Hamiltonian (57) are given by

$$
\begin{aligned}
|1;R_0\rangle &= \frac{1}{\mathcal{N}}\left[(1+G)|L\rangle + (1-G)|R\rangle\right] \\
|2;R_0\rangle &= \frac{1}{\mathcal{N}}\left[-(1-G)|L\rangle + (1+G)|R\rangle\right],
\end{aligned} \tag{58}
$$

where $\mathcal{N}(R_0) = \sqrt{2(1+G^2(R_0))}$ and

$$
G(R_0) = \gamma(R_0)^{-1}\left[-\Omega + \sqrt{\Omega^2 + \gamma^2(R_0)}\right]. \tag{59}
$$

The corresponding adiabatic energies are $E_{1,2}(R) = V_b(R) \mp \sqrt{\Omega^2 + \gamma^2(R_0)}$, where

$$
V_b(R) = V_n(R_0) + \sum_{j=0}^{N}\frac{P_j^2}{2M_j} + \sum_{j=1}^{N}\frac{M_j}{2}\omega_j^2\left(R_j^2 - \frac{c_j}{M_j\omega_j^2}R_0\right)^2. \tag{60}
$$

Insight into the nature of the quantum reaction dynamics can be gained by considering the ground and first excited adiabatic free energies along the R_0 coordinate, as given by

$$
\begin{aligned}
W_\alpha(R_0) &= -\beta^{-1}\ln\left(\int\prod_{j=1}^{N}\mathrm{d}R_j\ Z_\alpha^{-1}\mathrm{e}^{-\beta E_\alpha(R)}\right) \\
&= \beta^{-1}\ln Z_\alpha + V_n(R_0) \mp \sqrt{\Omega^2 + \gamma_0^2 R_0^2}\,,
\end{aligned} \tag{61}
$$

where $Z_\alpha = \int \mathrm{d}R\ \exp(-\beta E_\alpha(R))$ and $\alpha = 1,2$. They are plotted in Fig. 1 for both a small (left) and large (right) value of γ_0. Based on the structure of these profiles, we may choose $\theta(R_0)$ and $\theta(-R_0)$ for the A and B species variables, respectively.

For small values of γ_0, the potential in the reactant region is approximately harmonic, making the ground and excited free energy surfaces nearly parallel. As a result, the partition function for the reactant state, $n_A^{\mathrm{eq}}Z_Q$, can be approximated using the mean free energy surface and given by

$$
(n_A^{\mathrm{eq}}Z_Q)^{-1} \approx \mathrm{e}^{\beta V_r}\sinh(\beta\hbar\omega_r/2)\prod_{j=1}^{N}2\sinh(\beta\hbar\omega_j/2), \tag{62}
$$

where ω_r is the frequency in the reactant well and V_r is the bare potential at the bottom of it. Using the high temperature form of (62) in the transition state theory ($t = 0^+$) form of (53) we obtain

$$
k_{AB}^{TST} \approx \frac{\omega_r \mathrm{e}^{\beta V_r}}{2\pi}\frac{\mathrm{e}^{-\beta\Omega} + \mathrm{e}^{\beta\Omega}}{2}, \tag{63}
$$

which will be used to scale the results presented in the figures. When the coupling between the two-level system and the quartic oscillator is negligible, Ω is also negligible, and k_{AB}^{TST} becomes the well-known value of $\omega_r e^{\beta V_r}/2\pi$. In this regime, a symmetric oscillator has the frequency $\omega_r \approx \sqrt{2}\omega^{\ddagger}$. Using these results and (34) and (47) for $k_{AB}(t)$, the transmission coefficient, $\kappa_{AB}(t) = k_{AB}(t)/k_{AB}^{TST}$, takes the form

$$\kappa_{AB}(t) = -\sum_{\alpha}\sum_{\alpha' \geq \alpha}(2 - \delta_{\alpha'\alpha})\int \mathrm{d}X \mathrm{Re}[B_{\mathrm{W}}^{\alpha\alpha'}(X,t) w_{QRB}^{\alpha'\alpha}(X)], \qquad (64)$$

where

$$w_{QRB}^{\alpha'\alpha}(X) = \frac{2u}{\sin 2u}\frac{\sinh u_r}{u_r}\frac{P_0}{M_0}\sqrt{\frac{\pi M_0 \beta}{2u'}}\frac{F_{\alpha'\alpha}}{\sum_\alpha e^{-\beta\varepsilon_\alpha(0)}}$$
$$\times G_a\left(R_0; \frac{2M_0 u'}{\beta\hbar^2}\right) G_a\left(P_0; \frac{\beta}{2M_0 u'}\right)$$
$$\times \prod_{j=1}^{N} G_a\left(R_j - \frac{c_j R_0}{M_j \omega_j^2}; \frac{\beta}{2u_j''} M_j \omega_j^2\right) G_a\left(P_j; \frac{\beta}{2M_j u_j''}\right), \qquad (65)$$

the Gaussian function G_a is defined by $G_a(x;b) = \sqrt{\frac{b}{\pi}}\exp(-bx^2)$, and $u_j'' = u_j \coth u_j$ with $u_j = \beta\hbar\omega_j/2$. We label the results obtained using this formula, which treats the initial distribution of the reaction coordinate and bath quantum mechanically, by QRB.

When the initial distribution of the reaction coordinate and bath is treated classically, we can use (51) for $W_A^{\alpha'\alpha}$ to obtain

$$w_{CRB}^{\alpha\alpha'}(X) = \frac{P_0}{M_0}\sqrt{\frac{\pi M_0 \beta}{2}}\delta(R_0) G_a\left(P_0; \frac{\beta}{2M_0}\right)\frac{\mathcal{F}_{\alpha'\alpha}}{\sum_\alpha \mathrm{e}^{-\beta E_\alpha(0)}}$$
$$\times \prod_{j=1}^{N} G_a\left(R_j; \frac{\beta}{2} M_j \omega_j^2\right) G_a\left(P_j; \frac{\beta}{2M_j}\right). \qquad (66)$$

Results obtained using this formula are labeled by CRB.

We used a convenient set of dimensionless coordinates and parameters, which is given in [40]. The calculations were performed for a bath of $N = 100$ harmonic oscillators with the following values of the parameters: $\omega_{\max} = 3$, $\Omega = 0.1$, $\gamma_0 = 0.1$, $a = 0.05$, and $\omega^{\ddagger} = 1$. The simulation scheme for carrying out quantum–classical molecular dynamics has been described in detail earlier [40, 43, 58, 59], so only a few comments about the calculations are needed here. The initial distribution of X for the QRB and CRB results was sampled from weights determined by (65) and (66), respectively. The time evolution of the species variable, $B_W^{\alpha\alpha'}(X,t)$, is determined from constant energy quantum–classical trajectories generated using the sequential short-time propagation algorithm [58].

Results

First, we compare the QRB and CRB rate results for two temperatures in Fig. 2. For high temperatures ($\beta = 0.1$), both the QRB and CRB results are indistinguishable, except at very short times. At $t = 0$, the CRB result for the time-dependent transmission coefficient, $\kappa(t)$, is nonzero and equal to unity, which yields the transition state theory value of the rate constant. The QRB results for the time-dependent transmission coefficient are zero at $t = 0$, which is expected from quantum rate processes [46]. At lower temperatures ($\beta = 2$), where quantum effects are more pronounced, one sees that the QRB formula yields a larger rate constant than does the CRB one. This enhancement of the quantum rate has also been observed in other studies [15, 48, 60].

In Fig. 3, the QRB results for the transmission coefficient κ_{AB}, obtained from the plateau value of $\kappa_{AB}(t)$, are plotted as a function of the Kondo parameter ξ, which when increased, creates more friction in the bath. As the friction is increased from zero, the rate initially increases to a maximum and then continuously decreases, capturing the well-known turnover behavior [61]. This initial increase at low values of ξ is solely due to quantum effects.

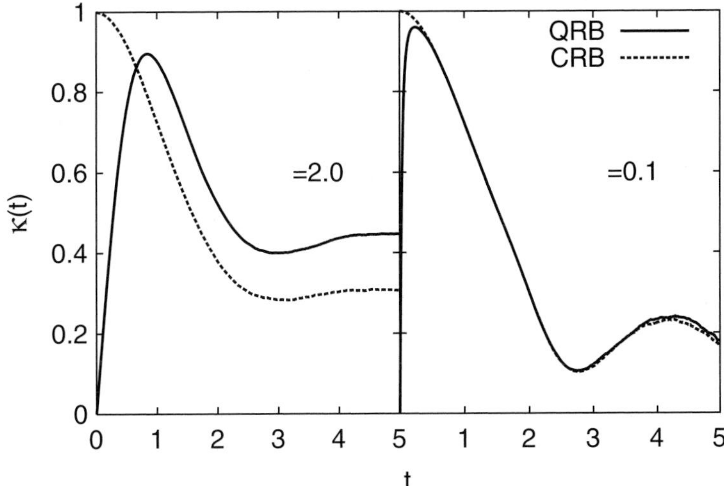

Fig. 2. Comparison between the time-dependent transmission coefficient of the case where the equilibrium structure of the reaction coordinate and bath is treated quantum mechanically (QRB) and that of the case where it is treated classically (CRB). Parameters values: $\beta = 2$ (left), $\beta = 0.1$ (right), $\gamma_0 = 0.1$, $\Omega = 0.1$, and $\xi = 3$

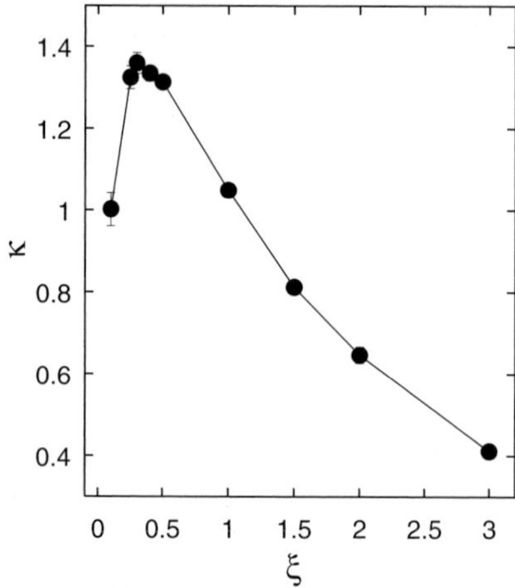

Fig. 3. Transmission coefficient (κ) vs. the Kondo parameter (ξ) for $\beta = 2, \gamma_0 = 0.1$, and $\Omega = 0.1$

4.2 Proton Transfer

Model

Proton transfer dynamics plays an important role in many chemical and biological systems. Therefore, an accurate picture of the global dynamics of these systems requires a careful treatment of the proton in the context of its environment. Since these systems are usually too complex and too large to simulate, one can resort to simplified models in order to gain valuable insights. In this connection, we studied a model for a proton transfer reaction (AH-$B \rightleftharpoons A^-$–H^+B) in a hydrogen-bonded complex (AHB) dissolved in a polar solvent. All the details of the model can be found in [43] and references therein, so we will only mention a few main aspects of it. This model has been used as a benchmark for testing a variety of techniques [62–68].

The potential energy describing the hydrogen bonding interaction within the complex in the absence of a solvent, which is a function of the protonic coordinate, models a slightly strongly hydrogen-bonded phenol (A) trimethylamine (B) complex. The parameters which control the strength of the $A - B$ bond were chosen to yield an equilibrium $A - B$ separation of $R_{AB} = 2.7$ Å. For this value of R_{AB}, the potential energy function has two minima, the deeper minimum corresponding to the stable covalent state and the shallower minimum corresponding to the metastable ionic state. We have constrained R_{AB} to be 2.7 Å in our simulations. The AHB complex is dissolved in a solvent composed of 255 polar, nonpolarizable model methyl chloride molecules. The temperature of the system in the simulations performed was approximately 250 K.

The time evolution of the system is determined using quantum–classical Liouville dynamics in which the complex and solvent are treated classically and the proton, quantum mechanically. The Hamiltonian operator, which is partially Wigner transformed over the solvent and A and B groups of the complex, can be found in [43].

Proton transfer dynamics in polar liquids is usually monitored [69, 70] by the solvent polarization, $\Delta E(R)$,

$$\Delta E(R) = \sum_{i,a} z_a \mathrm{e} \left(\frac{1}{|\mathrm{R}_i^a - s|} - \frac{1}{|\mathrm{R}_i^a - s'|} \right) , \tag{67}$$

where $z_a e$ ($e = 1.602 \times 10^{-19}$ C) is the charge on solvent atom a, and s and s' are two points within the complex, one at the center of mass and the other displaced by -0.56 Å from the center of mass, respectively, which correspond to the minima of the bare hydrogen bonding potential. The sums run over all solvent molecules i and atoms a. In essence, the solvent polarization is the difference between the solvent electrical potentials at points s and s' and drives the transfer of the proton, making it an ideal reaction coordinate.

In Fig. 4 we see that ΔE tracks the hops of the proton between the reactant/covalent state ($\Delta E \approx 0.005$ eC/Å) and the product/ionic state ($\Delta E \approx 0.0225$ eC/Å). The complex spends more time in the ionic configuration than in the covalent configuration since electrostatic interactions with the polar solvent preferentially stabilize the ionic configuration of the complex. In the absence of the polar solvent, the complex is primarily found in the covalent configuration.

The free energy profiles corresponding to adiabatic evolution on the ground, first and second excited state surfaces, are shown in Fig. 5. The free energy in the ground state has a double-well structure and a single-well structure in the first excited state. The second excited state free energy has a double-well structure with a relatively low barrier. Given the magnitude of the energy gap between the first and second excited state surfaces, the second excited state

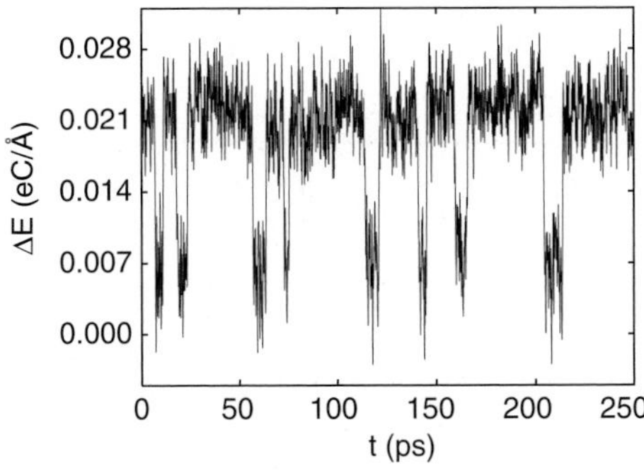

Fig. 4. Time series of the solvent polarization (ΔE) for a ground state adiabatic trajectory

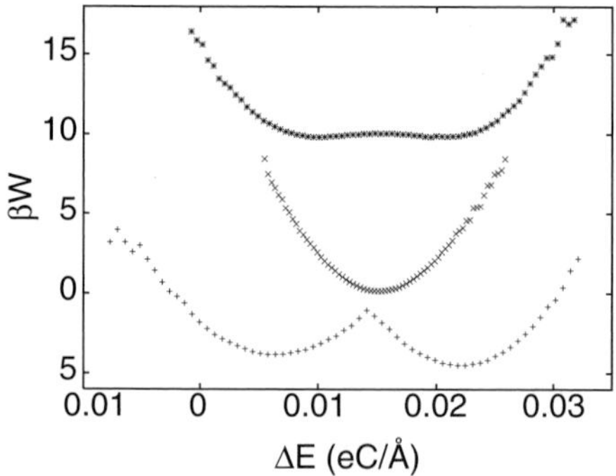

Fig. 5. Free energy (βW) profiles along the ΔE reaction coordinate for the system undergoing ground, first and second excited state adiabatic dynamics

is not expected to participate strongly in the nonadiabatic quantum-classical dynamics. It is evident from the ground state free energy profile that the minimum of the ionic state is lower in free energy than that of the covalent state as a result of the stabilizing effect of the polar solvent. The barrier top of the ground state surface is located at $\Delta E^{\ddagger} = 0.0141$ eC/Å.

Since the temperature of the system is fairly high and the dynamics of the solvent and complex atoms can be accurately captured using classical mechanics, a high temperature/classical approximation (analogous to the one which lead to (51)) is made to obtain a rate expression for this proton transfer reaction that employs $\Delta E(R)$ as the reaction coordinate. Based on the structure of the free energy profiles, we selected the A and B species variables as $\hat{N}_A = \theta(\Delta E(R) - \Delta E^{\ddagger})$ and $\hat{N}_B = \theta(\Delta E^{\ddagger} - \Delta E(R))$, respectively. The specific form of the diagonal part of the rate coefficient (which turns out to be the major contribution) for this choice of species variables is

$$k^d_{AB}(t) = -\frac{1}{n_A^{\text{eq}}} \sum_\alpha \int \mathrm{d}R\,\mathrm{d}P\, \Delta\dot{E}(R) N_B^{\alpha\alpha}(R,P,t) \times \delta(\Delta E(R) - \Delta E^{\ddagger}) \rho_{W_e}^{\alpha\alpha}\,, \tag{68}$$

where the equilibrium fraction of species A is

$$n_A^{\text{eq}} = \int \mathrm{d}\Delta E'\; \theta(\Delta E(R) - \Delta E^{\ddagger}) \mathrm{e}^{-\beta W(\Delta E')} P_u, \tag{69}$$

and the time derivative of the solvent polarization can be rewritten as $\Delta\dot{E}(R) = \frac{P}{M} \cdot \nabla_R \Delta E(R)$. The canonical equilibrium distribution is given by $\rho_{W_e}^{\alpha\alpha} = Z_0^{-1} \mathrm{e}^{-\beta H_W^\alpha}$, with $Z_0 = \sum_\alpha \int \mathrm{d}R\,\mathrm{d}P\, \mathrm{e}^{-\beta H_W^\alpha}$.

Equation (68) provides a well-defined formula involving initial sampling from the barrier top $\Delta E = \Delta E^{\ddagger}$. In addition, quantum–classical time evolution of $N_B^{\alpha\alpha}(R,P,t)$ must be carried out to compute the reaction rate.

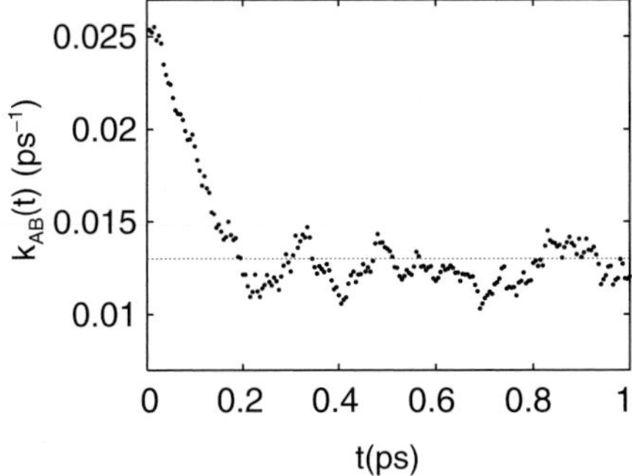

Fig. 6. The rate coefficient, $k_{AB}(t)$, as a function of time. The dotted line indicates the plateau value k_{AB}

Results

In Fig. 6, we present results for the time-dependent rate coefficient which were obtained from an average over 16,000 trajectories. As expected from the high temperature form of the rate coefficient, we see that it falls quickly from its initial transition state theory value in a few tenths of a picosecond to a plateau from which the rate constant can be extracted. The decrease in the rate coefficient from its transition state theory value is due to recrossing by the trajectory of the barrier top before the system reaches a stable state. The value of k_{AB} obtained from the plateau is $k_{AB} = 0.013$ ps^{-1}. This result is 32% lower than the adiabatic result, indicating a significant nonadiabatic quantum correction.

5 Concluding Remarks

The theory presented in this chapter shows how chemical reaction rates can be computed from time correlation function expressions that retain the quantum equilibrium structure of the system and employ a quantum–classical description of the dynamics of the species variables. Thus, the computational method combines a surface-hopping dynamics based on the quantum–classical Liouville equation, with initial sampling from a quantum equilibrium distribution. As such, the method differs from conventional surface-hopping schemes for reactive quantum–classical dynamics, both in the nature of the time evolution of operators and in the way the trajectories are sampled to compute the reaction rate.

The simulation results reported above utilized various approximate analytical expressions for the spectral density function that describes the quantum equilibrium structure. In some circumstances, especially for low temperatures, effects arising from the quantum equilibrium structure lead to important modifications of the reaction rate. To treat more general and complex molecular systems one could resort to numerical schemes for computing the equilibrium structure, similar to those based on the initial value representation [68, 71, 72] and linearization techniques [50, 73–75].

Different formulas for the time-dependent rate coefficient can be derived within this framework using other choices of the reaction coordinate and chemical species variables. These should allow one to effectively capture quantum effects in a variety of chemical rate processes occurring within a wide range of temperatures and in complex condensed phase environments.

References

1. A. Kohen, J. P. Klinman: Acc. Chem. Res. **31**, 397 (1998)
2. J. Lobaugh, G. A. Voth: J. Chem. Phys. **104**, 2056 (1996)
3. D. Marx, M. E. Tuckerman, J. Hutter, M. Parrinello: Nature **397**, 601 (1999)
4. T. Yamamoto: J. Chem. Phys. **33**, 281 (1960)
5. K. Thompson, N. Makri: J. Chem. Phys. **110**, 1343 (1999)
6. N. Makri: J. Phys. Chem. B **103**, 2823 (1999)
7. N. Makri, W. H. Miller: J. Chem. Phys. **89**, 2170 (1988)
8. C. H. Mak, D. Chandler: Phys. Rev. A **44**, 2352 (1991)
9. J. T. Stockburger, H. Grabert: Chem. Phys. **268**, 249 (2001)
10. J. S. Shao, C. Zerbe, P. Hanggi: Chem. Phys. **235**, 81 (1998)
11. E. Geva, Q. Shi, G. A. Voth: J. Chem. Phys. **115**, 9209 (2001)
12. R. I. Cukier, J. J. Zhu: J. Phys. Chem. **101**, 7180 (1997); R. I. Cukier: J. Chem. Phys. **88**, 5594 (1988)
13. D. R. Reichman, E. Rabani: Phys. Rev. Lett. **87**, 265702 (2001)
14. E. Rabani, D. R. Reichman: J. Chem. Phys. **120**, 1458 (2004)
15. H. B. Wang, X. Sun, W. H. Miller: J. Chem. Phys. **108**, 9726 (1998)
16. X. Sun, H. B. Wang, W. H. Miller: J. Chem. Phys. **109**, 7064 (1998)
17. X. Sun, H. B. Wang, and W. H. Miller: J. Chem. Phys. **109**, 4190 (1998)
18. X. Sun and W. H. Miller: J. Chem. Phys. **110**, 6635 (1999)
19. W. H. Miller: J. Phys. Chem. A **105**, 2942 (2001)
20. M. Thoss and H. B. Wang: Annu. Rev. Phys. Chem. **55**, 299 (2004)
21. K. G. Kay: Annu. Rev. Phys. Chem. **56**, 255 (2005)
22. M. A. Sepulveda, F. Grossmann: Adv. Chem. Phys. **96**, 191 (1996)
23. S. Bonella and D. F. Coker: J. Chem. Phys. **122**, 194102 (2005)
24. S. Bonella, D. Montemayor, D. F. Coker: Proc. Natl. Acad. Sci. **102**, 6715 (2005)
25. A. A. Neufeld: J. Chem. Phys. **122**, 164111 (2005)
26. J. C. Tully: J. Chem. Phys. **93**, 1061 (1990)
27. S. Hammes-Schiffer and J. C. Tully: J. Chem. Phys. **101**, 4657 (1994)
28. D. F. Coker, L. Xiao: J. Chem. Phys. **102**, 496 (1995)
29. F. Webster, E. T. Wang, P. J. Rossky, R. A. Friesner: J. Chem. Phys. **100**, 4835 (1994)
30. A. W. Jasper, S. N. Stechmann, D. G. Truhlar: J. Chem. Phys. **116**, 5424 (2002)
31. H. Wang, M. Thoss: J. Chem. Phys. **119**, 1289 (2003)
32. H. Wang, M. Thoss: Chem. Phys. Lett. **389**, 43 (2004)

33. I. V. Aleksandrov: Z. Naturforsch. **36**, 902 (1981)
34. A. Donoso, C. C. Martens: J. Phys. Chem. A **102**, 4291 (1998)
35. V. I. Gerasimenko: Theor. Math. Phys. **50**, 49 (1982)
36. I. Horenko, C. Salzmann, B. Schmidt, C. Schutte: J. Chem. Phys. **117**, 11075 (2002)
37. R. Kapral, G. Ciccotti: J. Chem. Phys. **110**, 8919 (1999)
38. C. Wan, J. Schofield: J. Chem. Phys. **113**, 7047 (2000)
39. S. Nielsen, R. Kapral, G. Ciccotti: J. Chem. Phys. **115**, 5805 (2001)
40. A. Sergi, R. Kapral: J. Chem. Phys. **118**, 8566 (2003)
41. J. L. Liao, E. Pollak: J. Chem. Phys. **116**, 2718 (2002)
42. A. Sergi, R. Kapral: J. Chem. Phys. **119**, 12776 (2003)
43. G. Hanna, R. Kapral: J. Chem. Phys. **122**, 244505 (2005)
44. A. Sergi, R. Kapral: J. Chem. Phys. **121**, 7565 (2004)
45. H. Kim, R. Kapral: J. Chem. Phys. **122**, 214105 (2005)
46. R. Kapral, S. Consta, L. McWhirter: Chemical rate laws and rate constants. In: *Classical and Quantum Dynamics in Condensed Phase Simulations*, ed by B. J. Berne, G. Ciccotti, D. F. Coker (World Scientific, Singapore 1998) pp 583–617
47. H. Kim, R. Kapral: J. Chem. Phys. **123**, 194108 (2005)
48. M. Topaler and N. Makri: J. Chem. Phys. **101**, 7500 (1994)
49. H. Kim, P. J. Rossky: J. Phys. Chem. B **106**, 8240 (2002)
50. J. A. Poulsen, G. Nyman, P. J. Rossky: J. Chem. Phys. **119**, 12179 (2003)
51. H. A. Kramers: Physica **6**, 284 (1940)
52. E. Pollak, J. L. Liao: J. Chem. Phys. **108**, 2733 (1998)
53. J. S. Shao, J. L. Liao, E. Pollak: J. Chem. Phys. **108**, 9711 (1998)
54. G. A. Voth, D. Chandler, W. H. Miller: J. Chem. Phys. **91**, 7749 (1989)
55. P. G. Wolynes: Phys. Rev. Lett. **47**, 968 (1981)
56. N. Makri, K. Thompson: Chem. Phys. Lett. **291**, 101 (1998); K. Thompson, N. Makri: J. Chem. Phys. **110**, 1343 (1999); N. Makri: J. Phys. Chem. B **103**, 2823 (1999)
57. D. MacKernan, R. Kapral, G. Ciccotti: J. Chem. Phys. **116**, 2346 (2002)
58. D. MacKernan, R. Kapral, G. Ciccotti: J. Phys.: Condens. Matter **14**, 9069 (2002)
59. A. Sergi, D. MacKernan, G. Ciccotti, R. Kapral: Theor. Chem. Acc. **110**, 49 (2003)
60. E. Rabani, G. Krilov, B. J. Berne: J. Chem. Phys. **112**, 2605 (2000)
61. P. Hanggi, P. Talkner, M. Borkovec: Rev. Mod. Phys. **62**, 251 (1990)
62. H. Azzouz, D. C. Borgis: J. Chem. Phys. **98**, 7361 (1993)
63. S. Hammes-Schiffer, J. C. Tully: J. Chem. Phys. **101**, 4657 (1994)
64. R. P. McRae, G. K. Schenter, B. C. Garrett, Z. Svetlicic, D. G. Truhlar: J. Chem. Phys. **115**, 8460 (2001)
65. D. Antoniou, S. D. Schwartz: J. Chem. Phys. **110**, 465 (1999)
66. D. Antoniou, S. D. Schwartz: J. Chem. Phys. **110**, 7359 (1999)
67. S. Y. Kim, S. Hammes-Schiffer: J. Chem. Phys. **119**, 4389 (2003)
68. T. Yamamoto, W. H. Miller: J. Chem. Phys. **122**, 044106 (2005)
69. P. M. Kiefer, J. T. Hynes: Solid State Ionics **168**, 219 (2004)
70. D. Laria, G. Ciccotti, M. Ferrario, R. Kapral: J. Chem. Phys. **97**, 378 (1992)
71. T. Yamamoto, W. H. Miller: J. Chem. Phys. **118**, 2135 (2003)
72. Y. Zhao, T. Yamamoto, W. H. Miller: J. Chem. Phys. **120**, 3100 (2004)
73. J. A. Poulsen, G. Nyman, P. J. Rossky: J. Phys. Chem. A **108**, 8743 (2004)
74. Q. Shi, E. Geva: J. Chem. Phys. **118**, 8173 (2003)
75. Q. Shi, E. Geva: J. Chem. Phys. **120**, 10647 (2004)

Linearized Nonadiabatic Dynamics in the Adiabatic Representation

D.F. Coker and S. Bonella

Summary. In this chapter we generalize a recently developed approximate method for computing quantum time correlation functions based on linearizing the phase of path integral expressions for these quantities in terms of the difference between paths representing the forward and backward propagators. The approach is designed with condensed phase applications in mind and involves partitioning the system into two subsystems: One best described by a few discrete quantum states, the other represented as a set of particle positions and momenta. In the original formulation, a diabatic basis was used to describe the quantum subsystem states. Here we extend the technique to allow for a description of the quantum subsystem in terms of adiabatic states. These can be more appropriate in certain dynamical regimes and have the formal advantage that they can be defined uniquely from the electronic Hamitonian. The linearized algorithm in the adiabatic basis is derived first, and its properties are then compared to those of alternative dynamical schemes.

1 Introduction

Statistical mechanics identifies time correlation functions as the fundamental link between the microscopic description of a system and its macroscopic behavior as observed in experiments. If classical mechanics provides an accurate description, it is possible to remedy the shortcomings of the analytical tools applicable to the calculation of time correlation functions for complex systems by using accurate, exact computer simulation schemes. Unfortunately, if quantum mechanics must be applied to the evolution of the system, exact numerical algorithms scale exponentially with the number of degrees of freedom and simulations become rapidly too expensive to study realistic models of condensed phase experiments. Therefore, the development of approximate methods to calculate quantum time dependent correlation functions by computer simulation is an active field of research in physics and chemistry.

In recent work [1, 2] we presented a new approximate method, called the Linearized Approach to Nonadiabatic Dynamics using the Mapping hamiltonian formulation or LAND-Map, for calculating nonadiabatic time correlation functions. When studying the properties of nonadiabatic systems, a fruitful strategy to simplify the situation with respect to the fully quantum propagation is to take the classical limit for the evolution of the nuclear degrees of freedom in the presence of a set of electronic states which maintain all their quantum characteristics. This results in a mixed quantum–classical picture of the world of the type described in Chap. 12 and the references therein.

A key property of the LAND-Map method, not shared by many of the alternative techniques for simulating nonadiabatic dynamics, is that the coupling mechanism between the classical and quantum evolution equations emerges naturally from a theoretical analysis of the time-correlation function. This is accomplished by combining the linearization ideas put forward by Poulsen [3–5] and Geva [6–9] with the mapping description of nonadiabatic transitions [10–14].

Linearization methods start from a path integral representation of the forward and backward propagators in the Heisenberg representation of a time correlation function, and combine them to describe the overall time evolution of the system in terms of a set of classical trajectories whose initial conditions are sampled from a quantity related to the Wigner transform of the density operator [15]. The linearized expression for a correlation function is a powerful tool for describing systems in the condensed phase since the rapid decay of correlation functions for such systems enables reliable results to be obtained using a representation of the dynamics strictly valid only for relatively short times. In order to extend the linearization scheme to nonadiabatic problems it is convenient to describe the electronic role in the dynamics in terms of operators with a continuous spectrum. A way to achieve this goal that has proved accurate in many situations is provided by the mapping formalism [16–19]. The method represents the electronic degrees of freedom and the transitions between the different states in terms of positions and momenta of a set of fictitious harmonic oscillators. The mapping formalism was originally applied to nonadiabatic dynamics in the context of semiclassical calculations. In that case, the continuous nature of the oscillator's spectra simplified the task of defining a classical analog for the total Hamiltonian of the system. In linearized approaches the mapping representation proves even more fruitful in that it gives an exact expression for the quantum system transition amplitude in the nuclear path integral representation. The expression depends parametrically on the nuclear path, but it is otherwise explicit, local in time, and computable by propagating a set of auxiliary classical equations that describe the evolution of the quadratic degrees of freedom that account for electronic transitions in the full quantum propagator exactly.

In LAND-Map the mixed quantum–classical evolution is obtained by applying the linearization procedure to the nuclear variables only. This results in a coupled nuclear and electronic evolution that is quite different from the propagations used in related methods and in other nonadiabatic techniques. For example, the exact solution of the mapping evolution by means of a propagation that is local in time avoids the self-consistent calculation of the nuclear trajectories and electronic transition amplitude that limits the usefulness of the Pechukas [20] path integral approach. Further the continuous description of the electronic states is reflected in the smooth changes in the forces experienced by the classical bath when a nonadiabatic transition takes place. This is to be contrasted, for example, with the situation in Surface Hopping [21–25] or Wigner mixed quantum–classical approaches [26–30]. In both cases, ensembles of trajectories that combine segments of classical evolution with "hops" from one state to another determined by some stochastic mechanism must be propagated. These hops introduce discontinuities, for example in the momenta of the classical particles, and, in some cases, ambiguities in the redistribution of the energy among the different degrees of freedom that are absent in our case.

LAND-Map was originally derived using a diabatic representation of the electronic states (i.e., an electronic basis set whose elements do not depend on the nuclear coordinates). Although the definition of such a basis for a given system is not unique, it has the convenient characteristic of describing the coupling between the different electronic surfaces *via* relatively smooth functions of the nuclear coordinates: the off-diagonal elements of the electronic Hamiltonian. Consequently the resulting evolution equations for the nuclei in a mixed quantum–classical scheme are relatively easy to integrate numerically. This description of the states is particularly advantageous, for example, when describing strongly vibrationally coupled systems. There are, however cases in which choosing an adiabatic basis set for the electronic degrees of freedom proves more convenient both conceptually and numerically. For instance, the adiabatic picture is an accurate representation of the dynamics of a system where the time scale separation of the nuclear and electronic motion is preserved throughout the evolution, so that, even when approaching an avoided crossing between adiabatic surfaces, the component of the nuclear momentum in the direction of the nonadiabatic coupling vector (see definition later) remains small [31]. The main benefit with this choice lies in the fact that the electronic states are determined by diagonalizing the electronic Hamiltonian at each nuclear configuration during the run. In this case, off-diagonal terms in the kinetic-energy operator, usually highly localized and rapidly varying functions of the nuclear coordinates, are responsible for the coupling.

Here we extend our linearized approach to nonadiabatic dynamics to situations in which an adiabatic electronic basis is more convenient.

The generalization is worth investigating for several reasons. Quantum mechanics prescribes that the result of an observation, whether it be the average value of an operator or a correlation function, does not depend on the choice of a particular basis set. This is manifest in the mathematical representation of an observable as a trace (i.e., an invariant with respect to basis choice). Approximate methods, however, can fail to preserve this fundamental property.

The reasons of this shortcoming fall into three broad categories. First, it might be difficult to maintain the physical picture underlying the prescriptions of a given dynamical scheme in a basis set different from the one in which the scheme was originally conceived. Surface hopping [21–25] is a prominent example of this kind. Second, even if it is possible to cast a given approach in a different basis, the approximations to the dynamics may lead to evolution equations that are not equivalent (i.e., not related by a canonical transformation) classically. In the self-consistent classical trajectory representation of nonadiabatic dynamics, such as for example the Ehrenfest method, this pathology manifests itself in the fact that the nuclear Hamiltonians derived from a diabatic or adiabatic representation of the electronic system contain identical first-order nonadiabatic coupling, but they differ in the form of the second-order coupling terms [32]. Finally, for methods that can be cast in more than one basis, choosing the most appropriate representation for a given problem might be crucial from a numerical point of view.

The chapter is organized as follows. A detailed derivation of the linearized nonadiabatic algorithm for calculating correlation functions within an adiabatic picture for the electronic states is presented and the features that differentiate it from other nonadiabatic techniques are mentioned. Some comments are offered on the method from a numerical viewpoint. The analogous linearized approach in the diabatic basis is presented in Chap. 12 in this book. Here we limit the discussion of the similarities and differences of the two formulations of LAND-Map to a few remarks. A detailed comparison on a specific application will be the object of a future publication [33].

2 Theory

The time-dependent quantum correlation function of operators $\hat{O}_1$ and $\hat{O}_2$ is defined as

$$\langle \hat{O}_1 \hat{O}_2(t) \rangle = \mathrm{Tr}\{\hat{\rho}\hat{O}_1 \mathrm{e}^{\frac{\mathrm{i}}{\hbar}\hat{H}t}\hat{O}_2 \mathrm{e}^{-\frac{\mathrm{i}}{\hbar}\hat{H}t}\}. \tag{1}$$

In this expression

$$\hat{H} = \frac{\hat{P}^2}{2M} + \hat{h}_{\mathrm{el}} \tag{2}$$

is the Hamiltonian of the interacting system of electrons and nuclei written in terms of the nuclear kinetic operator and an electronic contribution containing the kinetic energy of the electrons and the nuclear–nuclear, electron–electron, and electron–nuclear interaction potentials as well as any external field present. $\hat{\rho}$ is the density matrix for the system.

We begin by representing the total wavefunction in a basis chosen as the tensor product of the nuclear coordinates $|R\rangle$ and the adiabatic electronic basis set, i.e., $|\Psi_\lambda(R)\rangle$ such that

$$\hat{h}_{\mathrm{el}}|\Psi_\lambda(R)\rangle = E_\lambda(R)|\Psi_\lambda(R)\rangle. \tag{3}$$

The appropriate insertion of resolutions of the identity in this basis allows us to express the correlation function as

$$\langle \hat{O}_1 \hat{O}_2(t) \rangle = \sum_{\alpha\beta,\alpha'\beta'} \int \mathrm{d}R_0 \mathrm{d}R_N \mathrm{d}\tilde{R}_0 \mathrm{d}\tilde{R}_N \langle R_0 \Psi_\alpha(R_0)|\hat{\rho}\hat{O}_1|\tilde{R}_0 \Psi_{\alpha'}(\tilde{R}_0)\rangle$$
$$\times \langle \tilde{R}_0 \Psi_{\alpha'}(\tilde{R}_0)|\mathrm{e}^{\frac{\mathrm{i}}{\hbar}\hat{H}t}|\tilde{R}_N \Psi_{\beta'}(\tilde{R}_N)\rangle \langle \tilde{R}_N \Psi_{\beta'}(\tilde{R}_N)|\hat{O}_2|R_N \Psi_\beta(R_N)\rangle$$
$$\times \langle R_N \Psi_\beta(R_N)|\mathrm{e}^{-\frac{\mathrm{i}}{\hbar}\hat{H}t}|R_0 \Psi_\alpha(R_0)\rangle . \tag{4}$$

The quantum propagators in (4) are the transition amplitudes to move, for example, from nuclear position R_0 to position R_N in a time t while the electronic state changes from $\Psi_\alpha(R_0)$ to $\Psi_\beta(R_N)$. The fact that the amplitude is nondiagonal in the electronic states reflects the possibility of nonadiabatic transitions during the system's evolution.

Our first goal is to cast the propagators in a form suitable to be approximated by means of classical trajectories. To that end, we introduce a diadic representation of the Hamiltonian in (2)

$$\hat{H} = \frac{\hat{P}_\chi^2}{2M} + \sum_{\lambda,\mu} |\Psi_\lambda(R)\rangle \left(E_\lambda(R)\delta_{\lambda,\mu} + \hat{\Lambda}_{\lambda,\mu}(R) \right) \langle \Psi_\mu(R)| , \tag{5}$$

where

$$\hat{\Lambda}_{\lambda,\mu}(R) = - \left[\mathrm{i}D_{\lambda,\mu}(R)\frac{\hat{P}_\chi}{M} + \frac{1}{2M}G_{\lambda,\mu}(R) \right] \tag{6}$$

with

$$D_{\lambda,\mu}(R) = \langle \Psi_\lambda(R)|\frac{\partial}{\partial R}|\Psi_\mu(R)\rangle ,$$
$$G_{\lambda,\mu}(R) = \langle \Psi_\lambda(R)|\frac{\partial^2}{\partial R^2}|\Psi_\mu(R)\rangle . \tag{7}$$

Here and in the following $\hbar = 1$. This Hamiltonian acts on a general vibronic wave function of the form $|\Phi\rangle = \sum_\mu \chi_\mu(R)|\Psi_\mu(R)\rangle$ where $\chi_\mu(R)$ is the nuclear coefficient function. Note that $\hat{\Lambda}$ is an operator in $\hat{P}_\chi$ and a function of the nuclear coordinates. The differential operator $\hat{P}_\chi = \mathrm{i}(\partial/\partial R)$ acts on the nuclear coefficient functions, $\chi_\mu(R)$, only, i.e., it does not touch the parametric dependence on R of the adiabatic wavefunctions. $D_{\lambda,\mu}(R)$ is known as the nonadiabatic coupling vector and together with the other non Born–Oppenheimer term, $G_{\lambda,\mu}(R)$, is responsible for the nonadiabatic transitions.

2.1 Mapping Hamiltonian Representation of the Quantum Subsystem in the Adiabatic Basis

Let us now introduce the mapping representation of the system by proceeding in analogy to what is usually done in the case of a diabatic representation of the electronic states [10–14]. We replace the evolution of the electronic subsystem with the evolution of a system of fictitious harmonic oscillators to achieve a twofold objective. On the one hand, the replacement, defined by two mapping relations that we shall detail in a moment, is such that the propagation of the nuclear part of the problem is not changed. On the other, the introduction of this set of harmonic oscillators paves the way for a series of simplifications that will be exploited in the following.

The mapping relations act on the representations of the basis set and of the Hamiltonian. In the first case, the mapping is defined as

$$|\Psi_\lambda(R)\rangle \to |m_\lambda\rangle = |0_1, ..., 1_\lambda, ..0_n\rangle \tag{8}$$

and it transforms the Hilbert space spanned by the original n adiabatic states into one coinciding with a subspace of n-oscillators with at most one quantum of excitation in a single specific oscillator.

As for the Hamiltonian, we substitute the diadic operators as follows:

$$|\Psi_\lambda(R)\rangle\langle\Psi_\mu(R)| \to a^\dagger_\lambda a_\mu, \tag{9}$$

where a and $a^\dagger$ are creation and annihilation operators of mapping oscillators excitations such that, for example,

$$a^\dagger_\lambda a_\mu |0_1, ..., 1_\mu, ..0_n\rangle = |0_1, ..., 1_\lambda, ..0_n\rangle. \tag{10}$$

These operators can be expressed in terms of the positions and momenta of the n oscillators, for example

$$\hat{a}_\lambda = \frac{1}{\sqrt{2}}(\hat{q}_\lambda + i\hat{p}_\lambda). \tag{11}$$

Using this prescription for the creation operator, and the analogous for the annihilation operator, the Hamiltonian (5) becomes

$$\begin{aligned}\hat{H}_m = {} & \frac{\hat{P}_\chi^2}{2M} + \frac{1}{2}\sum_\lambda E_\lambda(R)(\hat{q}_\lambda^2 + \hat{p}_\lambda^2 - 1) \\ & + \frac{1}{2}\sum_{\lambda\mu}\left[\mathrm{Re}\hat{\Lambda}_{\lambda\mu}(R)(\hat{q}_\lambda\hat{q}_\mu + \hat{p}_\lambda\hat{p}_\mu) - \mathrm{Im}\hat{\Lambda}_{\lambda\mu}(R)(\hat{q}_\lambda\hat{p}_\mu - \hat{p}_\lambda\hat{q}_\mu)\right] \\ & - \frac{1}{2}\sum_\lambda \mathrm{Re}\hat{\Lambda}_{\lambda\lambda}(R),\end{aligned} \tag{12}$$

where we introduced the symbol $\hat{H}_m$ to indicate the Hamiltonian in the mapping representation. In deriving the equation above, we have used the fact that $\Lambda_{\lambda\mu} = \Lambda^*_{\mu\lambda}$. For future convenience we write the Hamiltonian as

$$\hat{H}_m = \frac{\hat{P}_\chi^2}{2M} + \hat{h}_m(R), \tag{13}$$

where $\hat{h}_m(R)$ is defined by comparing this equation with (12). Note that the mapping prescription takes us to a basis that, unlike the electronic states that appeared originally, has no parametric dependence on the nuclear coordinates. This is compatible with the diadic expression of the total Hamitonian in (5) since we took care of isolating the effect of the nuclear kinetic operator on the adiabatic states by introducing the nonadiabatic coupling terms $\hat{D}(R)$ and $G(R)$. As mentioned before, $\hat{P}_\chi$ acts on the nuclear coefficients only and these are left unchanged by the mapping.

It should be pointed out that the mapping defined in this section is not a conventional transformation of the electronic basis, in the sense that there is no unitary transformation that takes us from the adiabatic to the mapping states. So, for example, the dependence of $\hat{h}_m(R)$ on the nuclear coordinates is preserved by the prescriptions detailed above and it is determined by the original form of the non Born–Oppenheimer terms in the expression of the operator $\hat{\Lambda}(R)$. The mapping has, however, the remarkable property that it preserves the time evolution of the total system. Thus, since

$$\langle R_N \Psi_\beta(R_N)|\mathrm{e}^{-\frac{\mathrm{i}}{\hbar}\hat{H}t}|R_0\Psi_\alpha(R_0)\rangle = \langle R_N m_\beta|\mathrm{e}^{-\frac{\mathrm{i}}{\hbar}\hat{H}_m t}|R_0 m_\alpha\rangle \tag{14}$$

we can substitute the mapping quantum propagator in the expression for the correlation function without changing its value. Thus

$$\begin{aligned}\langle \hat{O}_1\hat{O}_2(t)\rangle = \sum_{\alpha\beta,\alpha'\beta'} \int & \mathrm{d}R_0\mathrm{d}R_N\mathrm{d}\tilde{R}_0\mathrm{d}\tilde{R}_N \langle R_0\alpha|\hat{\rho}\hat{O}_1|\tilde{R}_0\alpha'\rangle\langle\tilde{R}_N\beta'|\hat{O}_2|\beta R_N\rangle \\ & \times\langle\tilde{R}_0\alpha'|\mathrm{e}^{\frac{\mathrm{i}}{\hbar}\hat{H}_m t}|\tilde{R}_N\beta'\rangle\langle R_N\beta|\mathrm{e}^{-\frac{\mathrm{i}}{\hbar}\hat{H}_m t}|R_0\alpha\rangle \end{aligned} \tag{15}$$

The shorthand notation $\alpha \equiv m_\alpha$ etc. has been introduced in the sum.

2.2 Bath Subsystem Path Integral Representation

We then proceed by representing the nuclear part of the propagator as a path integral [34, 35]. This is slightly delicate, since the Λ operator contains a term, the nonadiabatic coupling vector, which features the product of the nuclear momentum operator with an operator function of the nuclear position. The situation is similar to that encountered when describing the motion of a charged quantum particle in a magnetic field. The coordinate representation of the discrete path integral expression of each propagator in the correlation function can therefore be obtained in analogy with that case [35]. For example,

$$\begin{aligned}\langle R_N m_\beta|\mathrm{e}^{-\mathrm{i}\hat{H}_m t}|R_0 m_\alpha\rangle = & \int \prod_{k=1}^{N-1} \mathrm{d}R_k \mathrm{e}^{\mathrm{i}S} \\ & \times\langle m_\beta|\mathrm{e}^{-\mathrm{i}\epsilon\hat{h}_m(R_N,R_{N-1})}....\mathrm{e}^{-\mathrm{i}\epsilon\hat{h}_m(R_1,R_0)}|m_\alpha\rangle,\end{aligned} \tag{16}$$

where

$$S = \frac{\epsilon M}{2}\sum_{j=0}^{N}\left(\frac{R_{j+1}-R_j}{\epsilon}\right)^2 \tag{17}$$

and

$$\hat{h}_m(R_{j+1}, R_j) = \frac{1}{2}\sum_{\lambda}(E_\lambda(R_j) - \mathrm{Re}\Lambda_{\lambda\lambda}(R_j))\,(\hat{q}_\lambda^2 + \hat{p}_\lambda^2 - 1)$$

$$+\frac{1}{2}\sum_{\lambda\neq\mu}[\mathrm{Re}\Lambda_{\lambda\mu}(R_j)(\hat{q}_\lambda\hat{q}_\mu + \hat{p}_\lambda\hat{p}_\mu) - \mathrm{Im}\Lambda_{\lambda\mu}(R_{j+1}, R_j)(\hat{q}_\lambda\hat{p}_\mu - \hat{p}_\lambda\hat{q}_\mu)]\,. \quad (18)$$

The propagator is thus written as the integral of a multidimensional function of the nuclear coordinates along the path connecting R_0 to R_N, but it is still an operator in the mapping subspace. However, an explicit expression of the mapping transition amplitude as a function of the nuclear coordinates can be obtained. The mapping Hamiltonian is a bilinear or quadratic operator in the subspace of the variables $\{q_\alpha, p_\alpha\}$, and this allows evaluation of the matrix element written above exactly, for example *via* a semiclassical expression. The result will depend parametrically on the nuclear path $\{R_j\}$ which, at this stage, plays the role of an external time dependent parameter.

2.3 Quantum Subsystem Mapping Transition Amplitude

A convenient exact representation for the mapping transition amplitude can be found using the Herman–Kluk expression [36, 37]. This is given by

$$\langle m_\beta|\mathrm{e}^{-\mathrm{i}\epsilon\hat{h}_m(R_N,R_{N-1})}....\mathrm{e}^{-\mathrm{i}\epsilon\hat{h}_m(R_1,R_0)}|m_\alpha\rangle =$$

$$\int \mathrm{d}q_0\mathrm{d}p_0\frac{(2\gamma q_{\beta t} + \mathrm{i}p_{\beta t})}{2\sigma_{pq}}\mathrm{e}^{-\frac{\mathrm{i}}{2\sigma_{pq}}\sum_\lambda q_{\lambda t}p_{\lambda t}}\mathrm{e}^{-\frac{1}{2}\sum_\lambda\left(\frac{q_{\lambda t}^2}{\sigma_q^2}+\frac{p_{\lambda t}^2}{\sigma_p^2}\right)}C_t\mathrm{e}^{\mathrm{i}s_t}$$

$$\times\frac{(2\gamma q_{\alpha 0} - \mathrm{i}p_{\alpha 0})}{2\sigma_{pq}}\mathrm{e}^{-\frac{1}{2}\sum_\lambda\left(\frac{q_{\lambda 0}^2}{\sigma_q^2}+\frac{p_{\lambda 0}^2}{\sigma_p^2}\right)}\mathrm{e}^{\frac{\mathrm{i}}{2\sigma_{pq}}\sum_\lambda q_{\lambda 0}p_{\lambda 0}}\,. \quad (19)$$

Here γ is the width of the coherent states used in the representation of the semiclassical propagator, $\sigma_q^2 = (\gamma + \frac{1}{2})/\gamma$, $\sigma_p^2 = 2(\gamma + \frac{1}{2})$, and $\sigma_{pq} = \gamma + \frac{1}{2}$. (q_t, p_t), where, for example, $q_t = (q_{1t}, \ldots, q_{nt})$, are the end points of classical trajectories starting at (q_0, p_0) and evolving according to the Hamiltonian

$$h(\mathbf{R}) = \frac{1}{2}\sum_\lambda A_{\lambda\lambda}(\mathbf{R})(q_\lambda^2 + p_\lambda^2 - 1)$$

$$+\frac{1}{2}\sum_{\lambda\neq\mu}[B_{\lambda\mu}(\mathbf{R})(q_\lambda q_\mu + p_\lambda p_\mu) + C_{\lambda\mu}(\mathbf{R})(p_\lambda q_\mu - q_\lambda p_\mu)]\,, \quad (20)$$

where (at nuclear time-slice j)

$$A_{\lambda\lambda}(\mathbf{R}) = E_\lambda(R_j) - \mathrm{Re}\Lambda_{\lambda\lambda}(R_j) = E_\lambda(R_j) + \frac{1}{2M}G_{\lambda\lambda}(R_j), \quad (21)$$

$$B_{\lambda\mu}(\mathbf{R}) = \mathrm{Re}\Lambda_{\lambda\mu}(R_j) = -\frac{1}{2M}G_{\lambda\mu}(R_j),$$

$$C_{\lambda\mu}(\mathbf{R}) = \mathrm{Im}\Lambda_{\lambda\mu}(R_j, R_{j+1}) = -\left[\frac{D_{\lambda\mu}(R_{j+1}) + D_{\lambda\mu}(R_j)}{2}\right]\frac{(R_{j+1} - R_j)}{\epsilon}$$

and the boldface indicates the dependence of h on the nuclear coordinates at more than one time-slice. s_t is the mapping action which we will compute in detail later, and

$$c_t = \left[\det \frac{1}{2}\left(\frac{\partial \mathbf{q}_t}{\partial \mathbf{q}_0} + \frac{\partial \mathbf{p}_t}{\partial \mathbf{p}_0} - 2\mathrm{i}\gamma \frac{\partial \mathbf{q}_t}{\partial \mathbf{p}_0} + \frac{\mathrm{i}}{2\gamma}\frac{\partial \mathbf{p}_t}{\partial \mathbf{q}_0}\right)\right]^{\frac{1}{2}} \tag{22}$$

is the square root of the determinant of the complex matrix which measures the stability of the mapping variable trajectories with respect to variations in the initial conditions.

The evolution of the mapping variables cannot, in general, be determined explicitly since it depends on the value of the nuclear coordinates at the various points along the path which has not been specified so far. The mapping transition amplitude can, however, be simplified by taking advantage of a number of properties of the evolution that can be derived without knowledge of the parametric dependence of (20) on nuclear coordinates.

The Hamiltonian dynamics of the mapping variables is determined by the following set of equations:

$$\frac{\mathrm{d}q_\beta}{\mathrm{d}t} = A_{\beta,\beta}(\mathbf{R})p_\beta + \sum_{\lambda(\neq\beta)} B_{\beta,\lambda}(\mathbf{R})p_\lambda + \sum_{\lambda(\neq\beta)} C_{\beta,\lambda}(\mathbf{R})q_\lambda, \tag{23}$$

$$\frac{\mathrm{d}p_\beta}{\mathrm{d}t} = -A_{\beta,\beta}(\mathbf{R})q_\beta - \sum_{\lambda(\neq\beta)} B_{\beta,\lambda}(\mathbf{R})q_\lambda + \sum_{\lambda(\neq\beta)} C_{\beta,\lambda}(\mathbf{R})p_\lambda.$$

Here, and in the following, $\sum_{\lambda(\neq\beta)}$ indicates a single sum over the index λ restricted to *not* include $\lambda = \beta$. Using these equations it is not difficult to show that the difference between the action s_t and the explicit phases arising from the initial and final mapping states in (19) can be manipulated as follows

$$\begin{aligned} s_t &- \frac{1}{2\sigma_{pq}}\sum_\lambda q_{\lambda t}p_{\lambda t} + \frac{1}{2\sigma_{pq}}\sum_\lambda q_{\lambda 0}p_{\lambda 0} \\ &= \int_0^t \mathrm{d}\tau \left\{\sum_\lambda p_{\lambda\tau}\frac{\mathrm{d}q_{\lambda\tau}}{\mathrm{d}\tau} - h(R_\tau)\right\} - \frac{1}{2\sigma_{pq}}\int_0^t \mathrm{d}\tau \frac{\mathrm{d}}{\mathrm{d}\tau}\sum_\lambda p_{\lambda\tau}q_{\lambda\tau}. \end{aligned} \tag{24}$$

Choosing $\sigma_{pq} = 1$, this reduces to

$$s_t - \frac{1}{2\sigma_{pq}}\sum_\lambda q_{\lambda t}p_{\lambda t} + \frac{1}{2\sigma_{pq}}\sum_\lambda q_{\lambda 0}p_{\lambda 0} = \frac{1}{2}\int_0^t \mathrm{d}\tau \sum_\lambda A_{\lambda,\lambda}(R_\tau). \tag{25}$$

The choice of σ_{pq}, which is equivalent to setting the value of the coherent state width to $\gamma = \frac{1}{2}$, does not compromise the generality of the method. The value of the width parameter does not affect the properties of the Gaussian basis set necessary to obtain the Herman–Kluk approximation for the quantum propagator and is usually fixed so as to optimize the numerical convergence of the calculations.

Let us then introduce the complex vector $\eta = \mathbf{q} + \mathrm{i}\mathbf{p}$ and observe that the determinant, in (22) can be expressed as

$$c_t = \det\left\{\frac{1}{2}\frac{\partial \eta_t}{\partial \eta_0}\right\}^{\frac{1}{2}}. \tag{26}$$

From the evolution equations of the mapping variables it follows that

$$\frac{\mathrm{d}\eta}{\mathrm{d}t} = M\eta, \tag{27}$$

where

$$M = -\mathrm{i}[A + B] + C. \tag{28}$$

A is the matrix with elements $\{A\}_{\alpha\beta}$, and similarly for B and C (see (23)). From (27) one gets

$$\frac{\partial \eta_t}{\partial \eta_0} = \mathrm{e}^{-\mathrm{i}\epsilon M(\mathbf{R_N})} \ldots \mathrm{e}^{-\mathrm{i}\epsilon M(\mathbf{R}_0)} \tag{29}$$

and

$$c_t = \frac{1}{\sqrt{2^d}}\mathrm{e}^{-\frac{\mathrm{i}}{2}\int_0^t d\tau \sum_\beta A_{\beta,\beta}(R_\tau)} \tag{30}$$

where d is the dimension of the stability matrix.

Further, from the mapping evolution equations it also follows that the quantity

$$\sum_\lambda (q_\lambda^2 + p_\lambda^2) \tag{31}$$

is a constant of the motion.

Using the results detailed above we can then rewrite the amplitude in (19) as

$$\begin{aligned}&\langle m_\beta|\mathrm{e}^{-\mathrm{i}\epsilon\hat{h}_m(R_N,R_{N-1})} \ldots \mathrm{e}^{-\mathrm{i}\epsilon\hat{h}_m(R_1,R_0)}|m_\alpha\rangle = \\ &\int \mathrm{d}q_0\mathrm{d}p_0(q_{\beta t} + \mathrm{i}p_{\beta t})(q_{\alpha 0} - \mathrm{i}p_{\alpha 0})\mathrm{e}^{-\frac{1}{2}\sum_\lambda (q_{\lambda 0}^2 + p_{\lambda 0}^2)}. \end{aligned} \tag{32}$$

An even more explicit expression for this object can be derived by introducing a polar representation of the complex polynomials. Thus defining

$$\begin{aligned} G_0 &= \mathrm{e}^{-\frac{1}{2}\sum_\lambda (q_{0,\lambda}^2 + p_{0,\lambda}^2)}, \\ r_{t,\beta}(\{R_k\}) &= \sqrt{q_{t,\beta}^2(\{R_k\}) + p_{t,\beta}^2(\{R_k\})}, \\ \Theta_{t,\beta}(\{R_k\}) &= \tan^{-1}\left(\frac{p_{t,\beta}(\{R_k\})}{q_{t,\beta}(\{R_k\})}\right). \end{aligned} \tag{33}$$

the transition amplitude becomes

$$\langle m_\beta | \mathrm{e}^{-\mathrm{i}\epsilon \hat{h}_m(R_N,R_{N-1})} .... \mathrm{e}^{-\mathrm{i}\epsilon \hat{h}_m(R_1,R_0)} | m_\alpha \rangle = \int \mathrm{d}q_0 \mathrm{d}p_0 r_{t,\beta}(\{R_k\}) \mathrm{e}^{-\mathrm{i}\Theta_{t\beta}(\{R_k\})} r_{0\alpha} \mathrm{e}^{\mathrm{i}\Theta_{0,\alpha}} G_0 . \tag{34}$$

The mapping evolution equations can also be used to show that

$$\begin{aligned} \Theta_{t,\beta}(\{R_k\}) &= \tan^{-1}\left(\frac{p_{0,\beta}}{q_{0,\beta}}\right) + \int_0^t \mathrm{d}\tau \left[\frac{\mathrm{d}}{\mathrm{d}\tau}\tan^{-1}\left(\frac{p_{\tau,\beta}(R_\tau)}{q_{\tau,\beta}(R_\tau)}\right)\right] \\ &= \tan^{-1}\left(\frac{p_{0,\beta}}{q_{0,\beta}}\right) - \int_0^t \mathrm{d}\tau A_{\beta,\beta}(R_\tau) \\ &\quad - \int_0^t \mathrm{d}\tau \sum_{\lambda(\neq\beta)} \left[B_{\beta,\lambda}(R_\tau)\frac{(p_{\tau\beta}p_{\tau\lambda} + q_{\tau\beta}q_{\tau\lambda})}{(p_{\tau\beta}^2 + q_{\tau\beta}^2)}\right] \\ &\quad + \int_0^t \mathrm{d}\tau \sum_{\lambda(\neq\beta)} \left[C_{\beta,\lambda}(R_\tau,\dot{R}_\tau)\frac{(p_{\tau\lambda}q_{\tau\beta} - p_{\tau\beta}q_{\tau\lambda})}{(p_{\tau\beta}^2 + q_{\tau\beta}^2)}\right] \end{aligned} \tag{35}$$

$$\begin{aligned} &= \tan^{-1}\left(\frac{p_{0,\beta}}{q_{0,\beta}}\right) + \int_0^t \theta_\beta(\tau)\mathrm{d}\tau \\ &\quad + \int_0^t \mathrm{d}\tau \sum_{\lambda(\neq\beta)} \left[C_{\beta,\lambda}(R_\tau,\dot{R}_\tau)\frac{(p_{\tau\lambda}q_{\tau\beta} - p_{\tau\beta}q_{\tau\lambda})}{(p_{\tau\beta}^2 + q_{\tau\beta}^2)}\right] . \end{aligned} \tag{36}$$

By using (34), and its counterpart for the backward propagated mapping variables, identified by a tilde, the correlation function can be expressed as

$$\begin{aligned} \langle \hat{O}_1 \hat{O}_2(t) \rangle = &\sum_{\alpha\beta,\alpha'\beta'} \int \mathrm{d}R_0 \mathrm{d}R_N \mathrm{d}\tilde{R}_0 \mathrm{d}\tilde{R}_N \int \prod_{k=1}^{N-1} \mathrm{d}R_k \int \prod_{k=1}^{N-1} \mathrm{d}\tilde{R}_k \\ &\times \mathrm{e}^{\mathrm{i}(S-\tilde{S})} \langle R_0\alpha | \hat{\rho}\hat{O}_1 | \tilde{R}_0\alpha' \rangle \langle \tilde{R}_N\beta' | \hat{O}_2 | \beta R_N \rangle \\ &\times \int \mathrm{d}\tilde{q}_0 \mathrm{d}\tilde{p}_0 \tilde{r}_{t,\beta'}(\{\tilde{R}_k\}) \mathrm{e}^{i\tilde{\Theta}_{t\beta'}(\{\tilde{R}_k\})} \tilde{r}_{0\alpha'} \mathrm{e}^{-\mathrm{i}\tilde{\Theta}_{0,\alpha'}} \tilde{G}_0 \\ &\times \int \mathrm{d}q_0 \mathrm{d}p_0 r_{t,\beta}(\{R_k\}) \mathrm{e}^{-\mathrm{i}\Theta_{t\beta}(\{R_k\})} r_{0\alpha} \mathrm{e}^{\mathrm{i}\Theta_{0,\alpha}} G_0 \end{aligned} \tag{37}$$

The forward and backward propagators are now represented as a combination of path integrals in the nuclear variables and integrals over the mapping oscillators phase space. We emphasize that, although the expression above has been obtained using semiclassical results, so far no approximation has been introduced: For any given pair of forward and backward paths in the nuclear configuration space, (37) is an exact rewriting of the definition of the correlation function from which we started. As such, its calculation is as problematic as that of any full quantum object. Before introducing the only approximation needed to obtain the expression of the correlation function in terms of sets of classical trajectories that we wish to discuss in this chapter, let us briefly summarize the steps performed so far. First, we obtained a nuclear path integral expression for the correlation function in which the effect of the nonadiabatic electronic transitions was accounted for by the fictitious set of harmonic oscillators introduced by the mapping method, see (16). We then took advantage of the quadratic nature of the operators in the mapping transition amplitude to derive an exact expression for this quantity. It should, however, be pointed out that the procedure used in this chapter is not the only possibility, since the harmonic nature of the mapping subspace opens up several opportunities for evaluating the amplitude exactly. Here we choose to evaluate the amplitude using a semiclassical representation for it that presented certain formal advantages. In particular, we exploited the properties of the Hamiltonian, classical evolution of the mapping variables in the Herman–Kluk expression of the amplitude to simplify the form of the phase in the nuclear path integral representation of the time correlation function (19–34).

2.4 Linearization of the Bath Subsystem Path Integrals: An approximate, Trajectory based Expression for the Correlation Function

To set the stage for the approximation we intend to perform to obtain a viable expression for the correlation function, it is convenient to modify our result further by introducing a set of momentum-like variables for the nuclear paths. This can be accomplished in the following way. Define, for example in the forward propagator of (16),

$$W_\beta = \sum_{\lambda(\neq\beta)} \frac{(D_{\beta\lambda}(R_{k+1}) + D_{\beta\lambda}(R_k))}{2} \left(\frac{p_{k\lambda}q_{k\beta} - q_{k\lambda}p_{k\beta}}{p_{k\beta}^2 + q_{k\beta}^2} \right) . \tag{38}$$

Then, using the results in (17) and (21), the terms containing a dependence on $(R_{k+1} - R_k)$ in the nuclear path integral are of the form

$$\mathrm{e}^{\mathrm{i}\epsilon\left[\frac{M(R_{k+1}-R_k)^2}{2\epsilon^2} - \frac{(R_{k+1}-R_k)}{\epsilon} W_\beta\right]} . \tag{39}$$

We can write this, barring constants, as the Fourier transform of a Gaussian

$$\int \mathrm{d}P_k \mathrm{e}^{\mathrm{i}P_k(R_{k+1}-R_k)} \mathrm{e}^{-\frac{\mathrm{i}\epsilon}{2M}[P_k+W_\beta]^2} , \tag{40}$$

where the Fourier variables are the nuclear momenta P_k. Using this result in the nuclear path integral along with the phase space expression for the mapping amplitude, the propagator becomes

$$\langle R_N m_\beta | \mathrm{e}^{-\mathrm{i}\hat{H}_m t} | R_0 m_\alpha \rangle = \int \prod_{k=1}^{N-1} \mathrm{d}R_k \prod_{k=1}^{N-1} \mathrm{d}P_k \mathrm{d}P_N \mathrm{e}^{\mathrm{i}S_P} \times \int \mathrm{d}q_0 \mathrm{d}p_0 r_{t,\beta}(\{R_k\}) \mathrm{e}^{-\mathrm{i}\Phi_{\beta t}(\{R_k\})} r_{0\alpha} \mathrm{e}^{\mathrm{i}\Phi_{0,\alpha}} G_0 \tag{41}$$

with

$$S_P = \epsilon \sum_k \left\{ P_k \frac{(R_{k+1} - R_k)}{\epsilon} - \frac{1}{2M} \left[P_k + W_\beta \right]^2 \right\} \tag{42}$$

and

$$\begin{aligned} \Phi_{\beta t} &= \tan^{-1}\left(\frac{p_{0,\beta}}{q_{0,\beta}}\right) - \epsilon \sum_k \left\{ A_{\beta,\beta}(R_k) + \sum_{\lambda(\neq\beta)} \left[B_{\beta,\lambda}(R_k) \frac{(p_{k\beta} p_{k\lambda} + q_{k\beta} q_{k\lambda})}{(p_{k\beta}^2 + q_{k\beta}^2)} \right] \right\} \\ &= \tan^{-1}\left(\frac{p_{0,\beta}}{q_{0,\beta}}\right) + \epsilon \sum_k \phi_{\beta,k} . \end{aligned} \tag{43}$$

Substituting (41) and its conterpart for the backward time evolution in the expression for the correlation function gives

$$\begin{aligned} \langle \hat{O}_1 \hat{O}_2(t) \rangle = \sum_{\alpha\beta,\alpha'\beta'} \int \mathrm{d}R_0 \mathrm{d}\tilde{R}_0 \mathrm{d} \int \prod_{k=1}^{N} \mathrm{d}R_k \mathrm{d}P_k \int \prod_{k=1}^{N} \mathrm{d}\tilde{R}_k \mathrm{d}\tilde{P}_k \mathrm{e}^{\mathrm{i}(S_P - \tilde{S}_P)} \\ \times \int \mathrm{d}q_0 \mathrm{d}p_0 \mathrm{d}\tilde{q}_0 \mathrm{d}\tilde{p}_0 \langle R_0 \alpha | \hat{\rho} \hat{O}_1 | \tilde{R}_0 \alpha' \rangle \langle \tilde{R}_N \beta' | \hat{O}_2 | \beta R_N \rangle \\ \times r_{t,\beta}(\{R_k\}) \mathrm{e}^{-\mathrm{i}\Phi_{t\beta}(\{R_k\})} r_{0\alpha} \mathrm{e}^{\mathrm{i}\Phi_{0,\alpha}} G_0 \\ \times \tilde{r}_{t,\beta'}(\{\tilde{R}_k\}) \mathrm{e}^{\mathrm{i}\tilde{\Phi}_{t\beta'}(\{\tilde{R}_k\})} \tilde{r}_{0\alpha'} \mathrm{e}^{-\mathrm{i}\tilde{\Phi}_{0,\alpha'}} \tilde{G}_0 . \end{aligned} \tag{44}$$

As a final step toward the linearized approximation of the correlation function, let us change variables to sum and difference in the nuclear paths

$$\begin{aligned} \bar{R}_k &= \frac{R_k + \tilde{R}_k}{2} \\ \Delta R_k &= R_k - \tilde{R}_k \end{aligned} \tag{45}$$

and similarly for P_k, $\tilde{P}_k$ for all k. We now proceed as indicated by Poulsen et al. [3–5] and Geva and co-workers [6–9]. After substituting the variable transformed propagators into the correlation function, the integrand is approximated by truncating the Taylor series expansion of its phase to linear order in the difference path. The terms appearing in the expansion of the phase can be classified as follows:

1. Terms of order zero in the difference path: They are of the form

$$\begin{aligned} \Omega = \epsilon \sum_k \frac{\bar{P}_k}{M}[W_\beta(\bar{R}_k, \{q_k, p_k\}) - \tilde{W}_{\beta'}(\bar{R}_k, \{\tilde{q}_k, \tilde{p}_k\})] \\ + \epsilon \sum_k \frac{1}{2M}[W_\beta^2(\bar{R}_k, \{q_k, p_k\}) - \tilde{W}_{\beta'}^2(\bar{R}_k, \{\tilde{q}_k, \tilde{p}_k\})] \\ + \epsilon \sum_k \left\{\phi_\beta(\bar{R}_k, \{q_k, p_k\}) - \tilde{\phi}_{\beta'}(\bar{R}_k, \{\tilde{q}_k, \tilde{p}_k\})\right\} . \end{aligned} \tag{46}$$

2. Linear terms in ΔP: They are of the form

$$\begin{aligned} f_{\Delta P} = \epsilon \sum_k \left[\frac{\bar{R}_k - \bar{R}_{k-1}}{\epsilon}\right] \Delta P_k \\ - \epsilon \sum_k \left[\frac{1}{M}\bar{P}_k + \frac{W_\beta(\bar{R}_{k-1}, \{q_{k-1}, p_{k-1}\}) + \tilde{W}_{\beta'}(\bar{R}_{k-1}, \{\tilde{q}_{k-1}, \tilde{p}_{k-1}\})}{2}\right] \Delta P_k \end{aligned} \tag{47}$$

3. Linear terms in ΔR: They are of the form

$$f_{\Delta R} = \epsilon \sum_k \left[\frac{\bar{P}_{k+1} - \bar{P}_k}{\epsilon}\right] \Delta R_k \tag{48}$$

$$\begin{aligned} + \epsilon \sum_k \left[\frac{1}{2M}\bar{P}_k \nabla_{\bar{R}_k}\{W_\beta(\bar{R}_k, \{q_k, p_k\}) + \tilde{W}_{\beta'}(\bar{R}_k, \{\tilde{q}_k, \tilde{p}_k\})\}\right] \Delta R_k \\ + \epsilon \sum_k \left[\mathrm{d}W_\beta(\bar{R}_k, \{q_k, p_k\}) + \mathrm{d}\tilde{W}_{\beta'}(\bar{R}_k, \{\tilde{q}_k, \tilde{p}_k\})\right] \Delta R_k \\ + \epsilon \sum_k \left[\frac{1}{2}\nabla_{\bar{R}_k}(\phi_\beta(\bar{R}_k, \{q_k, p_k\}) + \tilde{\phi}_{\beta'}(\bar{R}_k, \{\tilde{q}_k, \tilde{p}_k\})\right] \Delta R_k \end{aligned} \tag{49}$$

where, for example,

$$\begin{aligned} \mathrm{d}W_\beta(\bar{R}_k, \{q_k\}\{p_k\}) = \sum_{\lambda(\neq\beta)} D_{\beta\lambda}(\bar{R}_k) \left(\frac{p_{k\lambda}q_{k\beta} - q_{k\lambda}p_{k\beta}}{p_{k\beta}^2 + q_{k\beta}^2}\right) \\ \times \sum_{\mu(\neq\beta)} \nabla_{\bar{R}_k} D_{\beta\mu}(\bar{R}_k) \left(\frac{p_{k\mu}q_{k\beta} - q_{k\mu}p_{k\beta}}{p_{k\beta}^2 + q_{k\beta}^2}\right) . \end{aligned} \tag{50}$$

Once these expressions have been introduced in the exponent of the correlation function, the integrals over the difference path can be performed analytically. For $k > 0$, they are integral representation of δ-functions. As for the integral over ΔR_0 and ΔR_N, they define the partial Wigner transform with respect to the nuclear variables of the product $\hat{\rho}\hat{O}_1$, and of the operator $\hat{O}_2$. The linearized correlation function can therefore be written as

$$\langle \hat{O}_1 \hat{O}_2(t) \rangle = \sum_{\alpha\beta,\alpha'\beta'} \int \mathrm{d}\bar{R}_0 \mathrm{d}q_0 \mathrm{d}p_0 \mathrm{d}\tilde{q}_0 \mathrm{d}\tilde{p}_0 \int \prod_{k=1}^{N} \mathrm{d}\bar{R}_k \frac{\mathrm{d}\bar{P}_k}{2\pi}$$
$$\times \left[\hat{O}_2\right]^W_{\beta'\beta} (\bar{R}_N \bar{P}_N) \left[\hat{\rho}\hat{O}_1\right]^W_{\alpha,\alpha'} (\bar{R}_0, \bar{P}_1) \mathrm{e}^{-\mathrm{i}\Omega}$$
$$\times G_0 \tilde{G}_0 r_{0\alpha} \mathrm{e}^{\mathrm{i}\phi_{0,\alpha}} \tilde{r}_{0\alpha'} \mathrm{e}^{-\mathrm{i}\tilde{\phi}_{0,\alpha'}} r_{t,\beta}(\{\bar{R}_k\}) \tilde{r}_{t,\beta'}(\{\bar{R}_k\})$$
$$\times \prod_{k=1}^{N-1} \delta\left(\frac{\bar{P}_{k+1} - \bar{P}_k}{\epsilon} - F_k^{\beta,\beta'}\right) \prod_{k=1}^{N} \delta\left(\frac{\bar{R}_k - \bar{R}_{k-1}}{\epsilon} - \frac{\Pi_k^{\beta\beta'}}{M}\right), \quad (51)$$

where Ω is defined in (46),

$$\Pi_k^{\beta\beta'} = \left[\bar{P}_k + \frac{W_\beta(\bar{R}_{k-1}, \{q_{k-1}, p_{k-1}\}) + \tilde{W}_{\beta'}(\bar{R}_{k-1}, \{\tilde{q}_{k-1}, \tilde{p}_{k-1}\})}{2}\right] \quad (52)$$

and

$$F_k^{\beta,\beta'} = -\frac{1}{2M} \bar{P}_k \nabla_{\bar{R}_k} (W_\beta(\bar{R}_k, \{q_k, p_k\}) + \tilde{W}_{\beta'}(\bar{R}_k, \{\tilde{q}_k, \tilde{p}_k\}))$$
$$-\frac{1}{2M} (\mathrm{d}W_\beta(\bar{R}_k, \{q_k, p_k\}) + \mathrm{d}\tilde{W}_{\beta'}(\bar{R}_k, \{\tilde{q}_k, \tilde{p}_k\})$$
$$-\frac{1}{2} \nabla_{\bar{R}_k} (\phi_\beta(\bar{R}_k, \{q_k, p_k\}) + \tilde{\phi}_{\beta'}(\bar{R}_k, \{\tilde{q}_k, \tilde{p}_k\})) \quad (53)$$

The product of delta-functions in (51) amounts to a time-stepping prescription which forces the mean path to obey a classical evolution law. This, together with the classical motion of the mapping variables specified by (23), allows us to formulate the following algorithm for the evaluation of the integrals in the time correlation function calculation:

1. Assign the set of indexes $\alpha, \alpha', \beta, \beta'$
2. Sample initial values $\bar{R}_0, \bar{P}_1$ from the Wigner transform of the thermal density times the operator $\hat{O}_1$
3. Sample initial values of the forward and backward mapping variables in states α and α' respectively
4. Accumulate weights at the initial time
5. Perform integral over $\bar{R}_1$ advancing the nuclear position by one time step using

$$\bar{R}_1 = \bar{R}_0 + \frac{\epsilon}{M} \left[\bar{P}_1 + \frac{W_\beta(\bar{R}_0, \{q_0\}\{p_0\}) + \tilde{W}_{\beta'}(\bar{R}_0, \{\tilde{q}_0\}, \{\tilde{p}_0\})}{2}\right]$$

6. Calculate the matrix elements of the electronic Hamiltonian at the new nuclear positions and move mapping variables one step farther in the forward and backward propagation
7. Perform the integral over $\bar{P}_2$ advancing the nuclear momentum by one time step using

$$\bar{P}_2 = \bar{P}_1 + \epsilon F_1^{\beta,\beta'}$$

8. Accumulate weights and iterate from step 5. until the final time of the run is reached
9. Calculate $r_{t,\beta}(\{R_k\})\tilde{r}_{t,\beta'}(\{\bar{R}_k\})$ and the Wigner transform of operator $\hat{O}_2$, and multiply the weights
10. Iterate from step 1. to accumulate all terms in the correlation function until convergence is reached.

2.5 Some Comments on the Algorithm

All the propagations in the algorithm described above are classical and local in time, two features which simplify the numerical task considerably. The mapping variables evolve according to (23), while the classical trajectory for the nuclear variables is determined by the forces in (53). As such the overall evolution of the two coupled dynamical subsystems is not governed by a single Hamiltonian. This is in marked contrast to the usual semi-classical approach where the mapping Hamiltonian is differentiated to obtain a classically consistent system of equations of motion for all degrees of freedom [38–43]. The method we present here is also significantly different from the Pechukas semiclassical formulation of nonadiabatic dynamics [20]. The Pechukas transition amplitude and trajectory must be determined by self-consistent iteration. With the mapping formulation, however, we have an explicit form for the transition amplitude, (34), which can be integrated as the nuclear trajectory is advanced making the approach local in time and straightforward to implement.

The two distinct evolutions in the LAND-map approach are reminiscent of the situation in surface hopping or mean field nonadiabatic methods where different dynamical prescriptions apply to the quantum and classical subsystems (see for example [23]). However, in our approach the nuclear variables do not move in the mean field of the quantum subsystem nor vice-versa. The different terms in the expression for the correlation function in general involve trajectories moving with forces determined by linear combinations of different pairs of diabats and off-diagonal electronic Hamiltonian matrix elements. Since the motion does not happen on a single average potential surface our approach does not suffer from some of the limitations of mean field methods. Further the coupling between nuclear motion and electronic transitions is determined rigorously by the linearization procedure in contrast to the nonrigorous nature of the development of traditional surface hopping approaches [23].

Note also that the contributions to the force in (53) coming from the eigenvalues of the electronic matrix do not enter multiplied by mapping oscillator number operator terms as is found with the standard semiclassical implementation of the mapping formulation. As discussed in previous work for the case of a diabatic basis set [16, 17], this kind of terms can result in unstable trajectories due to inversion of the potentials which compromise the convergence of the standard algorithm for long times.

The use of the polar representation of the complex Hermite polynomials that project onto the final states β or β' complicates to some extent the implementation of the method. When the complex polynomial is zero, the phase is ill-defined. This is reflected in the expression of the force in (53) by the apparent singularity in the W contribution. The existence of a divergence in the force, however, depends on the behavior of the gradients of the nonadiabatic coupling vector. In fact, the analogous problem is also present in LAND-Map as derived using a diabatic representation of the electronic states. In that case the potential divergence is compensated by the behavior of the off-diagonal elements of the electronic Hamiltonian matrix that usually are, or go to, zero very rapidly in regions of zero population of the final state. As proved in applications [1, 2], it is therefore possible to remedy the problem by a careful implementation of the algorithm. Since, in this respect, the nonadiabatic coupling vector behaves similar to the diabatic off-diagonal elements of the electronic Hamiltonian, the numerical properties of the new algorithm are expected to resemble those observed in applications of the diabatic version of the scheme. Furthermore, like in the previous case, the weight of such trajectories is rigorously zero for zero population in the final state, thanks to the amplitudes r_β and $\tilde{r}_{\beta'}$ in (51). Thus, the effect of this apparent pathology in the phase affects the integration of the evolution equations but does not compromise the convergence of the average.

A second potential problem, that has affected the use of the mapping method to describe nonadiabatic processes in an adiabatic basis in the past [31], arises from the typically rapid variations in the coupling terms among the different states. The nonadiabatic coupling vector is usually strongly localized in the vicinity of avoided crossings or conical intersections. This results in very steep contributions to the forces that may lead to numerical instabilities in the integration algorithm for the evolution equations.

From a formal viewpoint, the main difference between the linearized approach for calculating quantum time correlation functions starting from an adiabatic representation of the electronic degrees of freedom and the result obtained from a diabatic representation lies in the structure of the evolution equations for the bath degrees of freedom. In the adiabatic picture, the force is not simply the gradient of a function of the nuclear coordinates times a term that depends on the mapping variables. Rather, it contains a multiplicative coupling of the "potential" to the momenta, in close analogy to the classical evolution equations of a charged particle moving in a magnetic field. The analogy is particularly evident in the form of the action, (42), where W plays the role of a vector potential. This structure does not come as a surprise, since it is a direct consequence of the nature of the nonadiabatic coupling in the chosen basis. Quantum mechanically, transitions among different electronic states are governed by the operator $\hat{\Lambda}_{\lambda\mu}$, introduced in (6), that contains the same kind of coupling between the nuclear momentum and position operators. It is interesting that, in spite of this fundamental difference in the structure of the quantum Hamiltonian, the formal manipulations necessary to derive the linearized expression of the correlation function can be performed in very close analogy to the procedure followed for the diabatic case. It also interesting to point out that, within the framework of a semiclassical implementation of the mapping method in the adiabatic representation, Sun and Miller [44], following earlier work of Meyer and Miller [10] obtained a similar result for the (single) Hamiltonian governing the motion of the coupled nuclear and electronic degrees of freedom. They followed a very different route, exploiting the correspondence between the similarity transformation leading from one basis set to another in quantum mechanics and a classical canonical transformation, to derive a form for the evolution equations in the adiabatic basis after taking the semiclassical limit for the propagator in the diabatic mapping representation. In this chapter, on the other hand, we have introduced the mapping of the electronic states in the adiabatic representation at the full quantum level and reduced the nuclear evolution to a classical prescription through linearization of correlation function expressions as a second step. Interestingly, their evolution equations for the bath variables obtained in these two different ways contain the same first order terms in the nonadiabatic coupling vector, but the second order terms, the function $G_{\lambda,\mu}(R)$ appears differently. It is known, see for example [31] and references therein, that different procedures to reduce the quantum propagation to classical motion can have such an effect. Further, different semiclassical or mixed quantum–classical approaches require to weight the contribution of the trajectories by different functions to account, at least approximately, for the coherence in the quantum propagation. It is the delicate balance of these ingredients that determines the performance of the available approximate methods and further investigation is required to build general criteria to assess their relative merits for a given class of applications.

3 Conclusions

The linearized nonadiabatic method for evaluating time quantum correlation functions, originally developed starting from a diabatic representation of the quantum subsystem's states, has been generalized to allow for an adiabatic representation of the electronic properties of the system. The generalization is interesting both from a formal and a numerical point of view. Not all available approximate methods for describing mixed quantum–classical nonadiabatic dynamics, in fact, can be cast in more than one representation. From a numerical viewpoint, the flexibility of LAND-Map with respect to the choice of the quantum subsystem's representation opens up, in principle, the possibility of choosing the most efficient electronic basis for any given problem. The practical usefulness of the algorithm described here will be tested in the future on the same set of model problems employed to assess the performance of the diabatic version of the method [33].

Acknowledgments

We acknowledge support for this research from the National Science Foundation under grant number CHE-0316856 as well a the Petroleum Research Fund administered by the American Chemical Society grant number 39180-AC6.

References

1. S. Bonella and D. F. Coker: J. Chem. Phys. **122**, 194102 (2005)
2. S. Bonella, D. Montemayor, and D.F. Coker: Proc. Natl. Acad. Sci. **102**, 6715 (2005)
3. J. A. Poulsen, G. Nyman, and P. J. Rossky: J. Chem. Phys. **119**, 12179 (2003)
4. J. A. Poulsen, G. Nyman, and P. J. Rossky: J. Phys. Chem A **108**, 8743 (2004)
5. J. A. Poulsen, G. Nyman, and P. J. Rossky: Proc. Natl. Acad. Sci. **102**, 6709 (2005)
6. Q. Shi and E. Geva: J. Chem. Phys. **118**, 8173 (2003)
7. Q. Shi and E. Geva: J. Chem. Phys. **120**, 10647 (2004)
8. Q. Shi and E. Geva: J. Phys. Chem. A **107**, 9059 (2003)
9. Q. Shi and E. Geva: J. Phys. Chem. A **107**, 9070 (2003)
10. H. D. Meyer and W. H. Miller: J. Chem. Phys. **70**, 3214 (1979)
11. W. H. Miller and C. W. McCurdy: J. Chem. Phys. **69**, 5163 (1978)
12. C. W. McCurdy, H. D. Meyer, and W. H. Miller: J. Chem. Phys. **70**, 3177 (1979)
13. G. Stock and M. Thoss: Phys. Rev. Lett. **78**, 578 (1997)
14. G. Stock and M. Thoss: Phys. Rev. A **59**, 64 (1999)
15. E. Wigner: Phys. Rev. **40**, 749 (1932)
16. S. Bonella and D. F. Coker: J. Chem. Phys. **114**, 7778 (2001)
17. S. Bonella and D. F. Coker: Chem. Phys. **268**, 323 (2001)
18. S. Bonella and D. F. Coker: J. Chem. Phys. **118**, 4370 (2003)
19. M. Thoss and W. H. Miller: J. Chem. Phys. **112**, 10282 (2000)
20. P. Pechukas: Phys. Rev. **181**, 174 (1969)
21. S. Hammes-Schiffer and J. C. Tully: J. Chem. Phys. **101**, 4657 (1994)
22. J. C. Burant and J. C. Tully: J. Chem. Phys. **112**, 6097 (2000)
23. J. C. Tully: In *Classical and Quantum Dynamics in Condensed Phase Simulations* Eds.: G. Ciccotti B. Berne and D. Coker. 489, World Scientific, Dordrecht (1998)
24. D. S. Sholl and J. C. Tully: J. Chem. Phys. **109**, 7702 (1998)
25. D. F. Coker and L. Xiao: J. Chem. Phys. **102**, 496 (1995)
26. R. Kapral and G. Ciccotti: J. Chem. Phys. **110**, 8919 (1999)
27. C. C. Martens and J.-Y. Fang: J. Chem. Phys. **106**, 4918 (1996); A. Donoso and C. C. Martens: J. Phys. Chem. **102**, 4291 (1998); D. Kohen and C. C. Martens: J. Chem. Phys. **111**, 4343 (1999); **112**, 7345 (2000)

28. I. Horenko, C. Salzmann, B. Schmidt, and C. Schütte: J. Chem. Phys. **117**, 11075 (2002)
29. R. Kapral and G. Ciccotti: *A Statistical Mechanical Theory of Quantum Dynamics in Classical Environments*, in *Bridging Time Scales: Molecular Simulations for the Next Decade*, 2001, eds. P. Nielaba, M. Mareschal and G. Ciccotti, (Springer, Berlin, 2003), p. 445
30. S. Nielsen, R. Kapral and G. Ciccotti: J. Chem. Phys. **115**, 5805 (2001)
31. G. Stock and M. Thoss: Adv. Chem. Phys. **131**, 243 (2005)
32. G. Stock and U. Muller: J. Chem. Phys. **108**, 7516 (1998)
33. D. Montemayor, S. Bonella, and D. F. Coker: Work in progress (2006)
34. R. P. Feynman and A. R. Hibbs: *Quantum Mechanics and Path Integrals* McGraw-Hill, New York (1965)
35. L. S. Shulman: *Techniques and Applications of Path Integration* Wiley, New York (1981)
36. M. F. Herman and E. Kluk: Chem. Phys. **91**, 27 (1984)
37. W. H. Miller: Mol. Phys. **100**, 397 (2002)
38. H. Wang, X. Sun, and W. H. Miller: J. Chem. Phys. **108**, 9726 (1998)
39. X. Sun, H. Wang, and W. H. Miller: J. Chem. Phys. **109**, 7064 (1998)
40. M. Thoss, H. Wang, and W. H. Miller: J. Chem. Phys. **114**, 47 (2001)
41. W. H. Miller: J. Phys. Chem. A **105**, 2942 (2001)
42. Q. Shi and E. Geva: J. Phys. Chem. A **108**, 6109 (2004)
43. Q. Shi and E. Geva: J. Chem. Phys. **121**, 3393 (2004)
44. X. Sun and W. H. Miller: J. Chem. Phys. **106**, 6346 (1997)

II.3 Quantum Trajectory Methods

Atom–Surface Diffraction: A Quantum Trajectory Description

A.S. Sanz and S. Miret-Artés

Summary. The trajectory-based formalism of Bohmian mechanics constitutes an alternative (but equivalent at a predictive level) approach to the standard or conventional formulation of quantum mechanics. Here we show the advantages of this formalism in providing both an accurate description and a novel interpretation when applied to different phenomena of interest in elastic atom–surface scattering, such as (a) diffraction by a "soft" double-slit, (b) surface rainbows and quantum–classical correspondence, (c) quantum vortical dynamics due to the presence of single adsorbates, and (d) selective adsorption resonances and classical vs quantum trapping. These problems illustrate fairly well how quantum trajectories are able to satisfactorily reproduce the main features of real scattering experiments as well as to provide a causal insight of the underlying dynamics.

1 Introduction

Particle diffraction experiments are playing a key role in the conceptual development of quantum theory since its inception [1,2]; they explicitly display the three main ingredients of this theory: *quantization*, *interference*, and *uncertainty*. Hence, since the pioneering electron diffraction experiments carried out by Davisson and Germer [3], experiments with heavier and more complex particles have been performed, like the most recent ones with fullerenes [4] and large biomolecules [5] by Zeilinger's group.

Nonetheless, apart from their obvious fundamental implications, diffraction experiments have always had an important practical interest. Thus, the first experimental evidence of atom diffraction presented by Stern and coworkers [6], based on the study of scattering of light (noble gas) atoms off surfaces, became one of the cornerstones of surface science. Low-energy He-atom diffraction is nowadays a well established and valuable tool to study the structure of periodic surfaces, probe gas–surface interaction potentials, or investigate the presence of defects and adsorbates on surfaces. Moreover, this technique has the advantage that it causes no damage to the surface when probing its outermost layers [7–9].

In order to extract useful information, provide correct interpretations, and make further predictions when dealing with diffraction experiments, one needs quantitative theories firmly relying on quantum mechanics [10–14]. These theories, however, lack a clear intuitive picture for the studied phenomena because of the intrinsically probabilistic nature of quantum mechanics in its standard formulation [15,16] – although there is an answer to the question of how probable it is for a scattered atom to be deflected at a certain angle, there is no information about the actual deflection angle or how the scattering process takes place (unless one runs classical trajectories to get an approximate insight). Hence it is also highly desirable to have trajectory-based theories that not only accurately reproduce a quantum observable [17,18], but at the same time offer a clear and intuitive picture of the diffraction process [19–22].

Among the different trajectory-based approaches that one might consider (some of them gathered in this book), Bohmian mechanics [23–26] is the most accurate one. In it the trajectory picture directly emerges from a reinterpretation of the *quantum state* without involving any kind of approximation. Within the classical-like Bohmian formalism, quantum systems are understood as consisting of a wave and a particle (thus breaking the "old" dichotomy wave vs particle), both evolving according to deterministic laws of motion. In this sense, accurate statistical predictions and a consistent theory of quantum motion are gathered within a more general conceptual framework than the standard quantum-mechanical one. This makes quantum trajectories to be an important working tool in the analysis of quantum phenomena from both an interpretative and a computational viewpoint.

In order to stress the relevance of Bohmian mechanics as the appropriate framework to interpret and predict quantum phenomena, this chapter is organized as follows. A brief survey of the Bohmian formalism is given in Sect. 2, where we emphasize those aspects that are more closely related to the problems that will be described later on. In Sect. 3, we present a description based on quantum trajectories of a number of phenomena of interest in elastic atom–surface scattering, providing a more physically, intuitive and novel interpretation for them. Finally, in Sect. 4 we briefly discuss the directions in which our work will be addressed in the near future.

2 Fundamentals of Bohmian Mechanics

According to the statistical interpretation of quantum mechanics [15,16], the *probability* for finding a particle at a position $\boldsymbol{r}$ at a time t is described by the probability density, $\varrho_t(\boldsymbol{r}) := |\Psi_t(\boldsymbol{r})|^2$. Unlike classical mechanics, the *state* of a particle is not specified by its position and momentum, but by a probability amplitude or wave function, $\Psi_t(\boldsymbol{r})$, whose time evolution is determined by the Schrödinger equation,

$$i\hbar \frac{\partial \Psi_t(\boldsymbol{r})}{\partial t} = \left[-\frac{\hbar^2}{2m} \nabla^2 + V(\boldsymbol{r}) \right] \Psi_t(\boldsymbol{r}) \ . \qquad (1)$$

Despite this fully probabilistic viewpoint of the quantum world, Bohm [23] realized[1] that (1) not only provides statistical information about the particle. Rearranging this equation conveniently, valuable information concerning individual properties of the particle can also be obtained. This constitutes the main idea behind Bohmian mechanics; *uncertainty* arises from the *unpredictability* in determining the particle initial conditions – distributed according to $\varrho_0(\boldsymbol{r})$ [24, 29] –, but not from the *impossibility* to know the actual (quantum) trajectory pursued during its evolution.

The (Bohmian) equations of motion for the particle can be easily derived by expressing the wave function in polar form,

$$\Psi_t(\boldsymbol{r}) = \varrho_t^{1/2}(\boldsymbol{r})\, \mathrm{e}^{\mathrm{i}S_t(\boldsymbol{r})/\hbar} , \tag{2}$$

and then substituting it into (1). Separating the real and imaginary parts from the resulting expression, two real coupled equations are obtained:

$$\partial_t \varrho_t + \nabla \cdot \left(\varrho_t \, \frac{\nabla S_t}{m} \right) = 0 , \tag{3a}$$

$$\partial_t S_t + \frac{(\nabla S_t)^2}{2m} + V_{\mathrm{eff}} = 0 , \tag{3b}$$

where

$$V_{\mathrm{eff}} := V - \frac{\hbar^2}{2m} \frac{\nabla^2 \varrho_t^{1/2}}{\varrho_t^{1/2}} \tag{4}$$

is an *effective potential* resulting from the sum of the "classical" contribution, V, and the so-called *quantum potential*, Q_t, which depends on the quantum state via ϱ_t. Equation (3a) is the continuity equation for the particle flow (or the probability density, from a conventional viewpoint) and (3b) is a *generalized* Hamilton–Jacobi equation. As in classical mechanics, the *characteristics* or solutions, S_t, of (3b) define the particle velocity field,

$$\boldsymbol{v} := \frac{\boldsymbol{p}}{m} = \frac{\nabla S_t}{m} , \tag{5}$$

from which the *quantum trajectories* are kown.

An alternative way to obtain the quantum trajectories is by formulating Bohmian mechanics as a Newtonian-like theory. Then, (5) gives rise to a *generalized* Newton's second law,

$$m\, d_t \boldsymbol{v} = -\nabla V_{\mathrm{eff}} . \tag{6}$$

This formulation results very insightful; according to (6), particles move under the action of an *effective force*, $-\nabla V_{\mathrm{eff}}$, responsible for effects that

[1] Madelung [27] and de Broglie [28] had previously worked on similar approaches.

have no classical analog (e.g., interference and tunneling). Furthermore, this formulation is also particularly useful to solve problems from a quantum-hydrodynamical viewpoint [26] (see later).

As seen earlier, the quantum potential depends on the instantaneous curvature of the wave function, thus representing the (instantaneous) action of "internal" quantum forces on the system. Hence, though classical and Bohmian mechanics look formally the same, the presence of the quantum potential breaks down a full formal and conceptual equivalence. This can be seen by means of the following example (see also Sect. 3). The diffraction pattern obtained experimentally arises from a statistical count[2] of particles arriving in the detector after being deflected by the target. Any of these particles is described by the same initial state, i.e., an identical initial wave function, Ψ_0, is associated to all of them. Despite that, each particle reaches the detector at a different angular position because they have different initial conditions (from a Bohmian viewpoint). Up to here there is no difference between a classical and a quantum experiment. The difference arises from the kind of space–time coupling that Q_t establishes between those *independent*[3] particles. Consequently, quantum trajectories evolving from different initial conditions (but identical initial state) can never cross at a time t, unlike what one observes in classical descriptions – classical avoided crossings happen only in phase space, but not in configuration space. Because of this *nonlocal* action, Q_t can be considered an agent for the transmission of quantum information [25, 30, 31].

The nonlocal nature of Q_t has two important consequences in problems related to scattering processes. First, relevant quantum effects can be observed in regions where the classical interaction (described by V) is negligible and, more important, where $\varrho_t(\boldsymbol{r}) \approx 0$. This happens because quantum particles respond to the "shape" of Ψ_t, but not to its "intensity," $\varrho_t(\boldsymbol{r})$, unlike the classical analog – notice that Q_t is scale-invariant under the multiplication of $\varrho_t(\boldsymbol{r})$ by a real constant. Second, quantum-mechanically the concept of *asymptotic* or *free motion* only holds *locally*. According to classical mechanics, this concept is defined by the condition:

$$m\, d_t \boldsymbol{v} \approx 0 \,, \tag{7}$$

which is satisfied whenever $V \approx 0$. Analogously, $V_{\text{eff}} \approx 0$ determines the condition for quantum free motion. This condition is fulfilled, for example, along the directions specified by the diffraction channels [19, 21]; in between, particles are still subjected to strong quantum forces (although $V \approx 0$). The asymptotic condition allows one to establish a distinction between two dynamically different regions in configuration space [32]: the *interaction* or *Fresnel region*,

[2] Remember that (1) refers to the evolution of a single particle of mass m, although a statistics has to be invoked in order to reproduce $\varrho_t(\boldsymbol{r})$.

[3] By "independent" we mean that there is *no* physical connection between particles, since each one represents a different experiment.

where the particle dynamics is very intense due to the combined action of both V and Q_t; and the *asymptotic* or *Fraunhofer region*, where the dynamics becomes stationary, and only Q_t influences the particle motion.

The role played by the quantum potential is also relevant within *quantum hydrodynamics* [26, 27, 33], the hydrodynamical picture of quantum mechanics. Here, two important concepts come into play: the *quantum pressure* and the *quantum vortices*, which we will borrow to apply them in the context of scattering. Quantum hydrodynamics constitutes nowadays the basis for different computational techniques applied in the study of quantum dynamics of a large number of physical and chemical complex processes [34–39]. In analogy to classical fluids, quantum ones are characterized by a velocity field, $\boldsymbol{v}$, the probability density, ϱ_t, and the quantum density current, $\boldsymbol{J}_t := \varrho_t \boldsymbol{v}$. These elements allow to rewrite (3) as

$$\partial_t \varrho_t + \nabla \cdot \boldsymbol{J}_t = 0 \ , \tag{8a}$$

$$\partial_t \boldsymbol{v} + (\boldsymbol{v} \cdot \nabla) \, \boldsymbol{v} = -\frac{1}{m} \, \nabla V_{\text{eff}} \ , \tag{8b}$$

respectively, with (8a) describing the conservation of the probability density and (8b) being a *generalized* Euler equation. Notice that, in correspondence with classical fluids, m is identified with the mass of a piece of fluid separated from the rest by a closed surface, $m\varrho_t$ with the fluid density, and $\boldsymbol{v}$ with the velocity field of the flow. Nevertheless, let us stress out that, unlike classical fluids, quantum ones are generated by probability flows that do *not* have any material structure.

To understand the concept of *quantum pressure*, it is interesting to compare (8b) with its classical counterpart,

$$\varrho_t \Big[\ \partial_t v_i + (\boldsymbol{v} \cdot \nabla) \ v_i \ \Big] = \varrho_t f_i + \partial_j (-P \delta_{ij}) \ . \tag{9}$$

Here, ∂_j denotes the partial derivative along the j-direction, f_i is the external force acting on the fluid along the i-direction, and P the fluid pressure. Equation (9) describes the evolution of an *ideal* fluid (an incompressible and nonviscous flow) in absence of thermal effects, and shows that the flow dynamics is determined by both an external and an internal force – as seen, the internal force, given by $\varrho_t^{-1} \partial_j (-P \delta_{ij})$, depends on the properties of the fluid (via P). If (8b) is rewritten as

$$\varrho_t \Big[\ \partial_t v_i + (\boldsymbol{v} \cdot \nabla) \ v_i \ \Big] = \varrho_t f_i + \partial_j \mathsf{T}_{ij} \ , \tag{10}$$

in analogy to (9), we observe that

$$\mathsf{T}_{ij} := \frac{\hbar^2}{4m^2} \ \varrho_t \, \partial_{ij} \ln \varrho_t \tag{11}$$

is the quantum analog of the classical stress tensor, $-P\delta_{ij}$. This term, called the *quantum stress tensor*, implies the existence of a *quantum pressure*. Here, nonetheless, we are going to use this concept in a fully qualitative, simpler manner when we will present the interpretations provided for the different scattering processes discussed in Sect. 3.

There is an important feature regarding the nature of the quantum pressure that we would like to emphasize. In classical mechanics the pressure is associated to the number of particles that constitutes a fluid. On the contrary, the quantum pressure is related to ϱ_t rather than to the *total* number of particles, since, as said earlier, quantum fluids do not possess material structure. For example, all the phenomena discussed here are problems of a single particle scattered off a potential V, and therefore one should not expect to observe any pressure in a classical sense. However, one observes that each quantum trajectory (which represents the time-evolution of a particle with a certain initial condition with probability ϱ_0) "feels" the effect of the quantum pressure through the information about the (time-dependent) particle distribution, ϱ_t, conveyed by the quantum potential.

As stated earlier, quantum hydrodynamics is also characterized by the presence of *quantum vortices*. Observe that, since Ψ_t is a complex function, it is always uniquely determined except for a constant phase factor, i.e.,

$$\Psi'_t[S'_t] = \Psi_t[S_t] \iff S'_t(\boldsymbol{r}) = S_t(\boldsymbol{r}) + 2\pi\hbar n \ , \tag{12}$$

where n is an integer number. Therefore, considering that Ψ_t is a smooth function (its first derivative is continuous), discontinuities in its phase ($n \neq 0$) can only occur in nodal regions, where $\Psi_t = 0$ and S_t can display discrete "jumps" because of the wave function multivaluedness. These discontinuities give rise to a *vortical dynamics*, with the particles avoiding to cross the nodes of the wave function and moving parallel along them. In the case of point-like nodes particles undergo permanent or transient loops around them depending on the node lifetime (in general, nodes are time-dependent).

As inferred from (4), the presence of nodes leads to singularities in the quantum potential. From a computational viewpoint, this can be inconvenient for those numerical algorithms based on the direct solution of the hydrodynamical equations, since those singularities can explicitly appear. In such cases, it is necessary not only to know *where* they emerge (as could happen when one considers classical singular potentials, e.g., Coulombic-like potentials), but also *when* they do it. It is important to note that this problem disappears when quantum trajectories are computed by using the wave function – i.e., obtaining S_t at each time from Ψ_t, and then solving for (5). In this case, given Ψ_0 and $\boldsymbol{r}_0$, the velocity vector is well-defined and the trajectory propagates avoiding to come into nodal regions – only in cases of loss of accuracy these trajectories can enter into such regions, which is equivalent to say that the algorithm provides a wrong solution.

To finish this section, we are going to make a few remarks on the quantum–classical correspondence from the Bohmian viewpoint. Although according to (4) it is apparent that Bohmian mechanics should approach classical mechanics in the limit $Q_t \to 0$, in general a gradual, smooth transition does not exist, as happens, for example, from relativistic to Newtonian mechanics – in this case, the latter emerges when the particle velocity is much smaller than the speed of light. In quantum mechanics only the expectation value of an observable (but not the observable itself) can be compared with its corresponding classical counterpart. Nonetheless, an important insight on the quantum–classical correspondence can be obtained by expressing (5) in terms of the well-known ansatz of the WKB approximation [19,21],

$$\Psi_t(\boldsymbol{r}) = \mathrm{e}^{\mathrm{i}\bar{S}_t(\boldsymbol{r})/\hbar} , \tag{13}$$

where $\bar{S}_t$ is a general complex function. Introducing (13) into (1) one obtains

$$\partial_t \bar{S}_t + \frac{(\nabla \bar{S}_t)^2}{2m} + V + \frac{\hbar}{2m\mathrm{i}} \nabla^2 \bar{S}_t = 0 , \tag{14}$$

which is a complex equation totally equivalent to the Schrödinger equation, though similar to (3b). If, like in the procedure followed to derive the WKB approximation, $\bar{S}_t$ is expanded in a series of $\hbar/\mathrm{i}$,

$$\bar{S}_t = \sum_{n=0}^{\infty} \left(\frac{\hbar}{\mathrm{i}}\right)^n \bar{S}_t^{(n)} , \tag{15}$$

with $\bar{S}_t^{(n)}$ being real functions, and substituted into (14), a set of couple equations is obtained [21]. The equation corresponding to the zeroth order in $\hbar$ is the classical Hamilton–Jacobi equation; the remaining equations describe the evolution of higher order contributions to $\bar{S}_t$, responsible for the quantum behavior undergone by the particle.

On the other hand, introducing (15) into (5) results in

$$\dot{\boldsymbol{r}} = \frac{1}{m} \sum_{n=0}^{\infty} (-1)^n \hbar^{2n} \nabla \bar{S}_t^{(2n)} = \dot{\boldsymbol{r}}^{(\mathrm{cl})} + \frac{1}{m} \sum_{n=1}^{\infty} (-1)^n \hbar^{2n} \, \nabla \bar{S}_t^{(2n)} , \tag{16}$$

where $\dot{\boldsymbol{r}}^{(\mathrm{cl})} := \nabla \bar{S}_t^{(0)}/m$ is the classical law of motion. A simple inspection of (16) leads to the conclusion that quantum trajectories can be interpreted as classical trajectories "dressed" with a series of interfering terms, this showing the capital difference between both types of trajectories. Thus, in principle, it is always possible to distinguish some "classical" features in quantum trajectories whenever the interference effects are *relatively* weak. As will be seen in Sect. 3, this is a general statement that does not necessarily require expansions in terms of $\hbar$.

3 Applications

3.1 The "Soft" Double-slit Model

Apart from the fundamental implications of experiments with slits, they can also be used to understand the concepts underlying the gas–surface scattering dynamics; multiple-slit arrangements are diffracting gratings *equivalent* to perfect periodic surfaces. The simplest slit system is the famous *double-slit*, which we describe in this section in order to advance concepts and ideas that will be further used when dealing with scattering of He atoms off different types of surfaces. In particular, we consider two cases, regarded as experiments A and B, describing the scattering (and subsequent diffraction) of electrons by soft potentials modeling the double-slit [21] – these models are more *realistic* than the typical textbook example of sharp-edged slits.

The double-slit potential for experiment A is given by

$$V(x,y) = \left(V_0 - \frac{1}{2}\, m\omega^2 y^2 + \frac{m^2\omega^4 y^4}{16V_0} \right) \mathrm{e}^{-x^2/\alpha^2} \,, \tag{17}$$

a model used in the literature [17,18] to show the advantages of the backward–forward semiclassical initial value representation into the study of *decoherence*. Here, $\alpha = 25$ bohr, $\omega = 600$ cm^{-1}, $V_0 = 8,000$ cm^{-1}, and m is the electron mass. Experiment B is described by a modified version of (17) that consists in a slight shift forward (with respect to the plane containing the slit) of the central barrier. The corresponding potential model, introduced by Guantes et al. [21] to study the effects of the central barrier on the electron dynamics, is

$$V(x,y) = \frac{m^2\omega^4 y^4}{16V_0}\, \mathrm{e}^{-x^2/\alpha^2} + V_0\, \mathrm{e}^{-(x-x_b)^2/\alpha^2 - y^2/\beta^2} \,, \tag{18}$$

with $\beta = 90$ bohr and $x_b = 125$ bohr. Classically, this model presents direct transmission only for high values of the incidence energy, E_i. For lower values of E_i, the two slits become transversal channels that frustrate such a transmission – the electrons can only pass laterally after bouncing several times over the three walls of the arrangement. The results shown here are for a quasiplane (or quasimonochromatic) initial wave function with energy $\langle E \rangle_i \simeq 500$ cm^{-1}, for which there is no direct transmission in experiment B. This wave function is launched perpendicular to the double-slit from a distance $\langle x \rangle_0 = -400$ bohr (far enough from the interaction region of the soft potential).

Figure 1 shows the probability density after the collision with the double-slit (top panels), the transmission function[4] (center panels), and the intensity pattern that would appear on a screen behind the slits (bottom panels) for experiments A (left) and B (right). From the transmission function, a certain delay in reaching the Fraunhofer regime in experiment B (0.55 ps vs 0.27 ps for experiment A) is noticeable. This delay is caused by the barrier; as the wave function gets into the region $\Sigma := \{0 \lesssim x \lesssim x_b\}$, it becomes highly peaked inside, giving rise to a *transient trapping* or *resonance*. The portion of the wave packet inside Σ reaches its maximum at $t \approx 0.18$ ps (see Fig. 1b′), and then the resonance begins to *dissipate*, with the probability either flowing backward or passing through the transversal channels. Observe that the decay

[4] The transmission function is defined here as the probability to localize the electron behind the double-slit: $\mathcal{T}_t = \int_{x_b}^{+\infty} |\Psi_t(x)|^2 \mathrm{d}x$ (in experiment A we assume $x_b = 0$ bohr).

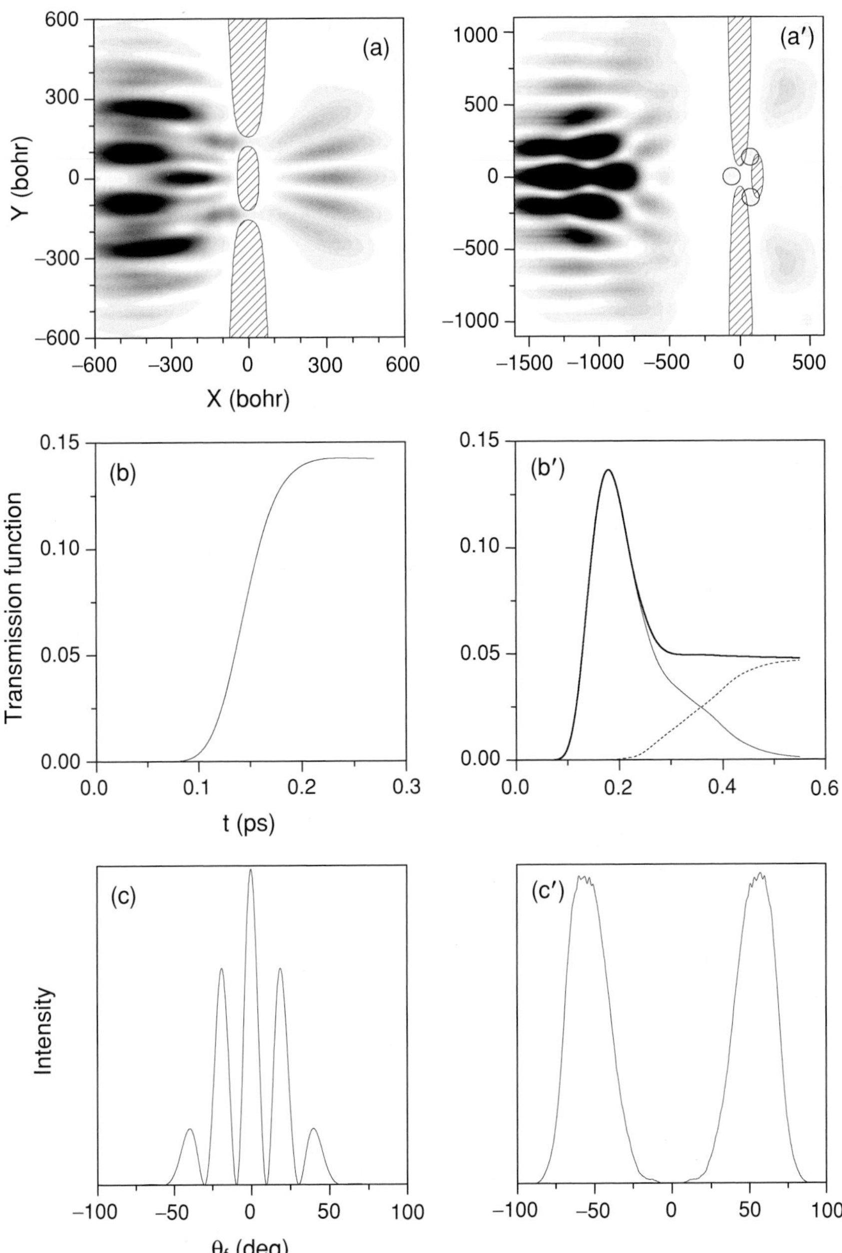

Fig. 1. Quantum results for the experiments A (*left panels*) and B (*right panels*): outgoing probability density (*top*), transmission function (*center*), and intensity pattern (*bottom*). In the upper panels, the initial propagation is from left to right with respect to the plane containing the slit. Circles in ($\mathbf{a}'$) enclose weak resonance peaks. The *thin solid line* in ($\mathbf{b}'$) refers to the probability inside Σ, and the *dotted line* denotes the probability behind $x = x_b$

of this resonance becomes asymptotically slower after $t \approx 0.28$ ps (thin solid line), thus still remaining for a relatively long time. This manifests as the appearance of a weak peak in each opening of Σ (the three corresponding peaks are enclosed by circles in Fig. 1a′).

Transient quantum trapping is intimately connected to the presence of *transient classical trapping* or *classical chaos* [21], what demonstrates a high quantum–classical correspondence. The classical values for the transmittance – the fraction of transmitted particles from an ensemble initially covering the same extension along the y-direction as ϱ_0 – are 10.48% for experiment A and 4.05% for experiment B. These values are comparable to those obtained quantum-mechanically – the quantum transmittance is the asymptotic value of the transmission function –, 14.24% and 4.78%, respectively. The slight difference is attributed to tunneling (see later) and diffractive effects.

The previous results have been explained by using the standard version of quantum mechanics. However, in our opinion, a deeper understanding of the dynamics can be gained by using Bohmian mechanics. As seen in Sect. 2, electrons undergo a motion similar to that of particles in a classical fluid, manifesting the action of an effective potential that is the sum of the classical potential plus the quantum one. The latter, which conveys information on the whole ensemble of particles, gives rise to the quantum pressure. In this way, the electrons with initial positions corresponding to the rear part of ϱ_0 (with respect to the direction of propagation) will not be able to reach regions that are accessible to those starting closer to the slits. Indeed, the latter will be "pressed" by those coming behind, being bounded to remain for a longer time in contact with the real double-slit potential. This is something with no analog in the classical problem of a single particle[5] passing through a double-slit.

The aforementioned statements are easily understood by looking at the different ensembles of quantum trajectories plotted in Fig. 2. Taking advantage of the reflection symmetry with respect to $y = 0$, only half of the trajectories (those corresponding to the upper slit) has been represented to make clearer the figures (moreover, the incident part is not shown either). The values of the initial y-coordinate for homologous trajectories in the different panels are the same, and only their initial x-coordinate changes. In particular, three different values of x_0 sampling the three parts of ϱ_0 (rear, middle, and front with respect to the direction of propagation, respectively) are considered: $x_0 = \langle x \rangle_0 - 100$, $x_0 = \langle x \rangle_0$, and $x_0 = \langle x \rangle_0 + 100$, with $\langle x \rangle_0 = -400$ (all units are given in bohr). As can be seen, the dynamical role of the quantum pressure is fundamental to understand the motion of the electrons. Notice how the trajectories starting at distances further from the double-slit potential (in both experiments) can not reach it, contrary to what happens in a purely classical situation, where the starting point (provided it is located at asymptotic distances from the potential) does not influence the behavior of subsequent groups of trajectories. Moreover, the distortion that the slits cause on the topology of the trajectories is also remarkable. If the potential was just a wall (i.e., no slits), electrons

[5] Single particle in the sense outlined in footnotes 2 and 3.

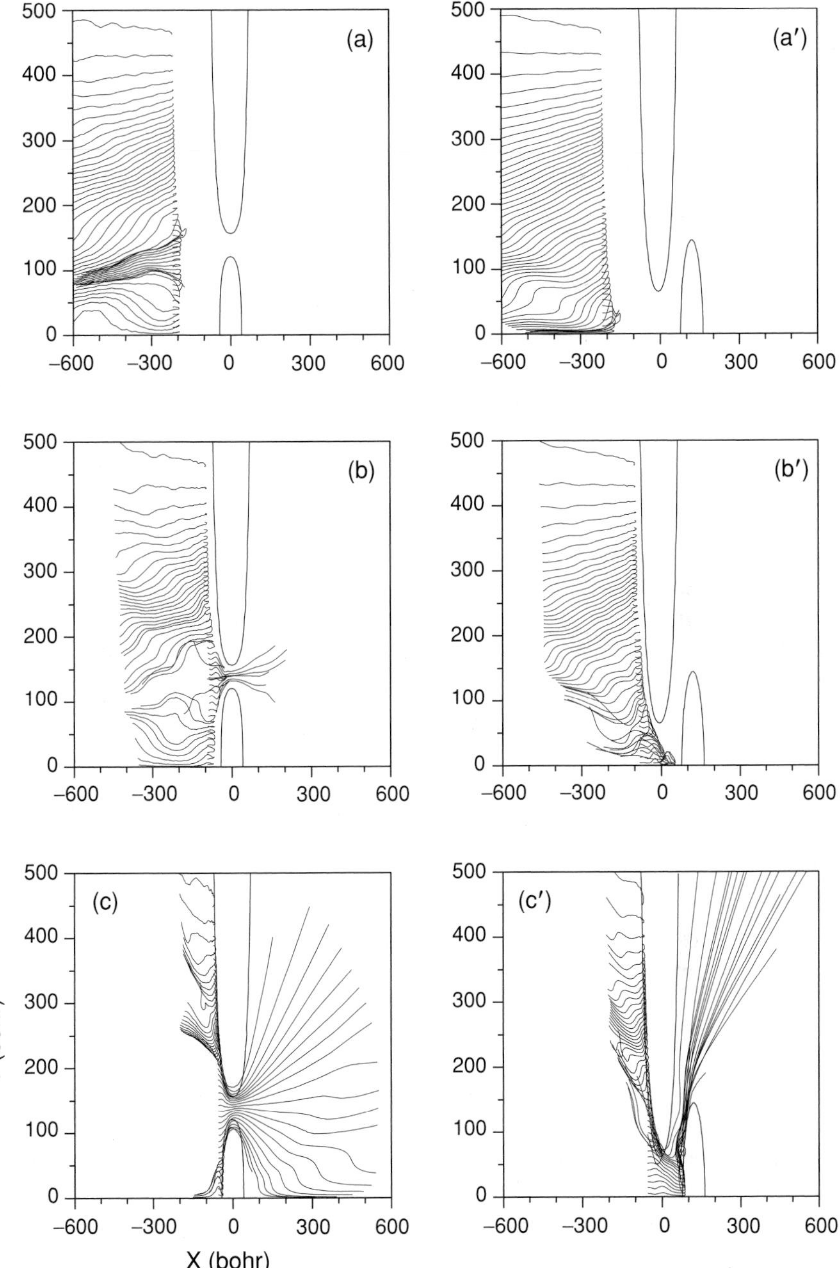

Fig. 2. Bohmian trajectories for experiments A (*left panels*) and B (*right panels*). The propagation is from left to right with respect to the plane containing the slit, launching the trajectories from: $x_0 = \langle x \rangle_0 - 100$ (*top*), $x_0 = \langle x \rangle_0$ (*center*), and $x_0 = \langle x \rangle_0 + 100$ (*bottom*), with $\langle x \rangle_0 = -400$ (all units are given in bohr). For clarity, only the scattered part of half of the trajectories is represented (see text for details)

would get diffracted backward giving rise to a diffraction pattern similar to that of a wave passing through a single-slit. However, the presence of slits leads to the appearance of some channels (two in experiment A and only one in experiment B) that disrupt the relatively smooth motion of the reflected electrons.

From a dynamical viewpoint, two interesting effects are worth discussing. First, observe that there are a number of electrons initially starting close to the potential that cannot pass through the slits, but that are pushed away by other particles coming behind. Due to quantum pressure, these electrons have two possibilities to "escape" when are reflected: either by going toward the borders of the incoming wave, or (in the case of experiment A; see Fig. 2c) toward the symmetry axis (i.e., the $y = 0$ axis). Thus, as happens in classical hydrodynamics, here the electrons also move toward those regions where the values of the quantum pressure are smaller. Second, notice the presence of tunneling mentioned earlier; Fig. 2c, c′ show how trajectories pass through regions that are classically forbidden. This is possible in Bohmian mechanics because quantum particles have an additional quantum energy arising from the quantum potential [36] which helps them to overcome regions that are classically forbidden. Regarding the conservation of the energy, this does not constitute a problem; quantum-mechanically, the magnitude that must be conserved is the average energy of the ensemble, $\langle E \rangle_t$, but *not* the energy of each individual particle.

Although the number of particles passing through the slits is a function of the energy $E_{\rm i}$ and the parameters defining the classical potential, it is clear that by studying the electron dynamics one can determine with no ambiguity which part of the initial wave packet is reflected and which one is transmitted. This is something unthinkable in standard quantum mechanics, where the wave function is a kind of "wholeness" from which such an information cannot be inferred. Here, we have seen that the electrons in the rear part of the ensemble do not cross the slits, while those initially closer to the potential do it. That is, the quantum transmittance has contributions from the front of the wave packet, but not from its rear part. Moreover, the electron quantum trajectories also indicates the part of the initial wave packet contributing to each diffraction peak. This fact, as we will see later, is of capital importance in characterizing of diffraction channels in atom–surface scattering.

Finally, let us stress the difference between Fig. 2c, c′ in relation to Fig. 1c, c′, respectively. In experiment A there is interference of the two diffracted electron beams. This manifests as a kind of "wiggly" behavior in the topology of the trajectories until the electrons reach a diffraction channel or Bragg direction; then, they move as free particles. The formation of these channels in the Fraunhofer region is a direct consequence of the information that the quantum potential transmits to the particles about the status of each slit (either open or close). In the case of experiment B, the electrons exiting from each slit behave like if such an information was not relevant, because the diffracted beams do not overlap. This makes electrons to display the free evolution – approximately, since there are still small disturbances produced by the remaining transiently trapped wave – that would correspond to motion under the guidance of a Gaussian wave packet.

3.2 Surface Rainbows

In classical scattering theory it is common to find that the scattering intensity displays singularities for certain deflection angles [40]. One of the effects responsible for such singularities is the so-called *rainbow* effect, which consists in a large accumulation of classical trajectories (*caustics*) as they approach the maximum/minimum deflection angle or *rainbow angle*, $\theta_{\rm r}$. An important tool to study this effect is the *classical deflection function*, i.e., the relationship between the deflection or final angle, θ_f, and the impact parameter. For $\theta_f = \theta_{\rm r}$, the deflection function presents local maxima and/or minima. This clearly explains the singularities in the classical intensity, since this magnitude is proportional to the inverse of the derivative of the deflection function with respect to the impact parameter.

The classical rainbow singularity gives rise in quantum mechanics to a certain modulation of the diffraction intensity patterns [41]. Strictly speaking, the quantum rainbow takes place when the rainbow angle is a Bragg or observable final angle. The general procedure followed to assign a feature from the intensity pattern to a rainbow is merely based on a direct correspondence between such a feature and the classical intensity [21,42,43]. By means of a semiclassical analysis, one also finds [44] that for (Bragg) diffraction peaks appearing at the place of classical rainbow angles, the (semiclassical) intensity comes from the contribution of the corresponding classical rainbow trajectories.

Despite the interest and accuracy that classical and semiclassical pictures might provide to this problem, the appropriate theoretical framework to establish a clear and unambiguous quantum–classical correspondence for the rainbow effect has to be of quantum nature. This working framework is precisely given by Bohmian mechanics. In analogy to classical scattering, one can define the *quantum deflection function* [45] with certain preventions. Since the initial wave packet has a finite width along the perpendicular direction to its propagation, the impact parameters have always to be chosen at different transversal (with respect to the propagation direction) cuts of the wave packet, as seen in Sect. 3.1. For periodic surfaces, this deflection function displays a seemingly step-ladder structure, with each step referring to a different diffraction channel that can be mapped backward onto a specific region of the initial probability density [32,45]. On the other hand, as also seen in Sect. 3.1, quantum dynamics is *global* unlike classical one, this meaning (within this context) that it involves the total number of unit cells illuminated by the incoming wave packet. Therefore, the quantum deflection function has to be built by considering the whole illuminated area in contrast to what happens in classical mechanics, where it is enough to study impact parameters covering a single unit cell.[6]

To study the quantum–classical correspondence in rainbow scattering, the classical limit is approached here by artificially increasing the mass (m) of the impinging particles. One can easily observe that the total number of diffraction orders, $n_{\rm t}$, scales with $\sqrt{m}$ for a fixed incidence energy, and leads to an increase of the complexity of the intensity pattern [45], which becomes highly oscillatory due to the large amount of emergent Bragg channels. On average, the intensity pattern resembles the one obtained classically; the diffraction intensity peaks become higher in the vicinity of those deflection angles that correspond to classical rainbow angles. From a Bohmian viewpoint, as m increases the trajectories get more complex and loose smoothness [21, 45]. Appealing to the quantum-hydrodynamical picture, this is equivalent to say from a qualitative point of view that light particles can be considered as moving in a laminar flow, while heavier ones undergo a more *turbulent* dynamics. In the latter case, crosses between trajectories can be observed (at different times), showing certain resemblance to the pattern formed by the classical trajectories (covering the same number of unit cells) in the Fresnel region. This makes clear the statement that a quantum trajectory can be understood as a classical trajectory "dressed" by a series of additional interference terms, as infers from (16). The presence of these additional terms is, nonetheless, very important regarding the properties of nonlocality and context-dependence provided by Bohmian trajectories, which still remain in the classical limit unless they are explicitly "removed"[7] from (16).

As an illustration of the previous statements, we are going to analyze the scattering of He atoms off Cu(110) at normal incidence and 21 meV. The quantum and classical deflection functions are plotted in Fig. 3. Quantum trajectories are run for two different masses, $m_{\rm He}$ and $m_{\rm He^*} = 500\, m_{\rm He}$, and their deflection functions are compared with the classical counterpart (for the same incidence conditions, the classical deflection function does not depend on the impinging particle mass). As commented earlier, the most remarkable feature is the transition from a (discrete) step-ladder structure, typical of a pure quantum regime, to another one that (on average) smoothly adapts to the classical deflection function. Observe that, effectively, the periodicity of the classical deflection function is absent in the Bohmian case; the corresponding deflection function only agrees with the classical one when the latter is rescaled to the range of impact parameters covered by the initial wave packet. Nonetheless, it is meaningful the fact that one can *locally* appreciate that the quantum deflection function displays an oscillating envelop with the periodicity of a

[6] Due to the translational symmetry of the potential surface, trajectories with impact parameters differing an integer amount of unit cells are identical.

[7] The interference terms can be very complicated in the classical limit, but it does *not* mean that they gradually disappear.

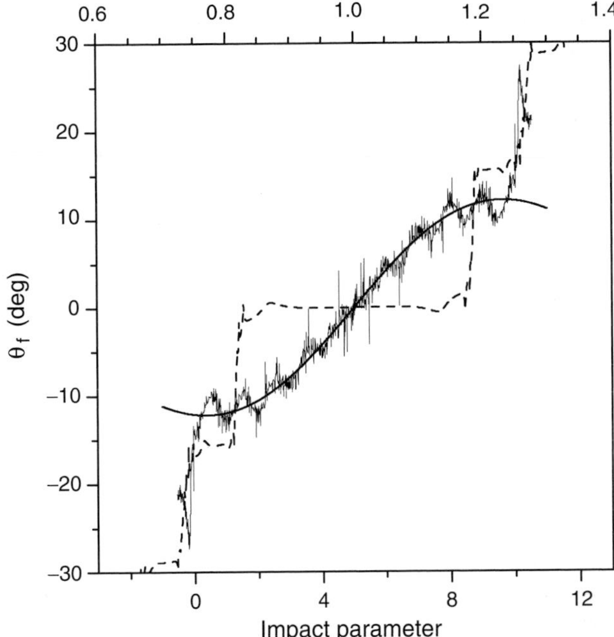

Fig. 3. Quantum deflection function for incident particles with masses m_{He} (*dashed line*) and $m^*_{\mathrm{He}} = 500\, m_{\mathrm{He}}$ (*thin solid line*); the classical deflection function is represented in thick solid line. The impact parameter is given in number of unit cells covered (1 = 1 u.c.). Observe the difference between the length covered by the quantum deflection function (lower horizontal axis) with respect to the set of classical impact parameters (upper horizontal axis)

single unit cell – notice that there are ten oscillations, which is in agreement with the number of unit cells illuminated.

As seen, the emergence of surface rainbows manifest in a completely different way in classical and quantum mechanics. In the former, rainbows are reproduced *independently* on each unit cell, while in the latter they appear as a *global* feature connected to the whole initially illuminated surface. This is a consequence of the type of information carried by the quantum potential, which does not fully vanish even for large values of the incident particle mass. As quantum trajectories show, observe that negligible values of this potential have still very important dynamical effects.

3.3 Vortical Dynamics

The presence of impurities, defects, and/or adsorbates on a surface greatly affects its physics and chemistry. They give rise to *diffusive scattering* [46], which is responsible for the formation of intensity peaks in between the Bragg angles. This phenomenon is characterized by the incoherent scattering of atoms from different defects [47], and is well illustrated by He scattering off CO adsorbed on Pt(111), for which there is a wealth of experimental data and theoretical work [42, 43, 46, 48, 49]. The Bohmian dynamics related to this system has been well characterized recently [50, 51], showing the key role played by quantized vortices as the dynamical origin of the different intensity peaks. As said in [21], vortical dynamics may lead to quantum chaos and can be of capital importance to understand a more complex situation like the He–Cu(117) system (see Sect. 3.4), which displays a strong classical chaotic behavior under certain conditions.

To simulate the He–CO/Pt(111) interaction, a simple two-dimensional, soft potential model has been used. This model, originally proposed by Yinnon et al. [42], has also been used to perform wave packet propagations [48]. It models a step defect (axially symmetric), rather than a point-like one (radially symmetric). The topological difference between both defects is only relevant from a statistical viewpoint, since the corresponding scattering intensities (computed either by means of the standard quantum mechanics or Bohmian mechanics) display certain differences in the relative height of their maxima [48]. However, here we are interested in the individual motion of the atoms involved in the scattering process for which both models are equivalent. This can be easily understood in the following intuitive way. Due to the noncrossing property of Bohmian trajectories, there will not be any "mixing" between trajectories with initial positions chosen to be contained in different planes transversal to the symmetry axis of the defect and perpendicular to the clean Pt surface. On the other hand, in the case of the punctual defect, such a choice is equivalent for trajectories with initial positions contained in different planes along the radial direction (with respect to the CO center-of-mass). Therefore, since the profile of both defects is the same, trajectories will also display the same features in both cases. To explain why the statistical results mentioned earlier are different, one must realize that this is a matter of diffusion; in the two-dimensional model (step defect) the diffusion of trajectories only takes place along the transversal direction (i.e., the diffusion can be seen basically as a one-dimensional motion), while in the three-dimensional model (point-like defect) their diffusion is along the radial direction (i.e., the diffusion takes place across the full two-dimensional plane because of the different orientation of the radial planes containing the trajectories).

As claimed in [50], the quantum dynamics associated to the perpendicular He–CO/Pt(111) scattering (i.e., the dynamical origin of the diffraction pattern) is strongly influenced by the presence of *quantum vortices*. These vortices appear because of the overlapping of semicircular and plane wavefronts; the semicircular fronts arise from the interaction of the wave function with the CO adsorbate, and the plane ones from its interaction with the clean Pt surface. In Fig. 4, a contour plot of the probability density when the interaction with the adsorbate/surface is more intense ($t \simeq 1.41$ ps) is shown for an incident energy of 10 meV. According to Sect. 2, the pattern of nodal lines translates into a pattern of quantum vortices that changes with time as the wave packet evolves.

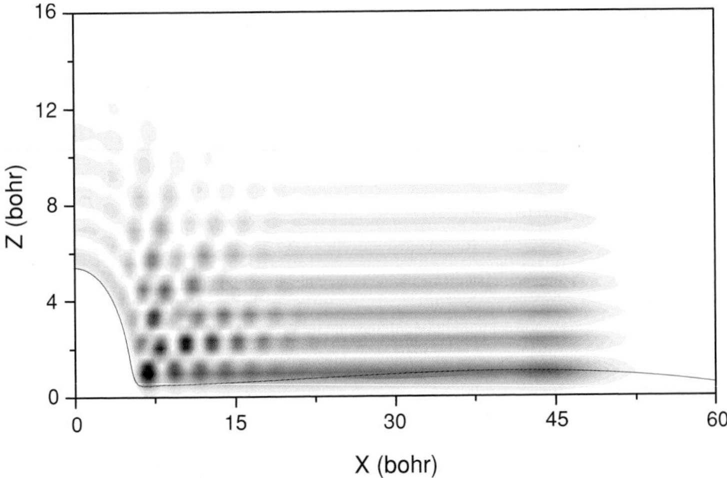

Fig. 4. Contour plot of the probability density during the instant of more intense interaction with the potential. The incidence is perpendicular to the surface, at an energy of 10 meV

This pattern of vortices is directly related to the de Broglie wavelength of the incident atoms – although $\lambda_{\rm dB} = 2\pi\hbar/\sqrt{2m(E_{\rm i} - V)}$, only a slight variation with respect to the initial wavelength ($\simeq$2.71 bohr) must be pointed out.

The web of vortices generates a highly organized, complex dynamics that Bohmian trajectories reflect; each trajectory manifests either a more laminar or a more turbulent behavior from a qualitative point of view depending on its initial position with respect to ϱ_0. This statement is illustrated by the three sets of trajectories shown in Fig. 5, with initial positions distributed along cuts perpendicular to the propagation direction and: $z_0 = \langle z \rangle_0 - 6$, $z_0 = \langle z \rangle_0$, and $z_0 = \langle z \rangle_0 + 6$ (from top to bottom), with $\langle z \rangle_0 = 19.4$ (all units are given in bohr). In the three right panels, enlargements of the region around the adsorbate are plotted for a better understanding of the corresponding quantum dynamics. In Fig. 5a it is apparent how trajectories covering regions of ϱ_0 closer to the surface mainly contribute to the peaks of the diffraction pattern with larger values of the momentum transfer, ΔK (see [50]). However, trajectories starting further away from the surface, as those shown in Fig. 5b, c, contribute to peaks with smaller values of ΔK. Again, this can be understood in terms of the quantum pressure; atoms closer to the surface suffer a higher pressure than those further away from it. In this sense, the latter manifest a classical-like motion, "bouncing" on the former, which act like an effective barrier (see enlargements of Fig. 5b, c). This explains why these atoms contribute to the central peaks of the diffraction pattern. On the contrary, those atoms starting closer to the surface remain "compressed" between the surface and the atoms coming from upper regions. This makes such atoms either to escape

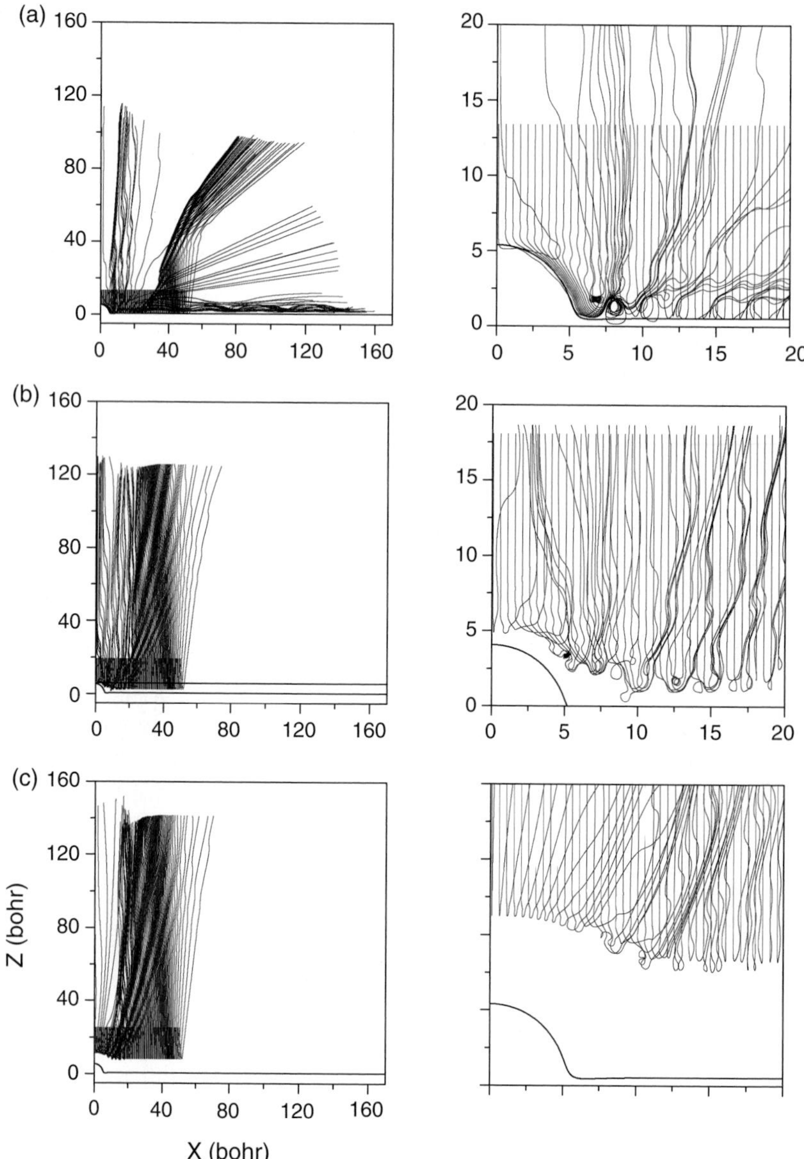

Fig. 5. Bohmian trajectories launched from: (**a**) $z_0 = \langle z \rangle_0 - 6$, (**b**) $z_0 = \langle z \rangle_0$, and (**c**) $z_0 = \langle z \rangle_0 + 6$, with $\langle z \rangle_0 = 19.4$ (all units are given in bohr). Right: Enlargements showing in more detail the particle dynamics in the region near the adsorbate

along the surface or to remain trapped permanently (see Fig. 5a), leading to the more marginal peaks of the diffraction pattern.

In a direct connection with the previous interpretation of the quantum motion, we observe that the He-atoms dynamics gets more complex as their initial positions are closer to the surface. This dynamics can be well described by recalling the concepts of laminarity and turbulence mentioned earlier. Within the hydrodynamical picture, there is a transition from laminar to more turbulent motions as initial positions are chosen closer to the surface (and mainly to the adsorbate), as inferred from Fig. 5. Thus, these regimes are clearly influenced by the presence of the nodal or vortical structure. Let us remark that the existence of a vortical regime leads to a *transient vortical trapping* [50, 51], different from the permanent trapping mentioned earlier, which is induced by the interaction with the surface. This temporal trapping is due to the action of the quantum pressure, and ends when the latter decreases sufficiently as for the atoms can escape from their confined motion within the vortical region. Given the complexity of the quantum motion, it is clear that the optical picture of this phenomenon [46] does not look like the Bohmian one at all, although it remains valid as any other semiclassical mechanism proposed to explain such diffraction patterns.

3.4 Surface Resonances

In Sect. 3.2, we have analyzed the quantum–classical correspondence in the rainbow effect, showing how Bohmian mechanics explains it in a way that goes beyond any classical or semiclassical approach. Here, we analyze the same correspondence in a more complicated phenomenon: the *selective adsorption*. It has been widely conjectured that the classical counterpart of a *selective adsorption resonance* (SAR) in atom–surface scattering is the temporal trapping of the incident atoms along the surface, with free parallel and bound perpendicular motions to the surface [52]. By means of a Bohmian analysis, we show that this classical picture has to be replaced within the quantum domain.

In order to study the elastic resonance effects, we have chosen the He–Cu(117) system [53] as a working model. Two types of quantum resonances of totally different nature are observable in this kind of systems [22, 54–56]: *threshold resonances* (TRs) and SARs. TRs occur when a diffraction channel just appears or disappears, i.e., the energy along the perpendicular direction to the surface vanishes. SARs, on the contrary, take place when the energy along the perpendicular direction to the surface becomes equal to any of the bound states of the attractive, surface-averaged potential. Here we are going to consider only SARs – discussion about TRs can be found elsewhere [21,56].

Based on classical and semiclassical arguments, it is common to think of these resonances as being connected to classical trapped trajectories that display a chaotic dynamics [55]. For example, if we consider the reciprocal lattice vector $\mathbf{B} = (3, 0)$ (in units of $2\pi/a$, a being the unit cell length of the assumed one-dimensional vicinal surface) and an incidence energy $E_\mathrm{i} = 21\,\mathrm{meV}$, three SARs appear at incidence angles $\theta^{(0)} \approx 51°$, $\theta^{(1)} \approx 46°$, and $\theta^{(2)} \approx 43°$ (superscripts refer to the bound states of the averaged potential), as seen in Fig. 6. Although these three resonance conditions are very close around the onset of

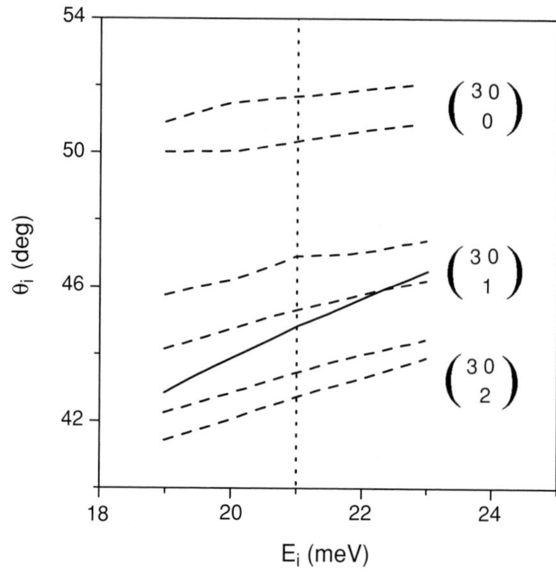

Fig. 6. SAR conditions (*dashed lines*) when the reciprocal lattice vector exchanged in the resonance process is $\mathbf{B} = (3, 0)$ [in units of $2\pi/a$, a being the unit cell length of the assumed one-dimensional Cu(117) surface], and onset of chaos (*solid line*) as a function of the incidence energy and angle for the He–Cu(117) system. Dashed lines represent here the positions plus/minus the angular halfwidths of the three bound states (ϵ_n, with $n = 0, 1, 2$) corresponding to each resonance. The vertical dotted line (at $E_i = 21$ meV) indicates the incidence energy for which the results shown in this work were obtained

classical chaos, $\theta_{\rm c} \approx 44.75°$, or multiple scattering regime, only two of them are earlier it, and therefore the classical picture for these two resonances should be applicable. However, the validity of such a picture breaks down when trying to explain the third resonance, which lays on the single scattering regime. Since it is later the vibrational trapping or classical chaos threshold, no (classical) trapped trajectory can provide an image of the resonance behavior. Even more, in this type of scattering, the transition from direct scattering (classical regularity) to trapping (classical chaos) can be easily controlled by the incident angle for a fixed incident energy.

To illustrate the behavior of the quantum trajectories in the multiple (or chaotic) and single scattering classical regimes, some calculations carried out at incidence angles 34.4° and 51.5°, and incident energy $E_{\rm i} = 21$ meV [56] are shown. As can be seen in Fig. 6, the first incidence angle lays on the classical regular region, while the second angle does it on the chaotic one. In Fig. 7, a sample of quantum trajectories for the two downhill (left) and uphill (right) incidence conditions, starting at cuts of ϱ_0 close to the surface (and perpendicular to the initial propagation of Ψ_0) are displayed. In Fig. 7a, we can

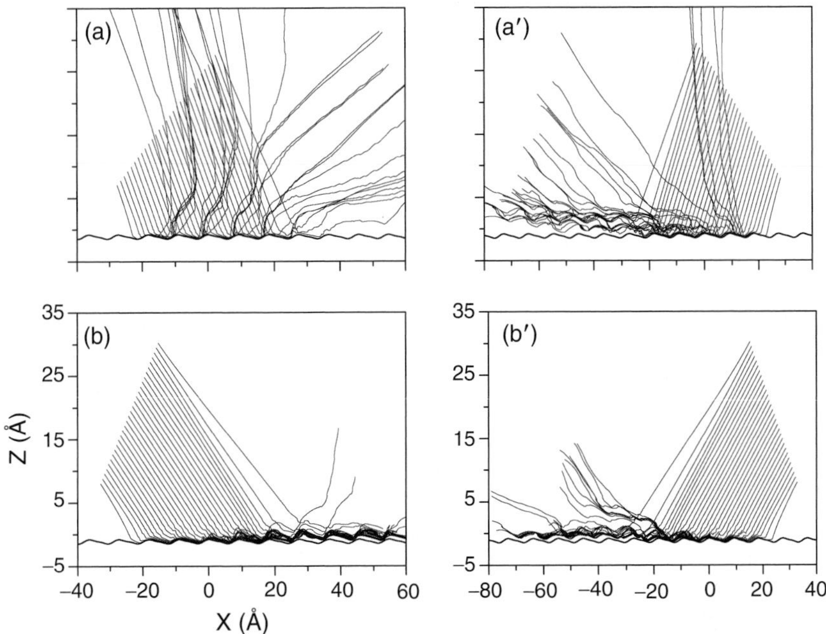

Fig. 7. *Left*: Quantum trajectories for downhill scattering at incidence conditions $E_{\rm i} = 21\,{\rm meV}$ and: (**a**) $\theta_{\rm i} = 34.4°$, and (**b**) $\theta_{\rm i} = 51.5°$. *Right*: Quantum trajectories for uphill scattering at incidence conditions $E_{\rm i} = 21\,{\rm meV}$ and: (**a′**) $\theta_{\rm i} = -34.4°$, and (**b′**) $\theta_{\rm i} = -51.5°$. The initial positions for the quantum trajectories are chosen along a cut in the front part of ϱ_0 transversal to the propagation direction

observe that quantum trajectories escape through several *exit* channels after remaining trapped along the surface a distance of one unit cell or less. On the contrary, in Fig. 7b, quantum trajectories remain trapped along the surface for a much longer time, thus covering a larger number of unit cells. These trapped quantum trajectories represent the direct analog of the (classical) *skipping orbits*, a bouncing motion with more than one turning point [21, 44]. The striking difference in the behavior of both types of trajectories (classical and quantum) comes from the presence of a kind of "sliding" motion in the region of stronger interaction for the quantum trajectories, which makes them to smoothly follow the potential contour. In this way, a SAR in the chaotic region should be interpreted as a bounded motion along the z-direction with a vibrational frequency given by the corresponding bound state of the surface average interaction potential, and a free motion parallel to the surface during a lifetime given by the inverse of the internal halfwidth covering a distance of two or more unit cells. On the contrary, a SAR in the regular or single scattering regime should be interpreted in a similar way, but in a shorter time scale and covering a length of one single unit cell or less. This is in sharp contrast to the classical idea of trapping, in which at least two consecutive unit cells are involved (one turning point at each unit cell). Moreover, notice the also apparent lack of vortical dynamics, unlike that observed in the He–CO/Pt(111) system [50, 51] (see also Sect. 3.3). Although quantum trapping is observed in the lower panel, due to the weak corrugation of the Cu surface and its periodicity, a well-defined structure of quantum vortices cannot be distinguished. Nonetheless, certain degree of vorticality should exist very close to the surface but, at the resolution level of the plots shown, vortices are not appreciated.

The nonparity and time-reversal invariance of this scattering process is manifested by the quantum trajectories for the two uphill conditions plotted in Fig. 7a′, b′. In both cases, the existence of a new type of *quantum skipping orbits* is apparent. Particles usually keep moving along the surface. However, as can be seen, some of them are now bouncing along a different axis, far from the surface. It could be said that they are feeling an effective corrugated (quantum) potential along such an axis. This effect is more clearly observed in Fig. 7a′ at final grazing angles.

As already mentioned in previous sections, an interesting and remarkable feature observed is the global (or nonlocal) character of Bohmian dynamics in contrast to the local character of classical dynamics. The quantum potential has the effect that atoms behave differently depending on their initial conditions with respect to ϱ_0 – observe that the classical dynamics does not show this dependence. The quantum dynamics is very strongly influenced by the initial distribution: particles coming from behind "know" (through the information transmitted by the quantum potential) that there are other particles reaching the surface in front of them, and therefore they cannot follow the same tracks. In this way, while the front trajectories reach the surface and move along it, those starting behind cannot approach it. Indeed, the trajectories starting in the outmost rear part of ϱ_0 can approach (on average) only at 6.4 Å. This behavior making the atoms to undergo a bounce when they are still far from the surface, arises from the different effective forces that they "feel" depending on their initial position relative to ϱ_0. The concept of quantum pressure previously introduced explains this observation; particles under a high pressure (and close to the surface) are constrained to keep moving along the surface until such a pressure decreases enough to let them escape. As inferred from Fig. 7a, the effects of quantum pressure are relatively small along the exit channels. On the other hand, quantum trapping also comes from the attractive dynamics governed by the interaction potential (which can coincide with conditions of classical trapping). In Fig. 7b it is apparent that the front trajectories follow a sliding motion along the potential surface and when the upper trajectories begin to leave, such a pressure decreases, and the quantum trajectories that remain trapped along the surface either keep moving in the same direction due to the attractive potential or simply escape.

Another feature worth mentioning is the *impulsive* character of classical trajectories in both the regular and the chaotic regime. In classical mechanics the collision can be considered, in a simplistic way, as an instantaneous kick of

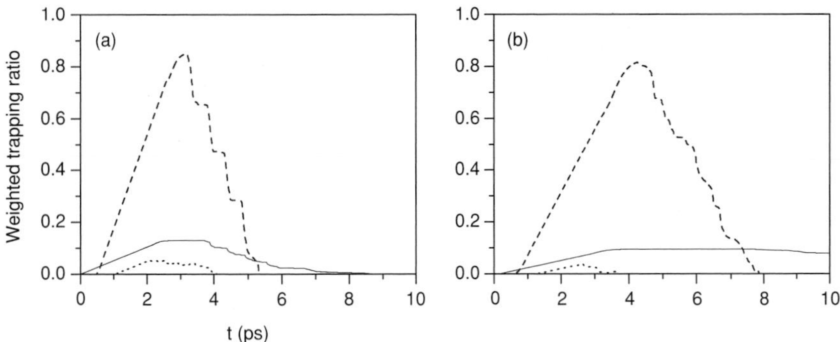

Fig. 8. Weighted fraction of trapped Bohmian trajectories inside Σ for incidence conditions $E_{\rm i} = 21\,{\rm meV}$ and: (**a**) $\theta_{\rm i} = 34.4°$, and (**b**) $\theta_{\rm i} = 51.5°$, for downhill scattering. The trajectories are started with initial positions covering the rear (*dotted line*), middle (*dashed line*), and front (*solid line*) parts of ϱ_0

the particle against the surface, thus changing its initial momentum. Depending on this initial momentum and how the kick takes place (i.e., depending on the particular orientation of the surface with respect to the direction of the incoming particle), the atom will get trapped (and move along a length of more than one unit cell) or not. This is something totally different with respect to what happens in Bohmian mechanics, where the concepts of collision and trapping acquire a more general physical meaning. Only those trajectories associated to the central and rear parts of ϱ_0 will display features typical of classical trajectories.

In the light of the previous statements, we note that for He atoms scattering off Cu(117), in particular, and any corrugated surface, in general, the quantum dynamics can be considered as an isomorphism of the classical one "plus interferences." Classical dynamics provides a kind of pattern ruling the different observable dynamics (regular or chaotic, in this case), and interferences determine the final motion displayed by the quantum trajectories, i.e., their global dynamics under the effects of the quantum pressure.

Finally, in order to complete our analysis, the fraction of trapped Bohmian trajectories (properly weighted) is given in Fig. 8 for three different sets of initial positions and downhill scattering. The (statistical) relevance of the contribution of each group of trajectories to the *restricted norm*,[8] $\mathcal{P}_t$ and the effects of the global quantum dynamics can be better understood by means of this plot. In particular, we define the weighted fraction of trapped quantum trajectories as:

$$\mathcal{W}_{\rm t} \propto \sum_i \rho_0(\boldsymbol{r}_i^0)\; \delta(z - z_i^t) \quad \text{for } z \in \Sigma := \{z \leq 8\,\text{Å}\}, \tag{19}$$

[8] The *restricted norm*, $\mathcal{P}_{\rm t}$, is the (time-dependent) probability inside Σ [56]. This definition is similar to that given in footnote 6 for the transmission function, $\mathcal{T}_{\rm t}$.

where the index i runs over the total number of trajectories chosen along a cut of ϱ_0, $\boldsymbol{r}_i^0$ is the initial position of each trajectory, and z_i^t is its z-coordinate at time t. The proportional relation comes from the fact that the $\mathcal{W}_\mathrm{t}$ function plotted is not exactly the value given by the r.h.s. of (19), but a renormalized one. This renormalization results from assuming the maximum value of $\mathcal{W}_\mathrm{t}$ equal to the maximum value of $\mathcal{P}_\mathrm{t}$. Let us remark that the restricted norm can be approximated by $\sum_n \mathcal{W}_\mathrm{t}^{(n)}$, where n runs over all cuts of ϱ_0 considered. Thus, while $\mathcal{W}_\mathrm{t}$ only accounts for the weighted number of trapped trajectories corresponding to one single cut of ϱ_0, $\mathcal{P}_\mathrm{t}$ stands for the (also weighted) total number of trapped trajectories with initial positions covering the whole spatial extension of ϱ_0. Hence, in Fig. 8a, b, it is observed that each $\mathcal{W}_\mathrm{t}$ function contributes differently to $\mathcal{P}_\mathrm{t}$. As is clearly seen, the main contribution to $\mathcal{P}_\mathrm{t}$ is given by the trajectories starting in the central and front cuts of the region covered by ϱ_0, while the rest contribute marginally. This is expected since the profile of the incoming plane wave along the direction parallel to the initial wave vector is a Gaussian, and then the number of trajectories in the central part of ϱ_0 will be proportionally larger than in any other. Moreover, the position of the maximum (in t) of $\mathcal{W}_\mathrm{t}$ agrees fairly well with that of the restricted norm. Other ensembles of trajectories initially located at the borders of the Gaussian profile will contribute only to small deviations of the maximum position of $\mathcal{W}_\mathrm{t}$, thus resulting a total agreement with the restricted norm. Notice, however, that the trajectories starting at the front part of ρ_0 determine the *long-range* behavior of $\mathcal{W}_\mathrm{t}$, and therefore are responsible for the tail displayed by the decay of the restricted norm, leading to higher residence times or SAR lifetimes.

4 Future Work

A natural extension of this type of studies is toward inelastic atom–surface scattering. In particular, the role played by the surface temperature as well as the surface phonons is very well established in the standard quantum mechanics, and we are convinced that new physical insight will be provided by Bohmian mechanics. Furthermore, if the surface is seen as a heat bath, this type of scattering can be seen as a dissipative, stochastic quantum process [22]. Work in both directions is now in progress.

References

1. G. Greenstein, A.G. Zajonc: *The Quantum Challenge* (Jones and Bartlett Publishers, Sudbury MA 1997)
2. L.E. Ballentine: *Quantum Mechanics: A Modern Development* (World Scientific, Singapore 1998)
3. C.J. Davisson, L.H. Germer: Phys. Rev. **30**, 705 (1927)

4. M. Arndt, O. Nairz, J. Vos-Andreae, C. Keller, G. van der Zouw, A. Zeilinger: Nature **401**, 680 (1999); O. Nairz, B. Brezger, M. Arndt, A. Zeilinger: Phys. Rev. Lett. **87**, 160401 (2001); B. Brezger, L. Hackermüller, S. Uttenthaler, J. Petschinka, M. Arndt, A. Zeilinger: Phys. Rev. Lett. **88**, 100404 (2002)
5. L. Hackermüller, S. Uttenthaler, K. Hornberger, E. Reiger, B. Brezger, A. Zeilinger, M. Arndt: Phys. Rev. Lett. **91**, 090408 (2003)
6. F. Knauer, O. Stern: Z. Physik **53**, 766 (1929); I. Estermann, O. Stern: Z. Physik **61**, 95 (1930); I. Estermann, R. Frisch, O. Stern: Z. Physik **73**, 348 (1931)
7. Helium Atom Scattering from Surfaces. In: *Springer Series in Surface Sciences*, vol 27, ed. by E. Hulpke (Springer, Berlin Heidelberg New York 1991)
8. D. Farías, K.H. Rieder: Rep. Prog. Phys. **61**, 1575 (1998)
9. D. Farías, E.G. Michel, S. Miret-Artés (Eds.): Special issue on Surface Dynamics, Phonons, Adsorbate Vibrations and Diffusion, J. Phys.: Condens Matter **14**, 5865 (2002)
10. N. Cabrera, V. Celli, J.R. Manson: Phys. Rev. Lett. **22**, 346 (1969); N. Cabrera, V. Celli, F.O. Goodman, J.R. Manson: Surf. Sci. **19**, 67 (1970)
11. A. Tsuchida: Surf. Sci. **14**, 375 (1969)
12. G. Wolken: J. Chem. Phys. **58**, 3047 (1973)
13. D. Kosloff, R. Kosllof: J. Comput. Phys. **52**, 35 (1983); A.T. Yinnon, R. Kosloff: Chem. Phys. Lett. **102**, 216 (1983)
14. S. Miret-Artés: Surf. Sci. **339**, 205 (1995)
15. M. Born: Z. Phys. **37**, 863 (1926); **38**, 803 (1926)
16. L.E. Ballentine: Rev. Mod. Phys. **42**, 358 (1970)
17. R. Gelabert, X. Giménez, M. Thoss, H. Wang, W.H. Miller: J. Chem. Phys. **114**, 2572 (2001)
18. S. Zhang, E. Pollak: Phys. Rev. Lett. **91**, 190201 (2003); J. Chem. Phys. **121**, 3384 (2004)
19. A.S. Sanz, F. Borondo, S. Miret-Artés: J. Phys.: Condens. Matter **14**, 6109 (2002)
20. E. Gindensperger, C. Meier, J.A. Beswick, and M.-C. Heitz: J. Chem. Phys. **116**, 10051 (2002)
21. R. Guantes, A.S. Sanz, J. Margalef-Roig, S. Miret-Artés: Surf. Sci. Rep. **53**, 199 (2004)
22. A.S. Sanz, S. Miret-Artés: Phys. Rep. (accepted, 2007)
23. D. Bohm: Phys. Rev. **85**, 166, 180 (1952)
24. D. Dürr, S. Goldstein, N. Zanghì: J. Stat. Phys. **67**, 843 (1992)
25. P.R. Holland: *The Quantum Theory of Motion* (Cambridge University Press, Cambridge 1993)
26. R.E. Wyatt: *Quantum Dynamics with Trajectories: Introduction to Quantum Hydrodynamics* (Springer, Berling Heidelberg New York 2005)
27. E. Madelung: Z. Phys. **40**, 332 (1926)
28. L. de Broglie: Compt. Rend. **184**, 273 (1927)
29. D. Bohm: Phys. Rev. **89**, 458 (1953)
30. D. Bohm, B.J. Hiley: *The Undivided Universe* (Routledge, London 1993)
31. O. Maroney, B.J. Hiley: Found. Phys. **29**, 1403 (1999)
32. A.S. Sanz, F. Borondo, S. Miret-Artés: Phys. Rev. B **61**, 7743 (2000)
33. J.O. Hirschfelder, A.C. Christoph, W.E. Palke: J. Chem. Phys. **61**, 5435 (1974); J.O. Hirschfelder, C.J. Goebel, L.W. Bruch: J. Chem. Phys. **61**, 5456 (1974); J.O. Hirschfelder, K.T. Tang: J. Chem. Phys. **64**, 760 (1976)

34. B.K. Dey, A. Askar, H. Rabitz: J. Chem. Phys. **109**, 8770 (1998); F. Sales-Mayor, A. Askar, H. Rabitz: J. Chem. Phys. **111**, 2423 (1999)
35. R.E. Wyatt: J. Chem. Phys. **111**, 4406 (1999); R.E. Wyatt: Chem. Phys. Lett. **313**, 189 (1999)
36. C.L. Lopreore, R.E. Wyatt: Phys. Rev. Lett. **82**, 5190 (1999); C.L. Lopreore, R.E. Wyatt: Chem. Phys. Lett. **325**, 73 (2000)
37. E.R. Bittner: J. Chem. Phys. **112**, 9703 (2000); R.E. Wyatt, E.R. Bittner: J. Chem. Phys. **113**, 8898 (2000)
38. C.J. Trahan, R.E. Wyatt: J. Chem. Phys. **119**, 7017 (2003)
39. I. Burghardt, L.S. Cederbaum: J. Chem. Phys. **115**, 10303, 10312 (2001); I. Burghardt, K.B. Møller: J. Chem. Phys. **117**, 7409 (2002)
40. M.S. Child: *Molecular Collision Theory* (Academic, London 1974)
41. J.D. McClure: J. Chem. Phys. **51**, 1687 (1969); **52**, 2712 (1970); **57**, 2810, 2823 (1972)
42. A.T. Yinnon, R. Kosloff, R.B. Gerber: J. Chem. Phys. **88** 7209 (1988)
43. B.H. Choi, K.T. Tang, J.P. Toennies: J. Chem. Phys. **107**, 9437 (1997)
44. R. Guantes, F. Borondo, C. Jaffé, S. Miret-Artés: Phys. Rev. B **53**, 14117 (1996) S. Miret-Artés, J. Margalef-Roig, R. Guantes, F. Borondo, C. Jaffé: Phys. Rev. B **54**, 10397 (1996); R. Guantes, F. Borondo, J. Margalef-Roig, S. Miret-Artés, J.R. Manson: Surf. Sci. **375**, L379 (1997)
45. A.S. Sanz, F. Borondo, S. Miret-Artés: Europhys. Lett. **55**, 303 (2001)
46. A.M. Lahee, J.R. Manson, J.P. Toennies, Ch. Wöll: Phys. Rev. Lett. **57**, 471 (1986); A.M. Lahee, J.R. Manson, J.P. Toennies, Ch. Wöll: J. Chem. Phys. **86**, 7194 (1987)
47. B. Poelsema, G. Comsa: Scattering of Thermal Energy Atoms from Disordered Surfaces. In: *Springer Tracts in Modern Physics* vol 115 (Springer, Berlin Heidelberg New York 1989)
48. M.N. Carré, D. Lemoine: J. Chem. Phys. **101**, 5305 (1994); D. Lemoine, Phys. Rev. Lett. **81**, 461 (1998)
49. G. Drolshagen, R. Vollmer: J. Chem. Phys. **87**, 4948 (1987)
50. A.S. Sanz, F. Borondo, S. Miret-Artés: Phys. Rev. B **69**, 115413 (2004)
51. A.S. Sanz, F. Borondo, S. Miret-Artés: J. Chem. Phys. **120**, 8794 (2004)
52. R.B. Gerber: Chem. Rev. **87**, 29 (1987)
53. D. Gorse, B. Salanon, F. Fabre, A. Kara, J. Perreau, G. Armand, J. Lapujoulade: Surf. Sci. **147**, 611 (1984); S. Miret-Artés, J.P. Toennies, G. Witte: Phys. Rev. B **54**, 5881 (1996)
54. M.I. Hernández, J. Campos-Martínez, S. Miret-Artés, R.D. Coalson: Phys. Rev. B **49**, 8300 (1994)
55. R. Guantes, S. Miret-Artés, F. Borondo: Phys. Rev. B **63**, 235401 (2001)
56. A.S. Sanz, S. Miret-Artés: J. Chem. Phys. **122**, 14702 (2005)

Hybrid Quantum/Classical Dynamics Using Bohmian Trajectories

C. Meier and J.A. Beswick

Summary. The mixed quantum–classical method based on Bohmian trajectories as introduced by Gindensperger, Meier and Beswick [J. Chem. Phys. **113**, 9369 (2000)] is reviewed, together with its basic properties. It is shown that this approximative method combining quantum and classical dynamics can be derived in a rigorous way from the hydrodynamic formulation of quantum mechanics. The quantum subsystem is described by a wave packet depending on the quantum variables and, via the total potential energy of the system, parametrically on the classical trajectories. The wave packet provides de Broglie–Bohm *quantum trajectories* which are used to calculate the force acting on the classical variables. Two examples are presented; the first one concerns molecule surface-scattering and the second one pump–probe spectroscopy of a molecule in a high pressure rare gas environment.

1 Introduction

The dynamics of systems containing a large number of degrees of freedom is one of the challenges in contemporary theoretical chemistry. Full quantum mechanical wave packet propagations in several degrees of freedom is a numerically demanding task, for which specialized methods like the multi-configuration time dependent Hartree (MCTDH) method [1, 2] has proven to be a unique and particularly efficient tool. However, for processes like proton and electron transfer in isolated polyatomic molecules, liquids, interfaces and biological systems, intramolecular energy redistribution and unimolecular fragmentation, as well as interactions of atoms and molecules with surfaces [3], a full quantum treatment of all degrees of freedom is still out of reach.

In many systems comprising a large number of particles, even though a detailed quantum treatment of all degrees of freedom is not necessary, there may exist subsets that have to be treated quantum mechanically under the influence of the rest of the system. If the typical timescales between system and bath dynamics are very different, Markovian models of quantum dissipation can succesfully mimic the influence of the bath onto the system dynamics [4]. However, in the femtosecond regime studied with ultrashort laser pulses, the so-called Markov approximation is not generally valid [5]. Furthermore, very often the bath operators are assumed to be of a special form (harmonic for instance) which are sometimes not realistic enough.

Another class of approximate methods are hybrid quantum/classical schemes in which only the essential degrees of freedom are treated quantum mechanically while all others are described classically. The most popular of these mixed quantum/classical methods are the mean-field approximation [6], the surface hopping trajectories [7] or methods based on quantum/classical Liouville space representations [8–14]. In the mean-field treatment the force for the classical motion is calculated by averaging over the quantum wavefunction. In the surface hopping scheme the classical trajectories move according to a force derived from a *single quantum state* with the possibility of transitions to other states.

An alternative treatment to mix quantum mechanics with classical mechanics, proposed in [15–21], is based on Bohmian quantum trajectories for the quantum/classical connection. Briefly, the quantum subsystem is described by a time-dependent Schrödinger equation that depends parametrically on classical variables. This is similar to the other approaches discussed earlier. The difference comes from the way the classical trajectories are calculated. In our approach, which was called (mixed quantum/classical bohmian MQCB) trajectories, the wave packet is used to define de Broglie–Bohm quantum *trajectories* [22–24] which in turn are used to calculate the force acting on the classical variables.

Recently, there has been a renewed interest in the de Broglie–Bohm formulation of quantum mechanics, both from a conceptual and numerical point of view [25]. As a numerical tool, it has been used to perform multidimensional wave packet calculations [26–34] In this context, one of the central problems is the accurate calculation of the quantum potential, especially in regions of small probability density ('node problem') [35–38].

It has also been used to visualize the motion of quantum mechanical wave packets by trajectories and to study the transition from quantum mechanics to classical mechanics [39–44]. Carlsen and Goscinsky [39] for instance, have studied fractional and full revivals of circular Rydberg wave packets in the hydrogen atoms using this formulation.

In what follows we shall show that the de Broglie–Bohm formulation can also be used to establish a hybrid quantum/classical scheme to treat the dynamics of systems with a large number of degrees of freedom in which a few need to be described quantum mechanically.

2 De Broglie–Bohm Formulation of Quantum Mechanics

Since the method to mix quantum and classical mechanics to be presented can be considered as an approximate method derived from the de Broglie–Bohm formulation of quantum mechanics, this completely equivalent perspective of quantum mechanics will be briefly reviewed. To this end, we consider a two dimensional Hilbert space.

Note that considering two dimensions is no restriction to what will be shown later, actually, x, X can be viewed as *collective variables* one of which will comprise all quantum degrees of freedom while the other all classical ones. Writing the wavefunction as $\psi(x, X, t) = R(x, X, t) \exp(\mathrm{i}S(x, X, t)/\hbar)$, with R, S being real, the Schrödinger equation can be recast in terms of a continuity equation,

$$\frac{\partial R^2}{\partial t} + \frac{1}{m}\frac{\partial S}{\partial x}\frac{\partial R^2}{\partial x} + \frac{1}{M}\frac{\partial S}{\partial X}\frac{\partial R^2}{\partial X} = -R^2\left(\frac{1}{m}\frac{\partial^2 S}{\partial x^2} + \frac{1}{M}\frac{\partial^2 S}{\partial X^2}\right) \tag{1}$$

and a quantum Hamilton–Jacobi equation

$$\frac{\partial S}{\partial t} + \frac{1}{2m}\left(\frac{\partial S}{\partial x}\right)^2 + \frac{1}{2M}\left(\frac{\partial S}{\partial X}\right)^2 + V(x, X) + Q(x, X, t) = 0, \tag{2}$$

where $Q(x, X, t)$ is the so-called *quantum potential* [22]

$$Q(x, X, t) = -\frac{\hbar^2}{2m}\frac{1}{R}\frac{\partial^2 R}{\partial x^2} - \frac{\hbar^2}{2M}\frac{1}{R}\frac{\partial^2 R}{\partial X^2}. \tag{3}$$

Hence one sees that the phase of $\psi(x, X, t)$ can be viewed as an action function, a solution to the Hamilton–Jacobi equation with an additional potential term, the quantum potential. This observation led to the definition of *trajectories* $(\mathbf{x}(t), \mathbf{X}(t))$ [22], the conjugate momenta of which are given by the derivative of $S(x, X, t)$:

$$\mathbf{p} = m\dot{\mathbf{x}} = \left.\frac{\partial S}{\partial x}\right|_{x=\mathbf{x}(t), X=\mathbf{X}(t)} \quad ; \quad \dot{\mathbf{p}} = -\frac{\partial}{\partial x}(V + Q), \tag{4}$$

$$\mathbf{P} = M\dot{\mathbf{X}} = \left.\frac{\partial S}{\partial X}\right|_{x=\mathbf{x}(t), X=\mathbf{X}(t)} \quad ; \quad \dot{\mathbf{P}} = -\frac{\partial}{\partial X}(V + Q). \tag{5}$$

Thus, within the Bohmian formulation of quantum mechanics, *quantum trajectories* move according to the usual Hamilton's equations, subject to the additional quantum potential defined in (3). An ensemble of *quantum particles* at positions $(\mathbf{x}(t), \mathbf{X}(t))$ distributed initially according to

$$P([x, x + \mathrm{d}x]; [X, X + \mathrm{d}X]) = |\psi_0(x, X)|^2 \,\mathrm{d}x\, \mathrm{d}X \tag{6}$$

and propagated alongside using (4,5), will represent the probability distribution of the quantum mechanical wavefunction at any time [22].

From (4) and (5) one sees that whenever the additional force due to the quantum potential is negligible, one has a purely classical motion. Thus, the limit from quantum theory to classical mechanics appears naturally within this theory.

3 From the de Broglie–Bohm Formulation of Quantum Mechanics to the MQCB Method

In order to establish the mixed quantum–classical method based on Bohmian trajectories (MQCB) [15], we take the same approach as in the de Broglie–Bohm formulation of quantum mechanics, as detailed earlier. Hence we start from the same, full dimensional initial wavefunction $\psi_0(x, X)$ alongside with an ensemble of trajectories at initial positions $(\mathbf{x}(t=0), \mathbf{X}(t=0))$ distributed acording to $R^2(x, X, t=0) = |\psi_0(x, X)|^2$. After taking derivatives with respect to x and X of (2) we neglect the term involving the second derivative of S with respect to X. In addition we neglect the second derivative of S with respect to X in (1) and the second derivative of R with respect to X in (3). Considering the simplest case of a free two-dimensional Gaussian wave packet, one sees that these terms describe the dispersion in X-direction. In the limit of large M the wave packet behaves classically and does *not* show much dispersion in the X-direction. Hence neglecting these terms should be a good approximation to the real quantum dynamics. In this sense, X will from now on be called the *classical* degree of freedom. Note that this approximation cannot be made in the original Schrödinger equation (1) but *only* in the equations for the amplitude and phase! We then have from (1):

$$\frac{\partial \widetilde{R}^2}{\partial t} + \frac{\partial}{\partial x}\left(\widetilde{R}^2 \frac{1}{m}\frac{\partial \widetilde{S}}{\partial x}\right) + \frac{1}{M}\frac{\partial \widetilde{S}}{\partial X}\frac{\partial \widetilde{R}^2}{\partial X} = 0 \tag{7}$$

and from (2)

$$\frac{\partial}{\partial t}\left(\frac{\partial \widetilde{S}}{\partial x}\right) + \left(\frac{1}{m}\frac{\partial \widetilde{S}}{\partial x}\right)\left(\frac{\partial^2 \widetilde{S}}{\partial x^2}\right) + \left(\frac{1}{M}\frac{\partial \widetilde{S}}{\partial X}\right)\left(\frac{\partial^2 \widetilde{S}}{\partial x \partial X}\right) = -\frac{\partial}{\partial x}(V + \widetilde{Q}), \tag{8}$$

$$\frac{\partial}{\partial t}\left(\frac{\partial \widetilde{S}}{\partial X}\right) + \left(\frac{1}{m}\frac{\partial \widetilde{S}}{\partial x}\right)\left(\frac{\partial^2 \widetilde{S}}{\partial X \partial x}\right) = -\frac{\partial}{\partial X}(V + \widetilde{Q}), \tag{9}$$

where $\widetilde{Q} = -(\hbar^2/2m\widetilde{R})\,\partial^2\widetilde{R}/\partial x^2$ and tilde quantities stand for the approximate solutions. As in the usual hydrodynamic formulation detailed earlier, the Bohmian trajectories associated with these approximate equations are

$$\mathbf{p} = m\dot{\mathbf{x}} = \left.\frac{\partial \widetilde{S}}{\partial x}\right|_{x=\mathbf{x}(t), X=\mathbf{X}(t)}, \tag{10}$$

$$\mathbf{P} = M\dot{\mathbf{X}} = \left.\frac{\partial \widetilde{S}}{\partial X}\right|_{x=\mathbf{x}(t), X=\mathbf{X}(t)}, \tag{11}$$

together with the *same initial conditions* as in the hydrodynamic formulation of quantum mechanics. This does not pose any problem, since within the MQCB method, the full-dimensional initial wavefunction is supposed to be known. Hence the initial values $(\mathbf{x}(t=0), \mathbf{X}(t=0))$ are chosen according to the distribution as earlier:

$$P([x, x+\mathrm{d}x]; [X, X+\mathrm{d}X]) = |\psi_0(x, X)|^2 \, \mathrm{d}x \, \mathrm{d}X. \tag{12}$$

The next step consists in evaluating (7) and (8) at $X = \mathbf{X}(t)$. Using (11) one gets

$$\frac{\mathrm{d}\widetilde{R}^2}{\mathrm{d}t} + \frac{\partial}{\partial x}\left(\widetilde{R}^2 \frac{1}{m}\frac{\partial \widetilde{S}}{\partial x}\right) = 0, \tag{13}$$

$$\frac{\mathrm{d}}{\mathrm{d}t}\left(\frac{\partial \widetilde{S}}{\partial x}\right) + \left(\frac{1}{m}\frac{\partial \widetilde{S}}{\partial x}\right)\left(\frac{\partial^2 \widetilde{S}}{\partial x^2}\right) = -\frac{\partial}{\partial x}(V + \widetilde{Q}), \tag{14}$$

where $\mathrm{d}/\mathrm{d}t$ stands for $\mathrm{d}\widetilde{f}/\mathrm{d}t = \partial \widetilde{f}/\partial t + \dot{\mathbf{X}} \cdot \left(\partial \widetilde{f}/\partial X\right)_{X=\mathbf{X}(t)}$.

The important observation at this point is that (13) and (14) are rigorously equivalent to a quantum problem in the x subspace with $\mathbf{X}$ being a time-dependent parameter. Thus the approximate wavefunction $\widetilde{\psi}(x, \mathbf{X}(t), t) = \widetilde{R}(x, \mathbf{X}(t), t) \exp(\mathrm{i}\widetilde{S}(x, \mathbf{X}(t), t)/\hbar)$ obeys the Schrödinger equation

$$\mathrm{i}\hbar \frac{\mathrm{d}\widetilde{\psi}(x, \mathbf{X}(t), t)}{\mathrm{d}t} = \left(-\frac{\hbar^2}{2m}\frac{\partial^2}{\partial x^2} + V(x, \mathbf{X}(t))\right)\widetilde{\psi}(x, \mathbf{X}(t), t), \tag{15}$$

Note the appearance of the total derivative in the left-hand side of this equation. As it will be shown later, this is important when an adiabatic basis set is used for solving the quantum problem. Since we have only approximated the *equations of motion* and supposed that the initial wavefunction $\psi_0(x, X)$ is known, we have

$$\widetilde{\psi}(x, \mathbf{X}(t=0), t=0) = \psi_0(x, X)|_{X=\mathbf{X}(t=0)} \tag{16}$$

as the initial wavefunction in the quantum subspace.

A consistent equation of motion for the classical degrees of freedom is obtained by taking the total derivative with respect to time of (11). Noting that this leads to a term $\frac{1}{M}\frac{\partial^2 S}{\partial X^2}$ which we assumed to be small we can use (9) to give:

$$\dot{\mathbf{P}} = -\frac{1}{M}\left.\frac{\partial\left(V(x, X) + \widetilde{Q}(x, X)\right)}{\partial X}\right|_{x=\mathbf{x}(t), X=\mathbf{X}(t)}. \tag{17}$$

The fact that at this level of approximation, the quantum potential corresponding to the quantum subsystem remains in the classical equation of motion, is somewhat reminescent of the Pechukas' force in the surface hopping method [7]. In both cases, the classical degrees of freedom are directly affected by changes in the quantum subspace. In practice, however, since we do not solve (8) and (9) directly but only following a specific trajectory $\mathbf{X}(t)$, we additionally neglect the quantum potential in the classical degree of freedom. At this level of approximation, the MQCB method is identical to the one proposed independently by Prezhdo et al. [19–21]. In [15] we have studied this approximation. In the example considered, its influence was negligible.

Equation (17), together with (15) and (10) provide the working equations of the MQCB method. According to their structure, the position of the classical degree of freedom, its momentum, the position of the Bohmian particle and the wavefunction in the quantum subspace need to be propagated simultaneously. For clarity, these quantities shall be combined as

$$\Gamma(t) = \left(\widetilde{\psi}(x, \mathbf{X}(t), t), \mathbf{x}(t), \mathbf{X}(t), \mathbf{P}(t)\right) . \tag{18}$$

Note that if one expands $\widetilde{\psi}(x, \mathbf{X}(t), t)$ in a basis set, the MQCB equations simply form a set of coupled, first-order ordinary differential equations and $\Gamma(t)$ is a vector of complex numbers, containing the classical position and momentum, the quantum trajectory and the expansion coefficients of the subspace wavefunction.

4 Structure of the MQCB Equations: Initial Conditions, Reversibility and Observables

The mathematical structure of the four MQCB equations (10), (11), (15), and (17) is such that the time evolution of the combined quantities $\Gamma(t)$ is uniquely determined by the MQCB equations, and reversibility is given in a strict mathematical sense. Upon propagation $\Gamma(0) \longrightarrow \Gamma(t)$, changing $\partial/\partial t \longrightarrow -\partial/\partial t$ and $\dot{\mathbf{X}}(t) \longrightarrow -\dot{\mathbf{X}}(t)$ propagates the state $\Gamma(t)$ backwards to yield the same initial state: $\Gamma(t) \longrightarrow \Gamma(0)$.

This simple *mathematical* structure of the quantities $\Gamma(t)$ together with the MQCB equations need also to be connected to *physical* quantities. For the initial conditions, we have shown above that $\Gamma(0)$ can be chosen in a consistent and physically sound way if the full-dimensional initial wavefunction is known.

However, due to the approximate nature of the MQCB equations, the full-dimensional wave packet is in general *not* known at later times. In this sense, the time zero is a special time at which, due to the physical situation considered, the full-dimensional wave packet must be known. However, this is not a severe restriction, since in many physically relevant situations the initial state is either a known asymtotic state like in all collisional processes or an eigenfunction (often the ground state).

The second point is to use the mathematical objects $\Gamma(t)$ to calculate physically measurable observables. Even though the definiton of observables is not unique within the MQCB scheme, in all cases considered so far, we use

$$\langle A \rangle = \sum_{\mathrm{i}} \left\langle \widetilde{\psi}_{\mathrm{i}} \left| A\left(x, \frac{\partial}{\partial x}, \mathbf{X}_{\mathrm{i}}, \mathbf{P}_{\mathrm{i}}\right) \right| \widetilde{\psi}_{\mathrm{i}} \right\rangle, \tag{19}$$

where the sum runs over all initially sampled trajectoires. Due to the approximate nature of the propagations, the observables obtained by the MQCB method are approximate as well. Especially, as compared to a full quantum wave packet calculation, the total energy is not a rigorously conserved quantity.

5 Applications

5.1 Molecule-Surface Scattering

The five-dimensional model chosen corresponds to molecular diffractive rotational scattering of N_2 from an LiF(001) surface. This problem still allows for a full quantum treatment and thus the MQCB results can be compared to reference calculations [18]. The interaction potential was chosen to be the dumbell model initially proposed by Gerber et al. [45]. Although this is a model surface it nevertheless has the same features of a realistic surface, in particular a two-dimensional corrugation with periodicity a. We considered energies up to 300 meV. Since at this energies, the first vibrational channel is closed, we approximate the diatomic as a rigid rotator, i.e. keep r fixed. Hence the Hamiltonian reads as

$$H(\mathbf{R}, \theta, \varphi) = \frac{\mathbf{P}_R^2}{2M} + \frac{\hbar^2 \mathbf{J}^2}{2\mu r^2} + V(\mathbf{R}, \theta, \varphi), \tag{20}$$

where M and μ are the total and reduced masses of N_2, respectively. In this expression, P_R is the total momentum of the center of mass of N_2 and $\mathbf{J}$ its angular momentum.

The process we are interested in is the rotational energy transfer during the collision as well as the diffraction of the diatomic from the surface that exhibits a two-dimensional corrugation. Since the diffraction is a quantum effect, we treat the whole system classically except for the directions X and Y parallel to the surface. The separation of the total Hamiltonian into a classical and quantum part reads as follows:

$$H(X, Y, Z, \theta, \varphi) = T_{\mathrm{q}}(X, Y) + T_{\mathrm{cl}}(Z, \theta, \varphi) + V(X, Y, Z, r, \theta, \varphi), \tag{21}$$

$$T_{\mathrm{q}}(X, Y) = -\frac{\hbar^2}{2M} \frac{\partial^2}{\partial X^2} - \frac{\hbar^2}{2M} \frac{\partial^2}{\partial Y^2}; \quad T_{\mathrm{cl}}(Z, \theta, \varphi) = \frac{P_Z^2}{2M} + \frac{\hbar^2 \mathbf{J}^2}{2\mu r^2}. \tag{22}$$

The MQCB equations for this problem are thus given by the Schrödinger equation in the (X, Y) subspace for $\widetilde{\psi}(X, Y, \boldsymbol{q}_\alpha(t), t)$ that depends parametrically on the classical variables $\boldsymbol{q}_\alpha(t) = \boldsymbol{Z}(t), \boldsymbol{\theta}(t), \boldsymbol{\varphi}(t)$, the corresponding equations for the quantum trajectories, as well as Hamilton's equations for the classical degrees of freedom and their conjugate momenta

$$\mathrm{i}\hbar\frac{\mathrm{d}\tilde{\psi}}{\mathrm{d}t} = \left(-\frac{\hbar^2}{2M}\frac{\partial^2}{\partial X^2} - \frac{\hbar^2}{2M}\frac{\partial^2}{\partial Y^2} + V(X,Y,\boldsymbol{q}_\alpha(t))\right)\tilde{\psi}, \tag{23}$$

$$\dot{\boldsymbol{X}}(t) = \frac{\hbar}{M}\,\mathrm{Im}\left(\frac{1}{\tilde{\psi}}\frac{\partial\tilde{\psi}}{\partial X}\right); \quad \dot{\boldsymbol{Y}}(t) = \frac{\hbar}{M}\,Im\left(\frac{1}{\tilde{\psi}}\frac{\partial\tilde{\psi}}{\partial Y}\right), \tag{24}$$

$$\dot{\boldsymbol{Z}}(t) = \frac{\boldsymbol{P}_Z(t)}{M}; \quad \dot{\boldsymbol{\theta}}(t) = \frac{\boldsymbol{p}_\theta(t)}{\mu r^2}; \quad \dot{\boldsymbol{\varphi}}(t) = \frac{\boldsymbol{p}_\varphi(t)}{\mu\sin^2\boldsymbol{\theta}(t)r^2}, \tag{25}$$

$$\dot{\boldsymbol{P}}_Z(t) = -\frac{\partial V}{\partial Z}; \quad \dot{\boldsymbol{p}}_\theta(t) = \frac{\boldsymbol{p}_\varphi^2(t)\cos\boldsymbol{\theta}(t)}{\mu r^2\sin^3\boldsymbol{\theta}(t)} - \frac{\partial V}{\partial\theta}; \quad \dot{\boldsymbol{p}}_\varphi(t) = -\frac{\partial V}{\partial\varphi} \tag{26}$$

evaluated at $X = \boldsymbol{X}(t)$, $Y = \boldsymbol{Y}(t)$, $q_\alpha = \boldsymbol{q}_\alpha(t)$. Starting with initial conditions to be specified later, these equations have to be solved simultaneously. In practice, the Schrödinger equation was solved in a basis of plane waves bearing the periodicity of the surface.

We are interested in the diffraction probability of the scattered molecule. To this end, we project the final asymptotic wave packet onto scattering states. When this projection is performed with the wave packet being in the asymptotic region, this is equivalent to projecting onto free waves, i.e. the scattering amplitudes are then simply given by the Fourier transform of the asymptotic wave packet [18]. Since in our case, the wavefunction depends parametrically on the classical variables, each trajectory yields a diffraction probability

$$P_\mathrm{i}(n,m) = \lim_{t\to\infty}\left|\int_0^a\int_0^a \mathrm{d}X\,\mathrm{d}Y\frac{1}{\sqrt{ab}}\,\mathrm{e}^{-\mathrm{i}(K_nX+K_mY)}\,\tilde{\psi}(X,Y,\boldsymbol{Z}_\mathrm{i},\boldsymbol{\theta}_\mathrm{i},\boldsymbol{\varphi}_\mathrm{i},t)\right|^2, \tag{27}$$

where $\mathrm{i} = 1, N$, $K_n = 2\pi n/a$, and $K_m = 2\pi m/a$. The energy dependence stems from the initial velocites of the classical trajectories. To obtain the total diffraction probability, we average over the whole ensemble of N trajectories:

$$P(n,m) = \frac{1}{N}\sum_{\mathrm{i}=1}^{N} P_\mathrm{i}(n,m). \tag{28}$$

The second quantity we are interested in is the rotational energy that is being transferred during the collision. If the initial state is taken to be a **J**=0 state, the energy transfer from the translational to the rotational degrees freedom is simply the rotational energy after the collision

$$E_{\mathrm{rot,i}} = \frac{1}{2\mu r^2}\left(\boldsymbol{p}_{\theta,\mathrm{i}}^2 + \frac{\boldsymbol{p}_{\varphi,\mathrm{i}}^2}{\sin^2\boldsymbol{\theta}_\mathrm{i}}\right) \qquad \mathrm{i} = 1, N \tag{29}$$

The *average rotational energy transfer* (ARET) is then given by

$$E_{\mathrm{rot}} = \frac{1}{N}\sum_{\mathrm{i}=1}^{N} E_{\mathrm{rot,i}} \tag{30}$$

With these definitions, $P(n,m)$ and E_{rot} can directly be compared to full-dimensional quantum wave packet results using the MCTDH method [1,2]. These quantities, which give a detailed information about the rotational diffractive scattering of a diatomic molecule, were calculated with the MQCB method and compared to full-dimensional quantum wave packet results based on the MCTDH method (for details of the calculations see [46]).

Results on the N_2/LiF(001) Molecule-Surface Scattering

As detailed earlier, we have treated the case of normal incidence with the molecule in an initial rotational state **J**=0.

In Fig. 1 we compare the results for the final diffraction probabilities of the orders $(0,0)$, $(1,1)$, $(2,2)$ as well as $(0,1)$, $(0,2)$ and $(2,1)$ for different collision energies between 0.1 and 0.3 eV. The solid lines correspond to the quantum result, and the dashed lines represent the results obtained by the approximate MQCB calculations. One clearly sees the very good agreement of our approximate method with the exact results for all orders considered.

At energies higher than about 0.17 eV, the diffraction into the order $(0,0)$, has decreased to only about 1%, while the other diffraction channels are being more and more populated. The diffraction probability $P(1,1)$ shows a maximum as about 0.15 eV. Taking the symmetry $n \longrightarrow -n, m \longrightarrow -m$ into account, we see that diffraction into the order $(\pm 1, \pm 1)$ accounts for about 30% of the whole diffraction probability.

Qualitatively and even quantitatively, the exact results are well reproduced. Note that these diffraction probabilities are a pure quantum effect and cannot be obtained directly (i.e. without boxing) by a classical calculation, as it is the case for the average rotational rotational energy transfer (ARET) to be shown later.

We now discuss the results for the average rotational energy transfer (ARET). Since this quantity can be obtained from a purely classical calculation, we can compare the ARET as a function of collision energy obtained by three methods: a purely classical treatment, the mixed MQCB method and the five-dimensional quantum treatment. To be able to rigorously test the MQCB method, in all three cases the same five-dimensional Hamiltonian (20) was used.

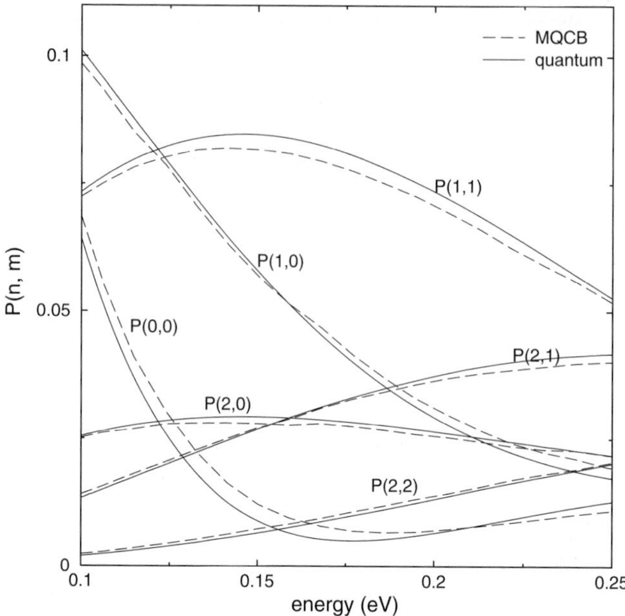

Fig. 1. Diffraction probability as a function of collision energy. *Full line*: exact quantum mechanical result obtained by the MCTDH method (from: [46]), *dashed line*: MQCB results (28)

The pure classical results are readily obtained by integrating the usual Hamilton's equations of classical mechanics, and a solution of the Schrödinger equation in (23) is not necessary any more. This again reflects that the MQCB method can be viewed as an extension of classical mechanics by adding quantum effects to certain degrees of freedom. The ARET is then calculated in exactly the same way with (30).

Figure 2 shows the ARET as a function of collision energy between 0.1 and 0.3 eV. The full line is the quantum result, and the dotted line correspond to the ARET obtained by a purely classical calculation. The MQCB result is shown as dashed lines. For this quantity, we find almost quantitative agreement between all three calculations, showing a monotonically increasing behaviour of the ARET as a function of the collision energy. The classical result is in very good agreement with the correct quantum one, and adding the quantum effects in the X- and Y-direction by the proposed MQCB scheme does not modify the already very good agreement between classical and quantum results.

However, one should keep in mind that even if the ARET for this system is well described by classical mechanics, the diffraction process, being of purely quantum mechanical nature, cannot be treated by classical mechanics. Thus

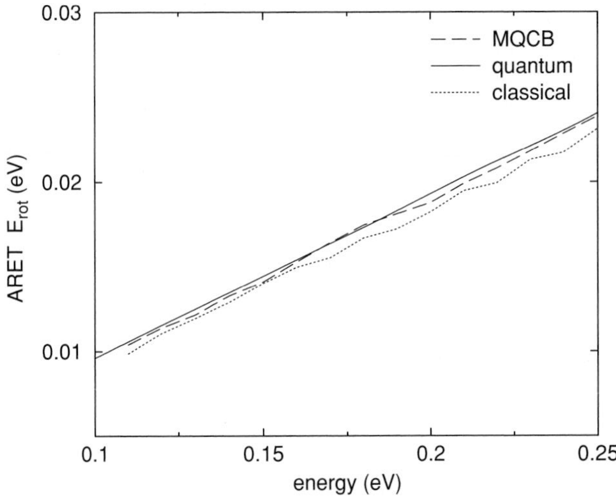

Fig. 2. Average rotational energy transfer (ARET) E_{rot} for different collision energies. *Full line*: quantum mechanical wave packet propagation using the MCTDH method (from: [46]); *dashed line*: MQCB method (30); *dotted line*: classical dynamics

one clearly sees how the MQCB method, as a mixed method, really combines classical mechanics with quantum mechanics in some degrees of freedom, when one is interested in effects that are of purely quantum mechanical nature. Comparing the MQCB method with the full quantum results presented in Figs. 1 and 2, one sees that both the diffraction probabilities as well as the average rotational energy transfer are extremely well described by the MQCB method.

5.2 Vibrational Decoherence of I_2 in a Dense Helium Environment

As second example, we consider the coherent vibrational dynamics of one diatomic molecule (I_2) after femtosecond laser pulse excitation to an excited state, while interacting with an environment of a high pressure rare gas [47]. After a well-defined delay time, a second laser pulse induces a transition to a final electronic state, from which the fluorescence is detected. The electronic ground, excited and final states are denoted by $|g\rangle$, $|e\rangle$ and $|f\rangle$, respectively.

We consider an iodine molecule and 40 rare gas atoms with periodic boundary conditions. This is justified, because the experiments we want to compare with are performed with a low I_2 concentration, and diatom–diatom interactions can be neglected.

The total Hamiltonian for the I_2/Rg system with the I_2 being in its electronic ground, excited, or final state are given by

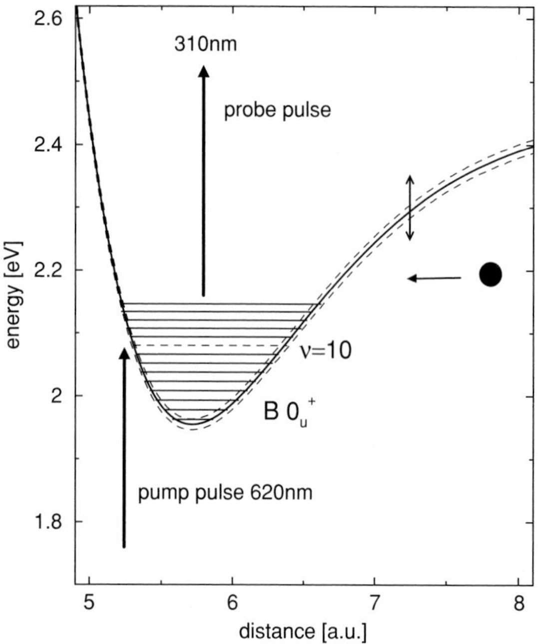

Fig. 3. Potential energy surface of the $B0_u^+$–state of I_2 together with the vibrational levels excited coherently be the pump pulse to create a vibrational wave packet centered around $\nu = 10$. The *arrows* indicate the pump- and probe pulses used for the calculations. The collisions with the buffer gas atoms lead to fluctuations of the PES, as indicated symbolically

$$H = H_{\mathrm{int}}^{(g,e,f)} + H_{\mathrm{trans}} + H_{\mathrm{rot}} + H_{\mathrm{int}} + H_{\mathrm{Rg}} \tag{31}$$

$$H_{\mathrm{r}}^{(g,e,f)} = \frac{1}{2\mu} p_{\mathrm{r}}^2 + V_{I-I}^{(g,e,f)}, \tag{32}$$

$$H_{\mathrm{trans}} = \frac{\boldsymbol{P}^2}{2M}, \tag{33}$$

$$H_{\mathrm{rot}} = \frac{1}{2\mu r^2}\left(p_\theta^2 + \frac{p_\phi^2}{\sin^2\theta}\right), \tag{34}$$

$$H_{\mathrm{int}}^{(g,e,f)} = \sum_{\mathrm{i}=1}^{N} V_{\mathrm{I_2-Rg}}^{(g,e,f)}, \tag{35}$$

$$H_{\mathrm{Rg}} = \sum_{\mathrm{i}=1}^{N} \frac{\boldsymbol{p}_{\mathrm{i}}}{2m} + \sum_{\mathrm{i}>j} V_{\mathrm{Rg-Rg}}. \tag{36}$$

In this expression, μ is the reduced mass of I_2, r its internuclear distance, θ and ϕ the polar angles describing its orientation in the laboratory frame of reference and $\boldsymbol{P}$ is its center-of-mass momentum. The N rare gas atoms are described by their positions $\boldsymbol{r}_\mathrm{i}$ and momenta $\boldsymbol{p}_\mathrm{i}$. The interaction potentials between the iodine molecule and helium atoms were taken to be the same as in the previous study by Ermoshin et al. [48]. The potential energy surfaces for the X- and B-states of I_2 are those of [49], and the f-state potential was taken from [50]. The I_2–He potentials were represented by a sum of atom–atom potentials with Morse functional form and parameters from [51]. The parameters of the Lennard-Jones parametrization for the He–He interactions were taken to reproduce the equilibrium distance and well depth of more sophisticated Hartree–Fock dispersion potentials from [52,53].

In the study presented in this work, which concerns the vibrational relaxation and decoherence of nuclear vibrational motion of I_2 in the electronic $B0_u^+$–state after femtosecond pulse excitation, we used for simplicity the same potentials $V_{\mathrm{I_2-Rg}}^{(g,e)}$ for the ground (X-) and excited (B-) state. Note however that with the methodology presented here this assumption is not necessary and more refined interaction potentials could be used.

The field-matter interaction

$$W_\alpha(t) = \boldsymbol{\mu}\cdot\boldsymbol{\epsilon}_\alpha\mathcal{E}_\alpha(t) \quad \alpha = \mathrm{pu},\mathrm{pr} \tag{37}$$

takes a different form for the pump and the probe transition, due to possibly different polarizations, central frequencies or pulse shapes. Following the experimental set-up described in [54,55], we consider linear polarizations for the two pulses, with an adjustable angle α bewteen the pump- and probe polarizations. Taking further into account, that the induced electronic transitions $I_2(B \leftarrow X)$ and $I_2(f \leftarrow B)$ are parallel transitions, the interaction terms for pump- and probe pulses are given by

$$W_{\mathrm{pu}}(t) = \cos\theta\;\mu_{eg}\;\mathcal{E}_{\mathrm{pu}}(t), \tag{38}$$

$$W_{\mathrm{pr}}(t) = [\sin\alpha\;\sin\theta\;\cos\varphi + \cos\alpha\;\cos\theta]\,\mu_{\mathrm{fe}}\;\mathcal{E}_{\mathrm{pr}}(t). \tag{39}$$

In writing these equations, we have assumed that the laboratory frame z-axis is defined by the pump–pulse polarization $\boldsymbol{\epsilon}_{\mathrm{pu}}$.

In what follows, we shall treat the internuclear distance r as quantum degree of freedom to account for the wave packet formation by the ultrafast laser excitation and for the vibrational revivals, as well as for the decoherence induced by the rotation and random collisions. All other degrees of freedom, like the center of mass motion, the rotation and the motion of a large number of rare gas atoms are treated classically within the spirit of a molecular dynamics simulation.

Prior to the pump–pulse excitation, the initial conditions of both the quantum as well as the classical degrees of freedom need to be defined. Since a wavefunction for the whole system is not known in this case, the initial conditions were chosen corresponding to the quantum wavefunction of the I_2 vibration and a statistical distribution for the rotation and rare gas atoms, according to a given pressure and temperature. This is similar as in a previous quantum-classical simulations that used the mean-field approach [48]. A full classical molecular dynamics equilibration was performed keeping the I–I distancne fixed to its ground state equilibrium value. To this end, the rare gas atoms are placed initially on a regular grid and the orientation of the diatomic was chosen at random. The rare gas momenta as well as the the values for p_θ and p_φ were chosen to correspond to the desired temperature. The initial quantum vibrational wavefunction $\psi_g(r, t = 0)$ was chosen randomly according to a Boltzmann distributon of vibrational levels. This vibrational wavefunction was then used to describe the femtosecond laser pulse interaction.

Between the pump and the probe pulse, we have to propagate the excited state vibrational wave packet alongside with the rotational and center of mass motion and their interactions with the colliding rare gas atoms. The time evolution of this coupled quantum/classical system is calculated by

$$\begin{aligned} \mathrm{i}\hbar\frac{\mathrm{d}}{\mathrm{d}t}\chi_e(r,t) = \Big[&-\frac{\hbar^2}{2\mu}\frac{\partial^2}{\partial r^2} + V^e_{I-I}(r) \\ &+\frac{1}{2\mu r^2}\Big(\mathbf{p}_\theta^2 + \frac{\mathbf{p}_\phi^2}{\sin^2\theta}\Big) + \sum_{\mathrm{i}=1}^{N} V_{\mathrm{I_2-Rg}}(r, \boldsymbol{R}, \boldsymbol{r}_\mathrm{i})\Big]\chi_e(r,t) \end{aligned} \tag{40}$$

$$\dot{\boldsymbol{r}} = \frac{\hbar}{\mu}\,\mathrm{Im}\Big(\frac{1}{\chi_e(r)}\frac{\partial\chi_e(r)}{\partial r}\Big)_{r=\mathbf{r}(t)} \tag{41}$$

$$\begin{aligned} &\dot{\boldsymbol{R}} = \frac{\boldsymbol{P}}{M} \qquad \dot{\boldsymbol{P}} = -\nabla_{\boldsymbol{R}} V_\mathrm{T}, \\ &\dot{\boldsymbol{r}}_\mathrm{i} = \frac{\boldsymbol{p}_\mathrm{i}}{m}, \qquad \dot{\boldsymbol{p}}_\mathrm{i} = -\nabla_{\boldsymbol{r}_\mathrm{i}} V_\mathrm{T}, \qquad \mathrm{i} = 1..N, \\ &\dot{\boldsymbol{\theta}}(t) = \frac{\boldsymbol{p}_\theta(t)}{\mu r^2} \qquad \dot{\boldsymbol{p}}_\theta(t) = \frac{\boldsymbol{p}_\varphi^2(t)\cos\boldsymbol{\theta}(t)}{\mu r^2\sin^3\boldsymbol{\theta}(t)} - \frac{\partial V_T}{\partial\theta}\Big|_{r=\mathbf{r}(t),q=\mathbf{q}(t)}, \\ &\dot{\boldsymbol{\varphi}}(t) = \frac{\boldsymbol{p}_\varphi^2(t)}{\mu\sin^2\boldsymbol{\theta}(t) r^2} \qquad \dot{\boldsymbol{p}}_\varphi(t) = -\frac{\partial V_\mathrm{T}}{\partial\varphi}\Big|_{r=\mathbf{r}(t),q=\mathbf{q}(t)}. \end{aligned} \tag{42}$$

For clarity, we have collectively denoted by $\mathbf{q}$ the classical variables $\boldsymbol{R}, \boldsymbol{r}_\mathrm{i}, \boldsymbol{\theta}, \boldsymbol{\varphi}$ and we have defined a classical potential energy $V_\mathrm{T} = \sum_{\mathrm{i}=1}^{N} V^e_{\mathrm{I_2-Rg}} + \sum_{\mathrm{i}>j} V_{\mathrm{Rg-Rg}}$.

This system of equations (40–42) consists of [15, 16]:

1. The Schrödinger equation (40) for the wavefunction $\chi_e(r, t)$ of the quantum subspace, which depends parametrically on the classical variables $\boldsymbol{R}, \boldsymbol{r}_\mathrm{i}, \boldsymbol{P}, \mathrm{p}_\mathrm{i}, \boldsymbol{\theta}, \boldsymbol{\varphi}, \boldsymbol{p}_\theta, \boldsymbol{p}_\varphi$
2. A *quantum trajectory* (41) that follows the quantum wave packet motion [22–24] and that is used in the classical equations
3. The classical Hamilton's equation, (42) for the classical degrees of freedom $\boldsymbol{R}, \boldsymbol{r}_\mathrm{i}, \boldsymbol{P}, \boldsymbol{p}_\mathrm{i}, \boldsymbol{\theta}, \boldsymbol{\varphi}, \boldsymbol{p}_\theta, \boldsymbol{p}_\varphi$

Starting from the inital values $\boldsymbol{R}_0, \boldsymbol{r}_{\mathrm{i},0}, \boldsymbol{P}_0, \boldsymbol{p}_{\mathrm{i},0}, \boldsymbol{\theta}_0, \boldsymbol{\varphi}_0, \boldsymbol{p}_{\theta,0}, \boldsymbol{p}_{\varphi,0}$ and χ_e $(r, t = 0)$, these quantities have to be propagated simultaneously for a large number of different initial conditions representing an ensemble of given pressure and temperature as detailed above.

In our implementation, each wavefunction is represented by one quantum trajectory. It is this feature which allows to treat the back-reaction for wavefunctions that split in several subpackets as it is the case in the example presented here. An extension of the method is possible, in which several quantum trajectories are associated with one wavefunction, and the back-reaction being calculated by an average of these quantum trajectories. In the limit of all trajectories associated with one single wavefunction, the method becomes identical to the mean-field method. Hence this idea of regrouping trajectories is an interesting direction for future work, since one can consider it to be a continuous interpolation from MQCB to mean-field. With only one quantum trajectory associated with one wavefunction, as it was used in this work, the total energy is not conserved. In principle, this could be fixed by velocity adjustment in a way similar to the surface hopping methods. However, in the context of the work presented here, this is not necessary since the parameters used in the calculations correspond to fairly low pressures where practically no helium atoms collide twice with the vibrating I_2 molecule.

Note that (41) can be re-written as

$$\ddot{\boldsymbol{r}} = -\frac{1}{\mu}\frac{\partial}{\partial r}\left(V^{e}_{\mathrm{I-I}}(r) + \frac{1}{2\mu r^2}\left(\mathbf{p}_\theta^2 + \frac{\mathbf{p}_\phi^2}{\sin^2\theta}\right) + \sum_{\mathrm{i}} V_{\mathrm{I_2-Rg}}(r, \boldsymbol{R}, \boldsymbol{r}_{\mathrm{i}}) + Q(r)\right) \tag{43}$$

with

$$Q(r) = -\frac{\hbar^2}{2\mu}\frac{1}{|\chi_e(r)|}\frac{\partial^2|\chi_e(r)|}{\partial r^2}. \tag{44}$$

In this way it becomes clear that the MQCB method can be viewed as an extension of purely classical mechanics by adding an approximate quantum potential to selected degrees of freedom to include quantum effects (here the internuclear distance r). Details of the derivation of the equations have been given elsewhere, together with an analysis of their structure, how to sample initial conditions and how to calculate observables. In this work, the question of reversibility and resampling at intermediate times is also addressed thoroughly [16].

After a well-defined delay time τ, the probe pulse interacts with the sample which induces a transition from the ground to a final state $|f\rangle$, the population of which is the detected pump–probe signal (e.g., fluorescence). In the experiment to compare with [54, 55], the probe polarization is rotated by an angle α with respect to the pump–pulse polarization. We assume pulses of weak intensity to allow for a time-dependent perturbative treatment.

The total population in the final electronic state, after both laser pulses have interacted with the sample, is given by:

$$P(\tau)=\Big[(\cos\alpha\ \sin\boldsymbol{\theta}_\tau\ \cos\boldsymbol{\varphi}_\tau+\cos\alpha\cos\boldsymbol{\theta}_\tau)\ \cos\boldsymbol{\theta}_0\Big]^2\int \mathrm{d}r\ |\chi_f(r,t=\tau)|^2 \tag{45}$$

This expression has a clear physical interpretation: the first term, depending on the relative orientation of the dipole at time $t=0$ and $t=\tau$, reflects the fact that a transition is favoured when the dipole is aligned with the polarization vectors. However, between the pump- and probe pulse the molecule continues to rotate, perturbed by the random collisions with the rare gas atoms.

The second term describes the interaction of the optical pulse with the internal degrees of freedom, i.e. the electronic and vibrational states. It reflects the ultrafast vibrational wave packet motion induced by the femtosecond laser pulse. The total population of the final electronic states depends on both, the rotational motion as well as the electronic/vibrational motion, the first process being described by classical mechanics, the latter by quantum mechanics.

In the pump–probe experiment we want to compare our results with [54, 55], the measured signal is the total fluorescence from the final exited electronic state as a function of delay time between the pump and probe pulse. This fluorescence signal is taken to be proportional to the total excited state population after the pump- and probe pulse have interacted with the sample. In the preceding sections we have shown how, starting from a well-defined initial state for the quantum and classical degrees of freedom, this quantity can be calculated within a mixed quantum/classical scheme.

To simulate a pump probe spectrum of an ensemble of molecules in a high pressure environment at a given tempertature T, the total experimentally measured signal is obtained by averaging $P(\tau)$ over many individual quantum/classical propagations, but with different initial conditions corresponding to a statistical ensemble of given pressure and temperature (p,T).

Due to the fact that we consider the rotational motion frozen on the time scale of the pulse *duration*, we were able to separate angular parts (treated classically) and internal degrees of freedom (treated quantum mechanically) that take pulse duration, central frequency or specific pulse forms into account. By taking the azimuthal symmetry into account $\boldsymbol{\varphi}_0$ is uniformely distributed, and so is $\boldsymbol{\varphi}_\tau$. Hence we find

$$P_{\text{total}}^{(p,T)}(\tau;\alpha)=\left\langle\big(\cos^2\alpha-\sin^2\boldsymbol{\theta}_\tau\ P_2(\cos\alpha)\big)\cos^2\boldsymbol{\theta}_0\int \mathrm{d}r\,|\chi_f(r,t=\tau)|^2\right\rangle_{(p,T)}, \tag{46}$$

where P_2 is the second Legendre polynomial. In this form, it becomes immediately clear that for $\alpha_0=54.7^\circ$ ('magic angle' detection) the effects of rotational motion are suppressed, and the pump–probe signal $S(p,\tau)$ reflects the internal vibrational motion only

$$S(p,\tau) \equiv P_{\text{total}}^{(p,T)}(\tau;\alpha_0) \sim \left\langle \int \mathrm{d}r\ |\chi_f(r,t=\tau)|^2 \right\rangle_{(p,T)} . \tag{47}$$

The mixed quantum–classical expression (21) generalizes the picture of polarization anisotropy of classical dipoles to the case where internal degrees of freedom are excited by the laser pulses: if the latter effect, expressed by $\int \mathrm{d}r\ |\chi_f(r,t=\tau)|^2$ is unimportant, one recovers a classical expression for a dipole–dipole correlation function. If on the other hand rotational motion is unimportant (for very short times or cold samples as in molecular beams), the pump–probe signal reflects the internal vibrational wave packet motion in the the different elecronic states involved.

Results: Pump–Probe Spectroscopy of I_2 in a High Pressure Rare Gas Environment

For the results to be presented later, we used 40 rare gas atoms in a periodic molecular dynamics box of length 31.7 a.u. for 5 bar simulations (10 bar: 25.2 a.u., 20 bar: 20 a.u). After the equilibration, the femtosecond excitation and mixed quantum/classical MQCB propagation was performed for 1,000 runs independently, with the pump–probe signal being averaged over. Increasing the number of trajectories did not alter the signals presented later. The wave packet propagation was performed using the Split–Operator technique of Feit and Fleck [56], with the classical equations (42) integrated simultaneously using Gear's algorithm. The time step for the quantum part was chosen to be 1 fs, while for the classical part a variable timestep integration in steps of 1 fs was found to be a fast and stable way of propagating the MQCB equations simultaneously. Equation (41) requires special attention due to the highly oscillatory nature of the complex wavefunction. To increase numerical stability, (41) is evaluated using a smoothing procedure detailed in [47]. By this method, the overall shape of the wavefunction is unchanged (dispersion, revival, fractional revival) while preventing the integrator from slowing down when the wave packet interferes with itself at the inner turning points.

Pump–probe experiments of iodine dimers in different high pressure rare gas environments at different pressures were performed by Lienau et al. [54,55]. We have performed simulations as described above using helium as buffer gas and using the laser parameters of the experiment (620 nm pump and 310 nm probe). Using the mixed quantum/classical method descibed in the theory section, we simulated pump–probe spectra with the results shown in Fig. 4. As a first result one sees how increasing the pressure leads to a strong decrease in the revival structure around 10 ps. These spectra can be directly compared to the experimental signals measured by Zewail and coworkers (see Fig. 20 in [55]) and a very good agreement is found. Note that the short-time

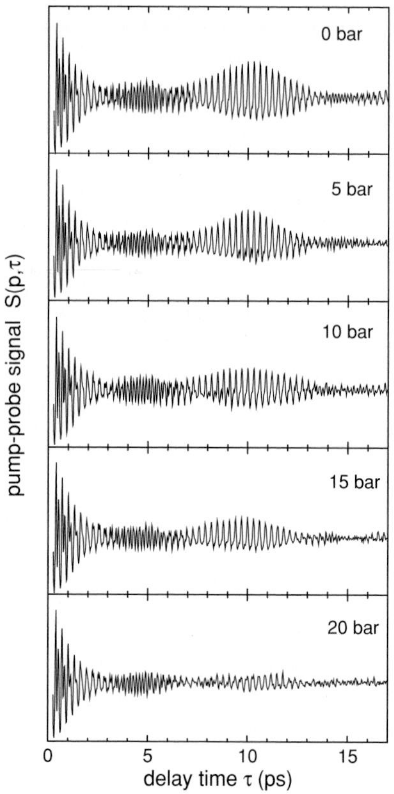

Fig. 4. Calculated pump–probe signals under *magic angle* conditions for different buffer gas pressures. The structure around 10 ps, originating from the phenomena of *fractional revivals*, vanishes gradually as pressure increases, due to dephasing induced by random collisions with the buffer gas atoms

signal, which reflects the first few vibrational periods, is almost unaffected for the pressures considered in this work. Second, for the delay times up to 20 ps, no considerable energy relaxation is found. The process can thus be considered as pure dephasing. This result is also documented in Fig. 5. It shows the vibrational populations and coherences of neighbouring vibrational levels calculated within the MQCB method as

$$\rho_{\nu',\nu} = \left\langle \langle \nu' | \psi_e \rangle \langle \psi_e | \nu \rangle \right\rangle_{(p,T)} , \tag{48}$$

where $|\psi_e\rangle$ is the wavefunction in the quantum subspace (depending also on the classical variables) and $|\nu\rangle$ the vibrational eigenfunctions of the purely quantum part $H_r^{(e)}$, which is the radial part of the Hamiltonian in the $B0_u^+$–state potential (see (40)).

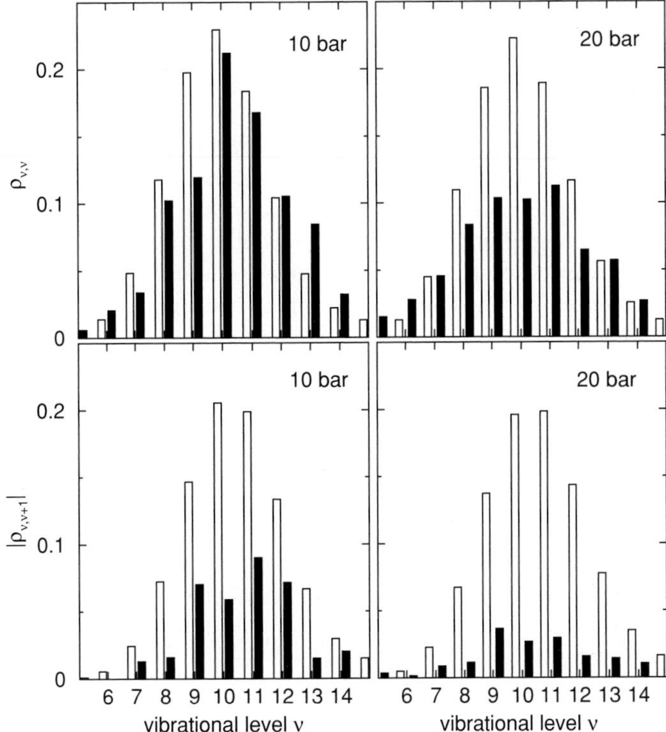

Fig. 5. Vibrational populations $\rho_{\nu,\nu}$ (*upper panel*) and coherences $\rho_{\nu,\nu+1}$ (*lower panel*) as defined in (23) for two different delay times (*white*: 300 fs, *black*: 10 ps) and pressures of 10 and 20 bar, as indicated. In the pressure/delay-time regime considered, the dominating process is dephasing, i.e. the decay of the coherences $\rho_{\nu,\nu+1}$ (*lower panel*), and not vibrational relaxation

The dynamical process of decoherence can be seen in Fig. 5: The upper panel shows the populations $\rho_{\nu,\nu}$ for two pressures, 10 bar and 20 bar, for two times, at 300 fs (open histograms) and at 10 ps (black histograms). The lower panel shows the coherences $\rho_{\nu,\nu+1}$ for two pressures and times.

Thes results have a clear physical significance. For both pressure conditions, the ultrafast laser pulse creates a wave packet around the vibrational state $\nu = 10$, comprised of states ranging from $\nu = 6$ up to $\nu = 14$ (shown as open histograms of the upper panel). After 10 ps, this distribution has not changed very much, it is still centered around $\nu = 10$. However, in the case of a buffer gas pressure of 20 bar the distribution is found to be enlarged. No significant vibrational relaxation is found for the pressures and delay times considered in this work. In contrast, the dephasing documented in the lower panel is clearly visible by a decay of the coherences between adjacent vibrational levels. While just after the femtosecond excitation the presence of coherences indicates a wave packet motion, which is detected by the pump–probe set-up, after 10 ps the coherences have significantly decayed. As a result, the system has evolved from one showing a coherent wave packet motion to one with statistically distributed population of vibrational levels. If, as it is the case for our system considered, the diagonal relaxation is not the dominating process, one usually speaks of *pure dephasing*. It is this process which prevents the wave packet from rephasing and showing the clear revival structure, which is detectable by pump–probe spectroscopy. Hence the pump–probe set-up considered in this work, which is able to detect fractional revivals of vibrational wave packets, is very sensitive to dephasing. This behaviour is clearly documented by the simulated pump–probe signals presented in Fig. 4.

6 Conclusions

In this paper we have given a review of the MQCB method as introduced by Gindensperger, Meier and Beswick [15], together with a discussion of its main properties. The method is illustrated by two examples, a five-dimensional model of a molecule–surface collision and a simulation of molecular vibrational decoherence, detected by pump–probe spectroscopy.

Since for the first example full quantum results can still be obtained by quantum wave packet propagation tecniques, we can give a direct comparison and thus show the accuracy of the (approximate) MQCB method for the model considered. We find an almost quantitative agreement, which is an encouraging results for possible further applications of the MQCB method in large systems, where inherent quantum effects require to treat at least parts of the full system quantum mechanically. However, whether the method would give good agreement with other systems, or a systematic study of its applicability to a large number of situations, remains to be studied.

The second example shows that the recently proposed scheme to mix quantum and classical mechanics can successfully be used to study the process of molecular vibrational decoherence induced by rotations and random collisions. It allows to employ realistic microscopic potentials to describe the system–environment interaction. Since dimensionality does not pose a major problem within the classical space, standard molecular dynamics methods can be employed to account for macroscopic parameters like pressure and temperature. As compared to the popular reduced-density matrix approaches based on master equations, the proposed method does not require a separation of time scales which in the femtosecond regime are often violated.

For this case, where ultrafast excitation and detection of I_2 in a helium environment is considered, we have shown that the phenomena of revivals of vibrational wave packets are a very sensitive to decoherence processes. Using the proposed theoretical approach of mixing quantum and classical dynamics, a very good agreement with experimental results was found. This method should thus be a promising tool to investigate the excitation and evolution of quantum systems with coherent light while interacting with a large number of other degrees of freedom, such as in a liquid or within a large molecule. This paves the way to the development of coherent control schemes to overcome or minimize decoherence effects induced by the environment in condensed phases.

Acknowledgments

We would like to thank E. Gindensperger for very helpful contributions at the early stage of this work.

References

1. H.-D. Meyer, U. Manthe, L. S. Cederbaum Chem. Phys. Lett. **165**, 73 (1990)
2. M. H. Beck, A. Jäckle, G. A. Worth, H.-D. Meyer Phys. Rep. **324**, 1 (2000)
3. *Classical and Quantum Dynamics in Condensed Phase Simulations* (B. Berne, G. Cicotti, D. Coker Eds.; World Scientific, Singapore, 1998)
4. *Quantum Dissipative Systems*, ed. U. Weiss (World Scientific, Singapore 1993)
5. C. Meier, D. Tannor J. Chem. Phys. **111**, 3365 (1999)
6. G. Billing Int. Rev. Phys. Chem. **13**, 309 (1994)
7. J. C. Tully, Int. J. Quant. Chem. **25**, 299 (1991)
8. D. A. Micha, B. Thorndyke Adv. Quantum Chem. **47**, 292–312 (2004)
9. B. Thorndyke, D. A. Micha Chem. Phys. Lett. **403**, 280–286 (2005)
10. A. Donoso, C. C. Martens J. Chem. Phys. **102**, 4291 (1998)
11. R. Kapral, G. Cicotti J. Chem. Phys. **110**, 8919 (1999)
12. S. Nielsen, R. Kapral, G. Cicotti J. Chem. Phys. **115**, 5805 (2001)
13. I. Horenko, C. Salzmann, B. Schmidt, C. Schütte J. Chem. Phys. **117**, 11075 (2002)
14. I. Horenko, M. Weiser, B. Schmidt, C. Schütte J. Chem. Phys. **120**, 8913 (2004)
15. E. Gindensperger, C. Meier, J. A. Beswick, J. Chem. Phys. **113**, 9369 (2000)
16. E. Gindensperger, C. Meier, J. A. Beswick, Adv. Quant. Chem. **47**, 331–346 (2004)
17. E. Gindensperger, C. Meier, J. A. Beswick, J. Chem. Phys. **116**, 8 (2002)
18. E. Gindensperger, C. Meier, J. A. Beswick, M.-C. Heitz, J. Chem. Phys. **116**, 10051 (2002)
19. O. V. Prezhdo, C. Brooksby Phys. Rev. Lett. **86**, 3215 (2001)
20. L. L. Salcedo Phys. Rev. Lett. **90**, 118901 (2003)
21. O. V. Prezhdo, C. Brooksby Phys. Rev. Lett. **90**, 118902 (2003)
22. *The quantum theory of motion* P. R. Holland (Cambridge University Press, Cambridge 1993)
23. L. de Broglie C. R. Acad. Sci. Paris **183**, 447 (1926); **184**, 273 (1927)
24. D. Bohm Phys. Rev. **85**, 166 (1952); **85**, 180 (1952)
25. R. E. Wyatt *Quantum Dynamics with Trajectories*, Springer, Berlin, Heidelberg New York (2005)
26. C. L. Lopreore and R. E. Wyatt Phys. Rev. Lett. **82**, 5190 (1999)
27. E. R. Bittner, J. Chem. Phys. **112**, 9703 (2000)
28. R. E. Wyatt, C. L. Lopreore, G. Parlant J. Chem. Phys. **114**, 5113 (2001)

29. D. Nerukh, J. H. Frederick Chem. Phys. Lett. **332**, 145 (2000)
30. F. Sales Mayor, A. Askar, H. Rabitz J. Chem. Phys. **111**, 2423 (1999)
31. C. Trahan, R. E. Wyatt J. Chem. Phys. **118**, 4784 (2003)
32. R. E. Wyatt, E. R. Bittner J. Chem. Phys. **113**, 8898 (2000)
33. R. E. Wyatt J. Chem. Phys. **117**, 9569 (2002)
34. R. E. Wyatt, C. L. Lopreore, G. Parlant J. Chem. Phys. **114**, 5113 (2001)
35. J. R. Maddox, E. R. Bittner J. Chem. Phys. **119**, 6465 (2003)
36. S. Garashchuk, V. A. Rassolov Chem. Phys. Lett. **364**, 562 (2002)
37. S. Garashchuk, V. A. Rassolov Chem. Phys. Lett. **376**, 358 (2003)
38. S. Garashchuk, V. A. Rassolov J. Chem. Phys. **120**, 1181 (2004)
39. H. Carlsen, O. Goscinski Phys. Rev. A **59**, 1063 (1999)
40. A. S. Sanz, F. Borondo, S. Miret-Artés, Phys. Rev. B **61**, 7743 (2000)
41. I. Burghardt, L. S. Cederbaum J. Chem. Phys. **115**, 10303 (2001)
42. A. Donoso, C. C. Martens Phys. Rev. Lett. **87**, 223202 (2001)
43. R. Bittner J. Chem. Phys. **119**, 1358 (2003)
44. I. Burghardt J. Chem. Phys. **122**, 094103 (2005)
45. R. B. Gerber, L. H. Beard, D. J. Kouri, J. Chem. Phys. **74**, 4709 (1981)
46. M.-C. Heitz, H.-D. Meyer J. Chem. Phys. **114**, 1382 (2001)
47. C. Meier, J. A. Beswick J. Chem. Phys. **121**, 4550 (2004)
48. V. A. Ermoshin, A. K. Kazansky, V. Engel, J. Chem. Phys. **111**, 7807 (1999)
49. J.-Y. Fang, C. C. Martens, J. Chem. Phys. **105**, 9072 (1996)
50. J. P. Perrot, A. J. Bouvier, A. Bouvier, B. Femelat, J. Chevaleyre, J. Mol. Spectrosc. **114**, 60 (1985)
51. A. Garcia-Vela, P. Villarreal, G. Delgado-Barrio, J. Chem. Phys. **92**, 6504 (1990)
52. R. A. Aziz, A. R. Janzen, M. R. Moldover, Phys. Rev. Lett. **74**, 1586 (1995)
53. R. A. Aziz, H. H. Chen, J. Chem. Phys. **67**, 5719 (1977)
54. C. Lienau, A. H. Zewail, J. Phys. Chem., **100**, 18629 (1996)
55. Q. Liu, C. Wan, A. H. Zewail J. Phys. Chem., **100**, 18666 (1996)
56. J. A. Fleck, J. R. Morris, M. D. Feit, Appl. Phys. **10**, 129 (1976); M. D. Feit, J. A. Fleck, A. Steiger J. Comput. Phys. **47**, 412 (1982)

Quantum Hydrodynamics and a Moment Approach to Quantum–Classical Theory

I. Burghardt, K.B. Møller, and K.H. Hughes

Summary. We review and extend the quantum hydrodynamic formulation for mixed states (density matrices), by which a hierarchy of coupled moment equations is derived from the quantum Liouville equation. The quantum hydrodynamic picture provides a complement, in a quantum–statistical context, to its pure-state "Bohmian mechanics" analogue, and also connects in a unique way to quantum phase-space distributions. This formulation is used to introduce a novel hybrid quantum–classical method based upon *partial* moments, which combine the hydrodynamic representation in the quantum subspace with a Liouvillian phase-space representation in the classical subspace. In the Lagrangian picture, this results in a *mixed quantum–classical molecular dynamics* scheme. The interleaved trajectory dynamics is guided by a quantum force which also depends upon the classical variables. The method is shown to be closely connected to the quantum–classical Liouville equation, but its deterministic trajectory evolution is specific to the hydrodynamic setting. Examples are given for the vibrationally nonadiabatic dynamics in harmonic and anharmonic oscillator systems coupled to a classical harmonic subspace, for pure-state vs. dissipative situations; for these systems, the method is exact.

1 Introduction

The hydrodynamic, or "Bohmian" representation of quantum mechanics [1–7] is of great appeal (and has led to much controversy) in that it postulates the existence of quantum-mechanical particle trajectories, derived from the analogy between the Schrödinger equation and the equations of motion of fluid dynamics. Independently of the de Broglie–Bohm particle interpretation [1–4], the fluid-dynamical analogy was recognized in 1926 by Madelung [8] – in the same year as Schrödinger's seminal papers and de Broglie's first "pilot wave" interpretation [1]. Madelung's perspective refers to fluid-dynamical trajectories to represent the quantum hydrodynamic fields, but does not necessitate the interpretation of the Lagrangian dynamics in terms of actual particle trajectories, as in the Bohmian picture.

The considerations of the present chapter are limited to this latter point of view, compatible with the conventional ("Copenhagen") interpretation of quantum mechanics. This is in line with the use of the fluid dynamical trajectory picture as an analysis tool [9–11], and with the extensive recent efforts aiming at the numerical implementation of quantum trajectory propagation, as an alternative to conventional wavepacket propagation methodology [7, 12–24].

While the majority of recent works – as well as most of the interpretational issues associated with Bohmian mechanics [5, 6, 25] – have focused on the representation of wavefunctions (pure quantum states) in the hydrodynamic picture [1, 3, 4, 8, 26], the main concern of the present chapter is quantum hydrodynamics for *mixed* states. Indeed, the quantum-*statistical* theory, suitable for nonequilibrium states and dissipative processes, can be entirely cast in the hydrodynamic language. This version of quantum hydrodynamics goes back to Moyal [27], Zwanzig [28], Takabayasi [26], and Fröhlich [29], followed more recently by others [30–36]. Starting from the phase space (Wigner) representation $\rho_{\mathrm{W}}(q,p)$ of the density operator, the associated hydrodynamic formulation can be derived as a "projection onto coordinate space" [26], in terms of *hydrodynamic moments*,

$$\langle \mathcal{P}^n \rho \rangle_q = \int \mathrm{d}p \, p^n \rho_{\mathrm{W}}(q,p), \tag{1}$$

obtained by integration over the phase-space momentum p only. A hierarchy of coupled moment equations can be derived, which have the form of the hydrodynamic equations of classical mechanics, i.e., coupled equations for the mass density, momentum density, kinetic energy density, etc. [27, 29, 31–35]. For pure states (wavefunctions), the hierarchy terminates with the first two members, thus yielding the equations of motion of Bohmian mechanics [26, 35, 37]. Even though pure-state hydrodynamics can thus be understood as a particular case of the quantum-statistical formulation, this relation has only been scarcely discussed in the literature, with the exception of [26, 35–37], see also the recent overview in [7].

Against this background, the purpose of the present chapter is twofold: First, to provide an overview of quantum-statistical hydrodynamic theory and its relation to quantum phase space theories and to pure-state Bohmian mechanics. Second, to address a hybrid quantum–classical formulation which is suitable for the quantum-statistical context. We have recently proposed such a formulation [38, 39], which provides a new framework for coupling hydrodynamic quantum trajectories with classical Hamiltonian trajectories. Using the connections between the phase-space (Wigner) representation and the hydrodynamic picture, we define a mixed hydrodynamic-Liouvillian representation based upon *partial hydrodynamic moments* [38, 39],

$$\langle \mathcal{P}^n \rho \rangle_{qQP} = \int \mathrm{d}p \, p^n \, \rho_{\mathrm{W}}(q,p;Q,P). \tag{2}$$

Here, the hydrodynamic projection is applied to the quantum (q, p) sector only. Exact equations of motion can be derived for the $\langle \mathcal{P}^n \rho \rangle_{qQP}$'s from the quantum Liouville equation (see Sect. 3.2) [39]. With a classical-limit approximation in the Liouvillian (Q, P) sector, and using the Lagrangian picture associated with the hydrodynamic representation, the following dynamical equations are obtained, which *couple quantum hydrodynamic trajectories with classical phase-space trajectories* (Sect. 3.4),

$$
\begin{aligned}
\dot{q} &= \frac{\bar{p}}{m}, \\
\dot{\bar{p}} &= -\frac{\partial}{\partial q}\Big(V_q(q) + V_{\text{int}}(q, Q)\Big) + F_{\text{hyd}}(q, Q, P), \\
\dot{Q} &= \frac{P}{M}, \\
\dot{P} &= -\frac{\partial}{\partial Q}\Big(V_Q(Q) + V_{\text{int}}(q, Q)\Big),
\end{aligned} \tag{3}
$$

where the hydrodynamic momentum $\bar{p} \equiv \bar{p}_{qQP}$ is a function of (q, Q, P), defined via the first moment $\langle \mathcal{P}\rho \rangle_{qQP} = \bar{p}_{qQP} \langle \rho \rangle_{qQP}$, and the hydrodynamic force $F_{\text{hyd}}(q, Q, P)$ acting within the quantum subspace is a function of q as well as of the classical phase-space variables (Q, P).

The approximation made in (3) is the same as in the quantum–classical Liouville equation [40–48], and (3) can in fact be derived from this equation (Sect. 3.3). Several aspects regarding the trajectory equations (3) are worth noting: (a) Equation (3) captures the details of the phase-space correlations between the quantum and classical sectors, thus going far beyond mean-field (Ehrenfest type) methods. (b) No hydrodynamic force arises in the classical subspace, which is described within a Liouvillian setting. (c) As is the case for the quantum–classical Liouville equation, (3) is exact if the classical subsystem is harmonic. (This is not the case for other mixed quantum–classical representations derived from the Bohmian picture [49–51].) (d) Contrary to a mixed quantum–classical representation in terms of phase-space ("Wigner") trajectories [52–55], the quantum correction terms do not "penetrate" into the classical sector, i.e., in (3), the classical sector obeys a Hamiltonian dynamics. (e) Contrary to the stochastic trajectory dynamics (i.e., surface hopping type trajectories) which is necessary in the classical sector if the quantum part of the mixed quantum–classical Liouville equation is expressed in a discretized representation [45,46], the coupled trajectory equations (3) are deterministic. These latter features are essentially due to the fact that the hydrodynamic representation "localizes" the quantum sector.

The remainder of the chapter is organized as follows: Section 2 gives an overview of the quantum-statistical formulation of quantum hydrodynamics, while Sect. 3 focuses on the mixed quantum–classical formulation using partial moments. Section 4 gives an application to pure-state and mixed-state situations in the context of vibrationally nonadiabatic dynamics, and Sect. 5 concludes.

2 Quantum Hydrodynamics for Mixed States

Even though largely disregarded from the viewpoint of de Broglie–Bohm theory [1, 3, 4, 6], which has traditionally been considered as a theory for pure quantum states,[1] quantum hydrodynamics for mixed states has a long history, going back to Moyal [27], Zwanzig [28], Takabayasi [26], Fröhlich [29], and Yvon [30], along with a number of more recent works [31–35]. Starting from either the coordinate representation or else the associated phase space (Wigner) representation of the density operator, the hydrodynamic formulation can be derived in terms of *hydrodynamic moments* – i.e., q-dependent momentum moments of the Wigner function, see (1). Especially noteworthy is the work by Takabayasi [26] – published in 1954, shortly after Bohm's seminal papers – which focuses on pure states while using the hydrodynamic moment language. Surprisingly, the pure/mixed-state connection has remained largely ignored, both in the Bohmian mechanics literature and in the mixed-state hydrodynamics literature, apart from few exceptions, notably [37].

In the following, a brief summary is given of the relevant relations, following [35, 36, 57]. These relations provide the basis for the mixed quantum–classical theory described in Sect. 3.

2.1 Hydrodynamic Moments: Projection of the Wigner Density upon Coordinate Space

The hydrodynamic moments in question can be derived from the density operator in the coordinate representation, $\rho(x, x')$, or else from its phase-space analog, the Wigner function $\rho_W(q, p)$ [58–60],

$$\rho_W(q,p) = \frac{1}{2\pi\hbar}\int_{-\infty}^{\infty} \mathrm{d}r\ \rho\left(q-\frac{r}{2}, q+\frac{r}{2}\right)\exp(ipr/\hbar) \tag{4}$$

with the sum and difference coordinates $q = 1/2(x + x')$ and $r = x - x'$. In the following, we address a single degree of freedom for simplicity, but all relations can be readily generalized to an arbitrary number of degrees of freedom.

In the hydrodynamic description, the quantum density is characterized by a set of moment functions obtained from $\rho_W(q, p)$ by integration over momentum only[2] [27, 29, 31–35], see also (1),

$$\langle \mathcal{P}^n \rho\rangle_q = \int_{-\infty}^{\infty} \mathrm{d}p\, p^n\, \rho_W(q,p). \tag{5}$$

The hydrodynamic moments correspond to the coefficients of the Taylor expansion of the coordinate space density with respect to the coordinate r

---

[1] See, however, the recent discussion on the role of density matrices in Bohmian mechanics in [56].

[2] Equivalently, the moments can be obtained by taking repeated derivatives with respect to the difference coordinate r of the coordinate space distribution [31, 35].

(noting that r and p are conjugate Fourier variables, see (4)),

$$\rho(q,r) = \sum_n \frac{1}{n!} \langle \mathcal{P}^n \rho \rangle_q \left(\frac{\mathrm{i}r}{\hbar} \right)^n , \tag{6}$$

where the coordinate space density was rewritten as a function of the sum and difference coordinates.

Depending on the structure of the quantum density, the moments $\langle \mathcal{P}^n \rho \rangle_q$ can carry redundant information. In particular, for a pure-state density, $\rho(x,x') = \psi(x)\psi^*(x')$, and its associated Wigner transform, the state is characterized by its first two moments (see Sect. 2.3). Similarly, a Gaussian mixed-state density is characterized by the first three moments [35,36,61]. In general, an infinite number of moments are necessary to describe the quantum state – or, to reconstruct the state according to (6).

2.2 Dynamical Equations: Moment Hierarchies

The dynamical equations for the hydrodynamic moments are derived from the Liouville–von Neumann (LvN) equation for the density operator:

$$\frac{\partial \hat{\rho}}{\partial t} = -\frac{\mathrm{i}}{\hbar} \hat{\hat{L}} \hat{\rho}, \tag{7}$$

with $\hat{\hat{L}}$ the Liouvillian superoperator and $\hat{\hat{L}}\hat{\rho} = [\hat{H}, \hat{\rho}]$ for a Hamiltonian system. In the Wigner phase-space representation, the LvN equation takes the following form ("quantum Liouville equation") [58–60]:

$$\begin{aligned} \frac{\partial \rho_{\mathrm{W}}}{\partial t} &= (\, H, \rho_{\mathrm{W}} \,)_{qp} \\ &= -\frac{\mathrm{i}}{\hbar} \Big(H \exp(\hbar \Lambda_{qp}/2\mathrm{i}) \rho_{\mathrm{W}} - \rho_{\mathrm{W}} \exp(\hbar \Lambda_{qp}/2\mathrm{i}) H \Big) \end{aligned} \tag{8}$$

where $(\, , \,)_{qp}$ denotes the Moyal bracket, and the Poisson bracket operator Λ_{qp} is given as

$$\Lambda_{qp} = \frac{\overleftarrow{\partial}}{\partial p} \cdot \frac{\overrightarrow{\partial}}{\partial q} - \frac{\overleftarrow{\partial}}{\partial q} \cdot \frac{\overrightarrow{\partial}}{\partial p} . \tag{9}$$

The linear approximation $\exp(\hbar\Lambda_{qp}/2\mathrm{i}) \simeq 1 + \hbar\Lambda_{qp}/2\mathrm{i}$ formally yields the classical Liouville equation, $\partial\rho_{\mathrm{W}}/\partial t = \{H, \rho_{\mathrm{W}}\}_{qp} = -1/2(H\Lambda_{qp}\rho_{\mathrm{W}} - \rho_{\mathrm{W}}\Lambda_{qp}H)$. In the following, we shall refer to the limiting procedure

$$(\, H, \rho_{\mathrm{W}} \,)_{qp} \longrightarrow \{\, H, \rho_{\mathrm{cl}} \,\}_{qp} \tag{10}$$

as the classical Liouville limit [58–60, 62].[3]

[3] The convergence properties associated with the limit equation (10) are in fact nontrivial, see, e.g., the discussion in [63, 64].

Equation (8) can be rewritten as a series expansion in $\hbar$ "correction terms," i.e., the so-called Wigner–Weyl series,

$$\frac{\partial \rho_W}{\partial t} = \{ H, \rho_W \}_{qp} + \sum_{\substack{k=3 \\ \text{odd}}}^{n} \frac{1}{k!} \left(\frac{\hbar}{2i} \right)^{k-1} \frac{\partial^k V}{\partial q^k} \frac{\partial^k \rho_W}{\partial p^k} . \tag{11}$$

Note that the quantum correction terms involve the third and higher order derivatives of the potential. For a harmonic potential, the quantum vs. classical phase space dynamical equations are thus identical.

From (11), and using prescription (5), one obtains the moment equations in a form which involves a classical part to which quantum correction terms are added,

$$\frac{\partial \langle \mathcal{P}^n \rho \rangle_q}{\partial t} = \langle \mathcal{P}^n \{ H, \rho_W \}_{qp} \rangle_q + \mathcal{C}_q \tag{12}$$

with the classical part

$$\langle \mathcal{P}^n \{ H, \rho_W \}_{qp} \rangle_q = -\frac{1}{m} \frac{\partial}{\partial q} \langle \mathcal{P}^{n+1} \rho \rangle_q - n \left(\frac{\partial V}{\partial q} \right) \langle \mathcal{P}^{n-1} \rho \rangle_q \tag{13}$$

and the quantum correction part

$$\mathcal{C}_q = - \sum_{\substack{k=3 \\ \text{odd}}}^{n} \binom{n}{k} \left(\frac{\hbar}{2i} \right)^{k-1} \frac{\partial^k V}{\partial q^k} \langle \mathcal{P}^{n-k} \rho \rangle_q . \tag{14}$$

Since the quantum correction terms only appear with the third order onwards, the first three moment equations are formally of classical appearance.[4] The first equation is the continuity equation, which reflects the conservation law for the local density $\langle \rho \rangle_q$. The second and third equations are the dynamical equations for the momentum density $\langle \mathcal{P} \rho \rangle_q$ and kinetic energy density $\langle \mathcal{P}^2 \rho \rangle_q / 2m$, respectively.

As detailed in [35, 36, 65], the formulation of (12)–(14) for the moment hierarchy can be readily generalized so as to include dissipation [31, 35, 36] and coupled electronic states [65].

In (13), the kinetic-energy contribution (first term on the rhs of the equation) generates a dynamical coupling between the nth hydrodynamic moment $\langle \mathcal{P}^n \rho \rangle_q$ and the $(n+1)$th moment $\langle \mathcal{P}^{n+1} \rho \rangle_q$. The coupling of successive orders implies that the hierarchy does not terminate, unless a given moment can be expressed in terms of the lower-order moments. For example, for Gaussian mixed-state densities, the hierarchy terminates with the equations for $\langle \mathcal{P}^n \rho \rangle_q$, $n = 0, 1, 2$, because the third moment can be represented in terms of the lower-order moments. Another special case are pure-state densities, for which the

[4] The classical or quantum nature of the lowest-order moment equations thus depends entirely on the (classical or quantum) nature of the density.

hierarchy closes with the first two equations. This case will be considered explicitly in Sect. 2.3.

The closure, or approximate closure, of the hydrodynamic hierarchy is a central problem in applying the moment method. Quoting from the context of plasma physics, where a similar problem occurs in connection with the Vlassov equation, the moment equations "*must be terminated somewhere by a flash of insight*" [66]. In classical hydrodynamics, a Gaussian ("Maxwellian") closure is invoked, in conjunction with an equilibrium assumption; thus one obtains closure at the level of the second moment [67]. For nonequilibrium states, systematic closure criteria at higher orders of the hierarchy have been developed [68], in particular involving maximum entropy estimates [69]. These approaches have recently been transposed to the quantum domain [70].

2.3 Pure States and the Connection to Bohmian Mechanics

For pure states, $\rho(x,x') = \psi(x)\psi^*(x')$, the hydrodynamic hierarchy closes with the first *two* equations since the second moment can be expressed in terms of the zeroth and first moments [35, 71]

$$\langle \mathcal{P}^2 \rho \rangle_q \Big|_{\text{pure}} = \bar{p}_q^2 \langle \rho \rangle_q - \frac{\hbar^2}{4} \langle \rho \rangle_q \frac{\partial^2}{\partial q^2} \ln \langle \rho \rangle_q , \tag{15}$$

where the hydrodynamic momentum field $\bar{p}_q = \bar{p}(q)$ was introduced, via the first moment $\langle \mathcal{P}\rho \rangle_q = \bar{p}_q \langle \rho \rangle_q$.

The pure-state closure relation (15) can be shown to lead to the conventional equations of pure-state hydrodynamics (Bohmian mechanics) [26, 35, 37]. That the coupled equations for the first two moments, i.e., the local density $\langle \rho \rangle_q$ and the hydrodynamic momentum field $\langle \mathcal{P}\rho \rangle_q$, form a closed subset in this case is not unexpected, since the first two moments entirely determine the wavefunction (apart from a piecewise constant phase factor [72]). Using the polar form of the wavefunction, $\psi(x) = R(x)\exp[\mathrm{i}S(x)/\hbar]$, the local density and momentum field are given as

$$\begin{aligned} \langle \rho \rangle_q &= R^2(q) , \\ \langle \mathcal{P}\rho \rangle_q &= \bar{p}_q \langle \rho \rangle_q = \frac{\partial S}{\partial q} \langle \rho \rangle_q \end{aligned} \tag{16}$$

and contain the amplitude and phase information, respectively.

The dynamical equations for the first two moments read as follows:

$$\begin{aligned} \frac{\partial \langle \rho \rangle_q}{\partial t} &= -\frac{1}{m} \frac{\partial}{\partial q} \langle \mathcal{P}\rho \rangle_q \\ \frac{\partial \langle \mathcal{P}\rho \rangle_q}{\partial t} &= -\frac{1}{m} \frac{\partial}{\partial q} \langle \mathcal{P}^2 \rho \rangle_q \Big|_{\text{pure}} - \frac{\partial V}{\partial q} \langle \rho \rangle_q . \end{aligned} \tag{17}$$

With (15) for the pure-state second moment, these two equations separate from the rest of the moment hierarchy.

Equation (17) represents the so-called Eulerian picture of hydrodynamics. In the associated Lagrangian ("moving with the flow") picture, one defines fluid particle trajectories $\dot{q} = v(q)$, with the velocity field $v(q) = \bar{p}(q)/m$. Using the total, or "material" time derivative, $\mathrm{d}/\mathrm{d}t = \partial/\partial t + v(q)\partial/\partial q$, the Lagrangian equations of Bohmian mechanics are obtained as follows [6,7],

$$
\begin{aligned}
\frac{\mathrm{d}\langle\rho\rangle_q}{\mathrm{d}t} &= -\frac{\langle\rho\rangle_q}{m}\frac{\partial\bar{p}}{\partial q} \\
\frac{\mathrm{d}}{\mathrm{d}t}\bar{p}(q,t) &= -\frac{\partial V}{\partial q} + F_{\mathrm{hyd}}\Big|_{\mathrm{pure}}
\end{aligned} \tag{18}
$$

where the hydrodynamic force F_{hyd} reads [26,35–37]

$$
F_{\mathrm{hyd}}\Big|_{\mathrm{pure}} = -\frac{1}{m}\langle\rho\rangle_q^{-1}\frac{\partial\sigma_q}{\partial q}\Big|_{\mathrm{pure}} \tag{19}
$$

i.e., F_{hyd} is the derivative of the q-dependent momentum variance

$$
\sigma_q\Big|_{\mathrm{pure}} = \langle\mathcal{P}^2\rho\rangle_q\Big|_{\mathrm{pure}} - \bar{p}_q^2\langle\rho\rangle_q \tag{20}
$$

For pure states, σ_q is obtained explicitly as follows, from (15) [35,37,71]:

$$
\sigma_q\Big|_{\mathrm{pure}} = -\frac{\hbar^2}{4}\langle\rho\rangle_q\frac{\partial^2}{\partial q^2}\ln\langle\rho\rangle_q \tag{21}
$$

As shown in [35], (19) is equivalent to the "quantum force" of the conventional Bohmian formulation

$$
F_{\mathrm{hyd}}\Big|_{\mathrm{pure}} = -\frac{\partial Q}{\partial q} \tag{22}
$$

i.e., the gradient of the Bohmian quantum potential [3,4,6,7]

$$
Q = -\frac{\hbar^2}{2m}\frac{1}{\langle\rho\rangle_q^{1/2}}\frac{\partial^2\langle\rho\rangle_q^{1/2}}{\partial q^2} \tag{23}
$$

The derivation of the pure-state hydrodynamic equations from the Liouville space perspective of densities and phase-space distributions sheds some light on the physical meaning of the fluid-dynamical quantities involved. In particular, the "Bohmian" momentum is found to correspond to the *average* momentum at a given value of the position variable, $p_{\mathrm{Bohm}}(q) = \bar{p}(q)$. In the de Broglie–Bohm formulation, this momentum is associated with an actual *particle* momentum [5,6].

2.4 Hydrodynamic Force for Mixed States and Classical Distributions

Equations (18)–(20) can be formally transposed to arbitrary mixed states. However, for nonpure states, the first two equations of the moment hierarchy, equation (18), do *not* form a closed set, since they are coupled to the higher orders of the hierarchy via the hydrodynamic force. Yet, the picture of fluid particle motion under the hydrodynamic force remains valid,

$$\frac{\mathrm{d}\langle\rho\rangle_q}{\mathrm{d}t} = -\frac{\langle\rho\rangle_q}{m}\frac{\partial\bar{p}}{\partial q}$$
$$\frac{\mathrm{d}}{\mathrm{d}t}\bar{p}(q,t) = -\frac{\partial V}{\partial q} + F_{\mathrm{hyd}}\Big|_{\mathrm{mixed}} \tag{24}$$

with the general, mixed-state hydrodynamic force

$$F_{\mathrm{hyd}}\Big|_{\mathrm{mixed}} = -\frac{1}{m}\langle\rho\rangle_q^{-1}\frac{\partial\sigma_q}{\partial q}\Big|_{\mathrm{mixed}} \tag{25}$$

obtained again as the spatial derivative of the variance

$$\sigma_q\Big|_{\mathrm{mixed}} = \langle\mathcal{P}^2\rho\rangle_q - \bar{p}_q^2\langle\rho\rangle_q, \tag{26}$$

where σ_q is now no longer constrained to the pure-state form equation (21).

The expression (25) is closely related to the "pressure force" of classical hydrodynamics [67, 73]. Indeed, the occurrence of this additional force is an intrinsic property of the hydrodynamic representation. It can be understood to *compensate for the fact that we consider the dynamical evolution of the average momentum* $\bar{p}_q$.

Importantly, the *same* relations hold for a purely classical phase-space distribution function, with the classical hydrodynamic force

$$F_{\mathrm{hyd}}^{\mathrm{cl}} = -\frac{1}{m}\langle\rho_{\mathrm{cl}}\rangle_q^{-1}\frac{\partial\sigma_q^{\mathrm{cl}}}{\partial q} \tag{27}$$

derived from the variance of the classical distribution $\rho_{\mathrm{cl}}(q,p)$, i.e., $\sigma_q^{\mathrm{cl}} = \langle\mathcal{P}^2\rho_{\mathrm{cl}}\rangle_q - \bar{p}_q^2\langle\rho_{\mathrm{cl}}\rangle_q$.

This implies that the hydrodynamic force is generally nonvanishing, *independently of whether the system is quantum or classical.* A pertinent example is the evolution of a quantum vs. classical harmonic oscillator: While the Liouville space evolution is governed by the same (classical) Liouville equation for both the quantum and classical oscillator (noting that the quantum correction terms $\mathcal{C}_q$ of (11) disappear for quadratic potentials), the hydrodynamic evolution involves an extra force in both cases, see also Sect. 2.6.[5]

[5] For illustration, consider the second hydrodynamic moment for the thermal equilibrium state of the harmonic oscillator, which is essentially of quantum origin at

2.5 Classical Limit Considerations

Following the arguments of the preceding Section, the classical limit of the hydrodynamic representation naturally corresponds to a transition from a *quantum–statistical* (mixed-state) description to a *classical–statistical* description. One thus obtains the purely classical force term (27) as the classical limit of the quantum–statistical hydrodynamic force (25),

$$F_{\mathrm{hyd}}\Big|_{\mathrm{mixed}} \longrightarrow F_{\mathrm{hyd}}^{\mathrm{cl}} \qquad \text{classical−statistical limit,} \tag{28}$$

where $F_{\mathrm{hyd}}^{\mathrm{cl}}$ derives from a classical distribution $\rho_{\mathrm{cl}}(q,p)$ which represents the classical phase-space limit of the reference mixed-state quantum distribution.

The classical limit equation (28) can be considered to follow from the classical Liouville limit equation (10), in conjunction with the approximation $\rho_{\mathrm{W}}(q,p) \longrightarrow \rho_{\mathrm{cl}}(q,p)$ for the phase-space density. This limit is valid for small mass ratios for the quantum vs. classical systems, $m/M \ll 1$ [45, 46], and translates to the usual semiclassical limit defined, e.g., in terms of the action $\hbar \ll A$. This limit also holds at high temperatures, where the width of the phase-space distribution is predominantly of thermal origin, and quantum interference effects are washed out.

While the presence of the classical hydrodynamic force (27) and (28) follows directly from the hydrodynamic moment construction, it conflicts, though, with the usual definition of the classical limit in Bohmian mechanics [6]. In the Bohmian, pure-state picture, the hydrodynamic, or "quantum" force vanishes in the classical limit

$$F_{\mathrm{hyd}}\Big|_{\mathrm{pure}} \longrightarrow 0 \qquad \text{"Bohmian" classical limit.} \tag{29}$$

This limiting process is formally a $\hbar \to 0$ limit (while disregarding the $\hbar$ dependence of the wavefunction component $R = \langle\rho\rangle_q^{1/2}$, see (22) and (23)).

---

low temperatures,

$$\langle \mathcal{P}^2 \rho\rangle_q\Big|_{kT \ll \hbar\omega} = \frac{1}{2} m\hbar\omega\, \langle\rho\rangle_q$$

and essentially of thermal origin at high temperatures

$$\langle \mathcal{P}^2 \rho\rangle_q\Big|_{kT \gg \hbar\omega} = mkT\, \langle\rho\rangle_q .$$

The quantum–statistical moment description naturally comprises these limits, and the intermediate case. The hydrodynamic force resulting from *both* of the above expressions is temperature-independent, $F_{\mathrm{hyd}} = m\omega^2 q$ – and exactly compensates for the effect of the external force on the oscillator. Hydrodynamics thus predicts stationary trajectory solutions for both the quantum ground state at zero Kelvin and the high-temperature thermal equilibrium state.

However, from the discussion of the previous sections, the limit equation (29) can only be strictly valid for states whose momentum variance σ_q – or spatial variation $(\mathrm{d}\sigma_q/\mathrm{d}q)$ – vanishes. This is the case for momentum eigenstates, i.e., in the "plane wave limit" [6, 25, 36]. Apart from this, (29) can be shown to be approximately valid for certain physical situations where the hydrodynamic force can be neglected as compared with the external forces, $F_{\mathrm{hyd}}/F_{\mathrm{ext}} \ll 1$. For example, this would apply to a coherent state peaked around high quantum numbers, and undergoing large-amplitude oscillations [6].

Overall, caution needs to be exercised when applying (29), see the detailed discussion in [6]. Difficulties of interpretation arise from the fact that the condition (29) is state-dependent and often cannot be directly related to the physical properties of the system. For the applications we are focusing on, in a quantum molecular dynamics context, the classical–statistical limit (28) is generally the appropriate limiting procedure.

We conclude this section with a remark on initial conditions. While the limiting procedure (28) implies the limit $\rho_{\mathrm{W}}(q,p) \longrightarrow \rho_{\mathrm{cl}}(q,p)$ for the distribution function, one often constructs the initial condition in a less consistent fashion. In particular, if one aims to *simulate quantum dynamics by classical dynamics*, the initial condition is chosen to be the Wigner distribution itself, rather than its classical approximant; this approach is generally denoted the "Wigner method" [74, 75]. While this method introduces inconsistencies, as a consequence of evolving a quantum initial condition under the classical Liouville equation, it is frequently used to approximate the true quantum evolution.

Both of the above schemes for constructing initial conditions are relevant for the present discussion: (a) the construction of the classical approximant distribution, $\rho_{\mathrm{W}}(q,p) \longrightarrow \rho_{\mathrm{cl}}(q,p)$, which guarantees a consistent classical limit; (b) the Wigner method, which preserves the quantum initial condition and remains quantum-mechanical in nature (simulation of quantum mechanics using classical mechanics). The examples discussed in Sect. 4 are in fact in line with the Wigner method, even though future applications aim at a classical-limit perspective in many-body systems, in accordance with scheme (a).

2.6 Hydrodynamic Phase Space

The Lagrangian, fluid-dynamical trajectories which are defined by the relation $\dot{q} = \bar{p}(q)/m$ in conjunction with (18) or (24), are fundamentally different from the Liouville phase space trajectories which represent a time-evolving classical distribution $\rho_{\mathrm{cl}}(q,p)$. This is so because the hydrodynamic momentum $\bar{p}(q) = \langle \mathcal{P}\rho\rangle_q/\langle\rho\rangle_q$ originates in an average with respect to the Liouville phase-space momentum p. Hence, the fluid–particle momentum is a *q-dependent average over Liouville-space momenta.*

To underscore the particular character of the hydrodynamic momentum, one can introduce the notion of a hydrodynamic phase space [26, 35, 36, 76], whose momentum variable corresponds to the hydrodynamic momentum field

$p = \bar{p}_q$. The distribution functions in this alternative phase space are of the form [26, 35, 36, 76]

$$\rho_{\text{hyd}}(q, p) = \langle \rho \rangle_q \, \delta \left(p - \bar{p}_q \right) , \tag{30}$$

i.e., they are single-valued in the momentum, with $p = \bar{p}_q$ a function of q. This is illustrated in Fig. 1, for a Gaussian distribution.

Information on the momentum-space width of the underlying Wigner phase-space distribution (along with all other higher-order moments) is thus not directly available in this alternative phase-space picture. However, the higher-order moments indirectly determine the time evolution, via the force F_{hyd}. An equation of motion can be formulated for the distribution $\rho_{\text{hyd}}(q, p)$, which is analogous to the classical Liouville equation but contains the additional, hydrodynamic force term [6, 36]. The associated trajectory equations are given as follows, in accordance with the Lagrangian equations (18) or (24) and using $p = \bar{p}_q$ [6, 36]:

$$\begin{aligned} \dot{q} &= \frac{p}{m}, \\ \dot{p} &= -\frac{\partial}{\partial q} V(q) + F_{\text{hyd}}(q). \end{aligned} \tag{31}$$

For illustration, two paradigm cases will be considered in the following: First, the evolution of a Gaussian free-particle distribution, and second, the evolution of a Gaussian wavepacket in a harmonic well.

For the free-particle distribution, the initial condition corresponds to a Gaussian wavepacket at rest, with parameters taken from [77]. As shown in Fig. 2 (left panel), the distribution tilts and elongates, thus giving rise to the observed spreading in coordinate space. While the average momentum is initially uniform, $\bar{p}_q = 0$, it acquires a pronounced dependence on q as the distribution spreads out. This is reflected in the $p = \bar{p}_q$ evolution of the hydrodynamic phase space distribution, see Fig. 2 (right panel).

Figure 3 illustrates the hydrodynamic force as a function of time, for the two trajectories $A(t)$ and $B(t)$ whose end points are shown in Fig. 2. F_{hyd} is

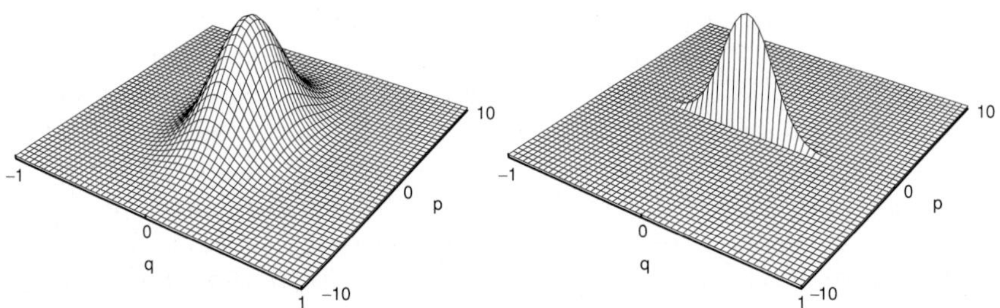

Fig. 1. Gaussian Wigner phase space distribution (left panel) and associated hydrodynamic phase space distribution (right panel), see (30)

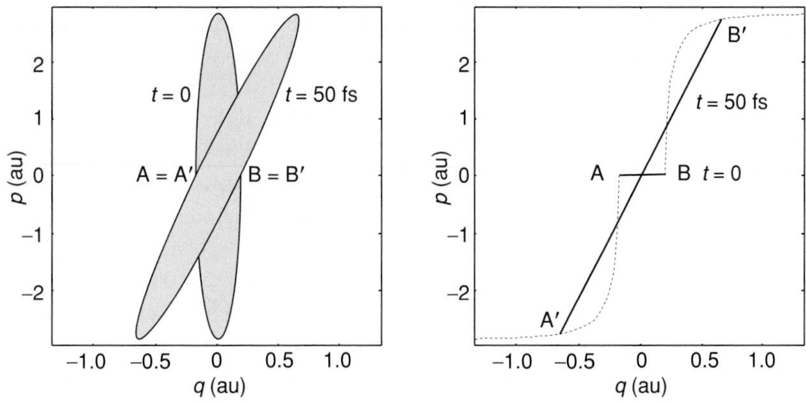

Fig. 2. Time-evolving Gaussian distribution for a free particle: Liouville phase space (left panel) vs. hydrodynamic phase space (right panel) representations. The two-dimensional Gaussian distribution is represented by a single contour line for the Liouville-space distribution, and by a line in the case of the associated hydrodynamic phase space distribution (see also Fig. 1). The distributions are shown for $t = 0$ and $t = 50$ fs. Two trajectories are indicated, starting at the points $(A,\ B)$, and evolving to $(A',\ B')$. At time $t = 0$, these points correspond to the average momentum at the given q values, and are therefore part of both the Wigner phase space distribution and of the hydrodynamic distribution. In the Wigner picture, these points remain stationary $(A = A', B = B')$, while they are nonstationary in the hydrodynamic picture. The dotted line (right panel) indicates the hydrodynamic phase space trajectory leading from $(A,\ B)$, at $t = 0$, to $(A',\ B')$ at $t = 50$ fs

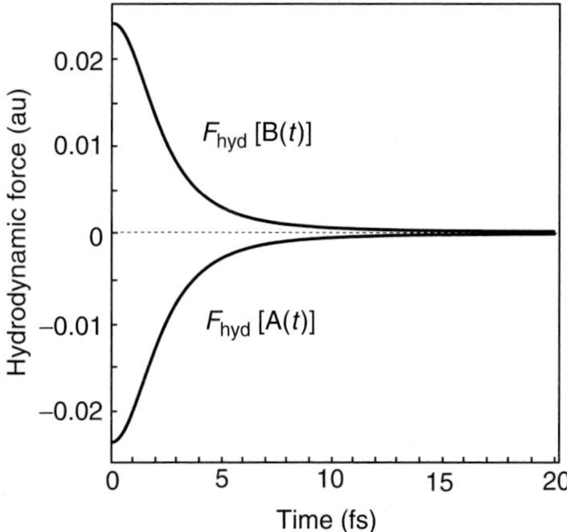

Fig. 3. Time-dependent hydrodynamic force for the free-particle evolution depicted in Fig. 2. F_{hyd} is shown for the trajectories $A(t)$ and $B(t)$, which include the points (A, A') and (B, B') indicated in the preceding figure. The hydrodynamic force tends to vanish as $t \to \infty$, as discussed in the text

seen to vanish with time; in fact, the hydrodynamic force is proportional to $q(0)[1+(t/\tau)^2]^{-3/2}$ [6] where τ is a constant that depends on the initial wave packet. This is due to the fact that as the Wigner distribution elongates, both the momentum width σ_q and its spatial variation decrease and eventually tend to zero.

In a harmonic potential (see Figs. 4 and 5), the hydrodynamic trajectories experience an oscillatory hydrodynamic force, reflecting the changing width of the wavepacket. (Only for a coherent state is the hydrodynamic force time independent [6,36].) The effect of the hydrodynamic force is to counter balance, to a certain extent, the distance dependence of the classical force, and thereby prevent the space–time trajectories from crossing. Figure 4 shows schematically a Gaussian Wigner function and the corresponding hydrodynamic distribution at various instants during the oscillation period in a harmonic well.

In both examples, the Wigner phase space distribution evolves under the classical Liouville equation, since quantum correction terms are absent for both the free particle and the harmonic oscillator. Moreover, the time-dependent Gaussian distributions in question are positive definite and can be taken to be either quantum or classical. From this follows that the hydrodynamic force could be either a "quantum force" (in the Bohmian sense), or else a classical–statistical hydrodynamic force.

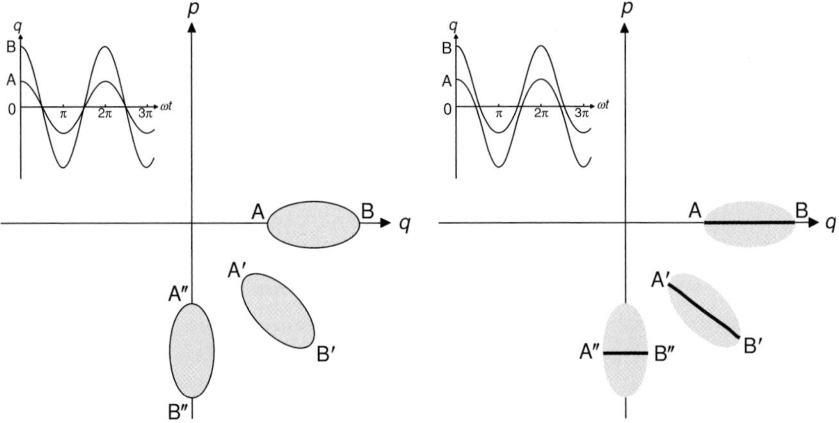

Fig. 4. Time-evolving Gaussian wavepacket for a harmonic potential: Liouville phase space (left panel) vs. hydrodynamic phase space (right panel) representations. For the initial distribution, the average momentum is q-independent, $\bar{p}_q = 0$, and the points (A, B) again coincide for the Liouvillian vs. hydrodynamic distributions. At later times, the trajectories (A, A', A'') and (B, B', B'') differ for the two phase space representations, as in the free particle case of Fig. 2. The inset shows the associated Liouvillian vs. hydrodynamic space–time trajectories

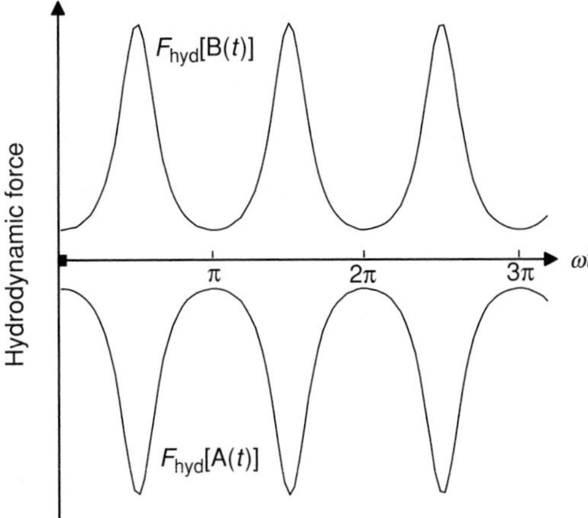

Fig. 5. Time-dependent hydrodynamic force for the harmonic-oscillator evolution, which determines the trajectories $A(t)$ and $B(t)$ that include the points (A, A', A'') and (B, B', B'') shown in the preceding figure. The hydrodynamic force is a periodic function of time

3 Quantum–classical dynamics via partial moments

Against the background of the earlier sections, we develop here a mixed quantum–classical formulation based upon the hydrodynamic picture. The dynamical scheme to be constructed should couple quantum hydrodynamic trajectories with classical (Hamiltonian) trajectories. This scheme should allow one to simulate potentially large systems containing many degrees of freedom, which can be subdivided into a quantum sector (with predominant quantum effects) and a classical sector (which is not strongly affected by quantum effects).

In view of the discussion of the earlier section, the following difficulty arises regarding the representation of the classical sector: While a classical limit can be defined within the hydrodynamic picture (see Sect. 2.5), this limit is associated with classical fluid-dynamical, *non-Hamiltonian* trajectories, evolving under the classical hydrodynamic force (27). Only in special cases, compatible with the Bohmian classical limit (29), does the additional force disappear. While a mixed quantum–classical formulation is thus feasible within the hydrodynamic representation – by taking into account both quantum and classical hydrodynamic forces – the disadvantage of this approach is that the simplicity of the classical Liouville-space dynamics, in terms of Hamiltonian trajectories, is lost. Even though the hydrodynamic representation is appropriate in order to define a trajectory representation in the quantum sector, it is thus desirable to retain a Liouville-space representation in the classical sector.

In the present development, we therefore consider a *hybrid* representation which is hydrodynamic with respect to the quantum subspace, and classical-Liouvillian with respect to the classical subspace. As a result, a *mixed quantum–classical molecular dynamics scheme* is obtained which couples quantum trajectories with classical Liouvillian trajectories. In order to construct this hybrid dynamics, it is necessary to introduce the partial moment quantities of equation (2).

3.1 Partial Moments: A Hybrid Hydrodynamic-Liouvillian Representation

Partial moments are constructed by introducing a hydrodynamic projection for *selected* degrees of freedom.[6] If one starts from the Wigner representation for two degrees of freedom, one may thus choose to integrate only over the phase space momentum variable p [38, 39],

$$\langle \mathcal{P}^n \rho \rangle_{qQP} = \int dp \; p^n \; \rho_{\mathrm{W}}(q, p; Q, P) \tag{32}$$

Given that the hybrid quantities $\langle \mathcal{P}^n \rho \rangle_{qQP}$ can be understood as hydrodynamic moments in q which are parameterized in the phase space variables (Q, P), many of the conclusions of the previous sections carry over to the mixed hydrodynamic-Liouvillian picture. In particular, for a pure-state density and its associated Wigner transform, the state is characterized by the first two partial moments; a brief discussion of the pure-state case is given in Appendix A. Similarly, a Gaussian mixed-state density is determined by the first three partial moments. In general, an infinite number of moments is required to characterize the system, and truncation schemes have to be designed by analogy with the considerations of Sect. 2.2.

Finally, one can introduce a hybrid hydrodynamic-Liouvillian phase space with distribution functions $\rho_{\mathrm{hybrid}}(q, p; Q, P) = \langle \rho \rangle_{qQP} \, \delta(p - \bar{p}_{qQP})$ which preserve the features of the underlying Liouville space distribution in the classical subspace but are single-valued in p in the quantum subspace [38, 39]. These distribution functions combine the hydrodynamic phase space picture (for the quantum subspace) and the Liouvillian phase space picture (for the classical subspace), see Figs. 1–5.

---

[6] We have previously applied this idea to the description of nonadiabatic processes where a hydrodynamic representation is chosen for the nuclear degrees of freedom while the electronic degrees of freedom remain in a Liouville-space setting (in a discretized representation of diabatic or adiabatic states [57, 65]). Note that the reverse strategy also appears promising, i.e., choosing a moment representation for the electronic degrees of freedom while treating the nuclear degrees of freedom by a classical phase space dynamics [78].

3.2 Exact Equations of Motion

Equations of motion for the partial moments can be derived from the LvN equation either in the coordinate space representation or else in the phase space Wigner representation. In the following, we will consider a Hamiltonian of the form $H = p^2/2m + P^2/2M + V(q,Q)$, with $V(q,Q) = V_q(q) + V_{\rm int}(q,Q) + V_Q(Q)$.

As shown in [39], exact equations of motion can be obtained in the following form:

$$\frac{\partial \langle \mathcal{P}^n \rho \rangle_{qQP}}{\partial t} = \langle \mathcal{P}^n \{ H, \rho_{\rm W} \}_{qp} \rangle_{qQP} + \{ H, \langle \mathcal{P}^n \rho \rangle_{qQP} \}_{QP} + \mathcal{C}_{qQP}. \tag{33}$$

As one would intuitively expect from the partial moment construction, (33) comprises (a) a classical hydrodynamic part in the (q,p) subspace

$$\langle \mathcal{P}^n \{ H, \rho_{\rm W} \}_{qp} \rangle_{qQP} = -\frac{1}{m}\frac{\partial}{\partial q} \langle \mathcal{P}^{n+1} \rho \rangle_{qQP} - n \frac{\partial [V_q(q) + V_{\rm int}(q,Q)]}{\partial q} \langle \mathcal{P}^{n-1} \rho \rangle_{qQP} \tag{34}$$

(b) a classical Liouvillian part in the (Q,P) subspace

$$\{ H, \langle \mathcal{P}^n \rho \rangle_{qQP} \}_{QP} = -\frac{P}{M} \frac{\partial \langle \mathcal{P}^n \rho \rangle_{qQP}}{\partial Q} + \frac{\partial [V_Q(Q) + V_{\rm int}(q,Q)]}{\partial Q} \frac{\partial \langle \mathcal{P}^n \rho \rangle_{qQP}}{\partial P} \tag{35}$$

and (c) a mixed hydrodynamic-Liouvillian quantum correction part,

$$\mathcal{C}_{qQP} = \sum_{l_1 + l_2 \geq 3} (-1)^{l_2 + 1} \frac{1}{l_2!} \binom{n}{l_1} \left(\frac{\hbar}{2i} \right)^{l_1 + l_2 - 1} \times \left(\frac{\partial^{l_1 + l_2} [V_q(q) + V_{\rm int}(q,Q) + V_Q(Q)]}{\partial q^{l_1} \partial Q^{l_2}} \right) \frac{\partial^{l_2}}{\partial P^{l_2}} \langle \mathcal{P}^{n - l_1} \rho \rangle_{qQP}, \tag{36}$$

where the summation runs over odd values of the sum of indices $l_1 + l_2$ and $l_1 \leq n$. The "quantum correction" part collects all terms that carry an explicit $\hbar$ dependence and involve third and higher order derivatives of the potential. Hence, this part is nonzero for moments of *all* orders, except for systems described by potentials that are at most second order polynomials. The fact that the equations of motion for the zeroth and first order moments carry explicit $\hbar$ contributions, which are absent in a pure hydrodynamic description, highlights the mixed hydrodynamic-Liouvillian nature of the partial moments. Indeed, for the 0th moment $\langle \rho \rangle_{qQP}$ ($l_1 = n = 0$), the explicit $\hbar$ terms originate entirely in the (Q,P) subspace, while the equation of motion for the first moment $\langle \mathcal{P} \rho \rangle_{qQP}$ ($l_1 = 0, 1$) contains correction terms involving mixed q/Q derivatives.

3.3 Quantum–classical Approximation and Connection to the Quantum–Classical Liouville Equation

In view of defining a mixed quantum–classical dynamics, a classical approximation is introduced in the (Q, P) subspace. To this end, only those quantum correction terms of (36) are retained which involve derivatives of order $l_2 = 0, 1$, i.e., we neglect in the equations of motion for the partial moments all terms involving multiple order derivatives with respect to the coordinate Q. This is the same approximation as the one made when obtaining the classical Poisson bracket from the Moyal bracket, see (10) and Sect. 2.5. The quantum–classical equations thus read as follows [39]:

$$\begin{aligned}\frac{\partial \langle \mathcal{P}^n \rho \rangle^c_{qQP}}{\partial t} &= \langle \mathcal{P}^n \left\{ H_q + V_{\text{int}}, \rho_{\text{W}} \right\}_{qp} \rangle^c_{qQP} \\ &\quad + \{ H_Q + V_{\text{int}}, \langle \mathcal{P}^n \, \rho \rangle^c \}_{QP} + \mathcal{C}^c_{qQP}\end{aligned} \tag{37}$$

with the *approximate* quantum correction part

$$\begin{aligned}\mathcal{C}^c_{qQP} = &- \sum_{\substack{l_1=3 \\ \text{odd}}}^{n} \binom{n}{l_1} \left(\frac{\hbar}{2i} \right)^{l_1 - 1} \frac{\partial^{l_1} V}{\partial q^{l_1}} \langle \mathcal{P}^{n-l_1} \, \rho \rangle^c_{qQP} \\ &+ \sum_{\substack{l_1=2 \\ \text{even}}}^{n} \binom{n}{l_1} \left(\frac{\hbar}{2i} \right)^{l_1} \frac{\partial^{l_1+1} V}{\partial q^{l_1} Q} \frac{\partial}{\partial P} \langle \mathcal{P}^{n-l_1} \, \rho \rangle^c_{qQP} .\end{aligned} \tag{38}$$

The index c indicates the approximate nature of the quantities evolving under the above equation of motion.

In contrast to (33)–(36), the approximation of (37) entails that the equations of motion for the *first two partial moments do not carry any quantum correction terms.* This will turn out to have important implications for the Lagrangian trajectory dynamics of (42), see Sect. 3.4. If the potentials in the classical subspace are harmonic, and the coupling between the quantum and classical subspaces is at most linear in the classical variables, (37) is exact.

Equation (37) is found to be identical to the partial moment equations one would obtain from the quantum–classical Liouville equation [40–43,45–48,79],

$$\begin{aligned}\frac{\partial \hat{\rho}^c_{\text{W}}}{\partial t} &= -\text{i}/\hbar \, [\hat{H}_{\text{W}}, \hat{\rho}^c_{\text{W}} \,] \\ &\quad + \frac{1}{2} \Big(\{ \hat{H}_{\text{W}}, \hat{\rho}^c_{\text{W}} \}_{QP} - \{ \hat{\rho}^c_{\text{W}}, \hat{H}_{\text{W}} \}_{QP} \Big) ,\end{aligned} \tag{39}$$

where $\hat{\rho}^c_{\text{W}}(Q, P)$ and $\hat{H}_{\text{W}}$ are partially Wigner transformed [45] operator quantities, i.e., operators with respect to the quantum subspace and functions of the classical phase-space variables (Q, P).

When constructing the moment equations, the characteristic commutator structure of (39) translates to nonclassical terms in the moment hierarchy for the local-in-q moments $\langle \mathcal{P}^n \rho \rangle_{qQP}$, starting from the second-order onwards. (These nonclassical terms will be affected by certain fundamental inconsistencies incurred at the level of the Jacobi identity [45, 46, 80]).

3.4 Lagrangian Picture, and Trajectory Representation

To obtain the trajectory equations (3), the Eulerian equations (37) for the mixed quantum–classical moments have to be translated to the Lagrangian frame. If the equation for the zeroth-order moment $\langle \rho \rangle^c_{qQP}$ is interpreted as a hybrid hydrodynamic-Liouvillian continuity equation, the fluid-particle dynamics follows from the definition of a three-component current $\mathbf{j}_{qQP}$ [39],

$$\begin{aligned}\frac{\partial \langle \rho \rangle^c_{qQP}}{\partial t} &= -\frac{1}{m}\frac{\partial \langle \mathcal{P} \rho \rangle^c_{qQP}}{\partial q} \\ &\quad + \{H_Q + V_{\text{int}}, \langle \rho \rangle^c_{qQP}\}_{QP} \\ &= -\nabla_{qQP} \cdot \mathbf{j}_{qQP}\end{aligned} \tag{40}$$

with $\nabla_{qQP} = (\partial/\partial q, \partial/\partial Q, \partial/\partial P)$ and the current

$$\frac{\mathbf{j}_{qQP}}{\langle \rho \rangle^c_{qQP}} = \begin{pmatrix} \dot{q} \\ \dot{Q} \\ \dot{P} \end{pmatrix} = \begin{pmatrix} \bar{p}_{qQP}/m \\ (\partial H/\partial P) \\ -(\partial H/\partial Q) \end{pmatrix}, \tag{41}$$

where the momentum field $\bar{p}_{qQP}$ was introduced via the first moment, $\langle \mathcal{P} \rho \rangle^c_{qQP} = \bar{p}_{qQP} \langle \rho \rangle^c_{qQP}$. The quantity $\bar{p}_{qQP}$ again represents the *average* momentum derived from the underlying Wigner distribution for a given combination of independent variables (q, Q, P).

In the Lagrangian picture, the hydrodynamic fields are evaluated *along the fluid particle trajectories* – or, more precisely, the characteristics of (40) [81] – as defined by (41). The temporal evolution in the Lagrangian frame is expressed via the total time derivative, $\mathrm{d}/\mathrm{d}t = \partial/\partial t + \mathbf{v}_{qQP} \cdot \nabla_{qQP}$. Thus, the continuity equation (40), which describes the local density balance at each point (q, Q, P), translates to the Lagrangian form $d\langle \rho \rangle^c_{qQP}/dt = -(\langle \rho \rangle^c_{qQP}/m)(\partial \bar{p}_{qQP}/\partial q)$.

In order to connect to a phase-space perspective similar to the one of (31), (41) is combined with an equation for the fluid particle acceleration $\mathrm{d}\bar{p}_{qQP}/\mathrm{d}t$ (obtained from the equation for the first moment, $\langle \mathcal{P} \rho \rangle^c_{qQP} = \bar{p}_{qQP} \langle \rho \rangle^c_{qQP}$) which involves a generalized hydrodynamic force term [39]. The overall picture is the one of a *correlated dynamics of the quantum hydrodynamic variables* $(q, p = \bar{p}_{qQP})$ *and classical variables* (Q, P), see (3)

$$\begin{aligned}\dot{q} &= \frac{\bar{p}}{m}, \\ \dot{\bar{p}} &= -\frac{\partial}{\partial q}\Big(V_q(q) + V_{\text{int}}(q, Q)\Big) + F_{\text{hyd}}(q, Q, P),\end{aligned}$$

$$\dot{Q} = \frac{P}{M},$$
$$\dot{P} = -\frac{\partial}{\partial Q}\Big(V_Q(Q) + V_{\mathrm{int}}(q,Q)\Big) \tag{42}$$

with $\bar{p} \equiv \bar{p}_{qQP}$ and the hydrodynamic force

$$F_{\mathrm{hyd}}(q,Q,P) = -\frac{1}{m\,\langle\rho\rangle^{c}_{qQP}}\,\frac{\partial\sigma_{qQP}}{\partial q} \tag{43}$$

obtained as the spatial derivative with respect to q of the generalized variance

$$\sigma_{qQP} = \langle\mathcal{P}^2\rho\rangle^{c}_{qQP} - \bar{p}^{2}_{qQP}\langle\rho\rangle^{c}_{qQP}. \tag{44}$$

The quantity σ_{qQP} reflects the width in p, for given $\mathbf{x} = (q,Q,P)$, of the (approximate) phase-space distribution $\rho^{c}_{\mathrm{W}}(q,p;Q,P)$, and it is the spatial variation of σ_{qQP} with respect to the hydrodynamic coordinate q which gives rise to F_{hyd}.

Apart from its dependence upon the classical phase-space variables (Q,P), (43) is entirely analogous to the quantum hydrodynamic equation obtained for a single quantum degree of freedom [35, 36] (see Sect. 2), and reduces to this equation in the absence of the classical subspace. Furthermore, if the isolated quantum subsystem corresponds to a pure state, one recovers the Bohmian quantum force $F_{\mathrm{hyd}} = -\partial\mathcal{Q}/\partial q$ of (22) [6, 35, 37].

The deterministic, Lagrangian trajectory representation equation (42) is a result of the classical nature of the first two moment equations, within the quantum–classical approximation. The representation equation (42) is rather unique in several respects. First, the dynamics of the coupled hydrodynamic and classical trajectories is nonstochastic, in contrast to the trajectory dynamics usually associated with the quantum–classical Liouville equation [45, 46]. Furthermore, the difficulty/ambiguity in formally defining and propagating quantum trajectories in Liouville space [52–55] is avoided. A Liouville space representation of both the quantum and the classical sector would, as a consequence of (39), most likely entail quantum correction terms in the classical sector. By "localizing" the quantum subsystem, the hydrodynamic representation leads to a remarkably simple form of the coupled trajectory equations (42).

4 Coupled Light-Heavy Oscillator Systems

For illustration, we consider the vibrationally nonadiabatic dynamics in a system of coupled light ("quantum") and heavy ("classical") oscillators. This type of system has been shown to pose a challenge for mixed quantum–classical methodology – notably the surface hopping and mean-field methods – even in the case of a harmonic potential in the classical subspace [82]. A quantum–classical method based upon the Bohmian picture (using the $F_{\mathrm{hyd}} \to 0$ approximation in the classical sector, see (29)) was also found to be subject to considerable error, of the same order as the mean-field approach [49]. A sensitive measure of the dynamics is the time-dependent probability for the light oscillator to be in a given adiabatic state, $p_n^{\text{ad}}(t) = \text{Tr}\{|\phi_n^{\text{ad}}\rangle\langle\phi_n^{\text{ad}}|\hat{\rho}(t)\}$ [82]; we therefore consider these survival probabilities in terms of the partial moment quantities introduced earlier (see Appendix B).

Within the quantum–classical picture developed here, the dynamics is described *exactly* if the classical subsystem is harmonic. This is exemplified by the cases to be discussed below: (a) a pure-state example where a double well system in the quantum subspace is coupled to a harmonic oscillator in the classical subspace, (b) a mixed-state case including dissipation, where a harmonic oscillator in the quantum subspace is coupled to a "Brownian" harmonic oscillator (undergoing Caldeira–Leggett type dissipation [83]) in the classical subspace. Both examples refer to cases where the moment hierarchy terminates at low orders, due to the pure-state vs. Gaussian closure conditions, respectively. Work in progress addresses suitable truncation schemes for more general situations.

While our mixed quantum–classical scheme is eventually meant for systems with a high-dimensional classical subspace for which the classical Liouville limit is applicable, we are discussing here low-dimensional situations with initial conditions which correspond to pure states. In the "classical" sector, these distributions thus remain quantum distributions, represented by classical trajectories. The present perspective in fact corresponds to the Wigner method rather than a rigorously defined classical limit situation, see Sect. 2.5.

4.1 Pure-State Case: Double Well Coupled to a Harmonic Oscillator

In this example, the quantum coordinate involves a quartic double well potential which is bilinearly coupled to a classical harmonic oscillator,

$$H = \frac{p^2}{2m} + \frac{P^2}{2M} + Aq^2 + Bq^4 + CqQ + \frac{K}{2}Q^2, \tag{45}$$

where $A < 0$ and $B > 0$.[7] We will focus upon the case where the overall system remains in a pure state.

We have used two complementary approaches to compute the hydrodynamic partial moments: First, conventional wavepacket propagation,[8] by which the partial moments are constructed from the relevant pure-state quantities, as described in Appendix A and [38, 39]. Second, propagation of the

[7] The parameters used in the calculation are as follows: for the masses, $m = 2,000$ a.u. and $M = 20,000$ a.u., and for the classical oscillator force constant $K = 0.118$ a.u. For the quantum part the parameters $A = -0.033$ a.u., $B = 0.030$ a.u. were used, along with a coupling constant of $C = 0.01$ a.u.

[8] The 2D wavepacket propagation was carried out by the Chebyshev propagator, combined with the grid based Fast Fourier Transform method to calculate the Hamiltonian operation on the wavefunction.

Wigner function,[9] with the partial moments obtained by integration over the quantum momentum variable as in (32). The numerical results obtained by these two methods were found to be in accurate agreement. In forthcoming work, we will report on a direct propagation of the partial moments using either (37) in a Eulerian frame, or else (42) in a Lagrangian frame.

The initial wavefunction is taken as a product of the ground adiabatic state $\phi_0^{\rm ad}(x|X)$ of the quantum subsystem multiplied by a Gaussian function in the classical sector,

$$\psi(x, X; t_0) = \left(\frac{2\beta}{\pi}\right)^{\frac{1}{4}} \phi_0^{\rm ad}(x|X) \exp(-\beta(X - X_e)^2), \tag{46}$$

where $X_e = 0.5$ a.u. defines the Gaussian maximum with a width characterized by the parameter $\beta = 44$ a.u. For the potential function parameters chosen in this study, this choice of X_e leads to a density which is predominantly localized in the left well. For comparison, $X_e = 0$ would delocalize the initial density in both potential wells, and $X_e < 0$ would localize the density in the right well.

Figure 6 depicts the first three hydrodynamic moments as a function of time, in the Eulerian picture. The moments were integrated over the classical variables (Q, P) so as to obtain the reduced quantities $\langle \mathcal{P}^n \rho \rangle_q = \int {\rm d}Q{\rm d}P \, \langle \mathcal{P}^n \rho \rangle_{qQP}$, $n = 0, 1, 2$. The figure illustrates the transfer of density between the two wells, by a dominant tunneling mechanism.

Figure 7 illustrates the time-dependent probability $p_{n=0}^{\rm ad}(t)$ for the "light", double well oscillator to remain in its adiabatic ground state. Appendix B provides details on the calculation of this probability in terms of the partial moment quantities, see (B.3). Nonadiabatic effects are extremely pronounced, since the time scales of the "light" (quantum) vs. "heavy" (classical) oscillators are not well separated. The density oscillates along Q with period 0.1 fs^{-1}, as manifested in the periodic revivals of $p_{n=0}^{\rm ad}(t)$ in Fig. 7. As density is transferred to the adjacent potential well, the magnitude of oscillations in $p_{n=0}^{\rm ad}(t)$ diminishes until at around $t = 800$ fs, when the density is fully delocalized in both wells.

We emphasize that the moment method is exact for the system considered here, where the classical coordinate is harmonic. This is also the case for the mixed quantum–classical phase space methods of [45, 46] (prior to approximations made in the propagation scheme), but is in contrast to other hybrid methods like the surface hopping and mean-field methods [49, 82], as well as the hybrid Bohmian-classical method of [49].

4.2 Quantum Oscillator Coupled to a Classical Brownian Oscillator

The second example addresses a mixed-state problem which can be solved analytically. We consider a harmonic "quantum" oscillator coupled bilinearly

[9] For the Wigner calculation in the four-dimensional (q, p, Q, P) phase space, the time integration was performed using a fourth-order Runge–Kutta method, and derivatives were calculated using the Fast Fourier Transform method.

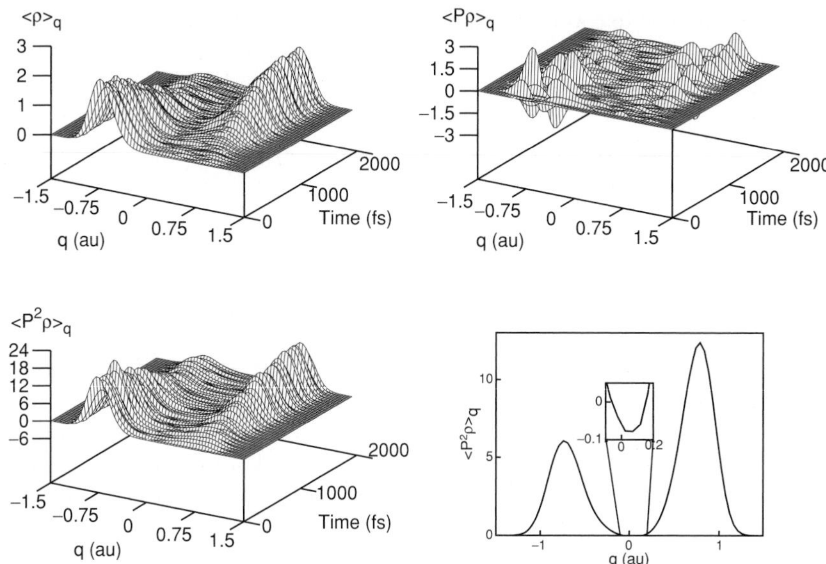

Fig. 6. The first three hydrodynamic moments are shown as a function of time (Eulerian picture), for the double well system coupled to a harmonic oscillator, (45). As explained in the text, the reduced q-dependent quantities $\langle \mathcal{P}^n \rho \rangle_q = \int \mathrm{d}Q\mathrm{d}P\, \langle \mathcal{P}^n \rho \rangle_{qQP}$ are shown. The moment evolution reflects the transfer of density between the two wells. Further, a one-dimensional cut is shown for the second moment $\langle \mathcal{P}^2 \rho \rangle_q$, illustrating that negative regions can occur. These can be traced back to the characteristic quantum interference (and tunneling) effects which give rise to negative regions of the Wigner function, typically in the barrier region between the two wells

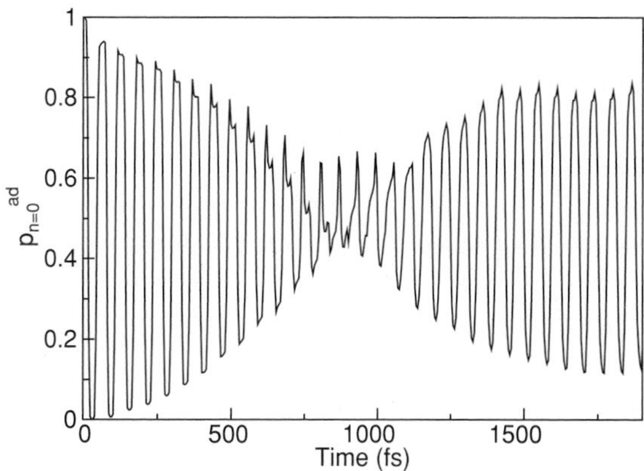

Fig. 7. For the double well system coupled to a harmonic oscillator, (45), the time-dependent probability $p^{\mathrm{ad}}_{n=0}$ for the quantum subsystem to be in the zeroth adiabatic state, is shown, calculated from (B.3) for the pure-state case. The marked oscillations in $p^{\mathrm{ad}}_{n=0}$ signal pronounced nonadiabatic effects which are generated by the coupling to the classical subsystem. Superimposed is a slow oscillation which corresponds to the tunneling period

to a "classical" oscillator which undergoes Caldeira–Leggett type dissipation [83]. The dynamics is described by a Fokker–Planck equation in the Wigner phase space representation,

$$\frac{\partial \rho_{\mathrm{W}}}{\partial t} = \{H, \rho_{\mathrm{W}}\}_{\mathbf{qp}} + \gamma \frac{\partial}{\partial P}(P\rho_{\mathrm{W}}) + \gamma M k_{\mathrm{B}} T \frac{\partial^2}{\partial P^2}\rho_{\mathrm{W}} \tag{47}$$

where the index denotes $\mathbf{qp} = (q, p, Q, P)$, and the Hamiltonian reads:

$$H = \frac{p^2}{2m} + \frac{P^2}{2M} + \frac{1}{2}k(q-Q)^2 + \frac{1}{2}KQ^2. \tag{48}$$

The quantum oscillator undergoes delayed (non-Markovian) dissipation due to its coupling to the classical "Brownian" oscillator. Given a Gaussian initial condition (which is again chosen to be pure-state), the system remains in a time-evolving Gaussian state, which is always mixed-state due to the effects of dissipation. The time evolution can be obtained analytically in terms of the time-evolving Gaussian parameters, i.e., the mean values of positions and momenta, as well as the matrix of variances [84].

The Gaussian density can at all times be represented in terms of the first three partial moments,

$$\rho_{\mathrm{W}}(q, p, Q, P) = \frac{\langle\rho\rangle_{qQP}}{(2\pi\tilde{\sigma}_{qQP})^{1/2}} \exp\left(-\frac{1}{2\tilde{\sigma}_{qQP}}(p - \bar{p}_{qQP})^2\right), \tag{49}$$

where $\tilde{\sigma}_{qQP} = \sigma_{qQP}/\langle\rho\rangle_{qQP}$. As detailed in [84], the momentum $\bar{p}_{qQP}$ and the hydrodynamic force (43) are linear functions of (q, Q, P). The time-dependent adiabatic probability $p^{\mathrm{ad}}_{n=0}(t)$ can also be constructed analytically in terms of the first three moments, see Appendix B.

Propagation in the Eulerian frame thus involves the time evolution of the first three partial moments [84]. In the Lagrangian frame, the fluid–particle dynamics is of stochastic character, due to the effects of dissipation in the classical subspace, so that (42) needs to be augmented by a Langevin term [84]

$$\begin{aligned}
\dot{q} &= \frac{\bar{p}}{m}, \\
\dot{\bar{p}} &= -\frac{\partial}{\partial q}\Big(V_q(q) + V_{\mathrm{int}}(q, Q)\Big) + F_{\mathrm{hyd}}(q, Q, P), \\
\dot{Q} &= \frac{P}{M}, \\
\dot{P} &= -\frac{\partial}{\partial Q}\Big(V_Q(Q) + V_{\mathrm{int}}(q, Q)\Big) - \gamma P + R(t)
\end{aligned} \tag{50}$$

with the Gaussian random force $R(t)$.

We have calculated the time-dependent adiabatic probability $p^{\mathrm{ad}}_{n=0}(t)$ (B.4) both in the Eulerian and Lagrangian representations [84], with parameters as specified in [38, 82, 84] and initial conditions analogous to (46). As

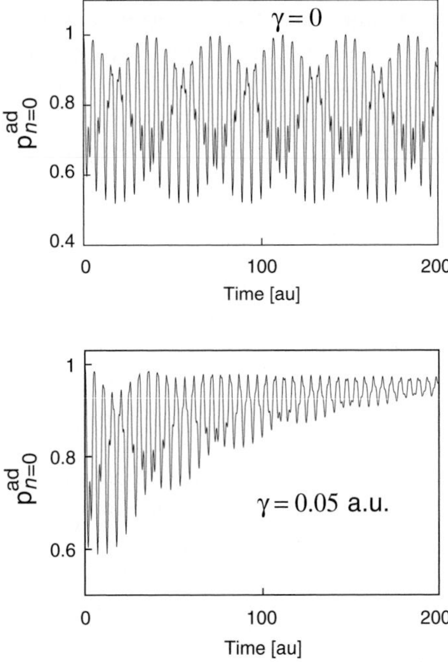

Fig. 8. The time-dependent adiabatic probability $p^{\mathrm{ad}}_{n=0}$ is shown for a system of coupled oscillators, (48), according to the mixed-state expression (B.4). Parameters were specified as in [38, 82, 84]. The system's behavior in the absence of dissipation (upper panel) is compared with the behavior in the presence of dissipation (lower panel). The friction coefficient was $\gamma = 0.05$ a.u. and the temperature was chosen as $T = \hbar\omega/(2\,k_{\mathrm{B}})$ (see text). The nonadiabatic oscillations are strongly reduced, i.e., an effect of "noise-induced adiabaticity" is observed

demonstrated in Fig. 8, a pronounced nonadiabatic behavior is again observed. As also illustrated by the figure, dissipation acts so as to make the dynamical behavior "more adiabatic": i.e., the characteristic oscillatory effects in the adiabatic probabilities subside as the stochastic dynamics affects the classical oscillator and, indirectly, the quantum oscillator. Dissipation slows down the classical oscillator (thus reducing the probability of nonadiabatic transitions) and, from an alternative viewpoint, leads to a loss of vibrational coherence in the quantum subsystem, thus impeding the population transfer. As a consequence of the high-temperature assumption inherent in the Caldeira–Leggett dynamics (47), the survival probability approaches $p^{\mathrm{ad}}_{n=0}(\infty) = (1/2 + k_{\mathrm{B}}T/\hbar\omega)^{-1}$ [84], here chosen as $p^{\mathrm{ad}}_{n=0}(\infty) \sim 1$.

5 Conclusions

The quantum–statistical moment formulation allows one to carry the hydrodynamic picture beyond the pure-state ("Bohmian") case, so as to include mixed states and dissipation. Complementary Eulerian and Lagrangian representations can be derived, by complete analogy with pure-state hydrodynamics (and classical hydrodynamics). The difference between the pure-state and mixed-state cases lies in the fact that *two* fluid-dynamical equations completely describe a pure quantum state, while an *infinite* number of coupled moment equations are necessary to fully characterize a general mixed state (Sect. 2.2). Special cases include, e.g., Gaussian mixed states, for which the moment hierarchy terminates with the first three equations. In general, the approximate closure of the moment hierarchy, by appropriate truncation schemes, is a key issue in applying the moment equations.

The hydrodynamic formulation provides a unique connection to the quantum phase space (Wigner) representation. The hydrodynamic moments in question are the moments of the Wigner function with respect to the phase space momentum variable p, i.e., obtained by a "projection onto coordinate space" [26]. Importantly, this perspective immediately leads one to identify the hydrodynamic momentum as an *average* momentum at given q, $p = \bar{p}_q$. The additional, hydrodynamic (or "pressure") force F_{hyd} which arises in the hydrodynamic picture can thus be taken to *compensate for the fact that the dynamics of the average momentum* $\bar{p}_q$ is considered [36, 37]. This aspect is underscored by introducing the concept of a hydrodynamic phase space (Sect. 2.6). In the Bohmian theory, the hydrodynamic momentum is identified with an actual particle momentum, $p_{\mathrm{Bohm}} \equiv \bar{p}_q$.

These observations raise several questions as far as the classical limit of quantum hydrodynamics is concerned (Sect. 2.5). While the "Bohmian" classical limit postulates the vanishing of the hydrodynamic, or "quantum" force, $F_{\mathrm{hyd}} \to 0$, the classical–*statistical* hydrodynamic limit implies that the force converges towards a purely classical form, $F_{\mathrm{hyd}} \to F_{\mathrm{hyd}}^{\mathrm{cl}}$, which is generally nonzero. The classical hydrodynamic force $F_{\mathrm{hyd}}^{\mathrm{cl}} = -(1/m)\,\langle\rho\rangle_q^{-1}\,\partial\sigma_q^{\mathrm{cl}}/\partial q$ can be deduced from the momentum variance σ_q^{cl} obtained in the classical Liouville limit, where the quantum Wigner distribution is approximately represented by a classical phase-space distribution. The classical-statistical hydrodynamic limit is thus compatible with – and can be taken to follow from – the classical Liouville limit. The two hydrodynamic classical limits – i.e., the classical-statistical limit (following from the classical Liouville limit) vs. the "Bohmian" $F_{\mathrm{hyd}} \to 0$ limit – generally differ substantially. (Exceptions are cases where F_{hyd} vanishes due to the form of the underlying phase-space distribution, e.g., in the free-particle case as discussed in Sect. 2.6).

Against this background, one of the central concerns of this chapter has been the construction of a hybrid quantum–classical scheme based upon the hydrodynamic picture (Sect. 3.1). We have chosen the dynamics in the classical subspace so as to be compatible with the classical Liouville limit (or, the statistical–hydrodynamic limit) rather than the Bohmian classical limit. This choice is motivated by the observation that the classical Liouville limit is most appropriate for the following two situations, which take a central role in a quantum molecular dynamics context: (a) the description of quantum–statistical phenomena in a regime where classical effects dominate, e.g., at high temperatures; (b) the simulation of quantum dynamical processes (even for pure states, at $T = 0$), by classical trajectory ensembles ("Wigner method").

The hybrid quantum–classical scheme in question is based upon the representation in terms of the *partial* moments of equation (3) [38, 39]. While quantum hydrodynamics (like classical hydrodynamics) is fundamentally a coordinate space formulation, its derivation from the Wigner picture makes it an ideal setting for combining the two representations. Thus, the essence of the partial moment construction is to introduce a hydrodynamic projection for a subset of selected degrees of freedom, while the remaining degrees of freedom remain in a Liouville space setting. In a further step, the classical Liouville limit is introduced in the classical subspace. As a result, one obtains the hybrid quantum–classical representation of (3) which *combines a hydrodynamic Lagrangian evolution in the quantum subspace with a Hamiltonian dynamics in the classical subspace.* Equation (3) can be considered as a hybrid quantum–classical molecular dynamics (MD) approach, where the sampling in the classical subspace is the same as in conventional MD simulations. This hybrid representation is distinct from other mixed quantum–classical schemes which are based upon the $F_{\rm hyd} \to 0$ limit in the classical subspace [49, 51].

The classical-limit description is obtained from the exact dynamical evolution for the partial moments (Sect. 3.2) by omitting the quantum correction terms pertaining to the classical sector. This approximation is the same as the one made in the quantum–classical Liouville equation (Sect. 3.3), and therefore shares all properties of the latter – in particular, the exactness for harmonic classical subsystems. The hybrid Lagrangian trajectory dynamics is unique, though, in that it describes deterministic, coupled quantum–classical trajectories according to (3). We have demonstrated this quantum–classical picture for coupled light-heavy oscillator systems.

The examples addressed in the present chapter have focused on cases for which the hydrodynamic hierarchy terminates with the second order (pure states) or with the third order (Gaussian mixed-state densities). Future work will focus on the closure problem for general densities, as briefly discussed in Sect. 2.1. In particular, maximum entropy methods [69, 70] will be applied and developed to construct approximate closure schemes. Once the closure problem is satisfactorily solved, the strength of the hydrodynamic description lies in (a) the possibility to selectively probe and propagate a limited number of moments, not the complete information contents of the quantum density, and (b) the use of the Lagrangian trajectory representation and its generalization to a hybrid hydrodynamic-Liouvillian setting. Quantum hydrodynamics for mixed quantum states could thus become a viable setting for dynamical calculations in a quantum–statistical setting.

Acknowledgments

Thanks are due to Giovanni Ciccotti, Ray Kapral, Gérard Parlant, Gert van der Zwan, and Eric Bittner for a number of stimulating conversations. Financial support was granted by the Centre National de la Recherche Scientifique (CNRS), the Danish Natural Science Research Council, the Carlsberg Foundation, the Ramsay Memorial Fellowhip Trust, and the Welsh Development Agency.

Appendix A Partial Moments for Pure States

For pure states, the partial moment hierarchy terminates with the first two equations. Since the fundamental representation for pure states is the coordinate space representation, the moments can be written as follows:

$$\langle \mathcal{P}^n \rho \rangle_{qQP} = \left(\frac{\hbar}{\mathrm{i}}\right)^n \frac{\partial^n}{\partial r^n} \int \mathrm{d}R \, \exp\left(-\frac{\mathrm{i}PR}{\hbar}\right) \rho(q,r;Q,R)\bigg|_{r=0} \tag{A.1}$$

with the pure state density matrix

$$\rho(q,r;Q,R) = \psi\left(q+\frac{r}{2}, Q+\frac{R}{2}\right)\psi^*\left(q-\frac{r}{2}, Q-\frac{R}{2}\right) \tag{A.2}$$

With the polar form of the wavefunction, $\psi(x,X) = R(x,X)\exp[\mathrm{i}S(x,X)/\hbar]$, the partial moments (A.1) can be entirely expressed in terms of the local-in-space quantities [38, 39]

$$\tilde{\rho}(q,Q) = R^2(q,Q), \tag{A.3}$$

$$\tilde{p}(q,Q) = \frac{\partial S(q,Q)}{\partial q}. \tag{A.4}$$

The explicit form of the pure-state partial moments is given in [39]. Note that the pure-state property is not generally conserved under the quantum–classical dynamics.

Appendix B Adiabatic Survival Probabilities

Within a Born–Oppenheimer zeroth-order description for a coupled light-heavy system [82], i.e., with a basis of functions $\{\phi_n^{\mathrm{ad}}(x|X)\,\varphi_{ln}(X)\}$, the projection onto the light-oscillator nth adiabatic state $\phi_n^{\mathrm{ad}}(x|X)$ is given as follows for a general coordinate-space density $\rho(x,X;x',X')$ [84]:

$$\begin{aligned} p_n^{\rm ad} &= {\rm Tr}\{|\phi_n^{\rm ad}\rangle\langle\phi_n^{\rm ad}|\hat\rho(t)\} \\ &= \int {\rm d}X\, {\rm d}x\, {\rm d}x'\ \phi_n^{{\rm ad}*}(x|X)\,\rho(x,X;x',X'=X)\,\phi_n^{\rm ad}(x'|X). \end{aligned} \tag{B.1}$$

This expression is a generalization of the corresponding pure-state expression given, e.g., in [82]. In the following we will consider, in particular, the time-dependent survival probabilities with respect to the $n=0$ adiabatic ground state. E.g., for a harmonic oscillator in the quantum subspace, the adiabatic ground state corresponds to a displaced Gaussian centered on the mean position of the classical oscillator [82].

In general, the representation of (B.1) in terms of moments would necessitate an expansion of $\rho(x,X;x',X')$ as in the Taylor expansion (6). Here, we will consider two special cases, for which the density itself, and hence, the expression (B.1) can be cast in terms of the lowest-order moments.

First, for pure states, the wavefunction $\psi(x,X) = R(x,X)\exp[{\rm i}S(x,X)/\hbar]$ can be rewritten in terms of the quantities $\tilde\rho(x,X)$ and $\tilde p(x,X)$ introduced in (A.3)–(A.4),

$$\psi(x,X;t) = \tilde\rho^{1/2}(x,X)\,\exp\Big(i/\hbar \int dq\,\tilde p(x,X) + f(X)\Big) \tag{B.2}$$

so as to obtain for the adiabatic probability

$$\begin{aligned} p_{n=0}^{\rm ad}(t)\Big|_{\rm pure} &= \int {\rm d}X \bigg| \int {\rm d}x\,\phi_0^{{\rm ad}*}(x|X) \\ &\qquad \tilde\rho^{1/2}(x,X;t)\,\exp\Big({\rm i}/\hbar \int {\rm d}x\,\tilde p(x,X;t)\Big)\bigg|^2, \end{aligned} \tag{B.3}$$

which we used in our analysis in [38]. When propagating the partial hydrodynamic moments, the local-in-space quantity $\tilde p(q,Q)$ is obtained by integration over the phase space momentum P, $\tilde p(q,Q) = \int {\rm d}P \langle \mathcal{P}\rho\rangle_{qQP} / \int {\rm d}P \langle\rho\rangle_{qQP}$.

Second, a Gaussian mixed-state density will be considered, for the specific case where both the quantum and classical parts are harmonic. In this case, the adiabatic probability can be expressed in the following analytical form, involving the first three partial moments [84]

$$p_{n=0}^{\rm ad}(t)\Big|_{\rm gauss} = \int {\rm d}q\,{\rm d}Q\,{\rm d}P\,\langle\rho\rangle_{qQP}(t)\,\mathcal{I}(q,Q,P;t) \tag{B.4}$$

with the kernel

$$\mathcal{I}(q,Q,P) = 2\left(\frac{\hbar^2\beta^2}{2\tilde\sigma_{qQP}+\hbar^2\beta^2}\right)^{1/2} \exp\left[-\beta^2(q-Q)^2 - \frac{\bar p_{qQP}^2}{2\tilde\sigma_{qQP}+\hbar^2\beta^2}\right], \tag{B.5}$$

where $\beta = (m\omega/\hbar)^{1/2}$ and $\tilde\sigma_{qQP} = \sigma_{qQP}/\langle\rho\rangle_{qQP}$.

References

1. L. de Broglie, Compt. Rend. **183**, 447 (1926)
2. L. de Broglie, *Tentative d'interprétation causale et non-linéaire de la mécanique ondulatoire*, Gauthier-Villars, Paris, (1956)
3. D. Bohm, Phys. Rev. **85**, 166 (1952)
4. D. Bohm, Phys. Rev. **85**, 180 (1952)
5. J. S. Bell, *Speakable and Unspeakable in quantum mechanics*, Cambridge University Press, Cambridge, (1989)
6. P. R. Holland, *The Quantum Theory of Motion*, Cambridge University Press, New York, (1993)
7. R. E. Wyatt, *Quantum Dynamics with Trajectories: Introduction to Quantum Hydrodynamics*, Springer, Berlin Heidelberg New York, (2005)
8. E. Madelung, Z. Phys. **40**, 322 (1926)
9. Z. S. Wang, G. R. Darling, and S. Holloway, J. Chem. Phys. **115**, 10373 (2001)
10. D. Babyuk, R. E. Wyatt, and J. H. Frederick, J. Chem. Phys. **119**, 6482 (2003)
11. A. Sanz, F. Borondo, and S. Miret-Artés, J. Chem. Phys. **120**, 8794 (2004)
12. J. H. Weiner and A. Askar, J. Chem. Phys. **54**, 3534 (1971)
13. B. K. Dey, A. Askar, and H. A. Rabitz, J. Chem. Phys. **109**, 8770 (1998)
14. F. Sales Mayor, A. Askar, and H. A. Rabitz, J. Chem. Phys. **111**, 2423 (1999)
15. R. E. Wyatt, Chem. Phys. Lett. **313**, 189 (1999)
16. R. E. Wyatt and E. R. Bittner, J. Chem. Phys. **113**, 8898 (2000)
17. E. R. Bittner, J. Chem. Phys. **112**, 9703 (2000)
18. C. Trahan, K. Hughes, and R. E. Wyatt, J. Chem. Phys. **118**, 9911 (2003)
19. E. R. Bittner, J. Chem. Phys. **119**, 1358 (2003)
20. Y. Zhao and N. Makri, J. Chem. Phys. **119**, 60 (2003)
21. V. A. Rassolov and S. Garashchuk, J. Chem. Phys. **120**, 6815 (2004)
22. B. Kendrick, J. Chem. Phys. **121**, 2471 (2004)
23. D. Babyuk and R. E. Wyatt, J. Chem. Phys. **121**, 9230 (2005)
24. C. J. Trahan, R. E. Wyatt, and B. Poirier, J. Chem. Phys. **122**, 164104 (2005)
25. V. Allori, V. Dürr, S. Goldstein, and N. Zanghì, J. Opt. B **4**, 482 (2002)
26. T. Takabayasi, Prog. Theor. Phys. **11**, 341 (1954)
27. J. E. Moyal, Proc. Camb. Philos. Soc. **45**, 99 (1949)
28. J. H. Irving and R. W. Zwanzig, J. Chem. Phys. **19**, 1173 (1951)
29. H. Fröhlich, Physica **37**, 215 (1967)
30. J. Yvon, J. Phys. Lett. **39**, 363 (1978)
31. W. R. Frensley, Rev. Mod. Phys. **62**, 745 (1990)
32. M. Ploszajczak and M. J. Rhoades-Brown, Phys. Rev. Lett. **55**, 147 (1985)
33. J. V. Lill, M. I. Haftel, and G. H. Herling, Phys. Rev. A **39**, 5832 (1989)
34. L. M. Johansen, Phys. Rev. Lett. **80**, 5461 (1998)
35. I. Burghardt and L. S. Cederbaum, J. Chem. Phys. **115**, 10303 (2001)
36. I. Burghardt and K. B. Møller, J. Chem. Phys. **117**, 7409 (2002)
37. J. G. Muga, R. Sala, and R. F. Snider, Phys. Scr. **47**, 732 (1993)
38. I. Burghardt and G. Parlant, J. Chem. Phys. **120**, 3055 (2004)
39. I. Burghardt, J. Chem. Phys. **122**, 094103 (2005)
40. O. V. Prezhdo and V. V. Kisil, Phys. Rev. A **56**, 162 (1997)
41. I. V. Aleksandrov, Z. Naturforsch. **36**, 902 (1981)
42. W. Boucher and J. Traschen, Phys. Rev. D **37**, 3522 (1988)
43. C. C. Martens and J.-Y. Fang, J. Chem. Phys. **106**, 4918 (1996)

44. E. Roman and C. C. Martens, J. Chem. Phys. **121**, 11572 (2004)
45. R. Kapral and G. Ciccotti, J. Chem. Phys. **110**, 8919 (1999)
46. S. Nielsen, R. Kapral, and G. Ciccotti, J. Chem. Phys. **115**, 5805 (2001)
47. I. Horenko, M. Weiser, B. Schmidt, and C. Schütte, J. Chem. Phys. **120**, 8913 (2004)
48. D. A. Micha and B. Thorndyke, Adv. Quant. Chem. **47**, 293 (2004)
49. E. Gindensperger, C. Meier, and J. A. Beswick, J. Chem. Phys. **113**, 9369 (2000)
50. E. Gindensperger, C. Meier, and J. A. Beswick, J. Chem. Phys. **116**, 8 (2002)
51. O. V. Prezhdo and C. Brooksby, Phys. Rev. Lett. **86**, 3215 (2001)
52. H. W. Lee and M. O. Scully, J. Chem. Phys. **77**, 4604 (1982)
53. H. W. Lee, Phys. Rep. **259**, 147 (1995)
54. A. Donoso and C. C. Martens, Phys. Rev. Lett. **87**, 223202 (2001)
55. J. Daligault, Phys. Rev. A **68**, 010501 (2003)
56. D. Dürr, S. Goldstein, R. Tumulka, and N. Zanghì, Found. Phys. **35**, 449 (2005)
57. I. Burghardt, K. B. Møller, G. Parlant, L. S. Cederbaum, and E. R. Bittner, Int. J. Quant. Chem. **100**, 1153 (2004)
58. E. Wigner, Phys. Rev. **40**, 749 (1932)
59. M. Hillery, R. F. O'Connell, M. O. Scully, and E. P. Wigner, Phys. Rep. **106**, 121 (1984)
60. W. P. Schleich, *Quantum Optics in Phase Space*, Wiley-VCH, Berlin, (2001)
61. J. B. Maddox, E. R. Bittner, and I. Burghardt, Int. J. Quant. Chem. **89**, 313 (2002)
62. L. E. Ballentine, Y. Yang, and J. P. Zibin, Phys. Rev. A **50**, 2854 (1994)
63. M. V. Berry and N. L. Balazs, J. Phys. A **12**, 625 (1979)
64. S. Habib, Phys. Rev. D **42**, 2566 (1990)
65. I. Burghardt and L. S. Cederbaum, J. Chem. Phys. **115**, 10312 (2001)
66. N. A. Krall and A. W. Trivelpiece, *Principles of Plasma Physics*, Mc-Graw Hill, New York, (1973)
67. L. D. Landau and E. M. Lifshitz, *Fluid Mechanics*, Pergamon Press, London, (1959)
68. C. D. Levermore, J. Stat. Phys. **83**, 1021 (1996)
69. M. Trovato and P. Falsaperla, Phys. Rev. B **57**, 4456 (1998)
70. P. Degond and C. Ringhofer, J. Stat. Phys. **112**, 587 (2003)
71. G. J. Iafrate, H. L. Grubin, and D. K. Ferry, J. Phys. (Paris) Colloq. C7, Suppl. 10, 42, 307 (1981)
72. S. Weigert, Phys. Rev. A **53**, 2078 (1996)
73. R. Balescu, *Equilibrium and Nonequilibrium Statistical Mechanics*, Wiley, New York, (1975)
74. E. J. Heller, J. Chem. Phys. **64**, 63 (1976)
75. K. B. Møller, J. P. Dahl, and N. E. Henriksen, J. Phys. Chem. **98**, 3272 (1994)
76. R. I. Sutherland, Found. Phys. **27**, 845 (1997)
77. C. L. Lopreore and R. E. Wyatt, Chem. Phys. Lett. **325**, 73 (2000)
78. R. Kapral, private communication
79. A. Donoso and C. C. Martens, J. Phys. Chem. **102**, 4291 (1998)
80. J. Caro and L. L. Salcedo, Phys. Rev. A **60**, 842 (1999)
81. S. J. Farlow, *Partial differential equations for scientists and engineers*, Dover, New York, (1982)
82. D. Kohen, F. H. Stillinger, and J. C. Tully, J. Chem. Phys. **109**, 4713 (1998)
83. A. O. Caldeira and A. J. Leggett, Physica A **121**, 587 (1983)
84. K. B. Møller, G. Parlant, and I. Burghardt, in preparation

Index